Solar System Data

Body	Mass (kg)	Mean Radius (m)	Period (s)	Distance from the Sun (m)
Mercury	3.18×10^{23}	2.43×10^{6}	7.60×10^{6}	5.79×10^{10}
Venus	4.88×10^{24}	6.06×10^{6}	1.94×10^{7}	1.08×10^{11}
Earth	5.98×10^{24}	6.37×10^{6}	3.156×10^{7}	1.496×10^{11}
Mars	6.42×10^{23}	3.37×10^{6}	5.94×10^{7}	2.28×10^{11}
Jupiter	1.90×10^{27}	6.99×10^{7}	3.74×10^{8}	7.78×10^{11}
Saturn	5.68×10^{26}	5.85×10^{7}	9.35×10^{8}	1.43×10^{12}
Uranus	8.68×10^{25}	2.33×10^{7}	2.64×10^{9}	2.87×10^{12}
Neptune	1.03×10^{26}	2.21×10^{7}	5.22×10^{9}	4.50×10^{12}
Pluto	$\approx 1.4 \times 10^{22}$	$\approx 1.5 \times 10^{6}$	7.82×10^{9}	5.91×10^{12}
Moon	7.36×10^{22}	1.74×10^{6}	—	—
Sun	1.991×10^{30}	6.96×10^{8}	—	—

Physical Data Often Used[a]

Average Earth–Moon distance	3.84×10^{8} m
Average Earth–Sun distance	1.496×10^{11} m
Average radius of the Earth	6.37×10^{6} m
Density of air (0°C and 1 atm)	1.29 kg/m^3
Density of water (20°C and 1 atm)	1.00×10^{3} kg/m^3
Free-fall acceleration	9.80 m/s^2
Mass of the Earth	5.98×10^{24} kg
Mass of the Moon	7.36×10^{22} kg
Mass of the Sun	1.99×10^{30} kg
Standard atmospheric pressure	1.013×10^{5} Pa

[a] These are the values of the constants as used in the text.

Some Prefixes for Powers of Ten

Power	Prefix	Abbreviation	Power	Prefix	Abbreviation
10^{-24}	yocto	y	10^{1}	deka	da
10^{-21}	zepto	z	10^{2}	hecto	h
10^{-18}	atto	a	10^{3}	kilo	k
10^{-15}	femto	f	10^{6}	mega	M
10^{-12}	pico	p	10^{9}	giga	G
10^{-9}	nano	n	10^{12}	tera	T
10^{-6}	micro	μ	10^{15}	peta	P
10^{-3}	milli	m	10^{18}	exa	E
10^{-2}	centi	c	10^{21}	zetta	Z
10^{-1}	deci	d	10^{24}	yotta	Y

W9-BNU-520

PHYSICS
For Scientists and Engineers
with Modern Physics

Fifth Edition

Raymond A. Serway

Robert J. Beichner
North Carolina State University

John W. Jewett, Jr., Contributing Author
California State Polytechnic University—Pomona

SAUNDERS COLLEGE PUBLISHING
A Division of Harcourt College Publishers

FORT WORTH   PHILADELPHIA   SAN DIEGO   NEW YORK   ORLANDO   AUSTIN   SAN ANTONIO   TORONTO   MONTREAL   LONDON   SYDNEY   TOKYO

Publisher: Emily Barrosse
Publisher: John Vondeling
Marketing Manager: Pauline Mula
Developmental Editor: Susan Dust Pashos
Project Editor: Frank Messina
Production Manager: Charlene Catlett Squibb
Manager of Art and Design: Carol Clarkson Bleistine
Text and Cover Designer: Ruth A. Hoover

Cover Image and Credit: Victoria Falls, Zimbabwe, at sunset. *(© Schafer & Hill/Tony Stone Images)*
Frontmatter Images and Credits: Title page: ripples on water. *(© Yagi Studio/Superstock, Inc.)*; water droplets on flower *(© Richard H. Johnson/FPG International Corp.)*; p. xi: Speed skaters *(Bill Bachman/Photo Researchers, Inc.)*; p. xii: Sky surfer *(Jump Run Productions/The Image Bank)*; p. xiii: jogger *(Jim Cummins/FPG International)*; p. xiv: Resistors on circuit board *(Superstock)*; p. xv: Fuel element of a nuclear reactor *(Courtesy of U.S. Department of Energy/Photo Researchers, Inc.)*; p. xvii: Long-jumper *(Chuck Muhlstock/FPG International)*; p.xviii: Penny-farthing bicycle race *(© Steve Lovegrove/Tasmanian Photo Library)*; p. xix: "Corkscrew" roller coaster *(Robin Smith/Tony Stone Images)*; p.xx: Cyclists pedaling uphill *(David Madison/Tony Stone Images)*; p. xxi: Twin Falls on the island of Kauai, Hawaii *(Bruce Byers/FPG)*; p. xxii: Bowling ball striking a pin *(Ben Rose/The Image Bank)*; p. xxiii: Sprinters at staggered starting positions *(© Gerard Vandystadt/Photo Researchers, Inc.)*; p. xxiv: U.S. Air Force F-117A stealth fighter in flight *(Courtesy of U.S. Air Force/Langley Air Force Base)*; p. xxv: Cheering crowd *(Gregg Adams/Tony Stone Images)*; p. xxviii: Welder *(© The Telegraph Colour Library/FPG)*; p. xxix: Basketball player dunking ball *(Ron Chapple/FPG International)*; p. xxx: Athlete throwing discus *(Bruce Ayres/Tony Stone Images)*; p. xxxi: Chum salmon *(Daniel J. Cox/Tony Stone Images)*; p. xxxiii: Bottle-nosed dolphin *(Stuart Westmorland/Tony Stone Images)*.

PHYSICS FOR SCIENTISTS AND ENGINEERS WITH MODERN PHYSICS, Fifth Edition, Volume 5
(Chapters 40–46)
ISBN 0-03-026952-0
Library of Congress Catalog Card Number: 99-61820

Address for domestic orders:
Saunders College Publishing, 6277 Sea Harbor Drive, Orlando, FL 32887-6777
1-800-782-4479
e-mail collegesales@harcourt.com

Address for international orders:
International Customer Service, Harcourt, Inc.
6277 Sea Harbor Drive, Orlando FL 32887-6777
(407) 345-3800
Fax (407) 345-4060
e-mail hbintl@harcourt.com

Address for editorial correspondence:
Saunders College Publishing, Public Ledger Building, Suite 1250
150 S. Independence Mall West, Philadelphia, PA 19106-3412

Web Site Address
http://www.harcourtcollege.com

Printed in the United States of America
9012345678 032 10 987654321

SAUNDERS COLLEGE PUBLISHING

soon to become

Harcourt
College Publishers

A Harcourt Higher Learning Company

Soon you will find Saunders College Publishing's distinguished innovation, leadership, and support under a different name . . . a new brand that continues our unsurpassed quality, service, and commitment to education.

We are combining the strengths of our college imprints into one worldwide brand: ◖Harcourt Our mission is to make learning accessible to anyone, anywhere, anytime—reinforcing our commitment to lifelong learning.

We'll soon be Harcourt College Publishers. Ask for us by name.

One Company
"Where Learning
Comes to Life."

PHYSICS

For Scientists and Engineers
with Modern Physics

Fifth Edition

Contents Overview

ix

Table of Contents

part IV   Electricity and Magnetism   707

Preface

*I*n writing this fifth edition of *Physics for Scientists and Engineers*, we have made a major effort to improve the clarity of presentation and to include new pedagogical features that help support the learning and teaching processes. Drawing on positive feedback from users of the fourth edition and reviewers' suggestions, we have made refinements in order to better meet the needs of students and teachers. We have also streamlined the supplements package, which now includes a CD-ROM containing student tutorials and interactive problem-solving software, as well as offerings on the World Wide Web.

This textbook is intended for a course in introductory physics for students majoring in science or engineering. The entire contents of the text could be covered in a three-semester course, but it is possible to use the material in shorter sequences with the omission of selected chapters and sections. The mathematical background of the student taking this course should ideally include one semester of calculus. If that is not possible, the student should be enrolled in a concurrent course in introductory calculus.

OBJECTIVES

This introductory physics textbook has two main objectives: to provide the student with a clear and logical presentation of the basic concepts and principles of physics, and to strengthen an understanding of the concepts and principles through a broad range of interesting applications to the real world. To meet these objectives, we have placed emphasis on sound physical arguments and problem-solving methodology. At the same time, we have attempted to motivate the student through practical examples that demonstrate the role of physics in other disciplines, including engineering, chemistry, and medicine.

CHANGES IN THE FIFTH EDITION

A large number of changes and improvements have been made in preparing the fifth edition of this text. Some of the new features are based on our experiences and on current trends in science education. Other changes have been incorporated in response to comments and suggestions offered by users of the fourth edition and by reviewers of the manuscript. The following represent the major changes in the fifth edition:

Improved Illustrations

- **Time-sequenced events** are represented by circled letters in selected mechanics illustrations. For example, Figure 2.1b (see page 25) shows such letters at the appropriate places on a position–time graph. This construction helps students "translate" the observed motion into its graphical representation.

- **Motion diagrams** are used early in the text to illustrate the difference between velocity and acceleration, concepts easily confused by beginning students. (For example, see Figure 2.9 on page 34, Figure 4.5 on page 81, and Figure 4.8 on page 84.) Students will benefit greatly from sketching their own motion diagrams, as they are asked to do in the Quick Quizzes found in Chapter 4.

- **Greater realism** is achieved with the superimposition of photographs and line art in selected figures (see pages 96 and 97). Also, the three-dimensional appearance of "blocks" in figures accompanying examples and problems in mechanics has been improved (see pages 142 and 143).

More Realistic Worked Examples   Readers familiar with the fourth edition may recall that Example 12.5 involved raising a cylinder onto a step of height *h*. In this idealized example, we calculated the minimum force **F** necessary to raise the cylinder, as well as the reaction force exerted by the step on the cylinder. In the fifth edition, we are pleased to present the revised Example 12.5 (see page 370), in which we calculate the force that a person must apply to a wheelchair's main wheel to roll it up over an uncut sidewalk curb. Although the revised Example 12.5 involves essentially the same calculation as its fourth-edition predecessor (with some changes in notation), we think that the increased realism makes the example more interesting and provides new motivation for studying physics.

Puzzlers   Every chapter begins with an interesting photograph and a caption that includes a Puzzler. Each Puzzler poses a thought-provoking question that is intended to motivate students' curiosity and enhance their interest in the chapter's subject matter. Part or all of the answer to each Puzzler is contained within the chapter text and is indicated by the ⚛ icon.

Chapter Outlines   The opening page of each chapter now includes an outline of the chapter's major headings. This outline gives students and instructors a preview of the chapter's content.

QuickLabs   This new feature encourages students to perform simple experiments on their own, thereby engaging them actively in the learning process. Most QuickLab experiments can be performed with low-cost items such as string, rubber bands, tape, a ruler, drinking straws, and balloons. In most cases, students are asked to observe the outcome of an experiment and to explain their results in terms of what they have learned in the chapter. When appropriate, students are asked to record data and to graph their results.

Quick Quizzes   Several Quick Quiz questions are included in each chapter to provide students with opportunities to test their understanding of the physical concepts presented. Many questions are written in multiple-choice format and require students to make decisions and defend them on the basis of sound reasoning. Some of them have been written to help students overcome common misconceptions. (Instructors should look to the Instructor's Notes in the margins of the Instructor's Annotated Edition for tips regarding certain Quick Quizzes.) Answers to all Quick Quiz questions are found at the end of each chapter.

Marginal Comments and Icons   To provide students with further guidance, common misconceptions and pitfalls are pointed out in comments in the margin of the text. Often, references to the *Saunders Core Concepts in Physics CD-ROM* and useful World Wide Web site addresses are given in these comments to encourage students to expand their understanding of physical concepts. The 💿 icon in the text margin refers students to the specific module and screen number(s) of the *Saunders Core*

Concepts in Physics CD-ROM that deals with the topic under discussion. A text illustration, example, Quick Quiz, or problem marked with the 🖱 icon indicates that it is accompanied by an Interactive Physics™ simulation that can be found on the *Student Tools CD-ROM*. See the Student Ancillaries section (page xviii) for descriptions of these two electronic learning packages.

Applications   Some chapters include Applications, which are about the same length as or slightly longer than worked examples. The Applications demonstrate to students how the physical principles covered in a chapter apply to practical problems of everyday life or engineering. For instance, Applications discuss antilock brakes within the context of static and kinetic friction (see Chapter 5); analyze the tension and compressional forces on the structural components of a truss bridge (see Chapter 12); explore the power delivered in automobile and diesel engine cycles (see Chapter 22); and discuss the construction and circuit wiring of holiday lightbulb strings (see Chapter 28).

Problems   A substantial revision of the end-of-chapter problems was made in an effort to improve their clarity and quality. Approximately 20 percent of the problems (about 650) are new, and most of these new problems are at the intermediate level (as identified by blue problem numbers). Many of the new problems require students to make order-of-magnitude calculations. All problems have been carefully edited and reworded when necessary. Solutions to approximately 20 percent of the end-of-chapter problems are included in the *Student Solutions Manual and Study Guide*. These problems are identified in the text by boxes around their numbers. A smaller subset of solutions are posted on the World Wide Web (**http://www.saunderscollege.com/physics/**) and are accessible to students and instructors using *Physics for Scientists and Engineers*. These problems are identified in the text by the **WEB** icon. See the next section for a complete description of other features of the problem set.

Line-by-Line Revision   The entire text has been carefully edited to improve clarity of presentation and precision of language. We believe that the result is a book that is both accurate and enjoyable to read.

Typographical and Notation Changes   The Text Features section (see page xvi) mentions the use of **boldface** type and screens for emphasizing important statements and definitions. Boldfaced passages in the text of the fifth edition replace the less legible passages appearing in italics in the fourth edition. Similarly, the symbols for vectors stand out very clearly from the surrounding text owing to the strong boldface type used in the fifth edition. As a step toward making equations more transparent and therefore more easily understood, the use of the subscripts "i" and "f" for initial and final values replaces the fourth edition's older notation, which makes use of subscript 0 (usually pronounced "naught") for an initial value and no subscript for a final value. In equations describing motion or direction, variables carry the subscripts x, y, or z whenever added clarity is needed.

Content Changes   Examination of the full Table of Contents might lead one to the impression that the content and organization of the textbook are essentially the same as those of the fourth edition. However, a number of subtle yet significant improvements in content have been made. Following are some examples:

• Section 16.8 contains a more complete and careful derivation of the power or rate of energy transfer for sinusoidal waves on strings. A similar development occurs in Section 17.3, which deals with the intensity of periodic sound waves.

- Section 18.2 contains an improved discussion of the envelope function of a standing wave.

- Chapter 20 contains an updated discussion of the distinction between heat and internal energy. Both heat and work are described and clarified as means of changing the energy of a system.

- Chapter 22 contains a new description of microstates and macrostates of a system, beginning with Section 22.6 on entropy and continuing through the end of the chapter.

- Section 24.3 contains a new list of guidelines for choosing a gaussian surface, allowing the student to take advantage of the symmetry of a charge distribution when determining the electric field.

- Chapter 25 contains new two- and three-dimensional graphs of the electric potential near a point charge and an electric dipole.

- In Chapter 27 and in following chapters, we use "Ohm's law" to refer only to the direct proportionality between current density and electric field seen in some (but not all) materials. See Section 27.2 and the corresponding Instructor's Note for a full explanation.

- Section 29.3 now makes explicit comparison between the potential energy of an electric dipole in an electric field and that of a magnetic dipole in a magnetic field. The section also contains new examples on satellite attitude control and the d'Arsonval galvanometer.

- Chapter 33 contains new information on rectifier circuits, including diodes. The material on rectifiers and filter circuits is now included in Optional Section 33.9, which follows the section on transformers and power transmission.

- In Chapter 35, reflection and refraction are now covered in separate sections, and discussion of Huygens's principle now precedes the section on dispersion and prisms. New Figure 35.8 illustrates retroreflection, which has many practical applications.

- Section 38.2 contains a new subsection considering two-slit diffraction patterns, in which the effects of diffraction and interference are combined.

- Within Section 39.4, new subsections cover space–time graphs and the relativistic Doppler effect. References to the concept of "rest mass" have been deleted.

Many sections in these and other chapters have been streamlined, deleted, or combined with other sections to allow for a more balanced presentation. In this extended version of the text, the former Chapter 44 on superconductivity in the fourth edition of *Physics for Scientists and Engineers with Modern Physics* has been deleted, and an abridged section on this topic has been added to Chapter 43. Some of the sections deleted from the fourth edition may be found on the textbook's Web sites for both instructors and students.

Instructor's Notes   For the first time, tips and comments are offered to instructors in blue marginal Instructor's Notes, which appear only in the Instructor's Annotated Edition. These annotations expand on common student misconceptions; call attention to certain worked examples, QuickLabs, and Quick Quizzes; or cite key physics education research literature that bears on the topic at hand. In some chapters, Instructor's Notes appear as footnotes in the end-of-chapter problem sets; these notes point out related groups of problems found in other chapters of the textbook. The Instructor's Annotated Edition includes Chapters 1 to 39.

CONTENT

The material in this book covers fundamental topics in classical physics and provides an introduction to modern physics. The book is divided into six parts. Part 1 (Chapters 1 to 15) deals with the fundamentals of Newtonian mechanics and the physics of fluids, Part 2 (Chapters 16 to 18) covers wave motion and sound, Part 3 (Chapters 19 to 22) addresses heat and thermodynamics, Part 4 (Chapters 23 to 34) treats electricity and magnetism, Part 5 (Chapters 35 to 38) covers light and optics, and Part 6 (Chapters 39 to 46) deals with relativity and modern physics. Each part opener includes an overview of the subject matter covered in that part, as well as some historical perspectives.

TEXT FEATURES

Most instructors would agree that the textbook selected for a course should be the student's primary guide for understanding and learning the subject matter. Furthermore, the textbook should be easily accessible and should be styled and written to facilitate instruction and learning. With these points in mind, we have included many pedagogical features in the textbook that are intended to enhance its usefulness to both students and instructors. These features are as follows:

Previews   Most chapters begin with a brief preview that includes a discussion of the chapters' objectives and content.

Important Statements and Equations   Most important statements and definitions are set in boldface type or are highlighted with a tan background screen for added emphasis and ease of review. Similarly, important equations are highlighted with a tan background screen to facilitate location.

Problem-Solving Hints   We have included general strategies for solving the types of problems featured both in the examples and in the end-of-chapter problems. This feature helps students to identify necessary steps in problem-solving and to eliminate any uncertainty they might have. Problem-Solving Hints are highlighted with a light blue-gray screen for emphasis and ease of location.

Marginal Notes   Comments and notes appearing in the margin can be used to locate important statements, equations, and concepts in the text.

Illustrations   The three-dimensional appearance of many illustrations has been improved in this fifth edition.

Mathematical Level   We have introduced calculus gradually, keeping in mind that students often take introductory courses in calculus and physics concurrently. Most steps are shown when basic equations are developed, and reference is often made to mathematical appendices at the end of the textbook. Vector products are introduced later in the text, where they are needed in physical applications. The dot product is introduced in Chapter 7, which addresses work and energy; the cross product is introduced in Chapter 11, which deals with rotational dynamics.

Worked Examples   A large number of worked examples of varying difficulty are presented to promote students' understanding of concepts. In many cases, the examples serve as models for solving the end-of-chapter problems. The examples are set off in boxes, and the answers to examples with numerical solutions are highlighted with a light blue-gray screen.

Worked Example Exercises   Many of the worked examples are followed immediately by exercises with answers. These exercises are intended to promote interactivity between the student and the textbook and to immediately reinforce the student's understanding of concepts and problem-solving techniques. The exercises represent extensions of the worked examples.

Conceptual Examples   As in the fourth edition, we have made a concerted effort to emphasize critical thinking and the teaching of physical concepts. We have accomplished this by including Conceptual Examples (for instance, see page 41). These examples provide students with a means of reviewing and applying the concepts presented in a section. Some Conceptual Examples demonstrate the connection between concepts presented in a chapter and other disciplines. The Conceptual Examples can serve as models for students when they are asked to respond to end-of-chapter questions, which are largely conceptual in nature.

Questions   Questions requiring verbal responses are provided at the end of each chapter. Over 1,000 questions are included in this edition. Some questions provide students with a means of testing their mastery of the concepts presented in the chapter. Others could serve as a basis for initiating classroom discussions. Answers to selected questions are included in the *Student Solutions Manual and Study Guide.*

Significant Figures   Significant figures in both worked examples and end-of-chapter problems have been handled with care. Most numerical examples and problems are worked out to either two or three significant figures, depending on the accuracy of the data provided.

Problems   An extensive set of problems is included at the end of each chapter; in all, over 3,000 problems are given throughout the text. Answers to odd-numbered problems are provided at the end of the book in a section whose pages have colored edges for ease of location. For the convenience of both the student and the instructor, about two thirds of the problems are keyed to specific sections of the chapter. The remaining problems, labeled "Additional Problems," are not keyed to specific sections.

Usually, the problems within a given section are presented so that the straightforward problems (those with black problem numbers) appear first; these straightforward problems are followed by problems of increasing difficulty. For ease of identification, the numbers of intermediate-level problems are printed in blue, and those of a small number of challenging problems are printed in magenta.

Review Problems   Many chapters include review problems that require the student to draw on numerous concepts covered in the chapter, as well as on those discussed in previous chapters. These problems could be used by students in preparing for tests and by instructors for special assignments and classroom discussions.

Paired Problems   Some end-of-chapter numerical problems are paired with the same problems in symbolic form. Two paired problems are identified by a common tan background screen.

Computer- and Calculator-Based Problems   Most chapters include one or more problems whose solution requires the use of a computer or graphing calculator. These problems are identified by the 🖥 icon. Modeling of physical phenomena enables students to obtain graphical representations of variables and to perform numerical analyses.

Units   The international system of units (SI) is used throughout the text. The British engineering system of units (conventional system) is used only to a limited extent in the chapters on mechanics, heat, and thermodynamics.

Summaries   Each chapter contains a summary that reviews the important concepts and equations discussed in that chapter.

Appendices and Endpapers   Several appendices are provided at the end of the textbook. Most of the appendix material represents a review of mathematical concepts and techniques used in the text, including scientific notation, algebra, geometry, trigonometry, differential calculus, and integral calculus. Reference to these appendices is made throughout the text. Most mathematical review sections in the appendices include worked examples and exercises with answers. In addition to the mathematical reviews, the appendices contain tables of physical data, conversion factors, atomic masses, and the SI units of physical quantities, as well as a periodic table of the elements. Other useful information, including fundamental constants and physical data, planetary data, a list of standard prefixes, mathematical symbols, the Greek alphabet, and standard abbreviations of units of measure, appears on the endpapers.

ANCILLARIES

The ancillary package has been updated substantially and streamlined in response to suggestions from users of the fourth edition. The most essential changes in the student package are a *Student Solutions Manual and Study Guide* with a tighter focus on problem-solving, the *Student Tools CD-ROM,* and the *Saunders Core Concepts in Physics CD-ROM* developed by Archipelago Productions. Instructors will find increased support for their teaching efforts with new electronic materials.

Student Ancillaries

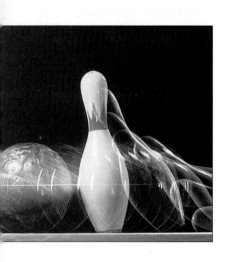

Student Solutions Manual and Study Guide by John R. Gordon, Ralph McGrew, and Raymond A. Serway, with contributions by Duane Deardorff. This two-volume manual features detailed solutions to 20 percent of the end-of-chapter problems from the textbook. Problems in the textbook whose complete solutions are found in the manual are identified by boxes around their numbers. The solutions to many problems follow the **GOAL** protocol described in the textbook (see page 47). The manual also features a list of important equations and concepts, as well as answers to selected end-of-chapter questions.

Pocket Guide by V. Gordon Lind. This 5″ × 7″ paperback is a section-by-section capsule of the textbook and serves as a handy guide for looking up important concepts, formulas, and problem-solving hints.

Student Tools CD-ROM   This CD-ROM contains tools that are designed to enhance the learning of physical concepts and train students to become better problem-solvers. It includes a textbook version of the highly acclaimed Interactive Physics™ software by MSC Working Knowledge and more than 100 Interactive Physics™ simulations keyed to appropriate figures, worked examples, Quick Quizzes, and selected end-of-chapter problems (as identified by the 🖳 icon).

Saunders Core Concepts in Physics CD-ROM   This CD-ROM package developed by Archipelago Productions applies the power of multimedia to the introductory physics course, offering full-motion animation and video, engaging interactive graphics, clear and concise text, and guiding narration. *Saunders Core Concepts in Physics CD-ROM* focuses on those concepts students usually find most difficult in the course, drawing from topics in mechanics, thermodynamics, electric fields, magnetic fields, and optics. The animations and graphics are presented to aid the student in developing accurate conceptual models of difficult topics—topics often too complex to be explained in words or chalkboard illustrations. The CD-ROM also presents step-by-step explorations of problem-solving strategies and provides animations of problems in order to promote conceptual understanding and sharpen problem-solving skills. Topics in the textbook that are further explored on the CD-ROM are identified by marginal  icons that give the appropriate module and screen number(s). Students should look to the CD-ROM to aid in their understanding of these topics.

Student Web Site   Students will have access to an abundance of material at **http://www.saunderscollege.com/physics/.** The Web Site features special topic essays by guest authors, practice problems with answers, and optional topics that accompany selected chapters of the textbook. Also included are selected solutions from the *Student Solutions Manual and Study Guide,* a sampling of the *Pocket Guide,* and a glossary that includes more than 300 physics terms. Students also can take practice quizzes in our Practice Exercises and Testing area.

Physics Laboratory Manual, Second Edition by David Loyd. Updated and redesigned, this manual supplements the learning of basic physical principles while introducing laboratory procedures and equipment. Each chapter includes a pre-laboratory assignment, objectives, an equipment list, the theory behind the experiment, step-by-step experimental procedures, and questions. A laboratory report form is provided for each experiment so that students can record data and make calculations. Students are encouraged to apply statistical analysis to their data so that they can develop their ability to judge the validity of their results.

So You Want to Take Physics: A Preparatory Course with Calculus by Rodney Cole. This introductory-level book is useful to those students who need additional preparation before or during a calculus-based course in physics. The friendly, straightforward style makes it easier to understand how mathematics is used in the context of physics.

Life Science Applications for Physics by Jerry Faughn. This book provides examples, readings, and problems from the biological sciences as they relate to physics. Topics include "Friction in Human Joints," "Physics of the Human Circulatory System," "Physics of the Nervous System," and "Ultrasound and Its Applications." This supplement is useful in those courses taken by a significant number of pre-med students.

Instructor's Ancillaries

Instructor's Manual with Solutions by Ralph McGrew, Jeff Saul, and Charles Teague, with contributions by Duane Deardorff and Rhett Allain. This manual contains chapter summaries, answers to even-numbered problems, and complete worked solutions to all the problems in the textbook. The solutions to problems new to the fifth edition are marked for easy identification by the instructor. New to this edition of the manual are suggestions on how to teach difficult topics and how to help students overcome common misconceptions. These suggestions are based on recent research in physics education.

Instructor's Web Site   The instructor's area at **http://www.saunderscollege.com/physics/** includes a listing of overhead transparencies; a guide to relevant experiments in David Loyd's *Physics Laboratory Manual, Second Edition*; a correlation guide between sections in *Physics for Scientists and Engineers* and modules in the *Saunders Core Concepts in Physics CD-ROM*; supplemental problems with answers; optional topics to accompany selected chapters of the textbook; and a problems correlation guide.

Instructor's Resource CD-ROM   This CD-ROM accompanying the fifth edition of *Physics for Scientists and Engineers* has been created to provide instructors with an exciting new tool for classroom presentation. The CD-ROM contains a collection of graphics files of line art from the textbook. These files can be opened directly, can be imported into a variety of presentation packages, or can be used in the presentation package included on the CD-ROM. The labels for each piece of art have been enlarged and boldfaced to facilitate classroom viewing. The CD-ROM contains electronic files of the *Instructor's Manual, Test Bank,* and *Practice Problems with Solutions.*

CAPA: A Computer-Assisted Personalized Approach   CAPA is a network system for learning, teaching, assessment, and administration. It provides students with personalized problem sets, quizzes, and examinations consisting of qualitative conceptual problems and quantitative problems, including problems from *Physics for Scientists and Engineers.* CAPA was developed through a collaborative effort of the Physics–Astronomy, Computer Science, and Chemistry Departments at Michigan State University. Students are given instant feedback and relevant hints via the Internet and may correct errors without penalty before an assignment's due date. The system records each student's participation and performance on assignments, quizzes, and examinations; and records are available on-line to both the individual student and to his or her instructor. For more information, visit the CAPA Web site at **http://www.pa.msu.edu/educ/CAPA/**

WebAssign: A Web-Based Homework System   WebAssign is a Web-based homework delivery, collection, grading, and recording service developed at North Carolina State University. Instructors who sign up for WebAssign can assign homework to their students, using questions and problems taken directly from *Physics for Scientists and Engineers.* WebAssign gives students immediate feedback on their homework that helps them to master information and skills, leading to greater competence and better grades. WebAssign can free instructors from the drudgery of grading homework and recording scores, allowing them to devote more time to meeting with students and preparing classroom presentations. Details about and a demonstration of WebAssign are available at **http://webassign.net/info**. For more information about ordering this service, contact WebAssign at **webassign@ncsu.edu**

Homework Service   With this service, instructors can reduce their grading workload by assigning thought-provoking homework problems using the World Wide Web. Instructors browse problem banks that include problems from *Physics for Scientists and Engineers,* select those they wish to assign to their students, and then let the Homework Service take over the delivery and grading. This system was developed and is maintained by Fred Moore at the University of Texas (**moore@physics.utexas.edu**). Students download their unique problems, submit their answers, and obtain immediate feedback; if students' answers are incorrect, they can resubmit them. This rapid grading feature facilitates effective learning. After the due date of their assignments, students can obtain the solutions to their problems. Minimal on-line connect time is required. The Homework Service uses algorithm-based problems: This means that each student solves sets of problems that are different from those given to other students. Details about and a demonstration of this service are available at **http://hw10.ph.utexas.edu/instInst.html**

Printed Test Bank by Edward Adelson. The *Printed Test Bank* contains approximately 2,300 multiple-choice questions. It is provided for the instructor who does not have access to a computer. About 20% of the old test items have been replaced with new, concept-based, thought-provoking questions.

Computerized Test Bank   Available in Windows™ and Macintosh® formats, the *Computerized Test Bank* contains more than 2,300 multiple-choice questions, representing every chapter of the text. The *Computerized Test Bank* enables the instructor to create many unique tests by allowing the editing of questions and the addition of new questions. The software program solves all problems and prints each answer on a separate grading key. All questions have been reviewed for accuracy.

Overhead Transparency Acetates   This collection of transparencies consists of 300 full-color figures from the text and features large print for easy viewing in the classroom.

Instructor's Manual for Physics Laboratory Manual by David Loyd. Each chapter contains a discussion of the experiment, teaching hints, answers to selected questions, and a post-laboratory quiz with short-answer and essay questions. It also includes a list of the suppliers of scientific equipment and a summary of the equipment needed for each of the laboratory experiments in the manual.

Saunders College Publishing, a division of Harcourt College Publishers, may provide complementary instructional aids and supplements or supplement packages to those adopters qualified under our adoption policy. Please contact your sales representative for more information. If as an adopter or potential user you receive supplements you do not need, please return them to your sales representative or send them to

Attn: Returns Department
Troy Warehouse
465 South Lincoln Drive
Troy, MO 63379

TEACHING OPTIONS

The topics in this textbook are presented in the following sequence: classical mechanics, mechanical waves, and heat and thermodynamics, followed by electricity and magnetism, electromagnetic waves, optics, and relativity. This presentation represents a more traditional sequence, with the subject of mechanical waves being presented before the topics of electricity and magnetism. Some instructors may prefer to cover this material after completing electricity and magnetism (i.e., after Chapter 34). The chapter on relativity was placed near the end of the text because this topic often is treated as an introduction to the era of "modern physics." If time permits, instructors may choose to cover Chapter 39 in the first semester after completing Chapter 14, as it concludes the material on Newtonian mechanics.

Instructors teaching a two-semester sequence can delete sections and chapters without any loss of continuity. We have labeled these as "Optional" in the Table of Contents and in the appropriate sections of the text. For student enrichment, instructors can assign some of these sections or chapters as extra reading.

ACKNOWLEDGMENTS

The fifth edition of this textbook was prepared with the guidance and assistance of many professors who reviewed part or all of the manuscript, the pre-revision text, or both. We wish to acknowledge the following scholars and express our sincere appreciation for their suggestions, criticisms, and encouragement:

Edward Adelson, *Ohio State University*
Roger Bengtson, *University of Texas at Austin*
Joseph Biegen, *Broome Community College*
Ronald J. Bieniek, *University of Missouri at Rolla*
Ronald Brown, *California Polytechnic State University—San Luis Obispo*
Michael E. Browne, *University of Idaho*
Tim Burns, *Leeward Community College*
Randall Caton, *Christopher Newport University*
Sekhar Chivukula, *Boston University*
Alfonso Diáz-Jiménez, *ADJOIN Research Center*
N. John DiNardo, *Drexel University*
F. Eugene Dunnum, *University of Florida*
William Ellis, *Cornell University*
F. Paul Esposito, *University of Cincinnati*
Paul Fahey, *University of Scranton*

Arnold Feldman, *University of Hawaii at Manoa*
Alexander Firestone, *Iowa State University*
Robert Forsythe, *Broome Community College*
Philip Fraundorf, *University of Missouri at St. Louis*
John Gerty, *Broome Community College*
John B. Gruber, *San Jose State University*
John Hubisz, *North Carolina State University*
Joey Huston, *Michigan State University*
Calvin S. Kalman, *Concordia University*
Natalie Kerr, M.D., *University of Tennessee, Memphis*
Peter Killen, *University of Queensland (Australia)*
Earl Koller, *Stevens Institute of Technology*
David LaGraffe, *U.S. Military Academy*
Ying-Cheng Lai, *University of Kansas*

Donald Larson, *Drexel University*
Robert Lieberman, *Cornell University*
Ralph McGrew, *Broome Community College*
David Mills, *Monash University (Australia)*
Clement J. Moses, *Utica College*
Peter Parker, *Yale University*
John Parsons, *Columbia University*
Arnold Perlmutter, *University of Miami*
Henry Schriemer, *Queen's University (Canada)*
Paul Snow, *University of Bath (U.K.)*
Edward W. Thomas, *Georgia Institute of Technology*
Charles C. Vuille, *Embry-Riddle Aeronautical University*
Xiaojun Wang, *Georgia Southern University*
Gail Welsh, *Salisbury State University*

This book was carefully checked for accuracy by James H. Smith *(University of Illinois at Urbana-Champaign)*, Gregory Snow *(University of Nebraska—Lincoln)*, Edward Gibson *(California State University—Sacramento)*, Ronald Jodoin *(Rochester Institute of Technology)*, Arnold Perlmutter *(University of Miami)*, Michael Paesler *(North Carolina State University)*, and Clement J. Moses *(Utica College)*.

We thank the following people for their suggestions and assistance during the preparation of earlier editions of this textbook:

George Alexandrakis, *University of Miami*
Elmer E. Anderson, *University of Alabama*
Wallace Arthur, *Fairleigh Dickinson University*
Duane Aston, *California State University—Sacramento*
Stephen Baker, *Rice University*
Richard Barnes, *Iowa State University*
Stanley Bashkin, *University of Arizona*
Robert Bauman, *University of Alabama*
Marvin Blecher, *Virginia Polytechnic Institute and State University*
Jeffrey J. Braun, *University of Evansville*
Kenneth Brownstein, *University of Maine*
William A. Butler, *Eastern Illinois University*
Louis H. Cadwell, *Providence College*

Ron Canterna, *University of Wyoming*
Bo Casserberg, *University of Minnesota*
Soumya Chakravarti, *California Polytechnic State University*
C. H. Chan, *The University of Alabama at Huntsville*
Edward Chang, *University of Massachusetts at Amherst*
Don Chodrow, *James Madison University*
Clifton Bob Clark, *University of North Carolina at Greensboro*
Walter C. Connolly, *Appalachian State University*
Hans Courant, *University of Minnesota*
Lance E. De Long, *University of Kentucky*
James L. DuBard, *Binghamton-Southern College*

F. Paul Esposito, *University of Cincinnati*
Jerry S. Faughn, *Eastern Kentucky University*
Paul Feldker, *Florissant Valley Community College*
Joe L. Ferguson, *Mississippi State University*
R. H. Garstang, *University of Colorado at Boulder*
James B. Gerhart, *University of Washington*
John R. Gordon, *James Madison University*
Clark D. Hamilton, *National Bureau of Standards*
Mark Heald, *Swarthmore College*
Herb Helbig, *Rome Air Development Center*
Howard Herzog, *Broome Community College*
Paul Holoday, *Henry Ford Community College*

Jerome W. Hosken, *City College of San Francisco*

Larry Hmurcik, *University of Bridgeport*

William Ingham, *James Madison University*

Mario Iona, *University of Denver*

Karen L. Johnston, *North Carolina State University*

Brij M. Khorana, *Rose-Hulman Institute of Technology*

Larry Kirkpatrick, *Montana State University*

Carl Kocher, *Oregon State University*

Robert E. Kribel, *Jacksonville State University*

Barry Kunz, *Michigan Technological University*

Douglas A. Kurtze, *Clarkson University*

Fred Lipschultz, *University of Connecticut*

Francis A. Liuima, *Boston College*

Robert Long, *Worcester Polytechnic Institute*

Roger Ludin, *California Polytechnic State University*

Nolen G. Massey, *University of Texas at Arlington*

Charles E. McFarland, *University of Missouri at Rolla*

Ralph V. McGrew, *Broome Community College*

James Monroe, *The Pennsylvania State University, Beaver Campus*

Bruce Morgan, *United States Naval Academy*

Clement J. Moses, *Utica College*

Curt Moyer, *Clarkson University*

David Murdock, *Tennessee Technological University*

A. Wilson Nolle, *University of Texas at Austin*

Thomas L. O'Kuma, *San Jacinto College North*

Fred A. Otter, *University of Connecticut*

George Parker, *North Carolina State University*

William F. Parks, *University of Missouri at Rolla*

Philip B. Peters, *Virginia Military Institute*

Eric Peterson, *Highland Community College*

Richard Reimann, *Boise State University*

Joseph W. Rudmin, *James Madison University*

Jill Rugare, *DeVry Institute of Technology*

Charles Scherr, *University of Texas at Austin*

Eric Sheldon, *University of Massachusetts—Lowell*

John Shelton, *College of Lake County*

Stan Shepard, *The Pennsylvania State University*

James H. Smith, *University of Illinois at Urbana-Champaign*

Richard R. Sommerfield, *Foothill College*

Kervork Spartalian, *University of Vermont*

Robert W. Stewart, *University of Victoria*

James Stith, *American Institute of Physics*

Charles D. Teague, *Eastern Kentucky University*

Edward W. Thomas, *Georgia Institute of Technology*

Carl T. Tomizuka, *University of Arizona*

Herman Trivilino, *San Jacinto College North*

Som Tyagi, *Drexel University*

Steve Van Wyk, *Chapman College*

Joseph Veit, *Western Washington University*

T. S. Venkataraman, *Drexel University*

Noboru Wada, *Colorado School of Mines*

James Walker, *Washington State University*

Gary Williams, *University of California, Los Angeles*

George Williams, *University of Utah*

Edward Zimmerman, *University of Nebraska, Lincoln*

Earl Zwicker, *Illinois Institute of Technology*

We are grateful to Ralph McGrew for organizing the end-of-chapter problems, writing many new problems, and his suggestions for improving the content of the textbook. The new end-of-chapter problems were written by Rich Cohen, John DiNardo, Robert Forsythe, Ralph McGrew, and Ronald Bieniek, with suggestions by Liz McGrew, Alexandra Héder, and Richard McGrew. We thank Laurent Hodges for permission to use selected end-of-chapter problems. We are grateful to John R. Gordon, Ralph McGrew, and Duane Deardorff for writing the *Student Solutions Manual and Study Guide,* and we thank Michael Rudmin for its attractive layout. Ralph McGrew, Jeff Saul, and Charles Teague have prepared an excellent *Instructor's Manual,* and we thank them. We thank Gloria Langer, Linda Miller, and Jennifer Serway for their excellent work in preparing the *Instructor's Manual* and the supplemental materials that appear on our Web site.

Special thanks and recognition go to the professional staff at Saunders College Publishing—in particular, Susan Pashos, Sally Kusch, Carol Bleistine, Frank Messina, Suzanne Hakanen, Ruth Hoover, Alexandra Buczek, Pauline Mula, Walter Neary, and John Vondeling—for their fine work during the development and production of this textbook. We are most appreciative of the intelligent line editing by Irene Nunes, the final copy editing by Sue Nelson and Mary Patton, the excellent artwork produced by Rolin Graphics, and the dedicated photo research efforts of Dena Digilio Betz.

Finally, we are deeply indebted to our wives and children for their love, support, and long-term sacrifices.

Raymond A. Serway
Chapel Hill, North Carolina

Robert J. Beichner
Raleigh, North Carolina

John W. Jewett, Jr.
Pomona, California

To the Student

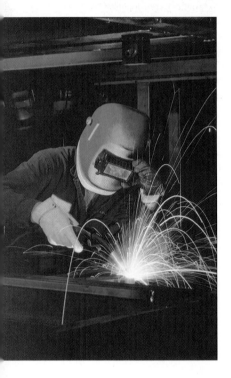

It is appropriate to offer some words of advice that should be of benefit to you, the student. Before doing so, we assume that you have read the Preface, which describes the various features of the text that will help you through the course.

HOW TO STUDY

Very often instructors are asked, "How should I study physics and prepare for examinations?" There is no simple answer to this question, but we would like to offer some suggestions that are based on our own experiences in learning and teaching over the years.

First and foremost, maintain a positive attitude toward the subject matter, keeping in mind that physics is the most fundamental of all natural sciences. Other science courses that follow will use the same physical principles, so it is important that you understand and are able to apply the various concepts and theories discussed in the text.

CONCEPTS AND PRINCIPLES

It is essential that you understand the basic concepts and principles before attempting to solve assigned problems. You can best accomplish this goal by carefully reading the textbook before you attend your lecture on the covered material. When reading the text, you should jot down those points that are not clear to you. We've purposely left wide margins in the text to give you space for doing this. Also be sure to make a diligent attempt at answering the questions in the Quick Quizzes as you come to them in your reading. We have worked hard to prepare questions that help you judge for yourself how well you understand the material. The QuickLabs provide an occasional break from your reading and will help you to experience some of the new concepts you are trying to learn. During class, take careful notes and ask questions about those ideas that are unclear to you. Keep in mind that few people are able to absorb the full meaning of scientific material after only one reading. Several readings of the text and your notes may be necessary. Your lectures and laboratory work supplement reading of the textbook and should clarify some of the more difficult material. You should minimize your memorization of material. Successful memorization of passages from the text, equations, and derivations does not necessarily indicate that you understand the material. Your understanding of the material will be enhanced through a combination of efficient study habits, discussions with other students and with instructors, and your ability to solve the problems presented in the textbook. Ask questions whenever you feel clarification of a concept is necessary.

STUDY SCHEDULE

It is important that you set up a regular study schedule, preferably one that is daily. Make sure that you to read the syllabus for the course and adhere to the schedule set by your instructor. The lectures will be much more meaningful if you read the corresponding textual material before attending them. As a general rule, you should devote about two hours of study time for every hour you are in class. If you are having trouble with the course, seek the advice of the instructor or other students who have taken the course. You may find it necessary to seek further instruction from experienced students. Very often, instructors offer review sessions in addition to regular class periods. It is important that you avoid the practice of delaying study until a day or two before an exam. More often than not, this approach has disastrous results. Rather than undertake an all-night study session, briefly review the basic concepts and equations and get a good night's rest. If you feel you need additional help in understanding the concepts, in preparing for exams, or in problem-solving, we suggest that you acquire a copy of the *Student Solutions Manual and Study Guide* that accompanies this textbook; this manual should be available at your college bookstore.

USE THE FEATURES

You should make full use of the various features of the text discussed in the preface. For example, marginal notes are useful for locating and describing important equations and concepts, and **boldfaced** type indicates important statements and definitions. Many useful tables are contained in the Appendices, but most are incorporated in the text where they are most often referenced. Appendix B is a convenient review of mathematical techniques.

Answers to odd-numbered problems are given at the end of the textbook, answers to Quick Quizzes are located at the end of each chapter, and answers to selected end-of-chapter questions are provided in the *Student Solutions Manual and Study Guide*. The exercises (with answers) that follow some worked examples represent extensions of those examples; in most of these exercises, you are expected to perform a simple calculation (see Example 4.7 on page 90). Their purpose is to test your problem-solving skills as you read through the text. Problem-Solving Hints are included in selected chapters throughout the text and give you additional information about how you should solve problems. The Table of Contents provides an overview of the entire text, while the Index enables you to locate specific material quickly. Footnotes sometimes are used to supplement the text or to cite other references on the subject discussed.

After reading a chapter, you should be able to define any new quantities introduced in that chapter and to discuss the principles and assumptions that were used to arrive at certain key relations. The chapter summaries and the review sections of the *Student Solutions Manual and Study Guide* should help you in this regard. In some cases, it may be necessary for you to refer to the index of the text to locate certain topics. You should be able to correctly associate with each physical quantity the symbol used to represent that quantity and the unit in which the quantity is specified. Furthermore, you should be able to express each important relation in a concise and accurate prose statement.

PROBLEM-SOLVING

R. P. Feynman, Nobel laureate in physics, once said, "You do not know anything until you have practiced." In keeping with this statement, we strongly advise that you develop the skills necessary to solve a wide range of problems. Your ability to solve problems will be one of the main tests of your knowledge of physics, and therefore you should try to solve as many problems as possible. It is good practice to try to find alternate solutions to the same problem. For example, you can solve problems in mechanics using Newton's laws, but very often an alternative method that draws on energy considerations is more direct. You should not deceive yourself into thinking that you understand a problem merely because you have seen it solved in class. You must be able to solve the problem and similar problems on your own.

The approach to solving problems should be carefully planned. A systematic plan is especially important when a problem involves several concepts. First, read the problem several times until you are confident you understand what is being asked. Look for any key words that will help you interpret the problem and perhaps allow you to make certain assumptions. Your ability to interpret a question properly is an integral part of problem-solving. Second, you should acquire the habit of writing down the information given in a problem and those quantities that need to be found; for example, you might construct a table listing both the quantities given and the quantities to be found. This procedure is sometimes used in the worked examples of the textbook. Finally, after you have decided on the method you feel is appropriate for a given problem, proceed with your solution. General problem-solving strategies of this type are included in the text and are highlighted with a light blue-gray screen. We have also developed the **GOAL** protocol (see page 47) to help guide you through complex problems. If you follow the steps of this procedure (**G**ather information, **O**rganize your approach, carry out your **A**nalysis, and finally **L**earn from your work), you will not only find it easier to come up with a solution, but you will also gain more from your efforts.

Often, students fail to recognize the limitations of certain formulas or physical laws in a particular situation. It is very important that you understand and remember the assumptions that underlie a particular theory or formalism. For example, certain equations in kinematics apply only to a particle moving with constant acceleration. These equations are not valid for describing motion whose acceleration is not constant, such as the motion of an object connected to a spring or the motion of an object through a fluid.

General Problem-Solving Strategy

Most courses in general physics require the student to learn the skills of problem-solving, and exams are largely composed of problems that test such skills. This brief section describes some useful ideas that will enable you to increase your accuracy in solving problems, enhance your understanding of physical concepts, eliminate initial panic or lack of direction in approaching a problem, and organize your work. One way to help accomplish these goals is to adopt a problem-solving strategy. Many chapters in this text include Problem-Solving Hints that should help you through the "rough spots."

In developing problem-solving strategies, five basic steps are commonly followed:

• Draw a suitable diagram with appropriate labels and coordinate axes (if needed).

- As you examine what is being asked in the problem, identify the basic physical principle (or principles) that are involved, listing the knowns and the unknowns.

- Select a basic relationship or derive an equation that can be used to find the unknown, and then solve the equation for the unknown symbolically.

- Substitute the given values along with the appropriate units into the equation.

- Obtain a numerical value for the unknown. The problem is verified and receives a check mark if the following questions can be properly answered: Do the units match? Is the answer reasonable? Is the plus or minus sign proper or meaningful?

One of the purposes of this strategy is to promote accuracy. Properly drawn diagrams can eliminate many sign errors. Diagrams also help to isolate the physical principles of the problem. Obtaining symbolic solutions and carefully labeling knowns and unknowns will help eliminate other careless errors. The use of symbolic solutions should help you think in terms of the physics of the problem. A check of units at the end of the problem can indicate a possible algebraic error. The physical layout and organization of your problem will make the final product more understandable and easier to follow. Once you have developed an organized system for examining problems and extracting relevant information, you will become a more confident problem-solver.

EXPERIMENTS

Physics is a science based on experimental observations. In view of this fact, we recommend that you try to supplement your reading of the text by performing various types of "hands-on" experiments, either at home or in the laboratory. Most chapters include one or two QuickLabs that describe simple experiments you can do on your own. These can be used to test ideas and models discussed in class or in the textbook. For example, the common Slinky™ toy is an excellent tool for studying traveling waves; a ball swinging on the end of a long string can be used to investigate pendulum motion; various masses attached to the end of a vertical spring or rubber band can be used to determine their elastic nature; an old pair of Polaroid sunglasses, some discarded lenses, and a magnifying glass are the components of various experiments in optics; and the approximate measure of the acceleration of gravity can be determined simply by measuring with a stopwatch the time it takes for a ball to drop from a known height. The list of such experiments is endless. When physical models are not available, be imaginative and try to develop models of your own.

NEW MEDIA

We strongly encourage you to use one or more of the following multimedia products that accompany this textbook. It is far easier to understand physics if you see it in action, and these new materials will enable you to become a part of that action.

Student Tools CD-ROM   The dual-platform (Windows™- and Macintosh®-compatible) *Student Tools CD-ROM* is available with each new copy of the textbook. This CD-ROM contains a textbook version of the Interactive Physics™ program by MSC Working Knowledge. Interactive Physics™ simulations are keyed to the following figures, worked examples, Quick Quizzes, and end-of-chapter problems (identified in the text with the  icon).

Help your students get to the "Core" of physics with Saunders Core Concepts in Physics CD-ROM!

Available with Serway & Beichner's *Physics for Scientists and Engineers, Fifth Edition*

Core Concepts in Physics CD-ROM covers the core principles of calculus-based physics through the use of on-screen simulations, animations, and videos. The combination of a conceptual presentation of physical principles and problem-solving helps students "see" material often too difficult for instructors to represent with chalkboard diagrams or with written and verbal descriptions.

Core Concepts in Physics CD-ROM contains more than 350 animated and live videos. "Real world" examples, graphics, models, and step-by-step explanations of mathematics help to spark students' interest and facilitate their exploration of the fascinating world of physics!

SERWAY • BEICHNER
PHYSICS
For Scientists and Engineers
FIFTH EDITION

PEDAGOGICAL COLOR CHART

Part 1 (Chapters 1–15)
Mechanics

Displacement and position vectors

Linear ($\mathbf{v}$) and angular ($\boldsymbol{\omega}$) velocity vectors

Velocity component vectors

Force vectors ($\mathbf{F}$)

Force component vectors

Acceleration vectors ($\mathbf{a}$)

Acceleration component vectors

Linear ($\mathbf{p}$) and angular ($\mathbf{L}$) momentum vectors

Torque vectors ($\boldsymbol{\tau}$)

Linear or rotational motion directions

Springs

Pulleys

Part 4 (Chapters 23–34)
Electricity and Magnetism

Electric fields

Magnetic fields

Positive charges

(continued on back)

SAUNDERS COLLEGE PUBLISHING

A Division of Harcourt College Publishers

SERWAY • BEICHNER

PHYSICS

For Scientists and Engineers
FIFTH EDITION

PEDAGOGICAL COLOR CHART

Part 4 (Chapters 23–34)
Electricity and Magnetism
(*continued*)

Negative charges

Batteries and other
  dc power supplies

Switches

Resistors

Capacitors

Inductors (coils)

Voltmeters

Ammeters

Galvanometers

ac Generators

Ground symbol

Part 5 (Chapters 35–38)
Light and Optics

Light rays

Lenses and
  prisms

Mirrors

Objects

Images

SAUNDERS COLLEGE
PUBLISHING

A Division of Harcourt College Publishers

Application,
Visualization, and Practice.
Student Tools CD-ROM helps
students understand the
natural forces and principles
of physics.

This CD-ROM contains tools that are designed to enhance the learning of physical concepts and to assist your students as they become better problem-solvers. It includes a textbook version of the highly acclaimed *Interactive Physics*™ software by Knowledge Revolution and more than 100 *Interactive Physics*™ simulations keyed to appropriate worked examples, figures, Quick Quizzes, and selected end-of-chapter problems. *The Student Tools CD-ROM* also features a graphing calculator and support for working the end-of-chapter problems that require a computer.

Saunders Core Concepts in Physics CD-ROM   In addition, you may purchase the *Saunders Core Concepts in Physics CD-ROM* developed by Archipelago Productions. This CD-ROM provides a complete multimedia presentation of selected topics in mechanics, thermodynamics, electromagnetism, and optics. It contains more than 350 movies—both animated and live video—that bring to life laboratory demonstrations, "real-world" examples, graphic models, and step-by-step explanations of essential mathematics. Those CD-ROM modules that supplement the material in *Physics for Scientists and Engineers* are identified in the margin of the text by the 💿 icon.

AN INVITATION TO PHYSICS

It is our sincere hope that you too will find physics an exciting and enjoyable experience and that you will profit from this experience, regardless of your chosen profession. Welcome to the exciting world of physics!

The scientist does not study nature because it is useful; he studies it because he delights in it, and he delights in it because it is beautiful. If nature were not beautiful, it would not be worth knowing, and if nature were not worth knowing, life would not be worth living.

—Henri Poincaré

About the Authors

Raymond A. Serway received his doctorate at Illinois Institute of Technology and is Professor Emeritus at James Madison University. In 1990, he received the Madison Scholar award at James Madison University, where he taught for 17 years. Dr. Serway began his teaching career at Clarkson University, where he conducted research and taught from 1967 to 1980. He was the recipient of the Distinguished Teaching Award at Clarkson University in 1977 and of the Alumni Achievement Award from Utica College in 1985. As Guest Scientist at the IBM Research Laboratory in Zurich, Switzerland, he worked with K. Alex Müller, 1987 Nobel Prize recipient. Dr. Serway also was a visiting scientist at Argonne National Laboratory, where he collaborated with his mentor and friend, Sam Marshall. In addition to earlier editions of this textbook, Dr. Serway is the author of *Principles of Physics, Second Edition,* and co-author of *College Physics, Fifth Edition,* and *Modern Physics, Second Edition*; he also is the author of the high-school textbook *Physics,* published by Holt, Rinehart, & Winston. In addition, Dr. Serway has published more than 40 research papers in the field of condensed matter physics and has given more than 60 presentations at professional meetings. Dr. Serway and his wife Elizabeth enjoy traveling, golfing, and spending quality time with their four children and four grandchildren.

Robert J. Beichner received his doctorate at the State University of New York at Buffalo. Currently, he is Associate Professor of Physics at North Carolina State University, where he directs the Physics Education Research and Development Group. He has more than 20 years of teaching experience at the community-college, four year–college, and university levels. His research interests are centered on improving physics instruction: In his work, he has published studies of video-based laboratories, collaborative learning, technology-supplemented learning environments, and the assessment of student understanding of various physics topics. Dr. Beichner has held several leadership roles in the field of physics education research and has given numerous talks and colloquia on his work. In addition to being an author of this textbook, he is the co-author of two CD-ROMs, several commercially available software packages, and two books for preservice elementary school teachers. Dr. Beichner is an avid sea kayaker and enjoys spending time with his wife Mary and their two daughters, Sarah and Julie.

John W. Jewett, Jr. earned his doctorate at Ohio State University, specializing in optical and magnetic properties of condensed matter. He is currently Professor of Physics at California State Polytechnic University—Pomona. Throughout his teaching career, Dr. Jewett has been active in promoting science education. In addition to receiving four National Science Foundation grants, he helped found and direct the Southern California Area Modern Physics Institute (SCAMPI). He also is the director of Science IMPACT (Institute for Modern Pedagogy and Creative Teaching), which works with teachers and schools to develop effective science curricula. Both organizations operate in the United States and abroad. Dr. Jewett's honors include four Meritorious Performance and Professional Promise awards, selection as Outstanding Professor at California State Polytechnic University for 1991–1992, and the Excellence in Undergraduate Physics Teaching Award from the American Association of Physics Teachers (AAPT) in 1998. He has given many presentations both domestically and abroad, including multiple presentations at national meetings of the AAPT. He will co-author the third edition of *Principles of Physics* with Dr. Serway. Dr. Jewett enjoys playing piano, traveling, and collecting antique quack medical devices, as well as spending time with his wife Lisa and their children.

Pedagogical Color Chart

Part 1 (Chapters 1–15) : Mechanics

Displacement and position vectors	→	Linear ($\mathbf{p}$) and angular ($\mathbf{L}$) momentum vectors	→
Linear ($\mathbf{v}$) and angular ($\boldsymbol{\omega}$) velocity vectors	→	Torque vectors (τ)	→
Velocity component vectors	→	Linear or rotational motion directions	
Force vectors ($\mathbf{F}$)	→	Springs	
Force component vectors	→		
Acceleration vectors ($\mathbf{a}$)	→	Pulleys	
Acceleration component vectors	→		

Part 4 (Chapters 23–34) : Electricity and Magnetism

Electric fields	→	Capacitors	
Magnetic fields	→	Inductors (coils)	
Positive charges	+	Voltmeters	V
Negative charges	−	Ammeters	A
Resistors		Galvanometers	G
Batteries and other dc power supplies		ac Generators	~
Switches		Ground symbol	

Part 5 (Chapters 35–38) : Light and Optics

Light rays	→	Objects	
Lenses and prisms		Images	
Mirrors			

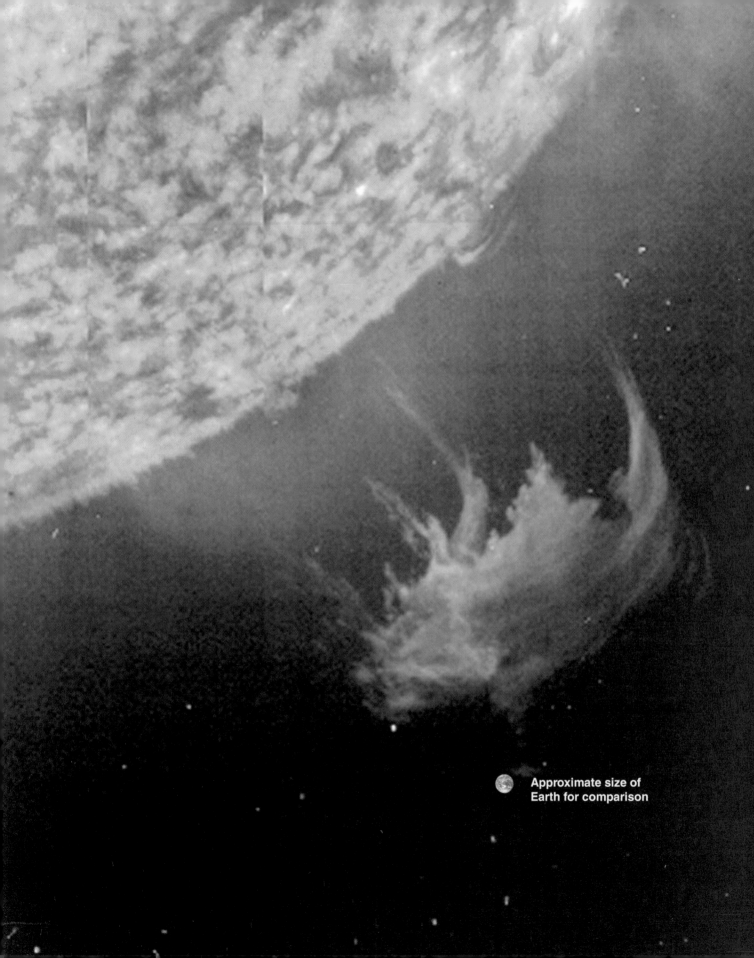

Approximate size of
Earth for comparison

part 6

Modern Physics

At the end of the 19th century, many scientists believed that they had learned most of what there was to know about physics. Newton's laws of motion and his theory of universal gravitation, Maxwell's theoretical work in unifying electricity and magnetism, and the laws of thermodynamics and kinetic theory were highly successful in explaining a variety of phenomena.

As the 19th century turned to the 20th, however, a major revolution shook the world of physics. In 1900 Planck provided the basic ideas that led to the formulation of the quantum theory, and in 1905 Einstein formulated his brilliant special theory of relativity. The excitement of the times is captured in Einstein's own words: "It was a marvelous time to be alive." Both ideas were to have a profound effect on our understanding of nature. Within a few decades, these two theories inspired new developments and theories in the fields of atomic physics, nuclear physics, and condensed-matter physics.

In Chapter 39 we introduce the special theory of relativity. The theory provides us with a new and deeper view of physical laws. Although the concepts underlying this theory often violate our common sense, the theory correctly predicts the results of experiments involving speeds near the speed of light. In the extended version of this textbook, *Physics for Scientists and Engineers with Modern Physics,* we cover the basic concepts of quantum mechanics and their application to atomic and molecular physics, and we introduce solid-state physics, nuclear physics, particle physics, and cosmology.

You should keep in mind that, although the physics developed during the 20th century has led to a multitude of important technological achievements, the story is still incomplete. Discoveries will continue to evolve during our lifetimes, and many of these discoveries will deepen or refine our understanding of nature and the world around us. It is still a "marvelous time to be alive."

◀ Courtesy of the SOHO-EIT Consortium.

c h a p t e r

39

Relativity

Most of our everyday experiences and observations have to do with objects that move at speeds much less than the speed of light. Newtonian mechanics was formulated to describe the motion of such objects, and this formalism is still very successful in describing a wide range of phenomena that occur at low speeds. It fails, however, when applied to particles whose speeds approach that of light.

Experimentally, the predictions of Newtonian theory can be tested at high speeds by accelerating electrons or other charged particles through a large electric potential difference. For example, it is possible to accelerate an electron to a speed of $0.99c$ (where c is the speed of light) by using a potential difference of several million volts. According to Newtonian mechanics, if the potential difference is increased by a factor of 4, the electron's kinetic energy is four times greater and its speed should double to $1.98c$. However, experiments show that the speed of the electron—as well as the speed of any other particle in the Universe—always remains less than the speed of light, regardless of the size of the accelerating voltage. Because it places no upper limit on speed, Newtonian mechanics is contrary to modern experimental results and is clearly a limited theory.

In 1905, at the age of only 26, Einstein published his special theory of relativity. Regarding the theory, Einstein wrote:

> The relativity theory arose from necessity, from serious and deep contradictions in the old theory from which there seemed no escape. The strength of the new theory lies in the consistency and simplicity with which it solves all these difficulties . . . . [1]

Although Einstein made many other important contributions to science, the special theory of relativity alone represents one of the greatest intellectual achievements of all time. With this theory, experimental observations can be correctly predicted over the range of speeds from $v = 0$ to speeds approaching the speed of light. At low speeds, Einstein's theory reduces to Newtonian mechanics as a limiting situation. It is important to recognize that Einstein was working on electromagnetism when he developed the special theory of relativity. He was convinced that Maxwell's equations were correct, and in order to reconcile them with one of his postulates, he was forced into the bizarre notion of assuming that space and time are not absolute.

This chapter gives an introduction to the special theory of relativity, with emphasis on some of its consequences. The special theory covers phenomena such as the slowing down of clocks and the contraction of lengths in moving reference frames as measured by a stationary observer. We also discuss the relativistic forms of momentum and energy, as well as some consequences of the famous mass–energy formula, $E = mc^2$.

In addition to its well-known and essential role in theoretical physics, the special theory of relativity has practical applications, including the design of nuclear power plants and modern global positioning system (GPS) units. These devices do not work if designed in accordance with nonrelativistic principles.

We shall have occasion to use relativity in some subsequent chapters of the extended version of this text, most often presenting only the outcome of relativistic effects.

---

[1] A. Einstein and L. Infeld, *The Evolution of Physics*, New York, Simon and Schuster, 1961.

39.1 THE PRINCIPLE OF GALILEAN RELATIVITY

To describe a physical event, it is necessary to establish a frame of reference. You should recall from Chapter 5 that Newton's laws are valid in all inertial frames of reference. Because an inertial frame is defined as one in which Newton's first law is valid, we can say that **an inertial frame of reference is one in which an object is observed to have no acceleration when no forces act on it.** Furthermore, any system moving with constant velocity with respect to an inertial system must also be an inertial system.

There is no preferred inertial reference frame. This means that the results of an experiment performed in a vehicle moving with uniform velocity will be identical to the results of the same experiment performed in a stationary vehicle. The formal statement of this result is called the **principle of Galilean relativity:**

Inertial frame of reference

> The laws of mechanics must be the same in all inertial frames of reference.

Let us consider an observation that illustrates the equivalence of the laws of mechanics in different inertial frames. A pickup truck moves with a constant velocity, as shown in Figure 39.1a. If a passenger in the truck throws a ball straight up, and if air effects are neglected, the passenger observes that the ball moves in a vertical path. The motion of the ball appears to be precisely the same as if the ball were thrown by a person at rest on the Earth. The law of gravity and the equations of motion under constant acceleration are obeyed whether the truck is at rest or in uniform motion.

Now consider the same situation viewed by an observer at rest on the Earth. This stationary observer sees the path of the ball as a parabola, as illustrated in Figure 39.1b. Furthermore, according to this observer, the ball has a horizontal component of velocity equal to the velocity of the truck. Although the two observers disagree on certain aspects of the situation, they agree on the validity of Newton's laws and on such classical principles as conservation of energy and conservation of linear momentum. This agreement implies that no mechanical experiment can detect any difference between the two inertial frames. The only thing that can be detected is the relative motion of one frame with respect to the other. That is, the notion of absolute motion through space is meaningless, as is the notion of a preferred reference frame.

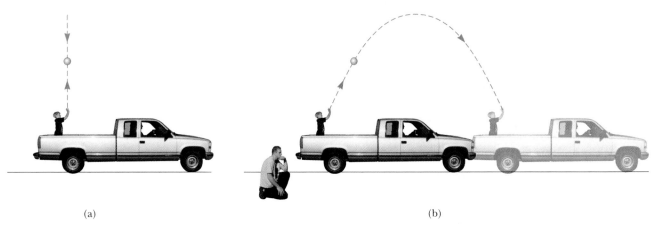

(a)   (b)

Figure 39.1   (a) The observer in the truck sees the ball move in a vertical path when thrown upward. (b) The Earth observer sees the path of the ball as a parabola.

Which observer in Figure 39.1 is right about the ball's path?

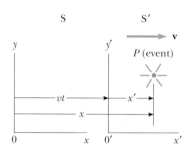

Suppose that some physical phenomenon, which we call an *event,* occurs in an inertial system. The event's location and time of occurrence can be specified by the four coordinates (x, y, z, t). We would like to be able to transform these coordinates from one inertial system to another one moving with uniform relative velocity.

Consider two inertial systems S and S′ (Fig. 39.2). The system S′ moves with a constant velocity **v** along the xx' axes, where **v** is measured relative to S. We assume that an event occurs at the point P and that the origins of S and S′ coincide at $t = 0$. An observer in S describes the event with space–time coordinates (x, y, z, t), whereas an observer in S′ uses the coordinates (x', y', z', t') to describe the same event. As we see from Figure 39.2, the relationships between these various coordinates can be written

Figure 39.2 An event occurs at a point P. The event is seen by two observers in inertial frames S and S′, where S′ moves with a velocity **v** relative to S.

$$x' = x - vt$$
$$y' = y$$
$$z' = z \tag{39.1}$$
$$t' = t$$

Galilean space–time transformation equations

These equations are the **Galilean space–time transformation equations.** Note that time is assumed to be the same in both inertial systems. That is, within the framework of classical mechanics, all clocks run at the same rate, regardless of their velocity, so that the time at which an event occurs for an observer in S is the same as the time for the same event in S′. Consequently, the time interval between two successive events should be the same for both observers. Although this assumption may seem obvious, it turns out to be incorrect in situations where v is comparable to the speed of light.

Now suppose that a particle moves a distance dx in a time interval dt as measured by an observer in S. It follows from Equations 39.1 that the corresponding distance dx' measured by an observer in S′ is $dx' = dx - v\,dt$, where frame S′ is moving with speed v relative to frame S. Because $dt = dt'$, we find that

$$\frac{dx'}{dt} = \frac{dx}{dt} - v$$

or

$$u'_x = u_x - v \tag{39.2}$$

Galilean velocity transformation equation

where u_x and u'_x are the x components of the velocity relative to S and S′, respectively. (We use the symbol **u** for particle velocity rather than **v**, which is used for the relative velocity of two reference frames.) This is the **Galilean velocity transformation equation.** It is used in everyday observations and is consistent with our intuitive notion of time and space. As we shall soon see, however, it leads to serious contradictions when applied to electromagnetic waves.

Applying the Galilean velocity transformation equation, determine how fast (relative to the Earth) a baseball pitcher with a 90-mi/h fastball can throw a ball while standing in a boxcar moving at 110 mi/h.

The Speed of Light

It is quite natural to ask whether the principle of Galilean relativity also applies to electricity, magnetism, and optics. Experiments indicate that the answer is no. Recall from Chapter 34 that Maxwell showed that the speed of light in free space is $c = 3.00 \times 10^8$ m/s. Physicists of the late 1800s thought that light waves moved through a medium called the *ether* and that the speed of light was c only in a special, absolute frame at rest with respect to the ether. The Galilean velocity transformation equation was expected to hold in any frame moving at speed v relative to the absolute ether frame.

Because the existence of a preferred, absolute ether frame would show that light was similar to other classical waves and that Newtonian ideas of an absolute frame were true, considerable importance was attached to establishing the existence of the ether frame. Prior to the late 1800s, experiments involving light traveling in media moving at the highest laboratory speeds attainable at that time were not capable of detecting changes as small as $c \pm v$. Starting in about 1880, scientists decided to use the Earth as the moving frame in an attempt to improve their chances of detecting these small changes in the speed of light.

As observers fixed on the Earth, we can say that we are stationary and that the absolute ether frame containing the medium for light propagation moves past us with speed v. Determining the speed of light under these circumstances is just like determining the speed of an aircraft traveling in a moving air current, or wind; consequently, we speak of an "ether wind" blowing through our apparatus fixed to the Earth.

A direct method for detecting an ether wind would use an apparatus fixed to the Earth to measure the wind's influence on the speed of light. If v is the speed of the ether relative to the Earth, then the speed of light should have its maximum value, $c + v$, when propagating downwind, as shown in Figure 39.3a. Likewise, the speed of light should have its minimum value, $c - v$, when propagating upwind, as shown in Figure 39.3b, and an intermediate value, $(c^2 - v^2)^{1/2}$, in the direction perpendicular to the ether wind, as shown in Figure 39.3c. If the Sun is assumed to be at rest in the ether, then the velocity of the ether wind would be equal to the orbital velocity of the Earth around the Sun, which has a magnitude of approximately 3×10^4 m/s. Because $c = 3 \times 10^8$ m/s, it should be possible to detect a change in speed of about 1 part in 10^4 for measurements in the upwind or downwind directions. However, as we shall see in the next section, all attempts to detect such changes and establish the existence of the ether wind (and hence the absolute frame) proved futile! (You may want to return to Problem 40 in Chapter 4 to see a situation in which the Galilean velocity transformation equation does hold.)

If it is assumed that the laws of electricity and magnetism are the same in all inertial frames, a paradox concerning the speed of light immediately arises. We can understand this by recognizing that Maxwell's equations seem to imply that the speed of light always has the fixed value 3.00×10^8 m/s in all inertial frames, a result in direct contradiction to what is expected based on the Galilean velocity transformation equation. According to Galilean relativity, the speed of light should not be the same in all inertial frames.

For example, suppose a light pulse is sent out by an observer S′ standing in a boxcar moving with a velocity **v** relative to a stationary observer standing alongside the track (Fig. 39.4). The light pulse has a speed c relative to S′. According to Galilean relativity, the pulse speed relative to S should be $c + v$. This is in contradiction to Einstein's special theory of relativity, which, as we shall see, postulates that the speed of the pulse is the same for all observers.

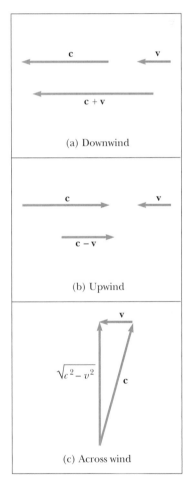

Figure 39.3 If the velocity of the ether wind relative to the Earth is **v** and the velocity of light relative to the ether is **c**, then the speed of light relative to the Earth is (a) $c + v$ in the downwind direction, (b) $c - v$ in the upwind direction, and (c) $(c^2 - v^2)^{1/2}$ in the direction perpendicular to the wind.

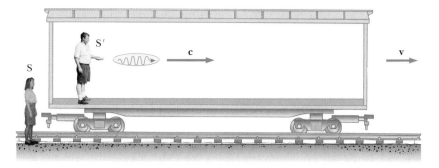

Figure 39.4   A pulse of light is sent out by a person in a moving boxcar. According to Galilean relativity, the speed of the pulse should be $c + v$ relative to a stationary observer.

To resolve this contradiction in theories, we must conclude that either (1) the laws of electricity and magnetism are not the same in all inertial frames or (2) the Galilean velocity transformation equation is incorrect. If we assume the first alternative, then a preferred reference frame in which the speed of light has the value c must exist and the measured speed must be greater or less than this value in any other reference frame, in accordance with the Galilean velocity transformation equation. If we assume the second alternative, then we are forced to abandon the notions of absolute time and absolute length that form the basis of the Galilean space–time transformation equations.

39.2 ▷ THE MICHELSON–MORLEY EXPERIMENT

The most famous experiment designed to detect small changes in the speed of light was first performed in 1881 by Albert A. Michelson (see Section 37.7) and later repeated under various conditions by Michelson and Edward W. Morley (1838–1923). We state at the outset that the outcome of the experiment contradicted the ether hypothesis.

The experiment was designed to determine the velocity of the Earth relative to that of the hypothetical ether. The experimental tool used was the Michelson interferometer, which was discussed in Section 37.7 and is shown again in Figure 39.5. Arm 2 is aligned along the direction of the Earth's motion through space. The Earth moving through the ether at speed v is equivalent to the ether flowing past the Earth in the opposite direction with speed v. This ether wind blowing in the direction opposite the direction of Earth's motion should cause the speed of light measured in the Earth frame to be $c - v$ as the light approaches mirror M_2 and $c + v$ after reflection, where c is the speed of light in the ether frame.

The two beams reflected from M_1 and M_2 recombine, and an interference pattern consisting of alternating dark and bright fringes is formed. The interference pattern was observed while the interferometer was rotated through an angle of 90°. This rotation supposedly would change the speed of the ether wind along the arms of the interferometer. The rotation should have caused the fringe pattern to shift slightly but measurably, but measurements failed to show any change in the interference pattern! The Michelson–Morley experiment was repeated at different times of the year when the ether wind was expected to change direction

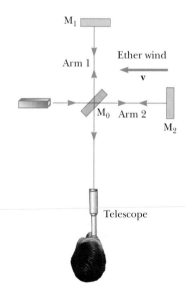

Figure 39.5   According to the ether wind theory, the speed of light should be $c - v$ as the beam approaches mirror M_2 and $c + v$ after reflection.

and magnitude, but the results were always the same: **no fringe shift of the magnitude required was ever observed.**[2]

The negative results of the Michelson–Morley experiment not only contradicted the ether hypothesis but also showed that it was impossible to measure the absolute velocity of the Earth with respect to the ether frame. However, as we shall see in the next section, Einstein offered a postulate for his special theory of relativity that places quite a different interpretation on these null results. In later years, when more was known about the nature of light, the idea of an ether that permeates all of space was relegated to the ash heap of worn-out concepts. **Light is now understood to be an electromagnetic wave, which requires no medium for its propagation.** As a result, the idea of an ether in which these waves could travel became unnecessary.

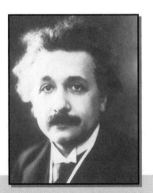

Albert Einstein (1879–1955)
Einstein, one of the greatest physicists of all times, was born in Ulm, Germany. In 1905, at the age of 26, he published four scientific papers that revolutionized physics. Two of these papers were concerned with what is now considered his most important contribution: the special theory of relativity.

In 1916, Einstein published his work on the general theory of relativity. The most dramatic prediction of this theory is the degree to which light is deflected by a gravitational field. Measurements made by astronomers on bright stars in the vicinity of the eclipsed Sun in 1919 confirmed Einstein's prediction, and as a result Einstein became a world celebrity.

Einstein was deeply disturbed by the development of quantum mechanics in the 1920s despite his own role as a scientific revolutionary. In particular, he could never accept the probabilistic view of events in nature that is a central feature of quantum theory. The last few decades of his life were devoted to an unsuccessful search for a unified theory that would combine gravitation and electromagnetism. *(AIP Niels Bohr Library)*

Optional Section

Details of the Michelson–Morley Experiment

To understand the outcome of the Michelson–Morley experiment, let us assume that the two arms of the interferometer in Figure 39.5 are of equal length L. We shall analyze the situation as if there were an ether wind, because that is what Michelson and Morley expected to find. As noted above, the speed of the light beam along arm 2 should be $c - v$ as the beam approaches M_2 and $c + v$ after the beam is reflected. Thus, the time of travel to the right is $L/(c - v)$, and the time of travel to the left is $L/(c + v)$. The total time needed for the round trip along arm 2 is

$$t_1 = \frac{L}{c + v} + \frac{L}{c - v} = \frac{2Lc}{c^2 - v^2} = \frac{2L}{c}\left(1 - \frac{v^2}{c^2}\right)^{-1}$$

Now consider the light beam traveling along arm 1, perpendicular to the ether wind. Because the speed of the beam relative to the Earth is $(c^2 - v^2)^{1/2}$ in this case (see Fig. 39.3), the time of travel for each half of the trip is $L/(c^2 - v^2)^{1/2}$, and the total time of travel for the round trip is

$$t_2 = \frac{2L}{(c^2 - v^2)^{1/2}} = \frac{2L}{c}\left(1 - \frac{v^2}{c^2}\right)^{-1/2}$$

Thus, the time difference between the horizontal round trip (arm 2) and the vertical round trip (arm 1) is

$$\Delta t = t_1 - t_2 = \frac{2L}{c}\left[\left(1 - \frac{v^2}{c^2}\right)^{-1} - \left(1 - \frac{v^2}{c^2}\right)^{-1/2}\right]$$

Because $v^2/c^2 \ll 1$, we can simplify this expression by using the following binomial expansion after dropping all terms higher than second order:

$$(1 - x)^n \approx 1 - nx \qquad \text{for } x \ll 1$$

In our case, $x = v^2/c^2$, and we find that

$$\Delta t = t_1 - t_2 \approx \frac{Lv^2}{c^3} \tag{39.3}$$

This time difference between the two instants at which the reflected beams arrive at the viewing telescope gives rise to a phase difference between the beams,

---

[2] From an Earth observer's point of view, changes in the Earth's speed and direction of motion in the course of a year are viewed as ether wind shifts. Even if the speed of the Earth with respect to the ether were zero at some time, six months later the speed of the Earth would be 60 km/s with respect to the ether, and as a result a fringe shift should be noticed. No shift has ever been observed, however.

producing an interference pattern when they combine at the position of the telescope. A shift in the interference pattern should be detected when the interferometer is rotated through 90° in a horizontal plane, so that the two beams exchange roles. This results in a time difference twice that given by Equation 39.3. Thus, the path difference that corresponds to this time difference is

$$\Delta d = c(2\,\Delta t) = \frac{2Lv^2}{c^2}$$

Because a change in path length of one wavelength corresponds to a shift of one fringe, the corresponding fringe shift is equal to this path difference divided by the wavelength of the light:

$$\text{Shift} = \frac{2Lv^2}{\lambda c^2} \qquad\qquad \textbf{(39.4)}$$

In the experiments by Michelson and Morley, each light beam was reflected by mirrors many times to give an effective path length L of approximately 11 m. Using this value and taking v to be equal to 3.0×10^4 m/s, the speed of the Earth around the Sun, we obtain a path difference of

$$\Delta d = \frac{2(11\text{ m})(3.0 \times 10^4\text{ m/s})^2}{(3.0 \times 10^8\text{ m/s})^2} = 2.2 \times 10^{-7}\text{ m}$$

This extra travel distance should produce a noticeable shift in the fringe pattern. Specifically, using 500-nm light, we expect a fringe shift for rotation through 90° of

$$\text{Shift} = \frac{\Delta d}{\lambda} = \frac{2.2 \times 10^{-7}\text{ m}}{5.0 \times 10^{-7}\text{ m}} \approx 0.44$$

The instrument used by Michelson and Morley could detect shifts as small as 0.01 fringe. However, **it detected no shift whatsoever in the fringe pattern.** Since then, the experiment has been repeated many times by different scientists under a wide variety of conditions, and no fringe shift has ever been detected. Thus, it was concluded that the motion of the Earth with respect to the postulated ether cannot be detected.

Many efforts were made to explain the null results of the Michelson–Morley experiment and to save the ether frame concept and the Galilean velocity transformation equation for light. All proposals resulting from these efforts have been shown to be wrong. No experiment in the history of physics received such valiant efforts to explain the absence of an expected result as did the Michelson–Morley experiment. The stage was set for Einstein, who solved the problem in 1905 with his special theory of relativity.

39.3 ▸ EINSTEIN'S PRINCIPLE OF RELATIVITY

In the previous section we noted the impossibility of measuring the speed of the ether with respect to the Earth and the failure of the Galilean velocity transformation equation in the case of light. Einstein proposed a theory that boldly removed these difficulties and at the same time completely altered our notion of space and time.[3] He based his special theory of relativity on two postulates:

[3] A. Einstein, "On the Electrodynamics of Moving Bodies," *Ann. Physik* 17:891, 1905. For an English translation of this article and other publications by Einstein, see the book by H. Lorentz, A. Einstein, H. Minkowski, and H. Weyl, *The Principle of Relativity*, Dover, 1958.

1. **The principle of relativity:** The laws of physics must be the same in all inertial reference frames.
2. **The constancy of the speed of light:** The speed of light in vacuum has the same value, $c = 3.00 \times 10^8$ m/s, in all inertial frames, regardless of the velocity of the observer or the velocity of the source emitting the light.

The first postulate asserts that *all* the laws of physics—those dealing with mechanics, electricity and magnetism, optics, thermodynamics, and so on—are the same in all reference frames moving with constant velocity relative to one another. This postulate is a sweeping generalization of the principle of Galilean relativity, which refers only to the laws of mechanics. From an experimental point of view, Einstein's principle of relativity means that any kind of experiment (measuring the speed of light, for example) performed in a laboratory at rest must give the same result when performed in a laboratory moving at a constant velocity past the first one. Hence, no preferred inertial reference frame exists, and it is impossible to detect absolute motion.

Note that postulate 2 is required by postulate 1: If the speed of light were not the same in all inertial frames, measurements of different speeds would make it possible to distinguish between inertial frames; as a result, a preferred, absolute frame could be identified, in contradiction to postulate 1.

Although the Michelson–Morley experiment was performed before Einstein published his work on relativity, it is not clear whether or not Einstein was aware of the details of the experiment. Nonetheless, the null result of the experiment can be readily understood within the framework of Einstein's theory. According to his principle of relativity, the premises of the Michelson–Morley experiment were incorrect. In the process of trying to explain the expected results, we stated that when light traveled against the ether wind its speed was $c - v$, in accordance with the Galilean velocity transformation equation. However, if the state of motion of the observer or of the source has no influence on the value found for the speed of light, one always measures the value to be c. Likewise, the light makes the return trip after reflection from the mirror at speed c, not at speed $c + v$. Thus, the motion of the Earth does not influence the fringe pattern observed in the Michelson–Morley experiment, and a null result should be expected.

If we accept Einstein's theory of relativity, we must conclude that relative motion is unimportant when measuring the speed of light. At the same time, we shall see that we must alter our common-sense notion of space and time and be prepared for some bizarre consequences. It may help as you read the pages ahead to keep in mind that our common-sense ideas are based on a lifetime of everyday experiences and not on observations of objects moving at hundreds of thousands of kilometers per second.

39.4 CONSEQUENCES OF THE SPECIAL THEORY OF RELATIVITY

Before we discuss the consequences of Einstein's special theory of relativity, we must first understand how an observer located in an inertial reference frame describes an event. As mentioned earlier, an event is an occurrence describable by three space coordinates and one time coordinate. Different observers in different inertial frames usually describe the same event with different coordinates.

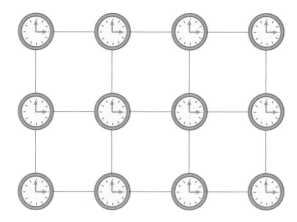

Figure 39.6 In studying relativity, we use a reference frame consisting of a coordinate grid and a set of synchronized clocks.

The reference frame used to describe an event consists of a coordinate grid and a set of synchronized clocks located at the grid intersections, as shown in Figure 39.6 in two dimensions. The clocks can be synchronized in many ways with the help of light signals. For example, suppose an observer is located at the origin with a master clock and sends out a pulse of light at $t = 0$. The pulse takes a time r/c to reach a clock located a distance r from the origin. Hence, this clock is synchronized with the master clock if this clock reads r/c at the instant the pulse reaches it. This procedure of synchronization assumes that the speed of light has the same value in all directions and in all inertial frames. Furthermore, the procedure concerns an event recorded by an observer in a specific inertial reference frame. An observer in some other inertial frame would assign different space–time coordinates to events being observed by using another coordinate grid and another array of clocks.

As we examine some of the consequences of relativity in the remainder of this section, we restrict our discussion to the concepts of simultaneity, time, and length, all three of which are quite different in relativistic mechanics from what they are in Newtonian mechanics. For example, in relativistic mechanics the distance between two points and the time interval between two events depend on the frame of reference in which they are measured. That is, **in relativistic mechanics there is no such thing as absolute length or absolute time.** Furthermore, **events at different locations that are observed to occur simultaneously in one frame are not observed to be simultaneous in another frame moving uniformly past the first.**

Simultaneity and the Relativity of Time

A basic premise of Newtonian mechanics is that a universal time scale exists that is the same for all observers. In fact, Newton wrote that "Absolute, true, and mathematical time, of itself, and from its own nature, flows equably without relation to anything external." Thus, Newton and his followers simply took simultaneity for granted. In his special theory of relativity, Einstein abandoned this assumption.

Einstein devised the following thought experiment to illustrate this point. A boxcar moves with uniform velocity, and two lightning bolts strike its ends, as illustrated in Figure 39.7a, leaving marks on the boxcar and on the ground. The marks on the boxcar are labeled A' and B', and those on the ground are labeled A and B. An observer O' moving with the boxcar is midway between A' and B', and a ground observer O is midway between A and B. The events recorded by the observers are the striking of the boxcar by the two lightning bolts.

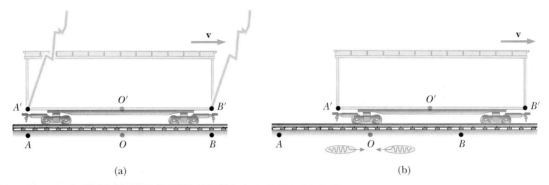

Figure 39.7 (a) Two lightning bolts strike the ends of a moving boxcar. (b) The events appear to be simultaneous to the stationary observer O, standing midway between A and B. The events do not appear to be simultaneous to observer O′, who claims that the front of the car is struck before the rear. Note that in (b) the leftward-traveling light signal has already passed O′ but the rightward-traveling signal has not yet reached O′.

The light signals recording the instant at which the two bolts strike reach observer O at the same time, as indicated in Figure 39.7b. This observer realizes that the signals have traveled at the same speed over equal distances, and so rightly concludes that the events at A and B occurred simultaneously. Now consider the same events as viewed by observer O′. By the time the signals have reached observer O, observer O′ has moved as indicated in Figure 39.7b. Thus, the signal from B′ has already swept past O′, but the signal from A′ has not yet reached O′. In other words, O′ sees the signal from B′ before seeing the signal from A′. According to Einstein, *the two observers must find that light travels at the same speed.* Therefore, observer O′ concludes that the lightning strikes the front of the boxcar before it strikes the back.

This thought experiment clearly demonstrates that the two events that appear to be simultaneous to observer O do not appear to be simultaneous to observer O′. In other words,

two events that are simultaneous in one reference frame are in general not simultaneous in a second frame moving relative to the first. That is, simultaneity is not an absolute concept but rather one that depends on the state of motion of the observer.

Quick Quiz 39.3

Which observer in Figure 39.7 is correct?

The central point of relativity is this: Any inertial frame of reference can be used to describe events and do physics. **There is no preferred inertial frame of reference.** However, observers in different inertial frames always measure different time intervals with their clocks and different distances with their meter sticks. Nevertheless, all observers agree on the forms of the laws of physics in their respective frames because these laws must be the same for all observers in uniform motion. For example, the relationship $F = ma$ in a frame S has the same form $F' = ma'$ in a frame S′ that is moving at constant velocity relative to frame S. It is the alteration of time and space that allows the laws of physics (including Maxwell's equations) to be the same for all observers in uniform motion.

Time Dilation

We can illustrate the fact that observers in different inertial frames always measure different time intervals between a pair of events by considering a vehicle moving to the right with a speed v, as shown in Figure 39.8a. A mirror is fixed to the ceiling of the vehicle, and observer O' at rest in this system holds a laser a distance d below the mirror. At some instant, the laser emits a pulse of light directed toward the mirror (event 1), and at some later time after reflecting from the mirror, the pulse arrives back at the laser (event 2). Observer O' carries a clock C' and uses it to measure the time interval Δt_p between these two events. (The subscript p stands for *proper*, as we shall see in a moment.) Because the light pulse has a speed c, the time it takes the pulse to travel from O' to the mirror and back to O' is

$$\Delta t_p = \frac{\text{Distance traveled}}{\text{Speed}} = \frac{2d}{c} \tag{39.5}$$

This time interval Δt_p measured by O' requires only a single clock C' located at the same place as the laser in this frame.

Now consider the same pair of events as viewed by observer O in a second frame, as shown in Figure 39.8b. According to this observer, the mirror and laser are moving to the right with a speed v, and as a result the sequence of events appears entirely different. By the time the light from the laser reaches the mirror, the mirror has moved to the right a distance $v\,\Delta t/2$, where Δt is the time it takes the light to travel from O' to the mirror and back to O' as measured by O. In other words, O concludes that, because of the motion of the vehicle, if the light is to hit the mirror, it must leave the laser at an angle with respect to the vertical direction. Comparing Figure 39.8a and b, we see that the light must travel farther in (b) than in (a). (Note that neither observer "knows" that he is moving. Each is at rest in his own inertial frame.)

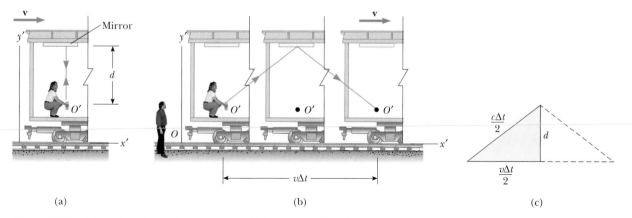

|(a)|(b)|(c)|

Figure 39.8 (a) A mirror is fixed to a moving vehicle, and a light pulse is sent out by observer O' at rest in the vehicle. (b) Relative to a stationary observer O standing alongside the vehicle, the mirror and O' move with a speed v. Note that what observer O measures for the distance the pulse travels is greater than $2d$. (c) The right triangle for calculating the relationship between Δt and Δt_p.

According to the second postulate of the special theory of relativity, both observers must measure c for the speed of light. Because the light travels farther in the frame of O, it follows that the time interval Δt measured by O is longer than the time interval Δt_p measured by O'. To obtain a relationship between these two time intervals, it is convenient to use the right triangle shown in Figure 39.8c. The Pythagorean theorem gives

$$\left(\frac{c\,\Delta t}{2}\right)^2 = \left(\frac{v\,\Delta t}{2}\right)^2 + d^2$$

Solving for Δt gives

$$\Delta t = \frac{2d}{\sqrt{c^2 - v^2}} = \frac{2d}{c\sqrt{1 - \dfrac{v^2}{c^2}}} \tag{39.6}$$

Because $\Delta t_p = 2d/c$, we can express this result as

Time dilation

$$\Delta t = \frac{\Delta t_p}{\sqrt{1 - \dfrac{v^2}{c^2}}} = \gamma\,\Delta t_p \tag{39.7}$$

where

$$\gamma = (1 - v^2/c^2)^{-1/2} \tag{39.8}$$

Because γ is always greater than unity, this result says that **the time interval Δt measured by an observer moving with respect to a clock is longer than the time interval Δt_p measured by an observer at rest with respect to the clock.** (That is, $\Delta t > \Delta t_p$.) This effect is known as **time dilation.** Figure 39.9 shows that as the velocity approaches the speed of light, γ increases dramatically. Note that for speeds less than one tenth the speed of light, γ is very nearly equal to unity.

The time interval Δt_p in Equations 39.5 and 39.7 is called the **proper time.** (In German, Einstein used the term *Eigenzeit*, which means "own-time.") In general, **proper time is the time interval between two events measured by an observer who sees the events occur at the same point in space.** Proper time is always the time measured with a single clock (clock C' in our case) at rest in the frame in which the events take place.

If a clock is moving with respect to you, it appears to fall behind (tick more slowly than) the clocks it is passing in the grid of synchronized clocks in your reference frame. Because the time interval $\gamma(2d/c)$, the interval between ticks of a moving clock, is observed to be longer than $2d/c$, the time interval between ticks of an identi-

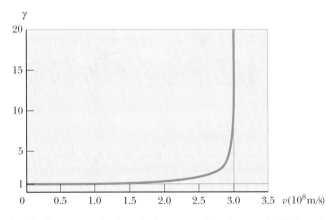

Figure 39.9 Graph of γ versus v. As the velocity approaches the speed of light, γ increases rapidly.

cal clock in your reference frame, it is often said that a moving clock runs more slowly than a clock in your reference frame by a factor γ. This is true for mechanical clocks as well as for the light clock just described. We can generalize this result by stating that all physical processes, including chemical and biological ones, slow down relative to a stationary clock when those processes occur in a moving frame. For example, the heartbeat of an astronaut moving through space would keep time with a clock inside the spaceship. Both the astronaut's clock and heartbeat would be slowed down relative to a stationary clock back on the Earth (although the astronaut would have no sensation of life slowing down in the spaceship).

Quick Quiz 39.4

A rocket has a clock built into its control panel. Use Figure 39.9 to determine approximately how fast the rocket must be moving before its clock appears to an Earth-bound observer to be ticking at one fifth the rate of a clock on the wall at Mission Control. What does an astronaut in the rocket observe?

Bizarre as it may seem, time dilation is a verifiable phenomenon. An experiment reported by Hafele and Keating provided direct evidence of time dilation.[4] Time intervals measured with four cesium atomic clocks in jet flight were compared with time intervals measured by Earth-based reference atomic clocks. In order to compare these results with theory, many factors had to be considered, including periods of acceleration and deceleration relative to the Earth, variations in direction of travel, and the fact that the gravitational field experienced by the flying clocks was weaker than that experienced by the Earth-based clock. The results were in good agreement with the predictions of the special theory of relativity and can be explained in terms of the relative motion between the Earth and the jet aircraft. In their paper, Hafele and Keating stated that "Relative to the atomic time scale of the U.S. Naval Observatory, the flying clocks lost 59 ± 10 ns during the eastward trip and gained 273 ± 7 ns during the westward trip . . . . These results provide an unambiguous empirical resolution of the famous clock paradox with macroscopic clocks."

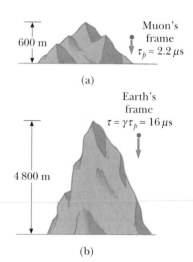

Another interesting example of time dilation involves the observation of *muons*, unstable elementary particles that have a charge equal to that of the electron and a mass 207 times that of the electron. Muons can be produced by the collision of cosmic radiation with atoms high in the atmosphere. These particles have a lifetime of $2.2\ \mu s$ when measured in a reference frame in which they are at rest or moving slowly. If we take $2.2\ \mu s$ as the average lifetime of a muon and assume that its speed is close to the speed of light, we find that these particles travel only approximately 600 m before they decay (Fig. 39.10a). Hence, they cannot reach the Earth from the upper atmosphere where they are produced. However, experiments show that a large number of muons do reach the Earth. The phenomenon of time dilation explains this effect. Relative to an observer on the Earth, the muons have a lifetime equal to $\gamma \tau_p$, where $\tau_p = 2.2\ \mu s$ is the lifetime in the frame traveling with the muons or the proper lifetime. For example, for a muon speed of $v = 0.99c$, $\gamma \approx 7.1$ and $\gamma \tau_p \approx 16\ \mu s$. Hence, the average distance traveled as measured by an observer on the Earth is $\gamma v \tau_p \approx 4\ 800$ m, as indicated in Figure 39.10b.

In 1976, at the laboratory of the European Council for Nuclear Research (CERN) in Geneva, muons injected into a large storage ring reached speeds of ap-

Figure 39.10 (a) Muons moving with a speed of $0.99c$ travel approximately 600 m as measured in the reference frame of the muons, where their lifetime is about $2.2\ \mu s$. (b) The muons travel approximately 4 800 m as measured by an observer on the Earth.

[4] J. C. Hafele and R. E. Keating, "Around the World Atomic Clocks: Relativistic Time Gains Observed," *Science*, 177:168, 1972.

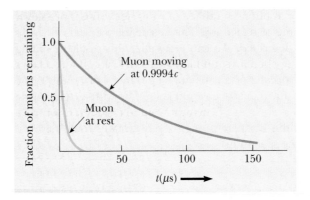

proximately 0.9994c. Electrons produced by the decaying muons were detected by counters around the ring, enabling scientists to measure the decay rate and hence the muon lifetime. The lifetime of the moving muons was measured to be approximately 30 times as long as that of the stationary muon (Fig. 39.11), in agreement with the prediction of relativity to within two parts in a thousand.

EXAMPLE 39.1 ▶ What Is the Period of the Pendulum?

The period of a pendulum is measured to be 3.0 s in the reference frame of the pendulum. What is the period when measured by an observer moving at a speed of 0.95c relative to the pendulum?

Solution   Instead of the observer moving at 0.95c, we can take the equivalent point of view that the observer is at rest and the pendulum is moving at 0.95c past the stationary observer. Hence, the pendulum is an example of a moving clock.

The proper time is $\Delta t_p = 3.0$ s. Because a moving clock runs more slowly than a stationary clock by a factor γ, Equation 39.7 gives

$$\Delta t = \gamma \Delta t_p = \frac{1}{\sqrt{1 - \dfrac{(0.95c)^2}{c^2}}} \Delta t_p = \frac{1}{\sqrt{1 - 0.902}} \Delta t_p$$

$$= (3.2)(3.0 \text{ s}) = \boxed{9.6 \text{ s}}$$

That is, a moving pendulum takes longer to complete a period than a pendulum at rest does.

EXAMPLE 39.2 ▶ How Long Was Your Trip?

Suppose you are driving your car on a business trip and are traveling at 30 m/s. Your boss, who is waiting at your destination, expects the trip to take 5.0 h. When you arrive late, your excuse is that your car clock registered the passage of 5.0 h but that you were driving fast and so your clock ran more slowly than your boss's clock. If your car clock actually did indicate a 5.0-h trip, how much time passed on your boss's clock, which was at rest on the Earth?

Solution   We begin by calculating γ from Equation 39.8:

$$\gamma = \frac{1}{\sqrt{1 - \dfrac{v^2}{c^2}}} = \frac{1}{\sqrt{1 - \dfrac{(3 \times 10^1 \text{ m/s})^2}{(3 \times 10^8 \text{ m/s})^2}}} = \frac{1}{\sqrt{1 - 10^{-14}}}$$

If you try to determine this value on your calculator, you will probably get $\gamma = 1$. However, if we perform a binomial expansion, we can more precisely determine the value as

$$\gamma = (1 - 10^{-14})^{-1/2} \approx 1 + \tfrac{1}{2}(10^{-14}) = 1 + 5.0 \times 10^{-15}$$

This result indicates that at typical automobile speeds, γ is not much different from 1.

Applying Equation 39.7, we find Δt, the time interval measured by your boss, to be

$$\Delta t = \gamma \Delta t_p = (1 + 5.0 \times 10^{-15})(5.0 \text{ h})$$

$$= 5.0 \text{ h} + 2.5 \times 10^{-14} \text{ h} = \boxed{5.0 \text{ h} + 0.09 \text{ ns}}$$

Your boss's clock would be only 0.09 ns ahead of your car clock. You might want to try another excuse!

The Twins Paradox

An intriguing consequence of time dilation is the so-called *twins paradox* (Fig. 39.12). Consider an experiment involving a set of twins named Speedo and Goslo. When they are 20 yr old, Speedo, the more adventuresome of the two, sets out on an epic journey to Planet X, located 20 ly from the Earth. Furthermore, his spaceship is capable of reaching a speed of $0.95c$ relative to the inertial frame of his twin brother back home. After reaching Planet X, Speedo becomes homesick and immediately returns to the Earth at the same speed $0.95c$. Upon his return, Speedo is shocked to discover that Goslo has aged 42 yr and is now 62 yr old. Speedo, on the other hand, has aged only 13 yr.

At this point, it is fair to raise the following question—which twin is the traveler and which is really younger as a result of this experiment? From Goslo's frame of reference, he was at rest while his brother traveled at a high speed. But from Speedo's perspective, it is he who was at rest while Goslo was on the high-speed space journey. According to Speedo, he himself remained stationary while Goslo and the Earth raced away from him on a 6.5-yr journey and then headed back for another 6.5 yr. This leads to an apparent contradiction. Which twin has developed signs of excess aging?

To resolve this apparent paradox, recall that the special theory of relativity deals with inertial frames of reference moving relative to each other at uniform speed. However, the trip in our current problem is not symmetrical. Speedo, the space traveler, must experience a series of accelerations during his journey. As a result, his speed is not always uniform, and consequently he is not in an inertial frame. He cannot be regarded as always being at rest while Goslo is in uniform motion because to do so would be an incorrect application of the special theory of relativity. Therefore, there is no paradox. During each passing year noted by Goslo, slightly less than 4 months elapsed for Speedo.

The conclusion that Speedo is in a noninertial frame is inescapable. Each twin observes the other as accelerating, but it is Speedo that actually undergoes dynamical acceleration due to the real forces acting on him. The time required to accelerate and decelerate Speedo's spaceship may be made very small by using large rockets, so that Speedo can claim that he spends most of his time traveling to Planet X at $0.95c$ in an inertial frame. However, Speedo must slow down, reverse his motion, and return to the Earth in an altogether different inertial frame. At the very best,

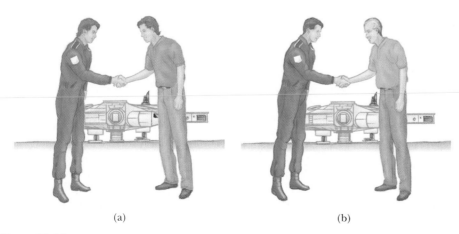

(a)   (b)

Figure 39.12 (a) As one twin leaves his brother on the Earth, both are the same age. (b) When Speedo returns from his journey to Planet X, he is younger than his twin Goslo.

Speedo is in two different inertial frames during his journey. Only Goslo, who is in a single inertial frame, can apply the simple time-dilation formula to Speedo's trip. Thus, Goslo finds that instead of aging 42 yr, Speedo ages only $(1 - v^2/c^2)^{1/2}(42 \text{ yr}) = 13$ yr. Conversely, Speedo spends 6.5 yr traveling to Planet X and 6.5 yr returning, for a total travel time of 13 yr, in agreement with our earlier statement.

Quick Quiz 39.5

Suppose astronauts are paid according to the amount of time they spend traveling in space. After a long voyage traveling at a speed approaching c, would a crew rather be paid according to an Earth-based clock or their spaceship's clock?

Length Contraction

The measured distance between two points also depends on the frame of reference. **The proper length L_p of an object is the length measured by someone at rest relative to the object.** The length of an object measured by someone in a reference frame that is moving with respect to the object is always less than the proper length. This effect is known as **length contraction.**

Consider a spaceship traveling with a speed v from one star to another. There are two observers: one on the Earth and the other in the spaceship. The observer at rest on the Earth (and also assumed to be at rest with respect to the two stars) measures the distance between the stars to be the proper length L_p. According to this observer, the time it takes the spaceship to complete the voyage is $\Delta t = L_p/v$. Because of time dilation, the space traveler measures a smaller time of travel by the spaceship clock: $\Delta t_p = \Delta t/\gamma$. The space traveler claims to be at rest and sees the destination star moving toward the spaceship with speed v. Because the space traveler reaches the star in the time Δt_p, he or she concludes that the distance L between the stars is shorter than L_p. This distance measured by the space traveler is

$$L = v \, \Delta t_p = v \frac{\Delta t}{\gamma}$$

Because $L_p = v \, \Delta t$, we see that

$$L = \frac{L_p}{\gamma} = L_p \left(1 - \frac{v^2}{c^2}\right)^{1/2} \tag{39.9}$$

where $(1 - v^2/c^2)^{1/2}$ is a factor less than unity. This result may be interpreted as follows:

> If an object has a proper length L_p when it is at rest, then when it moves with speed v in a direction parallel to its length, it contracts to the length $L = L_p(1 - v^2/c^2)^{1/2} = L_p/\gamma$.

For example, suppose that a stick moves past a stationary Earth observer with speed v, as shown in Figure 39.13. The length of the stick as measured by an observer in a frame attached to the stick is the proper length L_p shown in Figure 39.13a. The length of the stick L measured by the Earth observer is shorter than L_p by the factor $(1 - v^2/c^2)^{1/2}$. Furthermore, length contraction is a symmetrical effect: If the stick is at rest on the Earth, an observer in a moving frame would measure its length to be shorter by the same factor $(1 - v^2/c^2)^{1/2}$. Note that **length contraction takes place only along the direction of motion.**

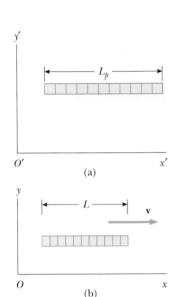

Figure 39.13 (a) A stick measured by an observer in a frame attached to the stick (that is, both have the same velocity) has its proper length L_p. (b) The stick measured by an observer in a frame in which the stick has a velocity **v** relative to the frame is shorter than its proper length L_p by a factor $(1 - v^2/c^2)^{1/2}$.

Length contraction

It is important to emphasize that proper length and proper time are measured in different reference frames. As an example of this point, let us return to the decaying muons moving at speeds close to the speed of light. An observer in the muon reference frame measures the proper lifetime (that is, the time interval τ_p), whereas an Earth-based observer measures a dilated lifetime. However, the Earth-based observer measures the proper height (the length L_p) of the mountain in Figure 39.10b. In the muon reference frame, this height is less than L_p, as the figure shows. Thus, in the muon frame, length contraction occurs but time dilation does not. In the Earth-based reference frame, time dilation occurs but length contraction does not. Thus, when calculations on the muon are performed in both frames, the effect of "offsetting penalties" is seen, and the outcome of the experiment in one frame is the same as the outcome in the other frame!

EXAMPLE 39.3   The Contraction of a Spaceship

A spaceship is measured to be 120.0 m long and 20.0 m in diameter while at rest relative to an observer. If this spaceship now flies by the observer with a speed of $0.99c$, what length and diameter does the observer measure?

Solution   From Equation 39.9, the length measured by the observer is

$$L = L_p \sqrt{1 - \frac{v^2}{c^2}} = (120.0 \text{ m}) \sqrt{1 - \frac{(0.99c)^2}{c^2}} = \boxed{17 \text{ m}}$$

The diameter measured by the observer is still 20.0 m because the diameter is a dimension perpendicular to the motion and length contraction occurs only along the direction of motion.

Exercise   If the ship moves past the observer with a speed of $0.100\,0c$, what length does the observer measure?

Answer   119.4 m.

EXAMPLE 39.4   How Long Was Your Car?

In Example 39.2, you were driving at 30 m/s and claimed that your clock was running more slowly than your boss's stationary clock. Although your statement was true, the time dilation was negligible. If your car is 4.3 m long when it is parked, how much shorter does it appear to a stationary roadside observer as you drive by at 30 m/s?

Solution   The observer sees the horizontal length of the car to be contracted to a length

$$L = L_p \sqrt{1 - \frac{v^2}{c^2}} \approx L_p \left(1 - \frac{1}{2}\frac{v^2}{c^2}\right)$$

where we have again used the binomial expansion for the factor $\sqrt{1 - \frac{v^2}{c^2}}$. The roadside observer sees the car's length as having changed by an amount $L_p - L$:

$$L_p - L \approx \frac{L_p}{2}\left(\frac{v^2}{c^2}\right) = \left(\frac{4.3 \text{ m}}{2}\right)\left(\frac{3.0 \times 10^1 \text{ m/s}}{3.0 \times 10^8 \text{ m/s}}\right)^2$$

$$= \boxed{2.2 \times 10^{-14} \text{ m}}$$

This is much smaller than the diameter of an atom!

EXAMPLE 39.5   A Voyage to Sirius

An astronaut takes a trip to Sirius, which is located a distance of 8 lightyears from the Earth. (Note that 1 lightyear (ly) is the distance light travels through free space in 1 yr.) The astronaut measures the time of the one-way journey to be 6 yr. If the spaceship moves at a constant speed of $0.8c$, how can the 8-ly distance be reconciled with the 6-yr trip time measured by the astronaut?

Solution   The 8 ly represents the proper length from the Earth to Sirius measured by an observer seeing both bodies

nearly at rest. The astronaut sees Sirius approaching her at $0.8c$ but also sees the distance contracted to

$$\frac{8 \text{ ly}}{\gamma} = (8 \text{ ly})\sqrt{1 - \frac{v^2}{c^2}} = (8 \text{ ly})\sqrt{1 - \frac{(0.8c)^2}{c^2}} = 5 \text{ ly}$$

Thus, the travel time measured on her clock is

$$t = \frac{d}{v} = \frac{5 \text{ ly}}{0.8c} = 6 \text{ yr}$$

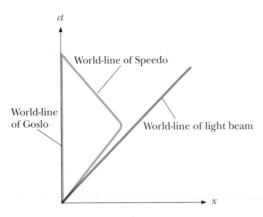

Figure 39.14 The twins paradox on a space–time graph. The twin who stays on the Earth has a world-line along the t axis. The path of the traveling twin through space–time is represented by a world-line that changes direction.

Space–Time Graphs

It is sometimes helpful to make a *space–time graph,* in which time is the ordinate and displacement is the abscissa. The twins paradox is displayed in such a graph in Figure 39.14. A path through space–time is called a **world-line.** At the origin, the world-lines of Speedo and Goslo coincide because the twins are in the same location at the same time. After Speedo leaves on his trip, his world-line diverges from that of his brother. At their reunion, the two world-lines again come together. Note that Goslo's world-line is vertical, indicating no displacement from his original location. Also note that it would be impossible for Speedo to have a world-line that crossed the path of a light beam that left the Earth when he did. To do so would require him to have a speed greater than c.

World-lines for light beams are diagonal lines on space–time graphs, typically drawn at 45° to the right or left of vertical, depending on whether the light beam is traveling in the direction of increasing or decreasing x. These two world-lines means that all possible future events for Goslo and Speedo lie within two 45° lines extending from the origin. Either twin's presence at an event outside this "light cone" would require that twin to move at a speed greater than c, which, as we shall see in Section 39.5, is not possible. Also, the only past events that Goslo and Speedo could have experienced occurred within two similar 45° world-lines that approach the origin from below the x axis.

Quick Quiz 39.6

How is acceleration indicated on a space–time graph?

The Relativistic Doppler Effect

Another important consequence of time dilation is the shift in frequency found for light emitted by atoms in motion as opposed to light emitted by atoms at rest. This phenomenon, known as the Doppler effect, was introduced in Chapter 17 as it pertains to sound waves. In the case of sound, the motion of the source with respect to the medium of propagation can be distinguished from the motion of the observer with respect to the medium. Light waves must be analyzed differently, however, because they require no medium of propagation, and no method exists for distinguishing the motion of a light source from the motion of the observer.

If a light source and an observer approach each other with a relative speed v, the frequency f_{obs} measured by the observer is

$$f_{obs} = \frac{\sqrt{1 + v/c}}{\sqrt{1 - v/c}} f_{source} \qquad \textbf{(39.10)}$$

where f_{source} is the frequency of the source measured in its rest frame. Note that this relativistic Doppler shift formula, unlike the Doppler shift formula for sound, depends only on the relative speed v of the source and observer and holds for relative speeds as great as c. As you might expect, the formula predicts that $f_{obs} > f_{source}$ when the source and observer approach each other. We obtain the expression for the case in which the source and observer recede from each other by replacing v with $-v$ in Equation 39.10.

The most spectacular and dramatic use of the relativistic Doppler effect is the measurement of shifts in the frequency of light emitted by a moving astronomical object such as a galaxy. Spectral lines normally found in the extreme violet region for galaxies at rest with respect to the Earth are shifted by about 100 nm toward the red end of the spectrum for distant galaxies—indicating that these galaxies are *receding* from us. The American astronomer Edwin Hubble (1889–1953) performed extensive measurements of this *red shift* to confirm that most galaxies are moving away from us, indicating that the Universe is expanding.

39.5 ▶ THE LORENTZ TRANSFORMATION EQUATIONS

We have seen that the Galilean transformation equations are not valid when v approaches the speed of light. In this section, we state the correct transformation equations that apply for all speeds in the range $0 \leq v < c$.

Suppose that an event that occurs at some point P is reported by two observers, one at rest in a frame S and the other in a frame S′ that is moving to the right with speed v, as in Figure 39.15. The observer in S reports the event with space–time coordinates (x, y, z, t), and the observer in S′ reports the same event using the coordinates (x', y', z', t'). We would like to find a relationship between these coordinates that is valid for all speeds.

The equations that are valid from $v = 0$ to $v = c$ and enable us to transform coordinates from S to S′ are the **Lorentz transformation equations:**

$$
\begin{aligned}
x' &= \gamma(x - vt) \\
y' &= y \\
z' &= z \\
t' &= \gamma\left(t - \frac{v}{c^2}x\right)
\end{aligned}
\qquad \textbf{(39.11)}
$$

Lorentz transformation equations for S → S′

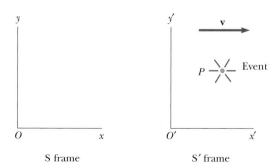

Figure 39.15 An event that occurs at some point P is observed by two persons, one at rest in the S frame and the other in the S′ frame, which is moving to the right with a speed v.

These transformation equations were developed by Hendrik A. Lorentz (1853–1928) in 1890 in connection with electromagnetism. However, it was Einstein who recognized their physical significance and took the bold step of interpreting them within the framework of the special theory of relativity.

Note the difference between the Galilean and Lorentz time equations. In the Galilean case, $t = t'$, but in the Lorentz case the value for t' assigned to an event by an observer O' standing at the origin of the S′ frame in Figure 39.15 depends both on the time t and on the coordinate x as measured by an observer O standing in the S frame. This is consistent with the notion that an event is characterized by four space–time coordinates (x, y, z, t). In other words, in relativity, space and time are not separate concepts but rather are closely interwoven with each other.

If we wish to transform coordinates in the S′ frame to coordinates in the S frame, we simply replace v by $-v$ and interchange the primed and unprimed coordinates in Equations 39.11:

Inverse Lorentz transformation equations for S′ → S

$$x = \gamma(x' + vt')$$
$$y = y'$$
$$z = z' \tag{39.12}$$
$$t = \gamma\left(t' + \frac{v}{c^2}\,x'\right)$$

When $v \ll c$, the Lorentz transformation equations should reduce to the Galilean equations. To verify this, note that as v approaches zero, $v/c \ll 1$ and $v^2/c^2 \ll 1$; thus, $\gamma = 1$, and Equations 39.11 reduce to the Galilean space–time transformation equations:

$$x' = x - vt \qquad y' = y \qquad z' = z \qquad t' = t$$

In many situations, we would like to know the difference in coordinates between two events or the time interval between two events as seen by observers O and O'. We can accomplish this by writing the Lorentz equations in a form suitable for describing pairs of events. From Equations 39.11 and 39.12, we can express the differences between the four variables x, x', t, and t' in the form

$$\left.\begin{aligned} \Delta x' &= \gamma(\Delta x - v\,\Delta t) \\ \Delta t' &= \gamma\left(\Delta t - \frac{v}{c^2}\,\Delta x\right) \end{aligned}\right\} \text{S} \rightarrow \text{S}' \tag{39.13}$$

$$\left.\begin{aligned} \Delta x &= \gamma(\Delta x' + v\,\Delta t') \\ \Delta t &= \gamma\left(\Delta t' + \frac{v}{c^2}\,\Delta x'\right) \end{aligned}\right\} \text{S}' \rightarrow \text{S} \tag{39.14}$$

where $\Delta x' = x_2' - x_1'$ and $\Delta t' = t_2' - t_1'$ are the differences measured by observer O' and $\Delta x = x_2 - x_1$ and $\Delta t = t_2 - t_1$ are the differences measured by observer O. (We have not included the expressions for relating the y and z coordinates because they are unaffected by motion along the x direction.[5])

---

[5] Although relative motion of the two frames along the x axis does not change the y and z coordinates of an object, it does change the y and z velocity components of an object moving in either frame, as we shall soon see.

EXAMPLE 39.6   **Simultaneity and Time Dilation Revisited**

Use the Lorentz transformation equations in difference form to show that (a) simultaneity is not an absolute concept and that (b) moving clocks run more slowly than stationary clocks.

Solution   (a) Suppose that two events are simultaneous according to a moving observer O', such that $\Delta t' = 0$. From the expression for Δt given in Equation 39.14, we see that in this case the time interval Δt measured by a stationary observer O is $\Delta t = \gamma v \, \Delta x'/c^2$. That is, the time interval for the same two events as measured by O is nonzero, and so the events do not appear to be simultaneous to O.

(b) Suppose that observer O' finds that two events occur at the same place ($\Delta x' = 0$) but at different times ($\Delta t' \neq 0$). In this situation, the expression for Δt given in Equation 39.14 becomes $\Delta t = \gamma \, \Delta t'$. This is the equation for time dilation found earlier (Eq. 39.7), where $\Delta t' = \Delta t_p$ is the proper time measured by a clock located in the moving frame of observer O'.

Exercise   Use the Lorentz transformation equations in difference form to confirm that $L = L_p/\gamma$ (Eq. 39.9).

Derivation of the Lorentz Velocity Transformation Equation

Once again S is our stationary frame of reference, and S′ is our frame moving at a speed v relative to S. Suppose that an object has a speed u'_x measured in the S′ frame, where

$$u'_x = \frac{dx'}{dt'} \qquad \text{(39.15)}$$

Using Equation 39.11, we have

$$dx' = \gamma(dx - v \, dt)$$

$$dt' = \gamma\left(dt - \frac{v}{c^2} \, dx\right)$$

Substituting these values into Equation 39.15 gives

$$u'_x = \frac{dx'}{dt'} = \frac{dx - v \, dt}{dt - \frac{v}{c^2} \, dx} = \frac{\frac{dx}{dt} - v}{1 - \frac{v}{c^2} \frac{dx}{dt}}$$

But dx/dt is just the velocity component u_x of the object measured by an observer in S, and so this expression becomes

$$u'_x = \frac{u_x - v}{1 - \frac{u_x v}{c^2}} \qquad \text{(39.16)}$$

Lorentz velocity transformation equation for S → S′

If the object has velocity components along the y and z axes, the components as measured by an observer in S′ are

$$u'_y = \frac{u_y}{\gamma\left(1 - \frac{u_x v}{c^2}\right)} \qquad \text{and} \qquad u'_z = \frac{u_z}{\gamma\left(1 - \frac{u_x v}{c^2}\right)} \qquad \text{(39.17)}$$

Note that u'_y and u'_z do not contain the parameter v in the numerator because the relative velocity is along the x axis.

The speed of light is the speed limit of the Universe. It is the maximum possible speed for energy transfer and for information transfer. Any object with mass must move at a lower speed.

Lorentz velocity transformation equations for S′ → S

When u_x and v are both much smaller than c (the nonrelativistic case), the denominator of Equation 39.16 approaches unity, and so $u'_x \approx u_x - v$, which is the Galilean velocity transformation equation. In the other extreme, when $u_x = c$, Equation 39.16 becomes

$$u'_x = \frac{c - v}{1 - \dfrac{cv}{c^2}} = \frac{c\left(1 - \dfrac{v}{c}\right)}{1 - \dfrac{v}{c}} = c$$

From this result, we see that an object moving with a speed c relative to an observer in S also has a speed c relative to an observer in S′—independent of the relative motion of S and S′. Note that this conclusion is consistent with Einstein's second postulate—that the speed of light must be c relative to all inertial reference frames. Furthermore, the speed of an object can never exceed c. That is, the speed of light is the ultimate speed. We return to this point later when we consider the energy of a particle.

To obtain u_x in terms of u'_x, we replace v by $-v$ in Equation 39.16 and interchange the roles of u_x and u'_x:

$$u_x = \frac{u'_x + v}{1 + \dfrac{u'_x v}{c^2}} \tag{39.18}$$

EXAMPLE 39.7 ▶ Relative Velocity of Spaceships

Two spaceships A and B are moving in opposite directions, as shown in Figure 39.16. An observer on the Earth measures the speed of ship A to be $0.750c$ and the speed of ship B to be $0.850c$. Find the velocity of ship B as observed by the crew on ship A.

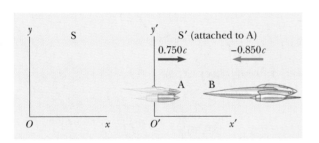

Figure 39.16   Two spaceships A and B move in opposite directions. The speed of B relative to A is *less* than c and is obtained from the relativistic velocity transformation equation.

Solution   We can solve this problem by taking the S′ frame as being attached to ship A, so that $v = 0.750c$ relative to the Earth (the S frame). We can consider ship B as moving with a velocity $u_x = -0.850c$ relative to the Earth. Hence, we can obtain the velocity of ship B relative to ship A by using Equation 39.16:

$$u'_x = \frac{u_x - v}{1 - \dfrac{u_x v}{c^2}} = \frac{-0.850c - 0.750c}{1 - \dfrac{(-0.850c)(0.750c)}{c^2}} = \boxed{-0.977c}$$

The negative sign indicates that ship B is moving in the negative x direction as observed by the crew on ship A. Note that the speed is less than c. That is, a body whose speed is less than c in one frame of reference must have a speed less than c in any other frame. (If the Galilean velocity transformation equation were used in this example, we would find that $u'_x = u_x - v = -0.850c - 0.750c = -1.60c$, which is impossible. The Galilean transformation equation does not work in relativistic situations.)

EXAMPLE 39.8 ▶ The Speeding Motorcycle

Imagine a motorcycle moving with a speed $0.80c$ past a stationary observer, as shown in Figure 39.17. If the rider tosses a ball in the forward direction with a speed of $0.70c$ relative to himself, what is the speed of the ball relative to the stationary observer?

Solution The speed of the motorcycle relative to the stationary observer is $v = 0.80c$. The speed of the ball in the frame of reference of the motorcyclist is $u'_x = 0.70c$. Therefore, the speed u_x of the ball relative to the stationary observer is

$$u_x = \frac{u'_x + v}{1 + \dfrac{u'_x v}{c^2}} = \frac{0.70c + 0.80c}{1 + \dfrac{(0.70c)(0.80c)}{c^2}} = \boxed{0.96c}$$

Exercise Suppose that the motorcyclist turns on the headlight so that a beam of light moves away from him with a speed c in the forward direction. What does the stationary observer measure for the speed of the light?

Answer c.

Figure 39.17 A motorcyclist moves past a stationary observer with a speed of $0.80c$ and throws a ball in the direction of motion with a speed of $0.70c$ relative to himself.

EXAMPLE 39.9 Relativistic Leaders of the Pack

Two motorcycle pack leaders named David and Emily are racing at relativistic speeds along perpendicular paths, as shown in Figure 39.18. How fast does Emily recede as seen by David over his right shoulder?

Solution Figure 39.18 represents the situation as seen by a police officer at rest in frame S, who observes the following:

David: $\quad u_x = 0.75c \quad\quad u_y = 0$

Emily: $\quad u_x = 0 \quad\quad u_y = -0.90c$

To calculate Emily's speed of recession as seen by David, we take S′ to move along with David and then calculate u'_x and u'_y for Emily using Equations 39.16 and 39.17:

$$u'_x = \frac{u_x - v}{1 - \dfrac{u_x v}{c^2}} = \frac{0 - 0.75c}{1 - \dfrac{(0)(0.75c)}{c^2}} = -0.75c$$

$$u'_y = \frac{u_y}{\gamma\left(1 - \dfrac{u_x v}{c^2}\right)} = \frac{\sqrt{1 - \dfrac{(0.75c)^2}{c^2}}\,(-0.90c)}{\left(1 - \dfrac{(0)(0.75c)}{c^2}\right)} = -0.60c$$

Thus, the speed of Emily as observed by David is

$$u' = \sqrt{(u'_x)^2 + (u'_y)^2} = \sqrt{(-0.75c)^2 + (-0.60c)^2} = \boxed{0.96c}$$

Note that this speed is less than c, as required by the special theory of relativity.

Exercise Use the Galilean velocity transformation equation to calculate the classical speed of recession for Emily as observed by David.

Answer $1.2c$.

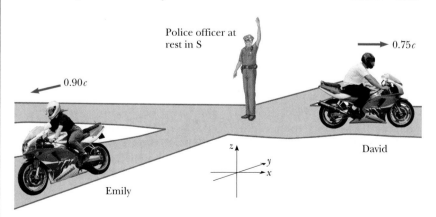

Figure 39.18 David moves to the east with a speed $0.75c$ relative to the police officer, and Emily travels south at a speed $0.90c$ relative to the officer.

39.6 RELATIVISTIC LINEAR MOMENTUM AND THE RELATIVISTIC FORM OF NEWTON'S LAWS

We have seen that in order to describe properly the motion of particles within the framework of the special theory of relativity, we must replace the Galilean transformation equations by the Lorentz transformation equations. Because the laws of physics must remain unchanged under the Lorentz transformation, we must generalize Newton's laws and the definitions of linear momentum and energy to conform to the Lorentz transformation equations and the principle of relativity. These generalized definitions should reduce to the classical (nonrelativistic) definitions for $v \ll c$.

First, recall that the law of conservation of linear momentum states that when two isolated objects collide, their combined total momentum remains constant. Suppose that the collision is described in a reference frame S in which linear momentum is conserved. If we calculate the velocities in a second reference frame S′ using the Lorentz velocity transformation equation and the classical definition of linear momentum, $\mathbf{p} = m\mathbf{u}$ (where $\mathbf{u}$ is the velocity of either object), we find that linear momentum is *not* conserved in S′. However, because the laws of physics are the same in all inertial frames, linear momentum must be conserved in all frames. In view of this condition and assuming that the Lorentz velocity transformation equation is correct, we must modify the definition of linear momentum to satisfy the following conditions:

- Linear momentum $\mathbf{p}$ must be conserved in all collisions.
- The relativistic value calculated for $\mathbf{p}$ must approach the classical value $m\mathbf{u}$ as $\mathbf{u}$ approaches zero.

For any particle, the correct relativistic equation for linear momentum that satisfies these conditions is

Definition of relativistic linear momentum

$$\mathbf{p} \equiv \frac{m\mathbf{u}}{\sqrt{1 - \dfrac{u^2}{c^2}}} = \gamma m\mathbf{u} \tag{39.19}$$

where $\mathbf{u}$ is the velocity of the particle and m is the mass of the particle. When u is much less than c, $\gamma = (1 - u^2/c^2)^{-1/2}$ approaches unity and $\mathbf{p}$ approaches $m\mathbf{u}$. Therefore, the relativistic equation for $\mathbf{p}$ does indeed reduce to the classical expression when u is much smaller than c.

The relativistic force $\mathbf{F}$ acting on a particle whose linear momentum is $\mathbf{p}$ is defined as

$$\mathbf{F} \equiv \frac{d\mathbf{p}}{dt} \tag{39.20}$$

where $\mathbf{p}$ is given by Equation 39.19. This expression, which is the relativistic form of Newton's second law, is reasonable because it preserves classical mechanics in the limit of low velocities and requires conservation of linear momentum for an isolated system ($\mathbf{F} = 0$) both relativistically and classically.

It is left as an end-of-chapter problem (Problem 63) to show that under relativistic conditions, the acceleration $\mathbf{a}$ of a particle decreases under the action of a constant force, in which case $a \propto (1 - u^2/c^2)^{3/2}$. From this formula, note that as the particle's speed approaches c, the acceleration caused by any finite force approaches zero. Hence, it is impossible to accelerate a particle from rest to a speed $u \geq c$.

EXAMPLE 39.10   Linear Momentum of an Electron

An electron, which has a mass of 9.11×10^{-31} kg, moves with a speed of $0.750c$. Find its relativistic momentum and compare this value with the momentum calculated from the classical expression.

Solution   Using Equation 39.19 with $u = 0.750c$, we have

$$p = \frac{m_e u}{\sqrt{1 - \dfrac{u^2}{c^2}}}$$

$$= \frac{(9.11 \times 10^{-31}\ \text{kg})(0.750 \times 3.00 \times 10^8\ \text{m/s})}{\sqrt{1 - \dfrac{(0.750c)^2}{c^2}}}$$

$$= 3.10 \times 10^{-22}\ \text{kg} \cdot \text{m/s}$$

The (incorrect) classical expression gives

$$p_{\text{classical}} = m_e u = 2.05 \times 10^{-22}\ \text{kg} \cdot \text{m/s}$$

Hence, the correct relativistic result is 50% greater than the classical result!

39.7   RELATIVISTIC ENERGY

We have seen that the definition of linear momentum and the laws of motion require generalization to make them compatible with the principle of relativity. This implies that the definition of kinetic energy must also be modified.

To derive the relativistic form of the work–kinetic energy theorem, let us first use the definition of relativistic force, Equation 39.20, to determine the work done on a particle by a force F:

$$W = \int_{x_1}^{x_2} F\, dx = \int_{x_1}^{x_2} \frac{dp}{dt}\, dx \tag{39.21}$$

for force and motion both directed along the x axis. In order to perform this integration and find the work done on the particle and the relativistic kinetic energy as a function of u, we first evaluate dp/dt:

$$\frac{dp}{dt} = \frac{d}{dt} \frac{mu}{\sqrt{1 - \dfrac{u^2}{c^2}}} = \frac{m(du/dt)}{\left(1 - \dfrac{u^2}{c^2}\right)^{3/2}}$$

Substituting this expression for dp/dt and $dx = u\, dt$ into Equation 39.21 gives

$$W = \int_0^t \frac{m(du/dt)u\, dt}{\left(1 - \dfrac{u^2}{c^2}\right)^{3/2}} = m \int_0^u \frac{u}{\left(1 - \dfrac{u^2}{c^2}\right)^{3/2}}\, du$$

where we use the limits 0 and u in the rightmost integral because we have assumed that the particle is accelerated from rest to some final speed u. Evaluating the integral, we find that

$$W = \frac{mc^2}{\sqrt{1 - \dfrac{u^2}{c^2}}} - mc^2 \tag{39.22}$$

Recall from Chapter 7 that the work done by a force acting on a particle equals the change in kinetic energy of the particle. Because of our assumption that the initial speed of the particle is zero, we know that the initial kinetic energy is zero. We therefore conclude that the work W is equivalent to the relativistic kinetic energy K:

Relativistic kinetic energy

$$K = \frac{mc^2}{\sqrt{1 - \dfrac{u^2}{c^2}}} - mc^2 = \gamma mc^2 - mc^2 \qquad \textbf{(39.23)}$$

This equation is routinely confirmed by experiments using high-energy particle accelerators.

At low speeds, where $u/c \ll 1$, Equation 39.23 should reduce to the classical expression $K = \frac{1}{2}mu^2$. We can check this by using the binomial expansion $(1 - x^2)^{-1/2} \approx 1 + \frac{1}{2}x^2 + \ldots$ for $x \ll 1$, where the higher-order powers of x are neglected in the expansion. In our case, $x = u/c$, so that

$$\frac{1}{\sqrt{1 - \dfrac{u^2}{c^2}}} = \left(1 - \frac{u^2}{c^2}\right)^{-1/2} \approx 1 + \frac{1}{2}\frac{u^2}{c^2}$$

Substituting this into Equation 39.23 gives

$$K \approx mc^2\left(1 + \frac{1}{2}\frac{u^2}{c^2}\right) - mc^2 = \frac{1}{2}mu^2$$

which is the classical expression for kinetic energy. A graph comparing the relativistic and nonrelativistic expressions is given in Figure 39.19. In the relativistic case, the particle speed never exceeds c, regardless of the kinetic energy. The two curves are in good agreement when $u \ll c$.

The constant term mc^2 in Equation 39.23, which is independent of the speed of the particle, is called the **rest energy** E_R of the particle (see Section 8.9). The term γmc^2, which does depend on the particle speed, is therefore the sum of the kinetic and rest energies. We define γmc^2 to be the **total energy** E:

Definition of total energy

$$\text{Total energy} = \text{kinetic energy} + \text{rest energy}$$
$$E = \gamma mc^2 = K + mc^2 \qquad \textbf{(39.24)}$$

or

$$E = \frac{mc^2}{\sqrt{1 - \dfrac{u^2}{c^2}}} \qquad \textbf{(39.25)}$$

This is Einstein's famous equation about mass–energy equivalence.

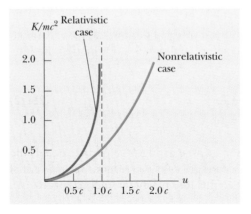

Figure 39.19 A graph comparing relativistic and nonrelativistic kinetic energy. The energies are plotted as a function of speed. In the relativistic case, u is always less than c.

The relationship $E = K + mc^2$ shows that **mass is a form of energy,** where c^2 in the rest energy term is just a constant conversion factor. This expression also shows that a small mass corresponds to an enormous amount of energy, a concept fundamental to nuclear and elementary-particle physics.

In many situations, the linear momentum or energy of a particle is measured rather than its speed. It is therefore useful to have an expression relating the total energy E to the relativistic linear momentum p. This is accomplished by using the expressions $E = \gamma mc^2$ and $p = \gamma mu$. By squaring these equations and subtracting, we can eliminate u (Problem 39). The result, after some algebra, is[6]

$$E^2 = p^2 c^2 + (mc^2)^2 \qquad \textbf{(39.26)}$$

Energy–momentum relationship

When the particle is at rest, $p = 0$ and so $E = E_R = mc^2$. For particles that have zero mass, such as photons, we set $m = 0$ in Equation 39.26 and see that

$$E = pc \qquad \textbf{(39.27)}$$

This equation is an exact expression relating total energy and linear momentum for photons, which always travel at the speed of light.

Finally, note that because the mass m of a particle is independent of its motion, m must have the same value in all reference frames. For this reason, m is often called the **invariant mass.** On the other hand, because the total energy and linear momentum of a particle both depend on velocity, these quantities depend on the reference frame in which they are measured.

Because m is a constant, we conclude from Equation 39.26 that the quantity $E^2 - p^2 c^2$ must have the same value in all reference frames. That is, $E^2 - p^2 c^2$ is invariant under a Lorentz transformation. (Equations 39.26 and 39.27 do not make provision for potential energy.)

When we are dealing with subatomic particles, it is convenient to express their energy in electron volts because the particles are usually given this energy by acceleration through a potential difference. The conversion factor, as you recall from Equation 25.5, is

$$1.00 \text{ eV} = 1.602 \times 10^{-19} \text{ J}$$

For example, the mass of an electron is 9.109×10^{-31} kg. Hence, the rest energy of the electron is

$$m_e c^2 = (9.109 \times 10^{-31} \text{ kg})(2.9979 \times 10^8 \text{ m/s})^2 = 8.187 \times 10^{-14} \text{ J}$$
$$= (8.187 \times 10^{-14} \text{ J})(1 \text{ eV}/1.602 \times 10^{-19} \text{ J}) = 0.511 \text{ MeV}$$

EXAMPLE 39.11 ▸ The Energy of a Speedy Electron

An electron in a television picture tube typically moves with a speed $u = 0.250c$. Find its total energy and kinetic energy in electron volts.

Solution Using the fact that the rest energy of the electron is 0.511 MeV together with Equation 39.25, we have

$$E = \frac{m_e c^2}{\sqrt{1 - \dfrac{u^2}{c^2}}} = \frac{0.511 \text{ MeV}}{\sqrt{1 - \dfrac{(0.250c)^2}{c^2}}}$$

$$= 1.03(0.511 \text{ MeV}) = \boxed{0.528 \text{ MeV}}$$

This is 3% greater than the rest energy.

We obtain the kinetic energy by subtracting the rest energy from the total energy:

$$K = E - m_e c^2 = 0.528 \text{ MeV} - 0.511 \text{ MeV} = \boxed{0.017 \text{ MeV}}$$

---

[6] One way to remember this relationship is to draw a right triangle having a hypotenuse of length E and legs of lengths pc and mc^2.

EXAMPLE 39.12   **The Energy of a Speedy Proton**

(a) Find the rest energy of a proton in electron volts.

Solution

$$E_R = m_p c^2 = (1.67 \times 10^{-27} \text{ kg})(3.00 \times 10^8 \text{ m/s})^2$$
$$= (1.50 \times 10^{-10} \text{ J})(1.00 \text{ eV}/1.60 \times 10^{-19} \text{ J})$$

$$= \boxed{938 \text{ MeV}}$$

(b) If the total energy of a proton is three times its rest energy, with what speed is the proton moving?

Solution   Equation 39.25 gives

$$E = 3m_p c^2 = \frac{m_p c^2}{\sqrt{1 - \dfrac{u^2}{c^2}}}$$

$$3 = \frac{1}{\sqrt{1 - \dfrac{u^2}{c^2}}}$$

Solving for u gives

$$1 - \frac{u^2}{c^2} = \frac{1}{9}$$

$$\frac{u^2}{c^2} = \frac{8}{9}$$

$$u = \frac{\sqrt{8}}{3}c = \boxed{2.83 \times 10^8 \text{ m/s}}$$

(c) Determine the kinetic energy of the proton in electron volts.

Solution   From Equation 39.24,

$$K = E - m_p c^2 = 3m_p c^2 - m_p c^2 = 2m_p c^2$$

Because $m_p c^2 = 938$ MeV, $K = \boxed{1\,880 \text{ MeV}}$

(d) What is the proton's momentum?

Solution   We can use Equation 39.26 to calculate the momentum with $E = 3m_p c^2$:

$$E^2 = p^2 c^2 + (m_p c^2)^2 = (3m_p c^2)^2$$

$$p^2 c^2 = 9(m_p c^2)^2 - (m_p c^2)^2 = 8(m_p c^2)^2$$

$$p = \sqrt{8}\,\frac{m_p c^2}{c} = \sqrt{8}\,\frac{(938 \text{ MeV})}{c} = \boxed{2\,650 \text{ MeV}/c}$$

The unit of momentum is written MeV/c for convenience.

39.8   **EQUIVALENCE OF MASS AND ENERGY**

To understand the equivalence of mass and energy, consider the following thought experiment proposed by Einstein in developing his famous equation $E = mc^2$. Imagine an isolated box of mass M_{box} and length L initially at rest, as shown in Figure 39.20a. Suppose that a pulse of light is emitted from the left side of the box, as depicted in Figure 39.20b. From Equation 39.27, we know that light of energy E carries linear momentum $p = E/c$. Hence, if momentum is to be conserved, the box must recoil to the left with a speed v. If it is assumed that the box is very massive, the recoil speed is much less than the speed of light, and conservation of momentum gives $M_{\text{box}}v = E/c$, or

$$v = \frac{E}{M_{\text{box}}c}$$

The time it takes the light pulse to move the length of the box is approximately $\Delta t = L/c$. In this time interval, the box moves a small distance Δx to the left, where

$$\Delta x = v\,\Delta t = \frac{EL}{M_{\text{box}}c^2}$$

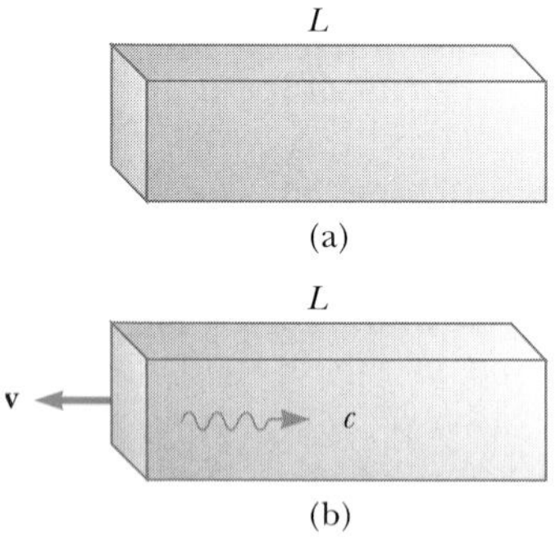

Figure 39.20   (a) A box of length L at rest. (b) When a light pulse directed to the right is emitted at the left end of the box, the box recoils to the left until the pulse strikes the right end.

The light then strikes the right end of the box and transfers its momentum to the box, causing the box to stop. With the box in its new position, its center of mass appears to have moved to the left. However, its center of mass cannot have moved because the box is an isolated system. Einstein resolved this perplexing situation by assuming that in addition to energy and momentum, light also carries mass. If M_{pulse} is the effective mass carried by the pulse of light and if the center of mass of the system (box plus pulse of light) is to remain fixed, then

$$M_{\text{pulse}}L = M_{\text{box}}\Delta x$$

Solving for M_{pulse}, and using the previous expression for Δx, we obtain

$$M_{\text{pulse}} = \frac{M_{\text{box}}\Delta x}{L} = \frac{M_{\text{box}}}{L}\frac{EL}{M_{\text{box}}c^2} = \frac{E}{c^2}$$

or

$$E = M_{\text{pulse}}c^2$$

Thus, Einstein reached the profound conclusion that "if a body gives off the energy E in the form of radiation, its mass diminishes by E/c^2, . . ."

Although we derived the relationship $E = mc^2$ for light energy, the equivalence of mass and energy is universal. Equation 39.24, $E = \gamma mc^2$, which represents the total energy of any particle, suggests that even when a particle is at rest ($\gamma = 1$) it still possesses enormous energy because it has mass. Probably the clearest experimental proof of the equivalence of mass and energy occurs in nuclear and elementary particle interactions, where large amounts of energy are released and the energy release is accompanied by a decrease in mass. Because energy and mass are related, we see that the laws of conservation of energy and conservation of mass are one and the same. Simply put, this law states that

the energy of a system of particles before interaction must equal the energy of the system after interaction, where energy of the ith particle is given by the expression

$$E_i = \frac{m_i c^2}{\sqrt{1 - \dfrac{u_i^2}{c^2}}} = \gamma m_i c^2$$

Conversion of mass–energy

The release of enormous quantities of energy from subatomic particles, accompanied by changes in their masses, is the basis of all nuclear reactions. In a conventional nuclear reactor, a uranium nucleus undergoes *fission*, a reaction that creates several lighter fragments having considerable kinetic energy. The combined mass of all the fragments is less than the mass of the parent uranium nucleus by an amount Δm. The corresponding energy Δmc^2 associated with this mass difference is exactly equal to the total kinetic energy of the fragments. This kinetic energy raises the temperature of water in the reactor, converting it to steam for the generation of electric power.

In the nuclear reaction called *fusion*, two atomic nuclei combine to form a single nucleus. The fusion reaction in which two deuterium nuclei fuse to form a helium nucleus is of major importance in current research and the development of controlled-fusion reactors. The decrease in mass that results from the creation of one helium nucleus from two deuterium nuclei is $\Delta m = 4.25 \times 10^{-29}$ kg. Hence, the corresponding excess energy that results from one fusion reaction is $\Delta mc^2 =$

3.83×10^{-12} J $= 23.9$ MeV. To appreciate the magnitude of this result, note that if 1 g of deuterium is converted to helium, the energy released is about 10^{12} J! At the current cost of electrical energy, this quantity of energy would be worth about $70 000.

CONCEPTUAL EXAMPLE 39.13

Because mass is a measure of energy, can we conclude that the mass of a compressed spring is greater than the mass of the same spring when it is not compressed?

Solution   Recall that when a spring of force constant k is compressed (or stretched) from its equilibrium position a distance x, it stores elastic potential energy $U = kx^2/2$. Ac-

cording to the special theory of relativity, any change in the total energy of a system is equivalent to a change in the mass of the system. Therefore, the mass of a compressed (or stretched) spring is greater than the mass of the spring in its equilibrium position by an amount U/c^2.

EXAMPLE 39.14   Binding Energy of the Deuteron

A deuteron, which is the nucleus of a deuterium atom, contains one proton and one neutron and has a mass of 2.013 553 u. This total deuteron mass is not equal to the sum of the masses of the proton and neutron. Calculate the mass difference and determine its energy equivalence, which is called the *binding energy* of the nucleus.

Solution   Using atomic mass units (u), we have

$$m_p = \text{mass of proton} = 1.007\ 276\ \text{u}$$

$$m_n = \text{mass of neutron} = 1.008\ 665\ \text{u}$$

$$m_p + m_n = 2.015\ 941\ \text{u}$$

The mass difference Δm is therefore 0.002 388 u. By defini-

tion, 1 u $= 1.66 \times 10^{-27}$ kg, and therefore

$$\Delta m = 0.002\ 388\ \text{u} = \boxed{3.96 \times 10^{-30}\ \text{kg}}$$

Using $E = \Delta mc^2$, we find that the binding energy is

$$E = \Delta mc^2 = (3.96 \times 10^{-30}\ \text{kg})(3.00 \times 10^8\ \text{m/s})^2$$

$$= 3.56 \times 10^{-13}\ \text{J} = \boxed{2.23\ \text{MeV}}$$

Therefore, the minimum energy required to separate the proton from the neutron of the deuterium nucleus (the binding energy) is 2.23 MeV.

39.9   RELATIVITY AND ELECTROMAGNETISM

Consider two frames of reference S and S′ that are in relative motion, and assume that a single charge q is at rest in the S′ frame of reference. According to an observer in this frame, an electric field surrounds the charge. However, an observer in frame S says that the charge is in motion and therefore measures both an electric field and a magnetic field. The magnetic field measured by the observer in frame S is created by the moving charge, which constitutes an electric current. In other words, electric and magnetic fields are viewed differently in frames of reference that are moving relative to each other. We now describe one situation that shows how an electric field in one frame of reference is viewed as a magnetic field in another frame of reference.

A positive test charge q is moving parallel to a current-carrying wire with velocity **v** relative to the wire in frame S, as shown in Figure 39.21a. We assume that the net charge on the wire is zero and that the electrons in the wire also move with velocity **v** in a straight line. The leftward current in the wire produces a magnetic field that forms circles around the wire and is directed into the page at the loca-

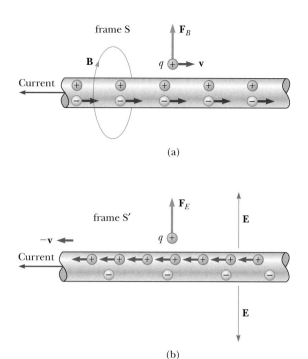

Figure 39.21 (a) In frame S, the positive charge q moves to the right with a velocity **v**, and the current-carrying wire is stationary. A magnetic field **B** surrounds the wire, and charge experiences a *magnetic* force directed away from the wire. (b) In frame S', the wire moves to the left with a velocity $-$**v**, and the charge q is stationary. The wire creates an electric field **E**, and the charge experiences an *electric* force directed away from the wire.

tion of the moving test charge. Therefore, a magnetic force $\mathbf{F}_B = q\mathbf{v} \times \mathbf{B}$ directed away from the wire is exerted on the test charge. However, no electric force acts on the test charge because the net charge on the wire is zero when viewed in this frame.

Now consider the same situation as viewed from frame S', where the test charge is at rest (Figure 39.21b). In this frame, the positive charges in the wire move to the left, the electrons in the wire are at rest, and the wire still carries a current. Because the test charge is not moving in this frame, $\mathbf{F}_B = q\mathbf{v} \times \mathbf{B} = 0$; there is no magnetic force exerted on the test charge when viewed in this frame. However, if a force is exerted on the test charge in frame S', the frame of the wire, as described earlier, a force must be exerted on it in any other frame. What is the origin of this force in frame S, the frame of the test charge?

The answer to this question is provided by the special theory of relativity. When the situation is viewed in frame S, as in Figure 39.21a, the positive charges are at rest and the electrons in the wire move to the right with a velocity **v**. Because of length contraction, the electrons appear to be closer together than their proper separation. Because there is no net charge on the wire this contracted separation must equal the separation between the stationary positive charges. The situation is quite different when viewed in frame S', shown in Figure 39.21b. In this frame, the positive charges appear closer together because of length contraction, and the electrons in the wire are at rest with a separation that is greater than that viewed in frame S. Therefore, there is a net positive charge on the wire when viewed in frame S′. This net positive charge produces an electric field pointing away from the wire toward the test charge, and so the test charge experiences an electric force directed away from the wire. Thus, what was viewed as a magnetic field (and a corresponding magnetic force) in the frame of the wire transforms into an electric field (and a corresponding electric force) in the frame of the test charge.

Optional Section

39.10 ▶ THE GENERAL THEORY OF RELATIVITY

Up to this point, we have sidestepped a curious puzzle. Mass has two seemingly different properties: a *gravitational attraction* for other masses and an *inertial* property that resists acceleration. To designate these two attributes, we use the subscripts *g* and *i* and write

$$\text{Gravitational property} \qquad F_g = m_g g$$

$$\text{Inertial property} \qquad \Sigma F = m_i a$$

The value for the gravitational constant G was chosen to make the magnitudes of m_g and m_i numerically equal. Regardless of how G is chosen, however, the strict proportionality of m_g and m_i has been established experimentally to an extremely high degree: a few parts in 10^{12}. Thus, it appears that gravitational mass and inertial mass may indeed be exactly proportional.

But why? They seem to involve two entirely different concepts: a force of mutual gravitational attraction between two masses, and the resistance of a single mass to being accelerated. This question, which puzzled Newton and many other physicists over the years, was answered when Einstein published his theory of gravitation, known as his *general theory of relativity,* in 1916. Because it is a mathematically complex theory, we offer merely a hint of its elegance and insight.

In Einstein's view, the remarkable coincidence that m_g and m_i seemed to be proportional to each other was evidence of an intimate and basic connection between the two concepts. He pointed out that no mechanical experiment (such as dropping a mass) could distinguish between the two situations illustrated in Figure 39.22a and b. In each case, the dropped briefcase undergoes a downward acceleration *g* relative to the floor.

Einstein carried this idea further and proposed that *no* experiment, mechanical or otherwise, could distinguish between the two cases. This extension to include all phenomena (not just mechanical ones) has interesting consequences. For example, suppose that a light pulse is sent horizontally across the elevator, as depicted in Figure 39.22c. During the time it takes the light to make the trip, the right wall of the elevator has accelerated upward. This causes the light to arrive at a location lower on the wall than the spot it would have hit if the elevator were not accelerating. Thus, in the frame of the elevator, the trajectory of the light pulse bends downward as the elevator accelerates upward to meet it. Because the accelerating elevator cannot be distinguished from a nonaccelerating one located in a gravitational field, Einstein proposed that a beam of light *should also be bent downward by a gravitational field.* Experiments have verified the effect, although the bending is small. A laser aimed at the horizon falls less than 1 cm after traveling 6 000 km. (No such bending is predicted in Newton's theory of gravitation.)

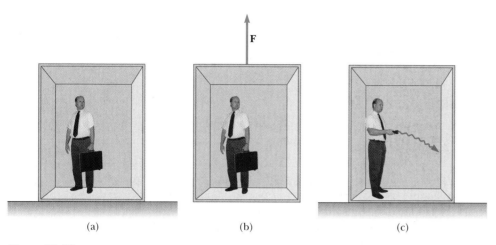

Figure 39.22 (a) The observer is at rest in a uniform gravitational field **g**. (b) The observer is in a region where gravity is negligible, but the frame of reference is accelerated by an external force **F** that produces an acceleration **g**. According to Einstein, the frames of reference in parts (a) and (b) are equivalent in every way. No local experiment can distinguish any difference between the two frames. (c) If parts (a) and (b) are truly equivalent, as Einstein proposed, then a ray of light should bend in a gravitational field.

The two postulates of Einstein's **general theory of relativity** are

- All the laws of nature have the same form for observers in any frame of reference, whether accelerated or not.
- In the vicinity of any point, a gravitational field is equivalent to an accelerated frame of reference in the absence of gravitational effects. (This is the *principle of equivalence.*)

The second postulate implies that gravitational mass and inertial mass are completely equivalent, not just proportional. What were thought to be two different types of mass are actually identical.

One interesting effect predicted by the general theory is that time scales are altered by gravity. A clock in the presence of gravity runs more slowly than one located where gravity is negligible. Consequently, the frequencies of radiation emitted by atoms in the presence of a strong gravitational field are *red-shifted* to lower frequencies when compared with the same emissions in the presence of a weak field. This gravitational red shift has been detected in spectral lines emitted by

This Global Positioning System (GPS) unit incorporates relativistically corrected time calculations in its analysis of signals it receives from orbiting satellites. These corrections allow the unit to determine its position on the Earth's surface to within a few meters. If the corrections were not made, the location error would be about 1 km. *(Courtesy of Trimble Navigation Limited)*

atoms in massive stars. It has also been verified on the Earth by comparison of the frequencies of gamma rays (a high-energy form of electromagnetic radiation) emitted from nuclei separated vertically by about 20 m.

Quick Quiz 39.7

Two identical clocks are in the same house, one upstairs in a bedroom and the other downstairs in the kitchen. Which clock runs more slowly?

The second postulate suggests that a gravitational field may be "transformed away" at any point if we choose an appropriate accelerated frame of reference—a freely falling one. Einstein developed an ingenious method of describing the acceleration necessary to make the gravitational field "disappear." He specified a concept, the *curvature of space–time,* that describes the gravitational effect at every point. In fact, the curvature of space–time completely replaces Newton's gravitational theory. According to Einstein, there is no such thing as a gravitational force. Rather, the presence of a mass causes a curvature of space–time in the vicinity of the mass, and this curvature dictates the space–time path that all freely moving objects must follow. In 1979, John Wheeler summarized Einstein's general theory of relativity in a single sentence: "Space tells matter how to move and matter tells space how to curve."

Consider two travelers on the surface of the Earth walking directly toward the North Pole but from different starting locations. Even though both say they are walking due north, and thus should be on parallel paths, they see themselves getting closer and closer together, as if they were somehow attracted to each other. The curvature of the Earth causes this effect. In a similar way, what we are used to thinking of as the gravitational attraction between two masses is, in Einstein's view, two masses curving space–time and as a result moving toward each other, much like two bowling balls on a mattress rolling together.

One prediction of the general theory of relativity is that a light ray passing near the Sun should be deflected into the curved space–time created by the Sun's mass. This prediction was confirmed when astronomers detected the bending of starlight near the Sun during a total solar eclipse that occurred shortly after World War I (Fig. 39.23). When this discovery was announced, Einstein became an international celebrity.

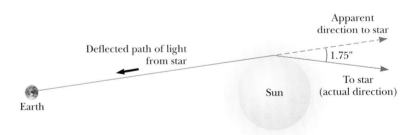

Figure 39.23 Deflection of starlight passing near the Sun. Because of this effect, the Sun or some other remote object can act as a *gravitational lens.* In his general theory of relativity, Einstein calculated that starlight just grazing the Sun's surface should be deflected by an angle of 1.75".

Einstein's cross. The four bright spots are images of the same galaxy that have been bent around a massive object located between the galaxy and the Earth. The massive object acts like a lens, causing the rays of light that were diverging from the distant galaxy to converge on the Earth. (If the intervening massive object had a uniform mass distribution, we would see a bright ring instead of four spots.) *(Courtesy of NASA)*

If the concentration of mass becomes very great, as is believed to occur when a large star exhausts its nuclear fuel and collapses to a very small volume, a **black hole** may form. Here, the curvature of space–time is so extreme that, within a certain distance from the center of the black hole, all matter and light become trapped.

SUMMARY

The two basic postulates of the special theory of relativity are

- The laws of physics must be the same in all inertial reference frames.
- The speed of light in vacuum has the same value, $c = 3.00 \times 10^8$ m/s, in all inertial frames, regardless of the velocity of the observer or the velocity of the source emitting the light.

Three consequences of the special theory of relativity are

- Events that are simultaneous for one observer are not simultaneous for another observer who is in motion relative to the first.
- Clocks in motion relative to an observer appear to be slowed down by a factor $\gamma = (1 - v^2/c^2)^{-1/2}$. This phenomenon is known as **time dilation.**
- The length of objects in motion appears to be contracted in the direction of motion by a factor $1/\gamma = (1 - v^2/c^2)^{1/2}$. This phenomenon is known as **length contraction.**

To satisfy the postulates of special relativity, the Galilean transformation equations must be replaced by the **Lorentz transformation equations:**

$$x' = \gamma(x - vt)$$
$$y' = y$$
$$z' = z$$
$$t' = \gamma\left(t - \frac{v}{c^2}\,x\right)$$

(39.11)

where $\gamma = (1 - v^2/c^2)^{-1/2}$.

The relativistic form of the **velocity transformation equation** is

$$u'_x = \frac{u_x - v}{1 - \dfrac{u_x v}{c^2}} \qquad\qquad \textbf{(39.16)}$$

where u_x is the speed of an object as measured in the S frame and u'_x is its speed measured in the S′ frame.

The relativistic expression for the **linear momentum** of a particle moving with a velocity **u** is

$$\mathbf{p} \equiv \frac{m\mathbf{u}}{\sqrt{1 - \dfrac{u^2}{c^2}}} = \gamma m\mathbf{u} \qquad\qquad \textbf{(39.19)}$$

The relativistic expression for the **kinetic energy** of a particle is

$$K = \gamma mc^2 - mc^2 \qquad\qquad \textbf{(39.23)}$$

where mc^2 is called the **rest energy** of the particle.

The total energy E of a particle is related to the mass m of the particle through the famous **mass–energy** equivalence expression:

$$E = \frac{mc^2}{\sqrt{1 - \dfrac{u^2}{c^2}}} \qquad\qquad \textbf{(39.25)}$$

The relativistic linear momentum is related to the total energy through the equation

$$E^2 = p^2 c^2 + (mc^2)^2 \qquad\qquad \textbf{(39.26)}$$

QUESTIONS

1. What two speed measurements do two observers in relative motion always agree on?

2. A spaceship in the shape of a sphere moves past an observer on the Earth with a speed $0.5c$. What shape does the observer see as the spaceship moves past?

3. An astronaut moves away from the Earth at a speed close to the speed of light. If an observer on Earth measures the astronaut's dimensions and pulse rate, what changes (if any) would the observer measure? Would the astronaut measure any changes about himself?

4. Two identical clocks are synchronized. One is then put in orbit directed eastward around the Earth while the other remains on Earth. Which clock runs slower? When the moving clock returns to Earth, are the two still synchronized?

5. Two lasers situated on a moving spacecraft are triggered simultaneously. An observer on the spacecraft claims to see the pulses of light simultaneously. What condition is necessary so that a second observer agrees?

6. When we say that a moving clock runs more slowly than a stationary one, does this imply that there is something physically unusual about the moving clock?

7. List some ways our day-to-day lives would change if the speed of light were only 50 m/s.

8. Give a physical argument that shows that it is impossible to accelerate an object of mass m to the speed of light, even if it has a continuous force acting on it.

9. It is said that Einstein, in his teenage years, asked the question, "What would I see in a mirror if I carried it in my hands and ran at the speed of light?" How would you answer this question?

10. Some distant star-like objects, called *quasars*, are receding from us at half the speed of light (or greater). What is the speed of the light we receive from these quasars?

11. How is it possible that photons of light, which have zero mass, have momentum?

12. With regard to reference frames, how does general relativity differ from special relativity?

13. Describe how the results of Example 39.7 would change if, instead of fast spaceships, two ordinary cars were approaching each other at highway speeds.

14. Two objects are identical except that one is hotter than the other. Compare how they respond to identical forces.

PROBLEMS

1, 2, 3 = straightforward, intermediate, challenging ☐ = full solution available in the *Student Solutions Manual and Study Guide*
WEB = solution posted at **http://www.saunderscollege.com/physics/** 🖥 = Computer useful in solving problem 🕹 = Interactive Physics
☐ = paired numerical/symbolic problems

Section 39.1   The Principle of Galilean Relativity

1. A 2 000-kg car moving at 20.0 m/s collides and locks together with a 1 500-kg car at rest at a stop sign. Show that momentum is conserved in a reference frame moving at 10.0 m/s in the direction of the moving car.

2. A ball is thrown at 20.0 m/s inside a boxcar moving along the tracks at 40.0 m/s. What is the speed of the ball relative to the ground if the ball is thrown (a) forward? (b) backward? (c) out the side door?

3. In a laboratory frame of reference, an observer notes that Newton's second law is valid. Show that it is also valid for an observer moving at a constant speed, small compared with the speed of light, relative to the laboratory frame.

4. Show that Newton's second law is *not* valid in a reference frame moving past the laboratory frame of Problem 3 with a constant acceleration.

Section 39.2   The Michelson–Morley Experiment
Section 39.3   Einstein's Principle of Relativity
Section 39.4   Consequences of the Special Theory of Relativity

5. How fast must a meter stick be moving if its length is observed to shrink to 0.500 m?

6. At what speed does a clock have to move if it is to be seen to run at a rate that is one-half the rate of a clock at rest?

7. An astronaut is traveling in a space vehicle that has a speed of $0.500c$ relative to the Earth. The astronaut measures his pulse rate at 75.0 beats per minute. Signals generated by the astronaut's pulse are radioed to Earth when the vehicle is moving in a direction perpendicular to a line that connects the vehicle with an observer on the Earth. What pulse rate does the Earth observer measure? What would be the pulse rate if the speed of the space vehicle were increased to $0.990c$?

8. The proper length of one spaceship is three times that of another. The two spaceships are traveling in the same direction and, while both are passing overhead, an Earth observer measures the two spaceships to have the same length. If the slower spaceship is moving with a speed of $0.350c$, determine the speed of the faster spaceship.

9. An atomic clock moves at 1 000 km/h for 1 h as measured by an identical clock on Earth. How many nanoseconds slow will the moving clock be at the end of the 1-h interval?

10. If astronauts could travel at $v = 0.950c$, we on Earth would say it takes $(4.20/0.950) = 4.42$ yr to reach Alpha

Centauri, 4.20 ly away. The astronauts disagree. (a) How much time passes on the astronauts' clocks? (b) What distance to Alpha Centauri do the astronauts measure?

WEB 11. A spaceship with a proper length of 300 m takes 0.750 μs to pass an Earth observer. Determine the speed of this spaceship as measured by the Earth observer.

12. A spaceship of proper length L_p takes time t to pass an Earth observer. Determine the speed of this spaceship as measured by the Earth observer.

13. A muon formed high in the Earth's atmosphere travels at speed $v = 0.990c$ for a distance of 4.60 km before it decays into an electron, a neutrino, and an antineutrino ($\mu^- \rightarrow e^- + \nu + \bar{\nu}$). (a) How long does the muon live, as measured in its reference frame? (b) How far does the muon travel, as measured in its frame?

14. **Review Problem.**   In 1962, when Mercury astronaut Scott Carpenter orbited the Earth 22 times, the press stated that for each orbit he aged 2 millionths of a second less than he would have had he remained on Earth. (a) Assuming that he was 160 km above the Earth in a circular orbit, determine the time difference between someone on Earth and the orbiting astronaut for the 22 orbits. You will need to use the approximation $\sqrt{1 - x} \approx 1 - x/2$ for small x. (b) Did the press report accurate information? Explain.

15. The pion has an average lifetime of 26.0 ns when at rest. In order for it to travel 10.0 m, how fast must it move?

16. For what value of v does $\gamma = 1.01$? Observe that for speeds less than this value, time dilation and length contraction are less-than-one-percent effects.

17. A friend passes by you in a spaceship traveling at a high speed. He tells you that his ship is 20.0 m long and that the identically constructed ship you are sitting in is 19.0 m long. According to your observations, (a) how long is your ship, (b) how long is your friend's ship, and (c) what is the speed of your friend's ship?

18. An interstellar space probe is launched from Earth. After a brief period of acceleration it moves with a constant velocity, 70.0% of the speed of light. Its nuclear-powered batteries supply the energy to keep its data transmitter active continuously. The batteries have a lifetime of 15.0 yr as measured in a rest frame. (a) How long do the batteries on the space probe last as measured by Mission Control on Earth? (b) How far is the probe from Earth when its batteries fail, as measured by Mission Control? (c) How far is the probe from Earth when its batteries fail, as measured by its built-in trip odometer? (d) For what total time after launch are data

received from the probe by Mission Control? Note that radio waves travel at the speed of light and fill the space between the probe and Earth at the time of battery failure.

19. **Review Problem.** An alien civilization occupies a brown dwarf, nearly stationary relative to the Sun, several lightyears away. The extraterrestrials have come to love original broadcasts of *The Ed Sullivan Show,* on our television channel 2, at carrier frequency 57.0 MHz. Their line of sight to us is in the plane of the Earth's orbit. Find the difference between the highest and lowest frequencies they receive due to the Earth's orbital motion around the Sun.

20. Police radar detects the speed of a car (Fig. P39.20) as follows: Microwaves of a precisely known frequency are broadcast toward the car. The moving car reflects the microwaves with a Doppler shift. The reflected waves are received and combined with an attenuated version of the transmitted wave. Beats occur between the two microwave signals. The beat frequency is measured. (a) For an electromagnetic wave reflected back to its source from a mirror approaching at speed v, show that the reflected wave has frequency

$$f = f_{source} \frac{c + v}{c - v}$$

where f_{source} is the source frequency. (b) When v is much less than c, the beat frequency is much less than the transmitted frequency. In this case, use the approximation $f + f_{source} \cong 2f_{source}$ and show that the beat frequency can be written as $f_b = 2v/\lambda$. (c) What beat frequency is measured for a car speed of 30.0 m/s if the microwaves have frequency 10.0 GHz? (d) If the beat frequency measurement is accurate to ± 5 Hz, how accurate is the velocity measurement?

21. *The red shift.* A light source recedes from an observer with a speed v_{source}, which is small compared with c. (a) Show that the fractional shift in the measured wavelength is given by the approximate expression

$$\frac{\Delta\lambda}{\lambda} \approx \frac{v_{source}}{c}$$

This phenomenon is known as the red shift because the visible light is shifted toward the red. (b) Spectroscopic measurements of light at $\lambda = 397$ nm coming from a galaxy in Ursa Major reveal a red shift of 20.0 nm. What is the recessional speed of the galaxy?

Section 39.5 The Lorentz Transformation Equations

22. A spaceship travels at $0.750c$ relative to Earth. If the spaceship fires a small rocket in the forward direction, how fast (relative to the ship) must it be fired for it to travel at $0.950c$ relative to Earth?

WEB **23.** Two jets of material from the center of a radio galaxy fly away in opposite directions. Both jets move at $0.750c$ relative to the galaxy. Determine the speed of one jet relative to that of the other.

24. A moving rod is observed to have a length of 2.00 m, and to be oriented at an angle of 30.0° with respect to the direction of motion (Fig. P39.24). The rod has a speed of $0.995c$. (a) What is the proper length of the rod? (b) What is the orientation angle in the proper frame?

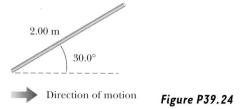

2.00 m
30.0°

Direction of motion **Figure P39.24**

25. A Klingon space ship moves away from the Earth at a speed of $0.800c$ (Fig. P39.25). The starship *Enterprise* pursues at a speed of $0.900c$ relative to the Earth. Observers on Earth see the *Enterprise* overtaking the Klingon ship at a relative speed of $0.100c$. With what speed is the *Enterprise* overtaking the Klingon ship as seen by the crew of the *Enterprise*?

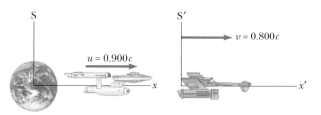

S
$u = 0.900c$
x

S′
$v = 0.800c$
x′

Figure P39.25

Figure P39.20 *(Trent Steffler/David R. Frazier Photolibrary)*

26. A red light flashes at position $x_R = 3.00$ m and time $t_R = 1.00 \times 10^{-9}$ s, and a blue light flashes at $x_B = 5.00$ m and $t_B = 9.00 \times 10^{-9}$ s (all values are measured in the S reference frame). Reference frame S′ has its origin at the same point as S at $t = t' = 0$; frame S′ moves constantly to the right. Both flashes are observed to occur at the same place in S′. (a) Find the relative velocity between S and S′. (b) Find the location of the two flashes in frame S′. (c) At what time does the red flash occur in the S′ frame?

Section 39.6 Relativistic Linear Momentum and the Relativistic Form of Newton's Laws

27. Calculate the momentum of an electron moving with a speed of (a) $0.010\ 0c$, (b) $0.500c$, (c) $0.900c$.

28. The nonrelativistic expression for the momentum of a particle, $p = mu$, can be used if $u \ll c$. For what speed does the use of this formula yield an error in the momentum of (a) 1.00 percent and (b) 10.0 percent?

29. A golf ball travels with a speed of 90.0 m/s. By what fraction does its relativistic momentum p differ from its classical value mu? That is, find the ratio $(p - mu)/mu$.

30. Show that the speed of an object having momentum p and mass m is

$$u = \frac{c}{\sqrt{1 + (mc/p)^2}}$$

WEB 31. An unstable particle at rest breaks into two fragments of *unequal* mass. The mass of the lighter fragment is 2.50×10^{-28} kg, and that of the heavier fragment is 1.67×10^{-27} kg. If the lighter fragment has a speed of $0.893c$ after the breakup, what is the speed of the heavier fragment?

Section 39.7 Relativistic Energy

32. Determine the energy required to accelerate an electron (a) from $0.500c$ to $0.900c$ and (b) from $0.900c$ to $0.990c$.

33. Find the momentum of a proton in MeV/c units if its total energy is twice its rest energy.

34. Show that, for any object moving at less than one-tenth the speed of light, the relativistic kinetic energy agrees with the result of the classical equation $K = mu^2/2$ to within less than 1%. Thus, for most purposes, the classical equation is good enough to describe these objects, whose motion we call *nonrelativistic.*

WEB 35. A proton moves at $0.950c$. Calculate its (a) rest energy, (b) total energy, and (c) kinetic energy.

36. An electron has a kinetic energy five times greater than its rest energy. Find (a) its total energy and (b) its speed.

37. A cube of steel has a volume of 1.00 cm³ and a mass of 8.00 g when at rest on the Earth. If this cube is now given a speed $u = 0.900c$, what is its density as measured by a stationary observer? Note that relativistic density is E_R/c^2V.

38. An unstable particle with a mass of 3.34×10^{-27} kg is initially at rest. The particle decays into two fragments that fly off with velocities of $0.987c$ and $-0.868c$. Find the masses of the fragments. (*Hint:* Conserve both mass–energy and momentum.)

39. Show that the energy–momentum relationship $E^2 = p^2c^2 + (mc^2)^2$ follows from the expressions $E = \gamma mc^2$ and $p = \gamma mu$.

40. A proton in a high-energy accelerator is given a kinetic energy of 50.0 GeV. Determine (a) its momentum and (b) its speed.

41. In a typical color television picture tube, the electrons are accelerated through a potential difference of 25 000 V. (a) What speed do the electrons have when they strike the screen? (b) What is their kinetic energy in joules?

42. Electrons are accelerated to an energy of 20.0 GeV in the 3.00-km-long Stanford Linear Accelerator. (a) What is the γ factor for the electrons? (b) What is their speed? (c) How long does the accelerator appear to them?

43. A pion at rest ($m_\pi = 270m_e$) decays to a muon ($m_\mu = 206m_e$) and an antineutrino ($m_{\bar{\nu}} \approx 0$). The reaction is written $\pi^- \rightarrow \mu^- + \bar{\nu}$. Find the kinetic energy of the muon and the antineutrino in electron volts. (*Hint:* Relativistic momentum is conserved.)

Section 39.8 Equivalence of Mass and Energy

44. Make an order-of-magnitude estimate of the ratio of mass increase to the original mass of a flag as you run it up a flagpole. In your solution explain what quantities you take as data and the values you estimate or measure for them.

45. When 1.00 g of hydrogen combines with 8.00 g of oxygen, 9.00 g of water is formed. During this chemical reaction, 2.86×10^5 J of energy is released. How much mass do the constituents of this reaction lose? Is the loss of mass likely to be detectable?

46. A spaceship of mass 1.00×10^6 kg is to be accelerated to $0.600c$. (a) How much energy does this require? (b) How many kilograms of matter would it take to provide this much energy?

47. In a nuclear power plant the fuel rods last 3 yr before they are replaced. If a plant with rated thermal power 1.00 GW operates at 80.0% capacity for the 3 yr, what is the loss of mass of the fuel?

48. A ^{57}Fe nucleus at rest emits a 14.0-keV photon. Use the conservation of energy and momentum to deduce the kinetic energy of the recoiling nucleus in electron volts. (Use $Mc^2 = 8.60 \times 10^{-9}$ J for the final state of the ^{57}Fe nucleus.)

49. The power output of the Sun is 3.77×10^{26} W. How much mass is converted to energy in the Sun each second?

50. A gamma ray (a high-energy photon of light) can produce an electron (e^-) and a positron (e^+) when

it enters the electric field of a heavy nucleus:
$\gamma \rightarrow e^+ + e^-$. What minimum γ-ray energy is
required to accomplish this task? (*Hint:* The masses of
the electron and the positron are equal.)

Section 39.9   Relativity and Electromagnetism

51. As measured by observers in a reference frame S, a particle having charge q moves with velocity **v** in a magnetic field **B** and an electric field **E**. The resulting force on the particle is then measured to be $\mathbf{F} = q(\mathbf{E} + \mathbf{v} \times \mathbf{B})$. Another observer moves along with the charged particle and also measures its charge to be q but measures the electric field to be $\mathbf{E}'$. If both observers are to measure the same force **F**, show that $\mathbf{E}' = \mathbf{E} + \mathbf{v} \times \mathbf{B}$.

ADDITIONAL PROBLEMS

52. An electron has a speed of $0.750c$. Find the speed of a proton that has (a) the same kinetic energy as the electron; (b) the same momentum as the electron.

WEB 53. The cosmic rays of highest energy are protons, which have kinetic energy on the order of 10^{13} MeV. (a) How long would it take a proton of this energy to travel across the Milky Way galaxy, having a diameter of $\sim 10^5$ ly, as measured in the proton's frame? (b) From the point of view of the proton, how many kilometers across is the galaxy?

54. A spaceship moves away from the Earth at $0.500c$ and fires a shuttle craft in the forward direction at $0.500c$ relative to the ship. The pilot of the shuttle craft launches a probe at forward speed $0.500c$ relative to the shuttle craft. Determine (a) the speed of the shuttle craft relative to the Earth and (b) the speed of the probe relative to the Earth.

55. The net nuclear fusion reaction inside the Sun can be written as $4^1\text{H} \rightarrow {}^4\text{He} + \Delta E$. If the rest energy of each hydrogen atom is 938.78 MeV and the rest energy of the helium-4 atom is 3 728.4 MeV, what is the percentage of the starting mass that is released as energy?

56. An astronaut wishes to visit the Andromeda galaxy (2.00 million lightyears away), making a one-way trip that will take 30.0 yr in the spaceship's frame of reference. If his speed is constant, how fast must he travel relative to the Earth?

57. An alien spaceship traveling at $0.600c$ toward the Earth launches a landing craft with an advance guard of purchasing agents. The lander travels in the same direction with a velocity $0.800c$ relative to the spaceship. As observed on the Earth, the spaceship is 0.200 ly from the Earth when the lander is launched. (a) With what velocity is the lander observed to be approaching by observers on the Earth? (b) What is the distance to the Earth at the time of lander launch, as observed by the aliens? (c) How long does it take the lander to reach the Earth as observed by the aliens on the mother ship? (d) If the lander has a mass of 4.00×10^5 kg, what is its

kinetic energy as observed in the Earth reference frame?

58. A physics professor on the Earth gives an exam to her students, who are on a rocket ship traveling at speed v relative to the Earth. The moment the ship passes the professor, she signals the start of the exam. She wishes her students to have time T_0 (rocket time) to complete the exam. Show that she should wait a time (Earth time) of

$$T = T_0\sqrt{\frac{1 - v/c}{1 + v/c}}$$

before sending a light signal telling them to stop. (*Hint:* Remember that it takes some time for the second light signal to travel from the professor to the students.)

59. Spaceship I, which contains students taking a physics exam, approaches the Earth with a speed of $0.600c$ (relative to the Earth), while spaceship II, which contains professors proctoring the exam, moves at $0.280c$ (relative to the Earth) directly toward the students. If the professors stop the exam after 50.0 min have passed on their clock, how long does the exam last as measured by (a) the students? (b) an observer on the Earth?

60. Energy reaches the upper atmosphere of the Earth from the Sun at the rate of 1.79×10^{17} W. If all of this energy were absorbed by the Earth and not re-emitted, how much would the mass of the Earth increase in 1 yr?

61. A supertrain (proper length, 100 m) travels at a speed of $0.950c$ as it passes through a tunnel (proper length, 50.0 m). As seen by a trackside observer, is the train ever completely within the tunnel? If so, with how much space to spare?

62. Imagine that the entire Sun collapses to a sphere of radius R_g such that the work required to remove a small mass m from the surface would be equal to its rest energy mc^2. This radius is called the *gravitational radius* for the Sun. Find R_g. (It is believed that the ultimate fate of very massive stars is to collapse beyond their gravitational radii into black holes.)

63. A charged particle moves along a straight line in a uniform electric field **E** with a speed of u. If the motion and the electric field are both in the x direction, (a) show that the acceleration of the charge q in the x direction is given by

$$a = \frac{du}{dt} = \frac{qE}{m}\left(1 - \frac{u^2}{c^2}\right)^{3/2}$$

(b) Discuss the significance of the dependence of the acceleration on the speed. (c) If the particle starts from rest at $x = 0$ at $t = 0$, how would you proceed to find the speed of the particle and its position after a time t has elapsed?

64. (a) Show that the Doppler shift $\Delta\lambda$ in the wavelength of light is described by the expression

$$\frac{\Delta\lambda}{\lambda} + 1 = \sqrt{\frac{c - v}{c + v}}$$

where λ is the source wavelength and v is the speed of relative approach between source and observer. (b) How fast would a motorist have to be going for a red light to appear green? Take 650 nm as a typical wavelength for red light, and one of 550 nm as typical for green.

65. A rocket moves toward a mirror at $0.800c$ relative to the reference frame S in Figure P39.65. The mirror is stationary relative to S. A light pulse emitted by the rocket travels toward the mirror and is reflected back to the rocket. The front of the rocket is 1.80×10^{12} m from the mirror (as measured by observers in S) at the moment the light pulse leaves the rocket. What is the total travel time of the pulse as measured by observers in (a) the S frame and (b) the front of the rocket?

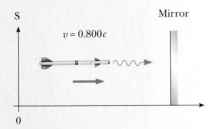

Figure P39.65 Problems 65 and 66.

66. An observer in a rocket moves toward a mirror at speed v relative to the reference frame labeled by S in Figure P39.65. The mirror is stationary with respect to S. A light pulse emitted by the rocket travels toward the mirror and is reflected back to the rocket. The front of the rocket is a distance d from the mirror (as measured by observers in S) at the moment the light pulse leaves the rocket. What is the total travel time of the pulse as measured by observers in (a) the S frame and (b) the front of the rocket?

67. Ted and Mary are playing a game of catch in frame S′, which is moving at $0.600c$, while Jim in frame S watches the action (Fig. P39.67). Ted throws the ball to Mary at $0.800c$ (according to Ted) and their separation (measured in S′) is 1.80×10^{12} m. (a) According to Mary,

how fast is the ball moving? (b) According to Mary, how long does it take the ball to reach her? (c) According to Jim, how far apart are Ted and Mary, and how fast is the ball moving? (d) According to Jim, how long does it take the ball to reach Mary?

68. A rod of length L_0 moving with a speed v along the horizontal direction makes an angle θ_0 with respect to the x' axis. (a) Show that the length of the rod as measured by a stationary observer is $L = L_0[1 - (v^2/c^2) \cos^2 \theta_0]^{1/2}$. (b) Show that the angle that the rod makes with the x axis is given by $\tan \theta = \gamma \tan \theta_0$. These results show that the rod is both contracted and rotated. (Take the lower end of the rod to be at the origin of the primed coordinate system.)

69. Consider two inertial reference frames S and S′, where S′ is moving to the right with a constant speed of $0.600c$ as measured by an observer in S. A stick of proper length 1.00 m moves to the left toward the origins of both S and S′, and the length of the stick is 50.0 cm as measured by an observer in S′. (a) Determine the speed of the stick as measured by observers in S and S′. (b) What is the length of the stick as measured by an observer in S?

70. Suppose our Sun is about to explode. In an effort to escape, we depart in a spaceship at $v = 0.800c$ and head toward the star Tau Ceti, 12.0 ly away. When we reach the midpoint of our journey from the Earth, we see our Sun explode and, unfortunately, at the same instant we see Tau Ceti explode as well. (a) In the spaceship's frame of reference, should we conclude that the two explosions occurred simultaneously? If not, which occurred first? (b) In a frame of reference in which the Sun and Tau Ceti are at rest, did they explode simultaneously? If not, which exploded first?

71. The light emitted by a galaxy shows a continuous distribution of wavelengths because the galaxy is composed of billions of different stars and other thermal emitters. Nevertheless, some narrow gaps occur in the continuous spectrum where light has been absorbed by cooler gases in the outer photospheres of normal stars. In particular, ionized calcium atoms at rest produce strong absorption at a wavelength of 394 nm. For a galaxy in the constellation Hydra, 2 billion lightyears away, this absorption line is shifted to 475 nm. How fast is the galaxy moving away from the Earth? (*Note:* The assumption that the recession speed is small compared with c, as made in Problem 21, is not a good approximation here.)

72. Prepare a graph of the relativistic kinetic energy and the classical kinetic energy, both as a function of speed, for an object with a mass of your choice. At what speed does the classical kinetic energy underestimate the relativistic value by 1 percent? By 5 percent? By 50 percent?

73. The total volume of water in the oceans is approximately 1.40×10^9 km³. The density of sea water is 1 030 kg/m³, and the specific heat of the water is 4 186 J/(kg·°C). Find the increase in mass of the oceans produced by an increase in temperature of 10.0°C.

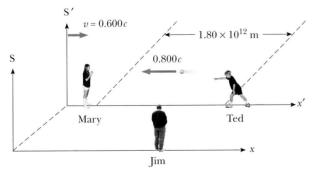

Figure P39.67

ANSWERS TO QUICK QUIZZES

39.1 They both are because they can report only what they see. They agree that the person in the truck throws the ball up and then catches it a bit later.

39.2 It depends on the direction of the throw. Taking the direction in which the train is traveling as the positive x direction, use the values $u'_x = +90$ mi/h and $v = +110$ mi/h, with u_x in Equation 39.2 being the value you are looking for. If the pitcher throws the ball in the same direction as the train, a person at rest on the Earth sees the ball moving at 110 mi/h + 90 mi/h = 200 mi/h. If the pitcher throws in the opposite direction, the person on the Earth sees the ball moving in the same direction as the train but at only 110 mi/h − 90 mi/h = 20 mi/h.

39.3 Both are correct. Although the two observers reach different conclusions, each is correct in her or his own reference frame because the concept of simultaneity is not absolute.

39.4 About 2.9×10^8 m/s, because this is the speed at which $\gamma = 5$. For every 5 s ticking by on the Mission Control clock, the Earth-bound observer (with a powerful telescope!) sees the rocket clock ticking off 1 s. The astronaut sees her own clock operating at a normal rate. To her, Mission Control is moving away from her at a speed of 2.9×10^8 m/s, and she sees the Mission Control clock as running slow. Strange stuff, this relativity!

39.5 If their on-duty time is based on clocks that remain on the Earth, they will have larger paychecks. Less time will have passed for the astronauts in their frame of reference than for their employer back on the Earth.

39.6 By a curved line. This can be seen in the middle of Speedo's world-line in Figure 39.14, where he turns around and begins his trip home.

39.7 The downstairs clock runs more slowly because it is closer to the Earth and hence experiences a stronger gravitational field than the upstairs clock does.

c h a p t e r

Introduction to Quantum Physics

40

In the preceding chapter, we discussed the fact that Newtonian mechanics must be replaced by Einstein's special theory of relativity when we are dealing with particle speeds comparable to the speed of light. As the 20th century progressed, many experimental and theoretical problems were resolved by the special theory. There were many other problems, however, for which classical physics could not provide a theoretical answer. Attempts to apply the laws of classical physics to explain the behavior of matter on the atomic scale were consistently unsuccessful. For example, blackbody radiation, the photoelectric effect, and the emission of sharp spectral lines from atoms could not be explained within the framework of classical physics.

As physicists sought new ways to solve these puzzles, another revolution took place in physics between 1900 and 1930. A new theory called *quantum mechanics* was highly successful in explaining the behavior of atoms, molecules, and nuclei. Like the special theory of relativity, the quantum theory requires a modification of our ideas concerning the physical world.

The basic ideas of quantum theory were introduced by Max Planck, but most of the subsequent mathematical developments and interpretations were made by a number of other distinguished physicists, including Einstein, Bohr, Schrödinger, de Broglie, Heisenberg, Born, and Dirac. Despite the great success of the quantum theory, Einstein frequently played the role of its critic, especially with regard to the manner in which the theory was interpreted. In particular, Einstein did not accept Heisenberg's interpretation of the uncertainty principle, which states that it is impossible to obtain a precise simultaneous measurement of the position and the velocity of a particle. According to this principle, the best we can do is predict the *probability* of the future of a system, contrary to the deterministic view held by Einstein.[1]

Because an extensive study of quantum theory is beyond the scope of this book, this chapter is simply an introduction to its underlying principles.

40.1 ▶ BLACKBODY RADIATION AND PLANCK'S HYPOTHESIS

An object at any temperature emits radiation sometimes referred to as **thermal radiation,** which we discussed in Section 20.7. The characteristics of this radiation depend on the temperature and properties of the object. At low temperatures, the wavelengths of thermal radiation are mainly in the infrared region of the electromagnetic spectrum, and hence the radiation is not observed by the eye. As the temperature of the object increases, the object eventually begins to glow red—in other words, enough visible radiation is emitted so that the object appears to glow. At sufficiently high temperatures, the object appears to be white, as in the glow of the hot tungsten filament of a lightbulb. Careful study shows that, as the temperature of an object increases, the thermal radiation it emits consists of a continuous distribution of wavelengths from the infrared, visible, and ultraviolet portions of the spectrum.

From a classical viewpoint, thermal radiation originates from accelerated charged particles in the atoms near the surface of the object; those charged particles emit radiation much as small antennas do. The thermally agitated particles can have a distribution of accelerations, which accounts for the continuous spec-

---

[1] Einstein's views on the probabilistic nature of quantum theory are expressed in his statement, "God does not play dice with the Universe."

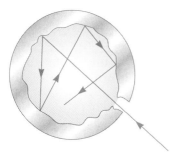

Figure 40.1 The opening to the cavity inside a hollow object is a good approximation of a black body. Light entering the small opening strikes the far wall, where some is absorbed and some is reflected at a random angle. The light continues to be reflected, and at each reflection a portion of the light is absorbed by the cavity walls. After many reflections, essentially all of the incident energy has been absorbed.

trum of radiation emitted by the object. By the end of the 19th century, however, it became apparent that the classical theory of thermal radiation was inadequate. The basic problem was in understanding the observed distribution of wavelengths in the radiation emitted by a black body. As we saw in Section 20.7, a black body is an ideal system that absorbs all radiation incident on it. A good approximation of a black body is a hole leading to the inside of a hollow object, as shown in Figure 40.1. The nature of the radiation emitted through the hole leading to the cavity depends only on the temperature of the cavity walls and not on the material of which the walls are made. The spaces between lumps of hot charcoal (Fig. 40.2) emit light that is very much like blackbody radiation.

Figure 40.3 shows how the energy of blackbody radiation varies with temperature and wavelength. As the temperature of the black body increases, two distinct behaviors are observed. The first effect is that the peak of the distribution shifts to shorter wavelengths. This is why the object described at the beginning of this section changes from not appearing to glow (peak in the infrared) to glowing red (peak in the near infrared with some visible at the red end of the spectrum) to glowing white (peak in the visible). This shift is found to obey the following rela-

QuickLab ➤

Use a black marker or pieces of black electrical tape to make a very dark area on the outside of a shoe box. Poke a hole in the center of the dark area with a pencil. Now put a lid on the box and compare the blackness of the hole with the blackness of the surrounding dark area. The hole acts like a black body.

Figure 40.2 The glow emanating from the spaces between these hot charcoal briquettes is, to a close approximation, blackbody radiation. The color of the light depends only upon the temperature of the briquettes. *(Corbis)*

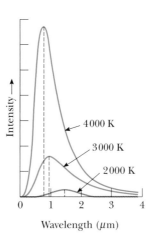

Figure 40.3 Intensity of blackbody radiation versus wavelength at three temperatures. Note that the amount of radiation emitted (the area under a curve) increases with increasing temperature.

tionship, called **Wien's displacement law:**

$$\lambda_{\max} T = 2.898 \times 10^{-3} \text{ m} \cdot \text{K} \qquad \textbf{(40.1)}$$

where $\lambda_{\max}$ is the wavelength at which the curve peaks and T is the absolute temperature of the object emitting the radiation. The wavelength at the peak of the curve is inversely proportional to the absolute temperature; that is, as the temperature increases, the peak is "displaced" to shorter wavelengths.

The second effect is that the total amount of energy the object emits increases with temperature. This is described by Stefan's law, given in Equation 20.18, which we wrote in the form $\mathcal{P} = \sigma A e T^4$. Recalling that $I = \mathcal{P}/A$ is the intensity of radiation at the surface of the object and that $e = 1$ for a black body, we can write Stefan's law in the form $I = \sigma T^4$.

To describe the distribution of energy from a black body, it is useful to define $I(\lambda, T) \, d\lambda$ to be the power per unit area emitted in the wavelength interval $d\lambda$. The result of a calculation based on a classical model of blackbody radiation known as the **Rayleigh–Jeans law** is

$$I(\lambda, T) = \frac{2\pi c k_B T}{\lambda^4} \qquad \textbf{(40.2)}$$

where k_B is Boltzmann's constant. In this classical model of blackbody radiation, the atoms in the cavity walls are treated as a set of oscillators that emit electromagnetic waves at all wavelengths. This model leads to an average energy per oscillator that is proportional to T.

An experimental plot of the blackbody radiation spectrum is shown in Figure 40.4, together with the theoretical prediction of the Rayleigh–Jeans law. At long wavelengths, the Rayleigh–Jeans law is in reasonable agreement with experimental data, but at short wavelengths major disagreement is apparent. We can understand the disagreement by noting that as λ approaches zero, the function $I(\lambda, T)$ given by Equation 40.2 approaches infinity. Hence, not only should short wavelengths predominate in a blackbody spectrum, but also the energy emitted by any black body should become infinite in the limit of zero wavelength. In contrast to this prediction, the experimental data plotted in Figure 40.4 show that as λ approaches zero, $I(\lambda, T)$ also approaches zero. This mismatch of theory and experiment was so disconcerting that scientists called it the *ultraviolet catastrophe*. (This name is a misnomer because the "catastrophe"—infinite energy—occurs as the wavelength approaches zero, not the ultraviolet wavelengths.)

Another discrepancy between theory and experiment concerns the total power emitted by the black body. Experimentally, the total power per unit area

QuickLab

On a clear night, go outdoors far from city lights and find the constellation Orion (visible from November through April in the evening sky). Look very carefully at the color of Betelgeuse and Rigel. Can you tell which star is hotter? If Orion is below the horizon, compare two of the brightest stars you can see, such as Vega in Lyra and Arcturus in Boötes. (*John Chumack/Photo Researchers, Inc.*)

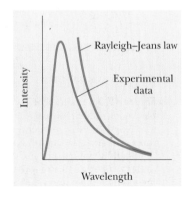

Figure 40.4 Comparison of experimental results and the curve predicted by the Rayleigh–Jeans law for the distribution of blackbody radiation.

given by $\int_0^\infty I(\lambda, T)\,d\lambda$ remains finite even though the Rayleigh–Jeans law (Eq. 40.2) says it should diverge to infinity.

In 1900, Planck derived a formula for blackbody radiation that was in complete agreement with experiment at all wavelengths. Planck's analysis led to the red curve in Figure 40.4. The function he proposed is

$$I(\lambda, T) = \frac{2\pi hc^2}{\lambda^5(e^{hc/\lambda k_B T} - 1)} \tag{40.3}$$

This function includes a parameter h, which Planck adjusted so that his curve matched the experimental data at all wavelengths. The value of this parameter is found to be independent of the material of which the black body is made and independent of the temperature. Rather than a variable parameter, it is a fundamental constant of nature. The value of h, Planck's constant, which we saw first in Chapter 11 and again in Chapter 35, is

$$h = 6.626 \times 10^{-34}\,\text{J} \cdot \text{s} \tag{40.4}$$

At long wavelengths, Equation 40.3 reduces to the Rayleigh–Jeans expression, Equation 40.2, and at short wavelengths it predicts an exponential decrease in $I(\lambda, T)$ with decreasing wavelength, in agreement with experimental results.

In his theory, Planck made two bold and controversial assumptions concerning the nature of the oscillating molecules at the surface of the black body:

1. The molecules can have only *discrete* values of energy E_n, given by

$$E_n = nhf \tag{40.5}$$

   where n is a positive integer called a **quantum number** and f is the natural frequency of oscillation of the molecules. This is quite different from the classical model of the harmonic oscillator, in which the energy of identical oscillators is related to the amplitude of the motion and unrelated to the frequency. Because the energy of a molecule can have only discrete values given by Equation 40.5, we say the energy is *quantized*. Each discrete energy value represents a different *quantum state* for the molecule, with each value of n representing a specific quantum state. When the molecule is in the $n = 1$ quantum state, its energy is hf; when it is in the $n = 2$ quantum state, its energy is $2hf$; and so on.

2. The molecules emit or absorb energy in discrete packets that later came to be called **photons.** The molecules emit or absorb these photons by "jumping" from one quantum state to another. If the jump is downward from one state to an adjacent lower state, Equation 40.5 shows that the amount of energy radi-

Max Planck (1858–1947)
Planck introduced the concept of "quantum of action" (Planck's constant, h) in an attempt to explain the spectral distribution of blackbody radiation, which laid the foundations for quantum theory. In 1918 he was awarded the Nobel Prize for this discovery of the quantized nature of energy. *(Photo courtesy of AIP Niels Bohr Library, W. F. Meggers Collection)*

Quantization of energy

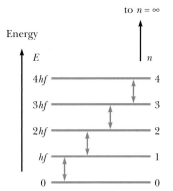

Figure 40.5 Allowed energy levels for a molecule that oscillates at its natural frequency f. Allowed transitions are indicated by the double-headed arrows.

ated by the molecule in a single photon equals *hf.* Hence, the energy of one photon corresponding to the energy difference between two adjacent quantum states is

| Energy of a photon |

$$E = hf \tag{40.6}$$

A molecule emits or absorbs energy only when it changes quantum states. If it remains in one quantum state, no energy is emitted or absorbed. Figure 40.5 shows the quantized energy levels and transitions between adjacent states.

The key point in Planck's theory is the radical assumption of quantized energy states. This development marked the birth of the quantum theory. When Planck presented his theory, most scientists (including Planck!) did not consider the quantum concept to be realistic. Hence, he and others continued to search for a more rational explanation of blackbody radiation. However, subsequent developments showed that a theory based on the quantum concept (rather than on classical concepts) had to be used to explain many other phenomena at the atomic level.

Quick Quiz 40.1

Which is more likely to cause sunburn (because more energy is absorbed by skin cells): (a) infrared light, (b) visible light, or (c) ultraviolet light?

EXAMPLE 40.1   Thermal Radiation from Different Objects

Find the peak wavelength of the radiation emitted by each of the following: (a) the human body when the skin temperature is 35°C.

Solution   From Wien's displacement law (Eq. 40.1), we have $\lambda_{max}T = 2.898 \times 10^{-3}$ m·K. Solving for λ_{max}, noting that 35°C corresponds to an absolute temperature of 308 K, we have

$$\lambda_{max} = \frac{2.898 \times 10^{-3} \text{ m·K}}{308 \text{ K}} = \boxed{9.4 \text{ } \mu\text{m}}$$

This radiation is in the infrared region of the spectrum and is invisible to the human eye. Some animals (pit vipers, for instance) are able to detect radiation of this wavelength and therefore can locate warm-blooded prey even in the dark.

(b) The tungsten filament of a lightbulb, which operates at 2 000 K.

Solution   Following the same procedure as in part (a), we find

$$\lambda_{max} = \frac{2.898 \times 10^{-3} \text{ m·K}}{2\,000 \text{ K}} = \boxed{1.4 \text{ } \mu\text{m}}$$

This also is in the infrared, meaning that most of the energy emitted by a lightbulb is not visible to us.

(c) The Sun, which has a surface temperature of about 5 800 K.

Solution   Again following the same procedure, we have

$$\lambda_{max} = \frac{2.898 \times 10^{-3} \text{ m·K}}{5\,800 \text{ K}} = \boxed{0.50 \text{ } \mu\text{m}}$$

This is near the center of the visible spectrum, about the color of a yellow-green tennis ball. Because it is the most prevalent color in sunlight, our eyes have evolved to be most sensitive to light of approximately this wavelength.

EXAMPLE 40.2   The Quantized Oscillator

A 2.0-kg block is attached to a massless spring that has a force constant of $k = 25$ N/m. The spring is stretched 0.40 m from its equilibrium position and released. (a) Find the total energy of the system and the frequency of oscillation according to classical calculations.

Solution   The total energy of a simple harmonic oscillator having an amplitude A is $\frac{1}{2}kA^2$ (Eq. 13.22). Therefore,

$$E = \tfrac{1}{2}kA^2 = \tfrac{1}{2}(25 \text{ N/m})(0.40 \text{ m})^2 = \boxed{2.0 \text{ J}}$$

The frequency of oscillation is (Eq. 13.19)

$$f = \frac{1}{2\pi}\sqrt{\frac{k}{m}} = \frac{1}{2\pi}\sqrt{\frac{25 \text{ N/m}}{2.0 \text{ kg}}} = \boxed{0.56 \text{ Hz}}$$

(b) Assuming that the energy is quantized, find the quantum number n for the system.

Solution If the energy is quantized, we have

$$E_n = nhf = n(6.626 \times 10^{-34} \text{ J·s})(0.56 \text{ Hz}) = 2.0 \text{ J}$$

$$n = \boxed{5.4 \times 10^{33}}$$

(c) How much energy is carried away in a one-quantum change?

Solution From Equation 40.6,

$$E = hf = (6.626 \times 10^{-34} \text{ J·s})(0.56 \text{ Hz}) = \boxed{3.7 \times 10^{-34} \text{ J}}$$

This energy carried away by a one-quantum change is such a small fraction of the total energy of the oscillator that we cannot detect it. Thus, even though the energy of a spring–block system is quantized and does indeed decrease by small quantum jumps, our senses perceive the decrease as continuous. Quantum effects become important and measurable only on the submicroscopic level of atoms and molecules.

40.2 THE PHOTOELECTRIC EFFECT

In the latter part of the 19th century, experiments showed that light incident on certain metal surfaces caused electrons to be emitted from the surfaces. This phenomenon, which we first met in Section 35.1, is known as the **photoelectric effect,** and the emitted electrons are called **photoelectrons.**

Figure 40.6 is a diagram of an apparatus in which the photoelectric effect can occur. An evacuated glass or quartz tube contains a metallic plate E connected to the negative terminal of a battery and another metallic plate C that is connected to the positive terminal of the battery. When the tube is kept in the dark, the ammeter reads zero, indicating no current in the circuit. However, when plate E is illuminated by light having a wavelength shorter than some particular wavelength that depends on the metal used to make plate E, a current is detected by the ammeter, indicating a flow of charges across the gap between plates E and C. This current arises from photoelectrons emitted from the negative plate (the emitter) and collected at the positive plate (the collector).

Figure 40.7 is a plot of photoelectric current versus potential difference ΔV between plates E and C for two light intensities. At large values of ΔV, the current reaches a maximum value. In addition, the current increases as the intensity of the incident light increases, as you might expect. Finally, when ΔV is negative—that is, when the battery in the circuit is reversed to make plate E positive and plate C negative—the current drops to a very low value because most of the emitted photoelectrons are repelled by the now negative plate C. In this situation, only those

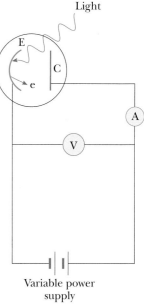

Figure 40.6 A circuit diagram for observing the photoelectric effect. When light strikes the plate E (the emitter), photoelectrons are ejected from the plate. Electrons moving from plate E to plate C (the collector) constitute a current in the circuit.

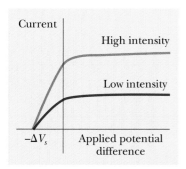

Figure 40.7 Photoelectric current versus applied potential difference for two light intensities. The current increases with intensity but reaches a saturation level for large values of ΔV. At voltages equal to or more negative than $-\Delta V_s$, the stopping potential, the current is zero.

photoelectrons having a kinetic energy greater than the magnitude of $e \, \Delta V$ reach plate C, where e is the charge on the electron.

When ΔV is equal to or more negative than $-\Delta V_s$, the **stopping potential,** no photoelectrons reach C and the current is zero. The stopping potential is independent of the radiation intensity. The maximum kinetic energy of the photoelectrons is related to the stopping potential through the relationship

$$K_{\text{max}} = e \, \Delta V_s \qquad \textbf{(40.7)}$$

Several features of the photoelectric effect could not be explained by classical physics or by the wave theory of light:

- No photoelectrons are emitted if the frequency of the incident light falls below some **cutoff frequency** f_c, which is characteristic of the material being illuminated. This is inconsistent with wave theory, which predicts that the photoelectric effect should occur at any frequency, provided the light intensity is sufficiently high.
- The maximum kinetic energy of the photoelectrons is independent of light intensity. According to wave theory, light of higher intensity should carry more energy into the metal per unit time and therefore eject photoelectrons having higher kinetic energies.
- The maximum kinetic energy of the photoelectrons increases with increasing light frequency. The wave theory predicts no relationship between photoelectron energy and incident light frequency.
- Photoelectrons are emitted from the surface almost instantaneously (less than 10^{-9} s after the surface is illuminated), even at low light intensities. Classically, we expect the photoelectrons to require some time to absorb the incident radiation before they acquire enough kinetic energy to escape from the metal.

A successful explanation of the photoelectric effect was given by Einstein in 1905, the same year he published his special theory of relativity. As part of a general paper on electromagnetic radiation, for which he received the Nobel Prize in 1921, Einstein extended Planck's concept of quantization to electromagnetic waves. He assumed that light (or any other electromagnetic wave) of frequency f can be considered a stream of photons. Each photon has an energy E, given by Equation 40.6, $E = hf$. A suggestive image of several photons, not to be taken too literally, is shown in Figure 40.8.

In Einstein's model, a photon is so localized that it gives *all* its energy hf to a single electron in the metal. According to Einstein, the maximum kinetic energy for these liberated photoelectrons is

Photoelectric effect equation

$$K_{\text{max}} = hf - \phi \qquad \textbf{(40.8)}$$

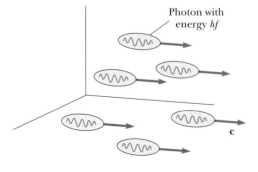

Photon with energy hf

c

Figure 40.8 A representation of photons. Each photon has a discrete energy hf.

where ϕ is called the **work function** of the metal. The work function represents the minimum energy with which an electron is bound in the metal and is on the order of a few electron volts. Table 40.1 lists work functions for various metals.

With the photon theory of light, we can explain the previously mentioned features of the photoelectric effect that we cannot understand using concepts of classical physics:

- That the effect is not observed below a cutoff frequency follows from the fact that the energy of the photon must be greater than or equal to ϕ. If the energy of the incoming photon does not satisfy this condition, the electrons are never ejected from the surface, regardless of the light intensity.
- That K_{max} is independent of light intensity can be understood by means of the following argument: If the light intensity is doubled, the number of photons is doubled, which doubles the number of photoelectrons emitted. However, their maximum kinetic energy, which equals $hf - \phi$, depends only on the light frequency and the work function, not on the light intensity.
- That K_{max} increases with increasing frequency is easily understood with Equation 40.8.
- That photoelectrons are emitted almost instantaneously is consistent with the particle theory of light, in which the incident energy arrives at the surface in small packets and there is a one-to-one interaction between photons and photoelectrons. In this interaction the photon's energy is imparted to an electron that then has enough energy to leave the metal. This is in contrast to the wave theory, in which the incident energy is distributed uniformly over a large area of the metal surface.

Experimental observation of a linear relationship between K_{max} and f would be a final confirmation of Einstein's theory. Indeed, such a linear relationship is observed, as sketched in Figure 40.9. The intercept on the horizontal axis gives the cutoff frequency below which no photoelectrons are emitted, regardless of light intensity. The frequency is related to the work function through the relationship $f_c = \phi/h$. The cutoff frequency corresponds to a **cutoff wavelength** of

$$\lambda_c = \frac{c}{f_c} = \frac{c}{\phi/h} = \frac{hc}{\phi} \qquad \text{(40.9)}$$

where we have used Equation 16.14 and c is the speed of light. Wavelengths greater than λ_c incident on a material having a work function ϕ do not result in the emission of photoelectrons.

One of the first practical uses of the photoelectric effect was as the detector in the light meter of a camera. Light reflected from the object to be photographed struck a photoelectric surface in the meter, causing it to emit photoelectrons that then passed through a sensitive ammeter. The magnitude of the current in the ammeter depended on the light intensity.

The phototube, another early application of the photoelectric effect, acts much like a switch in an electric circuit. It produces a current in the circuit when light of sufficiently high frequency falls on a metallic plate in the phototube but produces no current in the dark. Phototubes were used in burglar alarms and in the detection of the soundtrack on motion picture film. Modern semiconductor devices have now replaced older devices based on the photoelectric effect.

TABLE 40.1
Work Functions of Selected Metals

Metal	ϕ (eV)
Na	2.46
Al	4.08
Cu	4.70
Zn	4.31
Ag	4.73
Pt	6.35
Pb	4.14
Fe	4.50

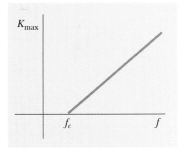

Figure 40.9 A plot of K_{max} of photoelectrons as a function of incident-light frequency in a typical photoelectric effect experiment. Photons having a frequency less than f_c do not have sufficient energy to eject an electron from the metal.

Quick Quiz 40.2

What does the slope of the line in Figure 40.9 represent? What does the y intercept represent? How would such a series of these graphs for different metals compare with one another?

>**Quick Quiz 40.3**

Make a sketch of how classical physicists expected Figure 40.9 to look.

EXAMPLE 40.3 **The Photoelectric Effect for Sodium**

A sodium surface is illuminated with light having a wavelength of 300 nm. The work function for sodium metal is 2.46 eV. Find (a) the maximum kinetic energy of the ejected photoelectrons and (b) the cutoff wavelength for sodium.

Solution (a) The energy of each photon in the illuminating light beam is

$$E = hf = \frac{hc}{\lambda} = \frac{(6.626 \times 10^{-34}\,\text{J·s})(3.00 \times 10^8\,\text{m/s})}{300 \times 10^{-9}\,\text{m}}$$

$$= 6.63 \times 10^{-19}\,\text{J} = \frac{6.63 \times 10^{-19}\,\text{J}}{1.60 \times 10^{-19}\,\text{J/eV}} = 4.14\,\text{eV}$$

Using Equation 40.8 gives

$$K_{\text{max}} = hf - \phi = 4.14\,\text{eV} - 2.46\,\text{eV} = \boxed{1.68\,\text{eV}}$$

(b) We can calculate the cutoff wavelength from Equation 40.9 after we convert ϕ from electron volts to joules:

$$\phi = (2.46\,\text{eV})(1.60 \times 10^{-19}\,\text{J/eV}) = 3.94 \times 10^{-19}\,\text{J}$$

$$\lambda_c = \frac{hc}{\phi} = \frac{(6.626 \times 10^{-34}\,\text{J·s})(3.00 \times 10^8\,\text{m/s})}{3.94 \times 10^{-19}\,\text{J}}$$

$$= 5.05 \times 10^{-7}\,\text{m} = \boxed{505\,\text{nm}}$$

This wavelength is in the yellow-green region of the visible spectrum.

Exercise Calculate the maximum speed of the photoelectrons under the conditions described in this example.

Answer 7.68×10^5 m/s.

40.3 THE COMPTON EFFECT

In 1919, Einstein concluded that a photon of energy E travels in a single direction (unlike a spherical wave) and carries a momentum equal to $E/c = hf/c$. In his own words, "If a bundle of radiation causes a molecule to emit or absorb an energy packet hf, then momentum of quantity hf/c is transferred to the molecule, directed along the line of the bundle for absorption and opposite the bundle for emission." In 1923, Arthur Holly Compton (1892–1962) and Peter Debye (1884–1966) independently carried Einstein's idea of photon momentum further.

Prior to 1922, Compton and his co-workers had accumulated evidence showing that the classical wave theory of light failed to explain the scattering of x-rays from electrons. According to classical theory, electromagnetic waves of frequency f_0 incident on electrons should have two effects, as shown in Figure 40.10a: (1) Radiation pressure (see Section 34.4) should cause the electrons to accelerate in the direction of propagation of the waves, and (2) the oscillating electric field of the incident radiation should set the electrons into oscillation at the apparent frequency f', where f' is the frequency in the frame of the moving electrons. This apparent frequency f' is different from the frequency f_0 of the incident radiation because of the Doppler effect (see Section 17.5): Each electron first absorbs as a moving particle and then reradiates as a moving particle, thereby exhibiting two Doppler shifts in the frequency of radiation.

Because different electrons will move at different speeds after the interaction, depending on the amount of energy absorbed from the electromagnetic waves, the scattered wave frequency at a given angle should show a distribution of Doppler-shifted values. Contrary to this prediction, Compton's experiments showed that, at a given angle, only *one* frequency of radiation was observed. Compton and his co-workers realized that they could explain these experiments by treat-

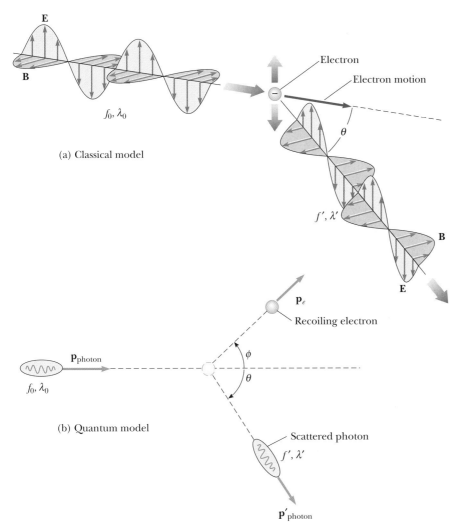

(a) Classical model

(b) Quantum model

Figure 40.10   X-ray scattering from an electron: (a) the classical model; (b) the quantum model.

ing photons not as waves but rather as point-like particles having energy hf and momentum hf/c and by assuming that the energy and momentum of any colliding photon–electron pair are conserved. Compton was adopting a particle model for something that was well known as a wave, and today this scattering phenomenon is known as the **Compton effect.** Figure 40.10b shows the quantum picture of the exchange of momentum and energy between an individual x-ray photon and an electron.

The second difference between the classical and quantum models is also shown in Figure 40.10b. In the classical model, the electron is pushed along the direction of propagation of the incident x-ray by radiation pressure. In the quantum model, the electron is scattered through an angle ϕ with respect to this direction, as if this were a billiard-ball type collision. (The symbol ϕ used here is not to be confused with work function, which was discussed in the previous section.)

Figure 40.11a is a schematic diagram of the apparatus used by Compton. The x-rays, scattered from a graphite target, were analyzed with a rotating crystal spectrometer, and the intensity was measured with an ionization chamber that generated a current proportional to the intensity. The incident beam consisted of mono-

Arthur Holly Compton
(1892–1962)

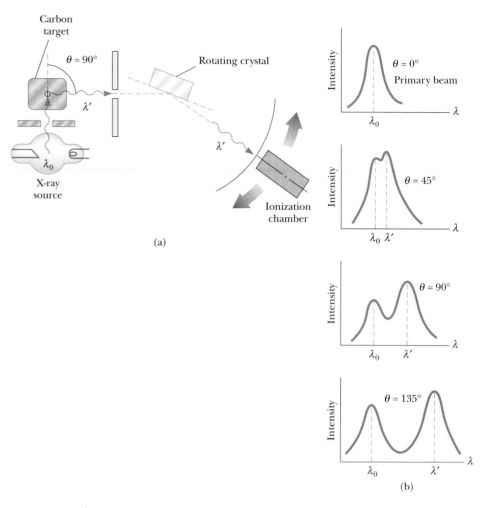

Figure 40.11   (a) Schematic diagram of Compton's apparatus. The wavelength was measured with a rotating crystal spectrometer using graphite (carbon) as the target. (b) Scattered x-ray intensity versus wavelength for Compton scattering at $\theta = 0°$, $45°$, $90°$, and $135°$.

chromatic x-rays of wavelength $\lambda_0 = 0.071$ nm. The experimental intensity-versus-wavelength plots observed by Compton for four scattering angles (corresponding to θ in Fig. 40.10) are shown in Figure 40.11b. The graphs for the three nonzero angles show two peaks, one at λ_0 and one at $\lambda' > \lambda_0$. The shifted peak at λ' is caused by the scattering of x-rays from free electrons, and it was predicted by Compton to depend on scattering angle as

Compton shift equation

$$\lambda' - \lambda_0 = \frac{h}{m_e c}(1 - \cos\theta) \qquad \textbf{(40.10)}$$

where m_e is the mass of the electron. This expression is known as the **Compton shift equation,** and the factor $h/m_e c$ is called the **Compton wavelength** λ_C of the electron. It has a currently accepted value of

Compton wavelength

$$\lambda_C = \frac{h}{m_e c} = 0.002\ 43 \text{ nm}$$

The unshifted peak at λ_0 in Figure 40.11b is caused by x-rays scattered from electrons tightly bound to the target atoms. This unshifted peak also is predicted by Equation 40.10 if the electron mass is replaced with the mass of a carbon atom, which is about 23 000 times the mass of the electron. Thus, there is a wavelength shift for scattering from an electron bound to an atom, but it is so small that it was undetectable in Compton's experiment.

Compton's measurements were in excellent agreement with the predictions of Equation 40.10. It is fair to say that these results were the first that really convinced many physicists of the fundamental validity of quantum theory!

Quick Quiz 40.4

Note that for any given scattering angle θ, Equation 40.10 gives the same value for the Compton wavelength shift for any wavelength. Keeping this in mind, explain why the experiment is normally done with x-rays rather than visible light.

Derivation of the Compton Shift Equation

We can derive the Compton shift equation by assuming that the photon behaves like a particle and collides elastically with a free electron initially at rest, as shown in Figure 40.12a. In this model, the photon is treated as a particle having energy $E = hf = hc/\lambda$ and mass zero. In the scattering process, the total energy and total linear momentum of the system must be conserved. Applying the principle of conservation of energy to this process gives

$$\frac{hc}{\lambda_0} = \frac{hc}{\lambda'} + K_e$$

where hc/λ_0 is the energy of the incident photon, hc/λ' is the energy of the scat-

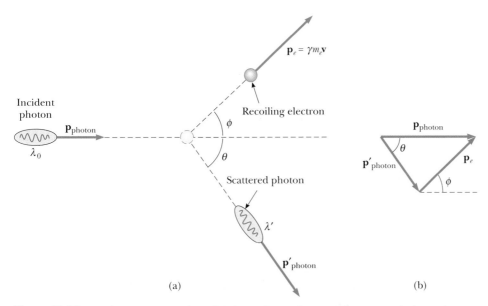

Figure 40.12 (a) Compton scattering of a photon by an electron. The scattered photon has less energy (longer wavelength) than the incident photon. (b) Momentum vectors for Compton scattering.

tered photon, and K_e is the kinetic energy of the recoiling electron. Because the electron may recoil at speeds comparable to the speed of light, we must use the relativistic expression $K_e = \gamma m_e c^2 - m_e c^2$ (Eq. 39.23). Therefore,

$$\frac{hc}{\lambda_0} = \frac{hc}{\lambda'} + \gamma m_e c^2 - m_e c^2 \qquad \textbf{(40.11)}$$

where $\gamma = 1/\sqrt{1 - v^2/c^2}$.

Next, we apply the law of conservation of momentum to this collision, noting that both the x and y components of momentum are conserved. Equation 39.27 shows that the momentum of a photon has a magnitude $p = E/c$, and we know from Equation 40.6 that $E = hf$. Therefore, $p = hf/c$. Substituting λf for c (Eq. 16.14) in this expression gives us $p = h/\lambda$. Because the relativistic expression for the momentum of the recoiling electron is $p_e = \gamma m_e v$ (Eq. 39.19), we obtain the following expressions for the x and y components of linear momentum, where the angles are as described in Figure 40.12b:

$$x\text{ component:} \qquad \frac{h}{\lambda_0} = \frac{h}{\lambda'}\cos\theta + \gamma m_e v \cos\phi \qquad \textbf{(40.12)}$$

$$y\text{ component:} \qquad 0 = \frac{h}{\lambda'}\sin\theta - \gamma m_e v \sin\phi \qquad \textbf{(40.13)}$$

By eliminating v and ϕ from Equations 40.11 to 40.13, we obtain a single expression that relates the remaining three variables (λ', λ_0, and θ). After some algebra (see Problem 68), we obtain the Compton shift equation:

$$\Delta\lambda = \lambda' - \lambda_0 = \frac{h}{m_e c}(1 - \cos\theta)$$

EXAMPLE *40.4* ▶ Compton Scattering at 45°

X-rays of wavelength $\lambda_0 = 0.200$ nm are scattered from a block of material. The scattered x-rays are observed at an angle of 45.0° to the incident beam. Calculate their wavelength.

Solution The shift in wavelength of the scattered x-rays is given by Equation 40.10:

$$\Delta\lambda = \lambda' - \lambda_0 = \frac{h}{m_e c}(1 - \cos\theta)$$

$$= \frac{6.626 \times 10^{-34}\,\text{J}\cdot\text{s}}{(9.11 \times 10^{-31}\,\text{kg})(3.00 \times 10^8\,\text{m/s})}(1 - \cos 45.0°)$$

$$= 7.10 \times 10^{-13}\,\text{m} = 0.000\ 710\ \text{nm}$$

Hence, the wavelength of the scattered x-ray at this angle is

$$\lambda' = \Delta\lambda + \lambda_0 = \boxed{0.200\ 710\ \text{nm}}$$

Exercise Find the fraction of energy lost by the photon in this collision.

Answer $\Delta E/E = 0.003\ 54$.

40.4 ▶ ATOMIC SPECTRA

As pointed out in Section 40.1, all objects emit thermal radiation characterized by a continuous distribution of wavelengths. In sharp contrast to this continuous distribution spectrum is the discrete **line spectrum** emitted by a low-pressure gas subject to an electric discharge. (Electric discharge occurs when the gas is subjected to a potential difference that creates an electric field larger than the dielec-

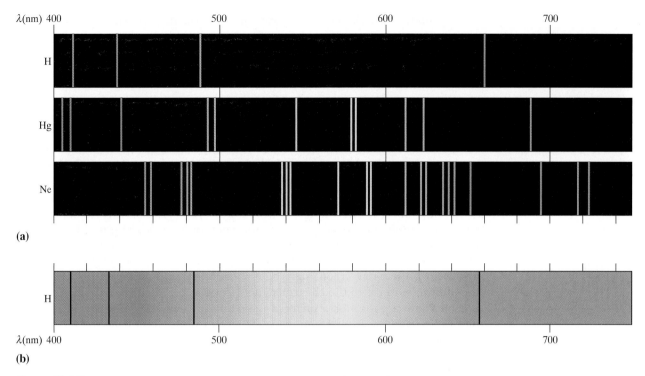

Figure 40.13 (a) Emission line spectra for hydrogen, mercury, and neon. (b) The absorption spectrum for hydrogen. Note that the dark absorption lines occur at the same wavelengths as the hydrogen emission lines in (a). *(K. W. Whitten, R. E. Davis, and M. L. Peck,* General Chemistry, *6th ed., Philadelphia, Saunders College Publishing, 2000.)*

tric strength of the gas.) Observation and analysis of this emitted light is called **emission spectroscopy.**

When the light from a gas discharge is examined with a spectroscope, it is found to consist of a few bright lines of color on a generally dark background. (The *lines* are due to the collimation of the light through a slit.) This discrete line spectrum contrasts sharply with the continuous rainbow of colors seen when a glowing solid is viewed through a spectroscope. Furthermore, as you can see from Figure 40.13a, the wavelengths contained in a given line spectrum are characteristic of the element emitting the light. The simplest line spectrum is that for atomic hydrogen, and we describe this spectrum in detail. Other atoms exhibit completely different line spectra. Because no two elements have the same line spectrum, this phenomenon represents a practical and sensitive technique for identifying the elements present in unknown samples.

Another form of spectroscopy very useful in analyzing substances is **absorption spectroscopy.** An absorption spectrum is obtained by passing light from a continuous source through a gas or dilute solution of the element being analyzed. The absorption spectrum consists of a series of dark lines superimposed on the continuous spectrum of the light source, as shown in Figure 40.13b for atomic hydrogen. In general, not all of the lines present in the emission spectrum of an element are present in the element's absorption spectrum.

The absorption spectrum of an element has many practical applications. For example, the continuous spectrum of radiation emitted by the Sun must pass through the cooler gases of the solar atmosphere and through the Earth's atmosphere. The various absorption lines observed in the solar spectrum have been used to identify elements in the solar atmosphere. In early studies of the solar spectrum, experimenters found some lines that did not correspond to any known element. A new element had been discovered! The new element was named helium, after the Greek word for Sun, *helios*. Helium was subsequently isolated from subterranean gas on the Earth.

Scientists are able to examine the light from stars other than our Sun in this fashion, but elements other than those present on the Earth have never been detected. Absorption spectroscopy has also been a useful technique in analyzing heavy-metal contamination of the food chain. For example, the first determination of high levels of mercury in tuna was made with the use of atomic absorption spectroscopy.

The discrete emissions of light from gas discharges are used in "neon" signs, as seen in the opening photograph of this chapter. Neon, the first gas used in these types of signs and the gas after which they are named, emits strongly in the red region. As a result, a glass tube filled with neon gas emits bright red light when an applied voltage causes a continuous discharge. Early signs used different gases to provide different colors, although the brightness of these signs was generally very low. Many present-day "neon" signs contain mercury vapor, which emits strongly in the ultraviolet range of the electromagnetic spectrum. The inside of the glass tube is coated with a phosphor, a material that emits a particular color when it absorbs ultraviolet radiation from the mercury. The color of the light from the tube is due to the particular phosphor chosen. A fluorescent light operates in the same manner, with a white-emitting phosphor coating the inside of the glass tube.

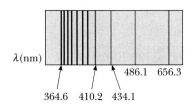

λ(nm)

486.1   656.3

364.6   410.2   434.1

Figure 40.14 The Balmer series of spectral lines for atomic hydrogen. The line labeled 364.6 is the shortest-wavelength line and is in the ultraviolet region of the electromagnetic spectrum. The other labeled lines are in the visible region.

From 1860 to 1885, scientists accumulated a great deal of data on atomic emissions using spectroscopic measurements. In 1885, a Swiss school teacher, Johann Jacob Balmer (1825–1898), found an empirical equation that correctly predicted the wavelengths of four visible emission lines of hydrogen: H_α(red), H_β(green), H_γ(blue), and H_δ(violet). Figure 40.14 shows these and other lines (in the ultraviolet) in the emission spectrum of hydrogen. The complete set of lines is called the **Balmer series.** The four visible lines occur at the wavelengths 656.3 nm, 486.1 nm, 434.1 nm, and 410.2 nm. The wavelengths of these lines can be described by the following equation, which is a modification of Balmer's original equation made by Johannes Rydberg (1854–1919):

Balmer series

$$\frac{1}{\lambda} = R_H \left(\frac{1}{2^2} - \frac{1}{n^2} \right) \tag{40.14}$$

where n may have integral values of 3, 4, 5, . . . and R_H is a constant now called the **Rydberg constant.** If the wavelength is in meters, R_H has the value $1.097\,373\,2 \times 10^7\,\text{m}^{-1}$. The line in the Balmer series at 656.3 nm corresponds to $n = 3$ in Equation 40.14; the line at 486.1 nm corresponds to $n = 4$, and so on. The measured spectral lines agree with this empirical formula to within 0.1%.

Rydberg constant

Other lines in the spectrum of hydrogen were found following Balmer's discovery. These spectra are called the Lyman, Paschen, and Brackett series after their discoverers. The wavelengths of the lines in these series can be calculated through the use of the following empirical formulas:

Lyman series

$$\frac{1}{\lambda} = R_H \left(1 - \frac{1}{n^2} \right) \qquad n = 2, 3, 4, \ldots \tag{40.15}$$

Paschen series

$$\frac{1}{\lambda} = R_H \left(\frac{1}{3^2} - \frac{1}{n^2} \right) \qquad n = 4, 5, 6, \ldots \tag{40.16}$$

Brackett series

$$\frac{1}{\lambda} = R_H \left(\frac{1}{4^2} - \frac{1}{n^2} \right) \qquad n = 5, 6, 7, \ldots \tag{40.17}$$

All of these equations were purely empirical; this means that no theoretical basis existed for them. They simply worked. In the next section, we discuss the remarkable achievement of a theory for the hydrogen atom that provided a theoretical basis for these equations.

40.5 ▷ BOHR'S QUANTUM MODEL OF THE ATOM

At the beginning of the 20th century, scientists were perplexed by the failure of classical physics to explain the characteristics of atomic spectra. Why did atoms of a given element exhibit only certain spectral lines? Furthermore, why did the atoms absorb only those wavelengths that they emitted? In 1913, Niels Bohr provided an explanation of atomic spectra that includes some features of the currently accepted theory. Bohr's theory contained a combination of ideas from Planck's original quantum theory, Einstein's photon theory of light, early models of the atom, and Newtonian mechanics. Using the simplest atom, hydrogen, Bohr described a model of what he thought must be the atom's structure. His model of the hydrogen atom contains some classical features, as well as some revolutionary postulates that could not be justified within the framework of classical physics.

The basic ideas of the Bohr theory as it applies to the hydrogen atom are as follows:

1. The electron moves in circular orbits around the proton under the influence of the Coulomb force of attraction, as shown in Figure 40.15.
2. Only certain electron orbits are stable. These stable orbits are ones in which the electron does not emit energy in the form of radiation. Hence, the total energy of the atom remains constant, and classical mechanics can be used to describe the electron's motion. Note that this representation is completely different from the classical model of an electron in a circular orbit. According to classical physics, the centripetally accelerated electron should continuously emit radiation, losing energy and eventually spiraling into the nucleus.
3. Radiation is emitted by the atom when the electron "jumps" from a more energetic initial orbit to a lower-energy orbit. This jump cannot be visualized or treated classically. In particular, the frequency f of the photon emitted in the jump is related to the change in the atom's energy and *is independent of the frequency of the electron's orbital motion*. The frequency of the emitted radiation is found from the energy-conservation expression

$$E_i - E_f = hf \qquad \textbf{(40.18)}$$

where E_i is the energy of the initial state, E_f is the energy of the final state, and $E_i > E_f$.
4. The size of the allowed electron orbits is determined by a condition imposed on the electron's orbital angular momentum: The allowed orbits are those for which the electron's orbital angular momentum about the nucleus is an integral multiple of $\hbar = h/2\pi$:

$$m_e vr = n\hbar \qquad n = 1, 2, 3, \ldots \qquad \textbf{(40.19)}$$

Using these four assumptions, we can calculate the allowed energy levels and emission wavelengths of the hydrogen atom. We can find the electric potential energy of the system shown in Figure 40.15 from Equation 25.13, $U = k_e q_1 q_2 / r = -k_e e^2 / r$, where k_e is the Coulomb constant and the negative sign arises from the charge $-e$ on the electron. Thus, the total energy of the atom,

Assumptions of the Bohr theory

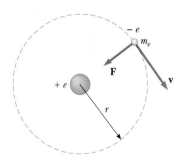

Figure 40.15 Diagram representing Bohr's model of the hydrogen atom, in which the orbiting electron is allowed to be only in specific orbits of discrete radii.

which contains both kinetic and potential energy terms, is

$$E = K + U = \tfrac{1}{2}m_e v^2 - k_e \frac{e^2}{r} \tag{40.20}$$

Applying Newton's second law to this system, we see that the Coulomb attractive force $k_e e^2/r^2$ exerted on the electron must equal the mass times the centripetal acceleration ($a = v^2/r$) of the electron:

$$\frac{k_e e^2}{r^2} = \frac{m_e v^2}{r}$$

From this expression, we see that the kinetic energy of the electron is

$$K = \frac{m_e v^2}{2} = \frac{k_e e^2}{2r} \tag{40.21}$$

Substituting this value of K into Equation 40.20, we find that the total energy of the atom is

$$E = -\frac{k_e e^2}{2r} \tag{40.22}$$

Note that the total energy is negative, indicating a bound electron–proton system. This means that energy in the amount of $k_e e^2/2r$ must be added to the atom to remove the electron and make the total energy of the system zero.

We can obtain an expression for r, the radius of the allowed orbits, by solving Equations 40.19 and 40.21 for v and equating the results:

$$v^2 = \frac{n^2 \hbar^2}{m_e^2 r^2} = \frac{k_e e^2}{m_e r}$$

$$r_n = \frac{n^2 \hbar^2}{m_e k_e e^2} \qquad n = 1, 2, 3, \ldots \tag{40.23}$$

This equation shows that the radii have discrete values—they are quantized. The result is based on the *assumption* that the electron can exist only in certain allowed orbits determined by the integer n.

The orbit with the smallest radius, called the **Bohr radius** a_0, corresponds to $n = 1$ and has the value

$$a_0 = \frac{\hbar^2}{m_e k_e e^2} = 0.052\ 9 \text{ nm} \tag{40.24}$$

We obtain a general expression for the radius of any orbit in the hydrogen atom by substituting Equation 40.24 into Equation 40.23:

Radii of Bohr orbits in hydrogen

$$r_n = n^2 a_0 = n^2 (0.052\ 9 \text{ nm}) \tag{40.25}$$

Bohr's theory gave a value of the right order of magnitude for the radius of a hydrogen atom from first principles rather than from any empirical assumption about orbit size. This result was a striking triumph for Bohr's theory. The first three Bohr orbits are shown to scale in Figure 40.16.

The quantization of orbit radii immediately leads to energy quantization. We can see this by substituting $r_n = n^2 a_0$ into Equation 40.22, obtaining for the allowed energy levels

Allowed energies of the hydrogen atom

$$E_n = -\frac{k_e e^2}{2 a_0}\left(\frac{1}{n^2}\right) \qquad n = 1, 2, 3, \ldots \tag{40.26}$$

Niels Bohr (1885–1962) Bohr, a Danish physicist, was an active participant in the early development of quantum mechanics and provided much of its philosophical framework. During the 1920s and 1930s, he headed the Institute for Advanced Studies in Copenhagen. The institute was a magnet for many of the world's best physicists and provided a forum for the exchange of ideas. When Bohr visited the United States in 1939 to attend a scientific conference, he brought news that the fission of uranium had been observed by Hahn and Strassman in Berlin. The results were the foundations of the atomic bomb developed in the United States during World War II. Bohr was awarded the 1922 Nobel Prize for his investigation of the structure of atoms and the radiation emanating from them. *(Photo courtesy of AIP Niels Bohr Library, Margarethe Bohr Collection)*

Inserting numerical values into this expression, we have

$$E_n = -\frac{13.606}{n^2}\,\text{eV} \qquad n = 1, 2, 3, \ldots \qquad \textbf{(40.27)}$$

Only energies satisfying this equation (called **energy levels**) are permitted. The lowest allowed energy level, called the **ground state,** has $n = 1$ and energy $E_1 = -13.606$ eV. The next energy level, the **first excited state,** has $n = 2$ and energy $E_2 = E_1/2^2 = -3.401$ eV. Figure 40.17 is an energy level diagram showing the energies of these discrete energy states and the corresponding quantum numbers n. The uppermost level, corresponding to $n = \infty$ (or $r = \infty$) and $E = 0$, represents the state for which the electron is removed from the atom. The minimum energy required to ionize the atom (that is, to completely remove an electron in the ground state from the proton's influence) is called the **ionization energy.** As can be seen from Figure 40.17, the ionization energy for hydrogen in the ground state, based on Bohr's calculation, is 13.6 eV. This constituted another major achievement for the Bohr theory because the ionization energy for hydrogen had already been measured to be 13.6 eV.

Equations 40.18 and 40.26 can be used to calculate the frequency of the photon emitted when the electron jumps from an outer orbit to an inner orbit:

$$f = \frac{E_i - E_f}{h} = \frac{k_e e^2}{2a_0 h}\left(\frac{1}{n_f^{\,2}} - \frac{1}{n_i^{\,2}}\right) \qquad \textbf{(40.28)}$$

Because the quantity measured experimentally is wavelength, it is convenient to use $c = f\lambda$ to convert frequency to wavelength:

$$\frac{1}{\lambda} = \frac{f}{c} = \frac{k_e e^2}{2a_0 hc}\left(\frac{1}{n_f^{\,2}} - \frac{1}{n_i^{\,2}}\right) \qquad \textbf{(40.29)}$$

The remarkable fact is that this expression, which is purely theoretical, is identical to the general form of the empirical relationships discovered by Balmer and Rydberg and given by Equations 40.14 to 40.17,

$$\frac{1}{\lambda} = R_{\text{H}}\left(\frac{1}{n_f^{\,2}} - \frac{1}{n_i^{\,2}}\right) \qquad \textbf{(40.30)}$$

provided the constant $k_e e^2/2a_0 hc$ is equal to the experimentally determined Rydberg constant $R_{\text{H}} = 1.097\,373\,2 \times 10^7\,\text{m}^{-1}$. After Bohr demonstrated that these two quantities agree to within approximately 1%, this work was soon recognized as the crowning achievement of his new theory of quantum mechanics. Furthermore, Bohr showed that all of the spectral series for hydrogen have a natural interpretation in his theory. Figure 40.17 shows these spectral series as transitions between energy levels.

Bohr immediately extended his model for hydrogen to other elements in which all but one electron had been removed. Ionized elements such as He$^+$, Li^{2+}, and Be^{3+} were suspected to exist in hot stellar atmospheres, where atomic collisions frequently have enough energy to completely remove one or more atomic electrons. Bohr showed that many mysterious lines observed in the spectra of the Sun and several other stars could not be due to hydrogen but were correctly predicted by his theory if attributed to singly ionized helium. In general, to describe a single electron orbiting a fixed nucleus of charge $+Ze$, where Z is the atomic number of the element (see Section 1.2), Bohr's theory

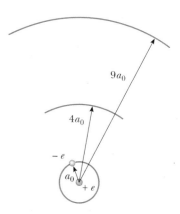

Figure 40.16 The first three circular orbits predicted by the Bohr model of the hydrogen atom.

Frequency of a photon emitted from hydrogen

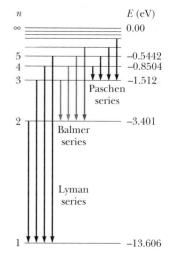

Figure 40.17 An energy level diagram for hydrogen. The discrete allowed energies are plotted on the vertical axis. Nothing is plotted on the horizontal axis, but the horizontal extent of the diagram is made large enough to show allowed transitions. Quantum numbers are given on the left and energies (in electron volts) on the right.

gives

$$r_n = (n^2) \frac{a_0}{Z} \tag{40.31}$$

$$E_n = -\frac{k_e e^2}{2a_0} \left(\frac{Z^2}{n^2} \right) \qquad n = 1, 2, 3, \ldots \tag{40.32}$$

EXAMPLE 40.5 Spectral Lines from the Star ξ-Puppis

Some mysterious lines observed in 1896 in the emission spectrum of the star ξ-Puppis (ξ is the Greek letter xi) fit the empirical formula

$$\frac{1}{\lambda} = R_H \left(\frac{1}{(n_f/2)^2} - \frac{1}{(n_i/2)^2} \right)$$

Show that these lines can be explained by the Bohr theory as originating from He$^+$.

Solution The ion He$^+$ has $Z = 2$. Thus, the allowed energy levels are given by Equation 40.32 as

$$E_n = -\frac{k_e e^2}{2a_0} \left(\frac{4}{n^2} \right)$$

Using Equation 40.28, we find

$$f = \frac{E_i - E_f}{h} = \frac{k_e e^2}{2a_0 h} \left(\frac{4}{n_f^2} - \frac{4}{n_i^2} \right)$$

$$= \frac{k_e e^2}{2a_0 h} \left(\frac{1}{(n_f/2)^2} - \frac{1}{(n_i/2)^2} \right)$$

$$\frac{1}{\lambda} = \frac{f}{c} = \frac{k_e e^2}{2a_0 hc} \left(\frac{1}{(n_f/2)^2} - \frac{1}{(n_i/2)^2} \right)$$

This is the desired solution when we recognize that $R_H \equiv k_e e^2/2a_0 hc$ (see text discussion immediately following Eq. 40.30).

EXAMPLE 40.6 Electronic Transitions in Hydrogen

(a) The electron in a hydrogen atom makes a transition from the $n = 2$ energy state to the ground state ($n = 1$). Find the wavelength and frequency of the emitted photon.

Solution We can use Equation 40.30 directly to obtain λ, with $n_i = 2$ and $n_f = 1$:

$$\frac{1}{\lambda} = R_H \left(\frac{1}{n_f^2} - \frac{1}{n_i^2} \right)$$

$$= R_H \left(\frac{1}{1^2} - \frac{1}{2^2} \right) = \frac{3R_H}{4}$$

$$\lambda = \frac{4}{3R_H} = \frac{4}{3(1.097 \times 10^7 \text{ m}^{-1})}$$

$$= 1.215 \times 10^{-7} \text{ m} = \boxed{121.5 \text{ nm}} \qquad \text{(ultraviolet)}$$

Because $c = f\lambda$, the frequency of the photon is

$$f = \frac{c}{\lambda} = \frac{3.00 \times 10^8 \text{ m/s}}{1.215 \times 10^{-7} \text{ m}} = \boxed{2.47 \times 10^{15} \text{ Hz}}$$

(b) In interstellar space, highly excited hydrogen atoms called Rydberg atoms have been observed. Find the wavelength to which radio astronomers must tune to detect signals from electrons dropping from the $n = 273$ level to $n = 272$.

Solution We can again use Equation 40.30, this time with $n_i = 273$ and $n_f = 272$:

$$\frac{1}{\lambda} = R_H \left(\frac{1}{n_f^2} - \frac{1}{n_i^2} \right)$$

$$\lambda = \boxed{0.922 \text{ m}}$$

(c) What is the radius of the electron orbit for a Rydberg atom for which $n = 272$?

Solution Using Equation 40.25, we find

$$r_{272} = (272)^2 (0.052\,9 \text{ nm}) = \boxed{3.91 \ \mu\text{m}}$$

This is large enough that the atom is on the verge of becoming macroscopic!

EXAMPLE 40.7 ▶ The Balmer Series for Hydrogen

The Balmer series for the hydrogen atom corresponds to electronic transitions that terminate in the $n = 2$ state, as shown in Figure 40.18. (a) Find the longest-wavelength photon emitted in this series and determine its energy.

Solution The longest-wavelength (lowest-energy) photon in the Balmer series results from the transition from $n = 3$ to $n = 2$. This is the lowest-energy photon in this series because it involves the smallest possible energy change. Equation 40.30 gives

$$\frac{1}{\lambda} = R_H \left(\frac{1}{n_f^2} - \frac{1}{n_i^2} \right)$$

$$\frac{1}{\lambda_{max}} = R_H \left(\frac{1}{2^2} - \frac{1}{3^2} \right) = \frac{5}{36} R_H$$

$$\lambda_{max} = \frac{36}{5R_H} = \frac{36}{5(1.097 \times 10^7 \, m^{-1})}$$

$$= \boxed{656.3 \, nm} \quad \text{(red)}$$

The energy of this photon is

$$E_{photon} = hf = \frac{hc}{\lambda_{max}}$$

$$= \frac{(6.626 \times 10^{-34} \, J \cdot s)(2.998 \times 10^8 \, m/s)}{656.3 \times 10^{-9} \, m}$$

$$= 3.03 \times 10^{-19} \, J = \boxed{1.89 \, eV}$$

We could also obtain the energy by using the expression $hf = E_3 - E_2$, where E_2 and E_3 can be calculated from Equation 40.26.

(b) Find the shortest-wavelength photon emitted in the Balmer series.

Solution The shortest-wavelength photon in the Balmer series is emitted when the electron makes a transition from $n = \infty$ to $n = 2$. Therefore,

$$\frac{1}{\lambda_{min}} = R_H \left(\frac{1}{2^2} - \frac{1}{\infty} \right) = \frac{R_H}{4}$$

$$\lambda_{min} = \frac{4}{R_H} = \frac{4}{1.097 \times 10^7 \, m^{-1}} = \boxed{364.6 \, nm}$$

This wavelength is in the ultraviolet region and corresponds to the series limit.

Exercise Find the energy of this shortest-wavelength photon.

Answer 3.40 eV.

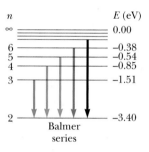

Figure 40.18 Transitions responsible for the Balmer series for the hydrogen atom. All transitions terminate at the $n = 2$ energy level. Energy levels are not drawn to scale.

Bohr's Correspondence Principle

In our study of relativity, we found that Newtonian mechanics is a special case of relativistic mechanics and is usable only when v is much less than c. Similarly, **quantum physics is in agreement with classical physics where the difference between quantized levels becomes vanishingly small.** This principle, first set forth by Bohr, is called the **correspondence principle.**

For example, consider an electron orbiting the hydrogen atom with $n > 10\,000$. For such large values of n, the energy differences between adjacent levels approach zero, and therefore the levels are nearly continuous. Consequently, the classical model is reasonably accurate in describing the system for large values of n. According to the classical picture, the frequency of the light emitted by the atom is equal to the frequency of revolution of the electron in its orbit about the nucleus. Calculations show that for $n > 10\,000$, this frequency is different from that predicted by quantum mechanics by less than 0.015%.

40.6 ▶ PHOTONS AND ELECTROMAGNETIC WAVES

Phenomena such as the photoelectric effect and the Compton effect offer ironclad evidence that when light (or other forms of electromagnetic radiation) and matter interact, the light behaves as if it were composed of particles having energy hf and momentum h/λ. An obvious question at this point is, "How can light be considered a photon (in other words, a particle) when we know it is a wave?" On the one hand, we describe light in terms of photons having energy and momentum. On the other hand, we recognize that light and other electromagnetic waves exhibit interference and diffraction effects, which are consistent only with a wave interpretation.

Which model is correct? Is light a wave or a particle? The answer depends on the phenomenon being observed. Some experiments can be explained either better or solely with the photon model, whereas others are explained better or solely with the wave model. The end result is that **we use both models and admit that the true nature of light is not describable in terms of any single classical picture.** However, you should recognize that the same light beam that can eject photoelectrons from a metal (meaning that the beam consists of photons) can also be diffracted by a grating (meaning that the beam is a wave). In other words, **the particle model and the wave model of light complement each other.**

The success of the particle model of light in explaining the photoelectric effect and the Compton effect raises many other questions. If light is a particle, what is the meaning of its "frequency" and "wavelength," and which of these two properties determines its energy and momentum? Is light *simultaneously* a wave and a particle? Although photons have no rest energy (a nonobservable quantity because a photon cannot be at rest!), is there a simple expression for the *effective mass* of a moving photon? If photons have effective mass, do they experience gravitational attraction? What is the spatial extent of a photon, and how does an electron absorb or scatter one photon? Although some of these questions are answerable, others are difficult to answer because our experiences from the everyday macroscopic world are far different from the behavior of microscopic particles. Many of these questions stem from classical analogies such as colliding billiard balls and water waves breaking on a shore. Quantum mechanics gives light a more fluid and flexible nature by incorporating both the particle model and the wave model as necessary and complementary. Hence,

light has a dual nature: It exhibits both wave and particle characteristics.

To understand why photons are compatible with electromagnetic waves, consider 2.5-MHz radio waves as an example. The energy of a photon having this frequency is only about 10^{-8} eV, too small to allow the photon to be detected. A sensitive radio receiver might require as many as 10^{10} of these photons to produce a detectable signal. Such a large number of photons would appear, on the average, as a continuous wave. With so many photons reaching the detector every second, it is unlikely that any graininess would appear in the detected signal. That is, with 2.5-MHz waves, one would not be able to detect the individual photons striking the antenna.

Now consider what happens as we go to higher frequencies. In the visible region, it is possible to observe both the particle characteristics and the wave characteristics of light. As we mentioned earlier, a beam of visible light shows interference phenomena (thus, it is a wave) and at the same time can produce photoelectrons (thus, it is a particle). At even higher frequencies, the momentum and energy of the photon increase. Consequently, the particle nature of light becomes more evident than its wave nature. For example, absorption of an x-ray photon is easily detected as a single event but wave effects are very difficult to observe.

40.7 ▶ THE WAVE PROPERTIES OF PARTICLES

Students introduced to the dual nature of light often find the concept difficult to accept. In the world around us, we are accustomed to regarding such things as baseballs solely as particles and such things as sound waves solely as forms of wave motion. Every large-scale observation can be interpreted by considering either a wave explanation or a particle explanation, but in the world of photons and electrons, such distinctions are not as sharply drawn. Even more disconcerting is the fact that, under certain conditions, the things we unambiguously call "particles" exhibit wave characteristics!

In 1923, in his doctoral dissertation, Louis de Broglie postulated that **because photons have both wave and particle characteristics, perhaps all forms of matter have both properties.** This was a highly revolutionary idea with no experimental confirmation at that time. According to de Broglie, electrons, just like light, have a dual particle-wave nature. Accompanying every electron is a wave (not an electromagnetic wave!). He explained the source of this assertion in his 1929 Nobel Prize acceptance speech:

> On the one hand the quantum theory of light cannot be considered satisfactory since it defines the energy of a light corpuscle by the equation $E = hf$ containing the frequency f. Now a purely corpuscular theory contains nothing that enables us to define a frequency; for this reason alone, therefore, we are compelled, in the case of light, to introduce the idea of a corpuscle and that of periodicity simultaneously. On the other hand, determination of the stable motion of electrons in the atom introduces integers, and up to this point the only phenomena involving integers in physics were those of interference and of normal modes of vibration. This fact suggested to me the idea that electrons too could not be considered simply as corpuscles, but that periodicity must be assigned to them also.

Louis de Broglie (1892–1987)
A French physicist, de Broglie was awarded the Nobel Prize in 1929 for his prediction of the wave nature of electrons. *(AIP Niels Bohr Library)*

In Section 39.7, we found that the relationship between the energy and the linear momentum of a photon, which has a rest energy of zero, is $p = E/c$. We also know that the energy of a photon is $E = hf = hc/\lambda$. Thus, the momentum of a photon can be expressed as

$$p = \frac{E}{c} = \frac{hc}{c\lambda} = \frac{h}{\lambda}$$

From this equation we see that the photon wavelength can be specified by its momentum: $\lambda = h/p$. De Broglie suggested that material particles of momentum p have a characteristic wavelength $\lambda = h/p$. Because the momentum of a particle of mass m and speed v is $p = mv$, the **de Broglie wavelength** of that particle is[2]

$$\lambda = \frac{h}{p} = \frac{h}{mv} \tag{40.33}$$

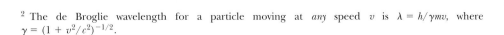

[2] The de Broglie wavelength for a particle moving at *any* speed v is $\lambda = h/\gamma mv$, where $\gamma = (1 + v^2/c^2)^{-1/2}$.

Furthermore, in analogy with photons, de Broglie postulated that the frequencies of **matter waves** (that is, waves associated with particles having nonzero rest energy) obey the Einstein relationship $E = hf$, where E is the total energy of the particle, so that

$$f = \frac{E}{h} \tag{40.34}$$

The dual nature of matter is apparent in these two equations because each contains both particle concepts (mv and E) and wave concepts (λ and f). The fact that these relationships are established experimentally for photons makes the de Broglie hypothesis that much easier to accept.

The Davisson–Germer Experiment

De Broglie's proposal in 1923 that matter exhibits both wave and particle properties was regarded as pure speculation. If particles such as electrons had wave properties, then under the correct conditions they should exhibit diffraction effects. Only three years later, C. J. Davisson (1881–1958) and L. H. Germer (1896–1971) of the United States succeeded in measuring the wavelength of electrons. Their important discovery provided the first experimental confirmation of the matter waves proposed by de Broglie.

Interestingly, the intent of the initial Davisson–Germer experiment was not to confirm the de Broglie hypothesis. In fact, their discovery was made by accident (as is often the case). The experiment involved the scattering of low-energy electrons (about 54 eV) from a nickel target in a vacuum. During one experiment, the nickel surface was badly oxidized because of an accidental break in the vacuum system. After the target was heated in a flowing stream of hydrogen to remove the oxide coating, electrons scattered by it exhibited intensity maxima and minima at specific angles. The experimenters finally realized that the nickel had formed large crystalline regions upon heating and that the regularly spaced planes of atoms in these regions served as a diffraction grating for electron matter waves.

Shortly thereafter, Davisson and Germer performed more extensive diffraction measurements on electrons scattered from single-crystal targets. Their results showed conclusively the wave nature of electrons and confirmed the de Broglie relationship $p = h/\lambda$. In the same year, G. P. Thomson (1892–1975) of Scotland also observed electron diffraction patterns by passing electrons through very thin gold foils. Diffraction patterns have since been observed for helium atoms, hydrogen atoms, and neutrons. Hence, the universal nature of matter waves has been established in various ways.

The problem of understanding the dual nature of matter and radiation is conceptually difficult because the two models seem to contradict each other. This problem as it applies to light was discussed earlier. Bohr helped to resolve this problem in his **principle of complementarity,** which states that the **wave and particle models of either matter or radiation complement each other.** Neither model can be used exclusively to describe matter or radiation adequately. Because humans can only generate mental images based on their experiences from the everyday world (baseballs, water waves, and so forth), we use both descriptions in a complementary manner to explain any given set of data from the quantum world.

EXAMPLE 40.8 ▶ The Wavelength of an Electron

Calculate the de Broglie wavelength for an electron ($m = 9.11 \times 10^{-31}$ kg) moving at 1.00×10^7 m/s.

Solution Equation 40.33 gives

$$\lambda = \frac{h}{mv} = \frac{6.63 \times 10^{-34}\,\text{J}\cdot\text{s}}{(9.11 \times 10^{-31}\,\text{kg})(1.00 \times 10^7\,\text{m/s})}$$

$$= \boxed{7.28 \times 10^{-11}\,\text{m}}$$

Exercise Find the de Broglie wavelength of a stone of mass 50 g thrown with a speed of 40 m/s.

Answer 3.3×10^{-34} m.

EXAMPLE 40.9 ▶ An Accelerated Charged Particle

A particle of charge q and mass m has been accelerated from rest through a potential difference ΔV. Find an expression for its de Broglie wavelength.

Solution When a charged particle is accelerated from rest through a potential difference ΔV, the gain in kinetic energy $\frac{1}{2}mv^2$ must equal the loss in potential energy $q\,\Delta V$:

$$\tfrac{1}{2}mv^2 = q\,\Delta V$$

Because $p = mv$, we can express this equation in the form

$$\frac{p^2}{2m} = q\,\Delta V$$

$$p = \sqrt{2mq\,\Delta V}$$

Substituting this expression for p into Equation 40.33 gives

$$\lambda = \frac{h}{p} = \boxed{\frac{h}{\sqrt{2mq\,\Delta V}}}$$

Exercise Calculate the de Broglie wavelength of an electron accelerated through a potential difference of 50 V.

Answer 0.174 nm.

SUMMARY

The characteristics of blackbody radiation cannot be explained using classical concepts. Planck introduced the quantum concept when he assumed that the atomic oscillators responsible for this radiation exist only in discrete energy states. Radiation is emitted in single quantized packets whenever an oscillator makes a transition between discrete energy states.

The **photoelectric effect** is a process whereby electrons are ejected from a metal surface when light is incident on that surface. Einstein provided a successful explanation of this effect by extending Planck's quantum hypothesis to electromagnetic radiation. In this model, light is viewed as a stream of light particles, or **photons,** each having energy $E = hf$, where f is the frequency and h is Planck's constant. The maximum kinetic energy of the ejected photoelectron is

$$K_{\max} = hf - \phi \qquad\qquad \textbf{(40.8)}$$

where ϕ is the **work function** of the metal.

X-rays are scattered at various angles by electrons in a target. In such a scattering event, a shift in wavelength is observed for the scattered x-rays, and the phenomenon is known as the **Compton effect.** Classical physics does not explain this effect. If the x-ray is treated as a photon, conservation of energy and linear momentum applied to the photon–electron collisions yields for the Compton shift:

$$\lambda' - \lambda_0 = \frac{h}{m_e c}(1 - \cos\theta) \qquad \textbf{(40.10)}$$

where m_e is the mass of the electron, c is the speed of light, and θ is the scattering angle.

The Bohr model of the atom is successful in describing the spectra of atomic hydrogen and hydrogen-like ions. One of the basic assumptions of the model is that the electron can exist only in discrete orbits such that the angular momentum mvr is an integral multiple of $h/2\pi = \hbar$. When we assume circular orbits and a simple Coulomb attraction between electron and proton, the energies of the quantum states for hydrogen are calculated to be

$$E_n = -\frac{k_e e^2}{2a_0}\left(\frac{1}{n^2}\right) \qquad n = 1, 2, 3, \ldots \qquad \textbf{(40.26)}$$

where k_e is the Coulomb constant, e is the electronic charge, n is an integer called the **quantum number,** and $a_0 = 0.052\,9$ nm is the **Bohr radius.**

If the electron in a hydrogen atom makes a transition from an orbit whose quantum number is n_i to one whose quantum number is n_f, where $n_f < n_i$, a photon is emitted by the atom, and the frequency of this photon is

$$f = \frac{k_e e^2}{2a_0 h}\left(\frac{1}{n_f^{\,2}} - \frac{1}{n_i^{\,2}}\right) \qquad \textbf{(40.28)}$$

Light has a dual nature in that it has both wave and particle characteristics. Some experiments can be explained either better or solely by the particle model, whereas others can be explained either better or solely by the wave model.

Every object of mass m and momentum $p = mv$ has wave properties, with a wavelength given by the de Broglie relationship:

$$\lambda = \frac{h}{p} = \frac{h}{mv} \qquad \textbf{(40.33)}$$

QUESTIONS

1. What assumptions were made by Planck in dealing with the problem of blackbody radiation? Discuss the consequences of these assumptions.
2. The classical model of blackbody radiation given by the Rayleigh–Jeans law has two major flaws. Identify them and explain how Planck's law deals with them.
3. If the photoelectric effect is observed for one metal, can you conclude that the effect will also be observed for another metal under the same conditions? Explain.
4. In the photoelectric effect, explain why the stopping potential depends on the frequency of light but not on the intensity.
5. Suppose the photoelectric effect occurs in a gaseous target rather than a solid plate. Will photoelectrons be produced at all frequencies of the incident photon? Explain.

6. How does the Compton effect differ from the photoelectric effect?
7. What assumptions did Compton make in dealing with the scattering of a photon from an electron?
8. The Bohr theory of the hydrogen atom is based upon several assumptions. Discuss these assumptions and their significance. Do any of them contradict classical physics?
9. Suppose that the electron in the hydrogen atom obeyed classical mechanics rather than quantum mechanics. Why should such a "hypothetical" atom emit a continuous spectrum rather than the observed line spectrum?
10. Can the electron in the ground state of hydrogen absorb a photon of energy (a) less than 13.6 eV and (b) greater than 13.6 eV?
11. Why would the spectral lines of diatomic hydrogen be different from those of monatomic hydrogen?

12. Explain why, in the Bohr model, the total energy of the atom is negative.
13. An x-ray photon is scattered by an electron. What happens to the frequency of the scattered photon relative to that of the incident photon?
14. Why does the existence of a cutoff frequency in the photoelectric effect favor a particle theory for light rather than a wave theory?
15. A student claims that he is going to eject electrons from a piece of metal by placing a radio transmitter antenna adjacent to the metal and sending a strong AM radio signal into the antenna. The work function of a metal is typically a few electron volts. Will this work?
16. All objects radiate energy. Why, then, are we not able to see all objects in a dark room?
17. Which has more energy, a photon of ultraviolet radiation or a photon of yellow light?
18. Why was the Davisson–Germer experiment involving the diffraction of electrons so important?

PROBLEMS

1, 2, 3 = straightforward, intermediate, challenging   ☐ = full solution available in the *Student Solutions Manual and Study Guide*
WEB = solution posted at **http://www.saunderscollege.com/physics/**   🖥 = Computer useful in solving problem   🖱 = Interactive Physics
☐ = paired numerical/symbolic problems

Section 40.1 Blackbody Radiation and Planck's Hypothesis

1. The human eye is most sensitive to 560-nm light. What is the temperature of a black body that would radiate most intensely at this wavelength?
2. (a) Lightning produces a maximum air temperature on the order of 10^4 K, whereas (b) a nuclear explosion produces a temperature on the order of 10^7 K. Use Wien's displacement law to find the order of magnitude of the wavelength of the thermally produced photons radiated with greatest intensity by each of these sources. Name the part of the electromagnetic spectrum where you would expect each to radiate most strongly.
3. (a) Assuming that the tungsten filament of a lightbulb is a black body, determine its peak wavelength if its temperature is 2 900 K. (b) Why does your answer to part (a) suggest that more energy from a lightbulb goes into infrared radiation than into visible light?
4. A black body at 7 500 K consists of an opening of diameter 0.050 0 mm, looking into an oven. Find the number of photons per second escaping the hole and having wavelengths between 500 nm and 501 nm.
5. Consider a black body of surface area 20.0 cm² and temperature 5 000 K. (a) How much power does it radiate? (b) At what wavelength does it radiate most intensely? Find the spectral power per wavelength at (c) this wavelength and at wavelengths of (d) 1.00 nm (an x- or γ-ray), (e) 5.00 nm (ultraviolet light or an x-ray), (f) 400 nm (at the boundary between UV and visible light), (g) 700 nm (at the boundary between visible and infrared light), (h) 1.00 mm (infrared light or a microwave), and (i) 10.0 cm (a microwave or radio wave). (j) About how much power does the object radiate as visible light?
6. The radius of our Sun is 6.96×10^8 m, and its total power output is 3.77×10^{26} W. (a) Assuming that the Sun's surface emits as a black body, calculate its surface temperature. (b) Using the result of part (a), find λ_{max} for the Sun.
7. Calculate the energy, in electron volts, of a photon whose frequency is (a) 620 THz, (b) 3.10 GHz, (c) 46.0 MHz. (d) Determine the corresponding wavelengths for these photons and state the classification of each on the electromagnetic spectrum.
8. A sodium-vapor lamp has a power output of 10.0 W. Using 589.3 nm as the average wavelength of this source, calculate the number of photons emitted per second.
9. An FM radio transmitter has a power output of 150 kW and operates at a frequency of 99.7 MHz. How many photons per second does the transmitter emit?
10. The average threshold of dark-adapted (scotopic) vision is 4.00×10^{-11} W/m² at a central wavelength of 500 nm. If light having this intensity and wavelength enters the eye and the pupil is open to its maximum diameter of 8.50 mm, how many photons per second enter the eye?
11. A simple pendulum has a length of 1.00 m and a mass of 1.00 kg. If the amplitude of oscillations of the pendulum is 3.00 cm, estimate the quantum number for the pendulum.
12. **Review Problem.** A star moving away from the Earth at $0.280c$ emits radiation that we measure to be most intense at the wavelength 500 nm. Determine the surface temperature of this star.
13. Show that at short wavelengths or low temperatures, Planck's radiation law (Eq. 40.3) predicts an exponential decrease in $I(\lambda, T)$ given by *Wien's radiation law:*

$$I(\lambda, T) = \frac{2\pi hc^2}{\lambda^5} \, e^{-hc/\lambda k_B T}$$

14. Show that at long wavelengths, Planck's radiation law (Eq. 40.3) reduces to the Rayleigh–Jeans law (Eq. 40.2).

Section 40.2 The Photoelectric Effect

15. Molybdenum has a work function of 4.20 eV. (a) Find the cutoff wavelength and cutoff frequency for the pho-

toelectric effect. (b) Calculate the stopping potential if the incident light has a wavelength of 180 nm.

16. Electrons are ejected from a metal surface with speeds ranging up to 4.60×10^5 m/s when light with a wavelength of $\lambda = 625$ nm is used. (a) What is the work function of the surface? (b) What is the cutoff frequency for this surface?

17. Lithium, beryllium, and mercury have work functions of 2.30 eV, 3.90 eV, and 4.50 eV, respectively. If 400-nm light is incident on each of these metals, determine (a) which metals exhibit the photoelectric effect and (b) the maximum kinetic energy for the photoelectrons in each case.

18. A student studying the photoelectric effect from two different metals records the following information: (i) the stopping potential for photoelectrons released from metal 1 is 1.48 V larger than that for metal 2, and (ii) the cutoff frequency for metal 1 is 40.0% smaller than that for metal 2. Determine the work function for each metal.

19. Two light sources are used in a photoelectric experiment to determine the work function for a particular metal surface. When green light from a mercury lamp ($\lambda = 546.1$ nm) is used, a stopping potential of 0.376 V reduces the photocurrent to zero. (a) Based on this measurement, what is the work function for this metal? (b) What stopping potential would be observed when using the yellow light from a helium discharge tube ($\lambda = 587.5$ nm)?

20. When 445-nm light strikes a certain metal surface, the stopping potential is 70.0% of that which results when 410-nm light strikes the same metal surface. Based on this information and the following table of work functions, identify the metal involved in the experiment.

Metal	Work Function (eV)
Cesium	1.90
Potassium	2.23
Silver	4.73
Tungsten	4.58

21. From the scattering of sunlight, Thomson calculated the classical radius of the electron as having a value of 2.82×10^{-15} m. If sunlight with an intensity of 500 W/m^2 falls on a disk with this radius, calculate the time required to accumulate 1.00 eV of energy. Assume that light is a classical wave and that the light striking the disk is completely absorbed. How does your result compare with the observation that photoelectrons are emitted promptly (within 10^{-9} s)?

22. **Review Problem.** An isolated copper sphere of radius 5.00 cm, initially uncharged, is illuminated by ultraviolet light of wavelength 200 nm. What charge will the photoelectric effect induce on the sphere? The work function for copper is 4.70 eV.

23. **Review Problem.** A light source emitting radiation at 7.00×10^{14} Hz is incapable of ejecting photoelectrons from a certain metal. In an attempt to use this source to eject photoelectrons from the metal, the source is given a velocity toward the metal. (a) Explain how this procedure produces photoelectrons. (b) When the speed of the light source is equal to $0.280c$, photoelectrons just begin to be ejected from the metal. What is the work function of the metal? (c) When the speed of the light source is increased to $0.900c$, determine the maximum kinetic energy of the photoelectrons.

Section 40.3 The Compton Effect

24. Calculate the energy and momentum of a photon of wavelength 700 nm.

25. X-rays having an energy of 300 keV undergo Compton scattering from a target. If the scattered rays are detected at 37.0° relative to the incident rays, find (a) the Compton shift at this angle, (b) the energy of the scattered x-ray, and (c) the energy of the recoiling electron.

26. A 0.110-nm photon collides with a stationary electron. After the collision, the electron moves forward and the photon recoils backwards. Find the momentum and kinetic energy of the electron.

WEB 27. A 0.001 60-nm photon scatters from a free electron. For what (photon) scattering angle does the recoiling electron have kinetic energy equal to the energy of the scattered photon?

28. In a Compton scattering experiment, a photon is scattered through an angle of 90.0°, and the electron is scattered through an angle of 20.0°. Determine the wavelength of the scattered photon.

29. A 0.880-MeV photon is scattered by a free electron initially at rest such that the scattering angle of the scattered electron equals that of the scattered photon ($\theta = \phi$ in Fig. 40.10b). (a) Determine the angles θ and ϕ. (b) Determine the energy and momentum of the scattered photon. (c) Determine the kinetic energy and momentum of the scattered electron.

30. A photon having energy E_0 is scattered by a free electron initially at rest such that the scattering angle of the scattered electron equals that of the scattered photon ($\theta = \phi$ in Fig. 40.10b). (a) Determine the angles θ and ϕ. (b) Determine the energy and momentum of the scattered photon. (c) Determine the kinetic energy and momentum of the scattered electron.

31. A 0.700-MeV photon scatters off a free electron such that the scattering angle of the photon is twice the scattering angle of the electron (Fig. P40.31). (a) Determine the scattering angle for the electron and (b) the final speed of the electron.

32. A photon having wavelength λ scatters off a free electron at A (Fig. P40.32), producing a second photon having wavelength λ'. This photon then scatters off another free electron at B, producing a third photon

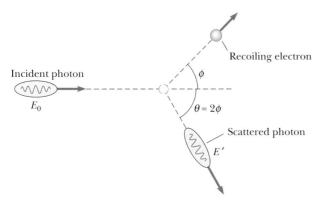

Figure P40.31

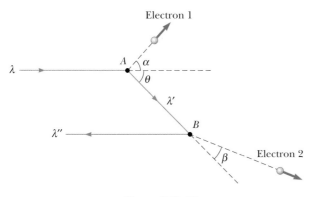

Figure P40.32

having wavelength λ'' and moving in a direction directly opposite the original photon as shown in Figure P40.32. Determine the numerical value of $\Delta\lambda = \lambda'' - \lambda$.

33. After a 0.800-nm x-ray photon scatters from a free electron, the electron recoils at 1.40×10^6 m/s. (a) What was the Compton shift in the photon's wavelength? (b) Through what angle was the photon scattered?

34. Find the maximum fractional energy loss for a 0.511-MeV gamma ray that is Compton scattered from a free (a) electron and (b) proton.

Section 40.4 Atomic Spectra

35. Show that the wavelengths for the Balmer series satisfy the equation

$$\lambda = \frac{364.5 n^2}{n^2 - 4} \text{ nm} \qquad \text{where } n = 3, 4, 5, \ldots$$

36. (a) Suppose that the Rydberg constant were given by $R_H = 2.00 \times 10^7 \text{ m}^{-1}$. In what part of the electromagnetic spectrum would the Balmer series lie? (b) Repeat for $R_H = 0.500 \times 10^7 \text{ m}^{-1}$.

37. (a) What value of n is associated with the 94.96-nm line in the Lyman hydrogen series? (b) Could this wavelength be associated with the Paschen or Brackett series?

38. (a) Compute the shortest wavelength in each of these hydrogen spectral series: Lyman, Balmer, Paschen, and Brackett. (b) Compute the energy (in electron volts) of the highest-energy photon produced in each series.

39. Liquid oxygen has a bluish color, meaning that it preferentially absorbs light toward the red end of the visible spectrum. Although the oxygen molecule (O_2) does not strongly absorb visible radiation, it does absorb strongly at 1 269 nm, which is in the infrared region of the spectrum. Research has shown that it is possible for two colliding O_2 molecules to absorb a single photon, sharing its energy equally. The transition that both molecules undergo is the same transition that results when they absorb 1 269-nm radiation. What is the wavelength of the single photon that causes this double transition? What is the color of this radiation?

Section 40.5 Bohr's Quantum Model of the Atom

40. For a hydrogen atom in its ground state, use the Bohr model to compute (a) the orbital speed of the electron, (b) the kinetic energy of the electron, and (c) the electric potential energy of the atom.

WEB 41. A hydrogen atom is in its first excited state ($n = 2$). Using the Bohr theory of the atom, calculate (a) the radius of the orbit, (b) the linear momentum of the electron, (c) the angular momentum of the electron, (d) the kinetic energy, (e) the potential energy, and (f) the total energy.

42. Four possible transitions for a hydrogen atom are as follows:

(A) $n_i = 2$; $n_f = 5$ (B) $n_i = 5$; $n_f = 3$

(C) $n_i = 7$; $n_f = 4$ (D) $n_i = 4$; $n_f = 7$

(a) Which transition emits the shortest wavelength photon? (b) In which transition does the atom gain the most energy? (c) In which transition(s) does the atom lose energy?

43. A photon is emitted as a hydrogen atom undergoes a transition from the $n = 6$ state to the $n = 2$ state. Calculate (a) the energy, (b) the wavelength, and (c) the frequency of the emitted photon.

44. How much energy is required to ionize hydrogen (a) when it is in the ground state? (b) when it is in the state for which $n = 3$?

45. Show that the speed of the electron in the nth Bohr orbit in hydrogen is given by

$$v_n = \frac{k_e e^2}{n\hbar}$$

46. (a) Calculate the angular momentum of the Moon due to its orbital motion about the Earth. In your calculation, use 3.84×10^8 m as the average Earth–Moon distance and 2.36×10^6 s as the period of the Moon in its orbit. (b) Determine the corresponding quantum number if the Moon's angular momentum is given by the

Bohr assumption $mvr = n\hbar$. (c) By what fraction would the Earth–Moon distance have to be increased to increase the quantum number by 1?

47. A monochromatic beam of light is absorbed by a collection of ground-state hydrogen atoms in such a way that six different wavelengths are observed when the hydrogen relaxes back to the ground state. What is the wavelength of the incident beam?

48. Two hydrogen atoms collide head-on and end up with zero kinetic energy. Each then emits a 121.6-nm photon ($n = 2$ to $n = 1$ transition). At what speed were the atoms moving before the collision?

49. (a) Construct an energy level diagram for the He$^+$ ion, for which $Z = 2$. (b) What is the ionization energy for He$^+$?

50. What is the radius of the first Bohr orbit in (a) He$^+$, (b) Li^{2+}, and (c) Be^{3+}?

51. A particle of charge q and mass m, moving with a constant speed v and perpendicular to a constant magnetic field B, follows a circular path. If the angular momentum about the center of this circle is quantized so that $mvr = n\hbar$, show that the allowed radii for the particle are

$$r_n = \sqrt{\frac{n\hbar}{qB}}$$

where $n = 1, 2, 3, \ldots$.

52. An electron is in the nth Bohr orbit of the hydrogen atom. (a) Show that the period of the electron is $T = t_0 n^3$, and determine the numerical value of t_0. (b) On the average, an electron remains in the $n = 2$ orbit for about 10 μs before it jumps down to the $n = 1$ (ground-state) orbit. How many revolutions does the electron make before it jumps to the ground state? (c) If one revolution of the electron is defined as an "electron year" (analogous to an Earth year being one revolution of the Earth around the Sun), does the electron in the $n = 2$ orbit "live" very long? Explain.

Section 40.6   Photons and Electromagnetic Waves
Section 40.7   The Wave Properties of Particles

53. Calculate the de Broglie wavelength for a proton moving with a speed of 1.00×10^6 m/s.

54. Calculate the de Broglie wavelength for an electron that has kinetic energy (a) 50.0 eV and (b) 50.0 keV.

55. (a) An electron has kinetic energy 3.00 eV. Find its wavelength. (b) A photon has energy 3.00 eV. Find its wavelength.

56. In the Davisson–Germer experiment, 54.0-eV electrons were diffracted from a nickel lattice. If the first maximum in the diffraction pattern was observed at $\phi = 50.0°$ (Fig. P40.56), what was the lattice spacing a?

WEB 57. The nucleus of an atom is on the order of 10^{-14} m in diameter. For an electron to be confined to a nucleus, its de Broglie wavelength would have to be of this order of magnitude or smaller. (a) What would be the kinetic energy of an electron confined to this region? (b) On

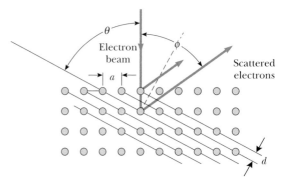

Figure P40.56

the basis of this result, would you expect to find an electron in a nucleus? Explain.

58. Robert Hofstadter won the 1961 Nobel Prize in physics for his pioneering work in scattering 20-GeV electrons from nuclei. (a) What is the γ-factor for a 20.0-GeV electron, where $\gamma = (1 - v^2/c^2)^{-1/2}$? (b) What is the momentum of the electron in kg·m/s? (c) What is the wavelength of a 20.0-GeV electron and how does it compare with the size of a nucleus?

59. (a) Show that the frequency f and wavelength λ of a freely moving particle are related by the expression

$$\left(\frac{f}{c}\right)^2 = \frac{1}{\lambda^2} + \frac{1}{\lambda_C^2}$$

where $\lambda_C = h/mc$ is the Compton wavelength of the particle. (b) Is it ever possible for a particle having nonzero mass to have the same wavelength *and* frequency as a photon? Explain.

60. After learning about de Broglie's hypothesis that particles of momentum p have wave characteristics with wavelength $\lambda = h/p$, an 80.0-kg student has grown concerned about being diffracted when passing through a 75.0-cm-wide doorway. Assume that significant diffraction occurs when the width of the diffraction aperture is less than 10.0 times the wavelength of the wave being diffracted. (a) Determine the maximum speed at which the student can pass through the doorway in order to be significantly diffracted. (b) With that speed, how long will it take the student to pass through the doorway if it is 15.0 cm thick? Compare your result to the currently accepted age of the Universe, which is 4×10^{17} s. (c) Should this student worry about being diffracted?

61. What is the speed of an electron if its de Broglie wavelength equals its Compton wavelength? (*Hint:* If you get an answer of c, see Problem 71.)

ADDITIONAL PROBLEMS

62. Figure P40.62 shows the stopping potential versus incident photon frequency for the photoelectric effect for

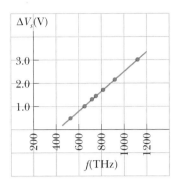

Figure P40.62

sodium. Use the graph to find (a) the work function, (b) the ratio h/e, and (c) the cutoff wavelength. (Data taken from R. A. Millikan, *Phys. Rev.* 7:362, 1916.)

63. Photons of wavelength 450 nm are incident on a metal. The most energetic electrons ejected from the metal are bent into a circular arc of radius 20.0 cm by a magnetic field with a magnitude of 2.00×10^{-5} T. What is the work function of the metal?

64. Photons of wavelength λ are incident on a metal. The most energetic electrons ejected from the metal are bent into a circular arc of radius R by a magnetic field having a magnitude B. What is the work function of the metal?

WEB 65. The table below shows data obtained in a photoelectric experiment. (a) Using these data, make a graph similar to Figure 40.9 that plots as a straight line. From the graph, determine (b) an experimental value for Planck's constant (in joule–seconds) and (c) the work function (in electron volts) for the surface. (Two significant figures for each answer are sufficient.)

Wavelength (nm)	Maximum Kinetic Energy of Photoelectrons (eV)
588	0.67
505	0.98
445	1.35
399	1.63

66. A 200-MeV photon is scattered at 40.0° by a free proton initially at rest. (a) Find the energy (in MeV) of the scattered photon. (b) What kinetic energy (in MeV) does the proton acquire?

67. Positronium is a hydrogen-like atom consisting of a positron (a positively charged electron) and an electron revolving around each other. Using the Bohr model, find the allowed radii (relative to the center of mass of the two particles) and the allowed energies of the system.

68. Derive the formula for the Compton shift (Eq. 40.10) from Equations 40.11, 40.12, and 40.13.

69. *An example of the correspondence principle.* Use Bohr's model of the hydrogen atom to show that when the electron moves from the state n to the state $n - 1$, the frequency of the emitted light is

$$f = \frac{2\pi^2 m_e k_e^2 e^4}{h^3} \left[\frac{2n - 1}{(n - 1)^2 n^2} \right]$$

Show that as $n \to \infty$, this expression varies as $1/n^3$ and reduces to the classical frequency one expects the atom to emit. (*Hint:* To calculate the classical frequency, note that the frequency of revolution is $v/2\pi r$, where r is given by Eq. 40.25.)

70. Show that a photon cannot transfer all of its energy to a free electron. (*Hint:* Note that energy and momentum must be conserved.)

71. Show that the speed of a particle having de Broglie wavelength λ and Compton wavelength $\lambda_C = h/(mc)$ is

$$v = \frac{c}{\sqrt{1 + (\lambda/\lambda_C)^2}}$$

72. The Lyman series for a (new?) one-electron atom is observed in the light from a distant galaxy. The wavelengths of the first four lines and the short-wavelength limit of this series are given by the energy-level diagram in Figure P40.72. Based on this information, calculate (a) the energies of the ground state and first four excited states for this one-electron atom and (b) the wavelengths of the first three lines and the short-wavelength limit in the Balmer series for this atom. (c) Show that the wavelengths of the first four lines and the short-wavelength limit of the Lyman series for the hydrogen atom are all 60.0% of the wavelengths for the Lyman se-

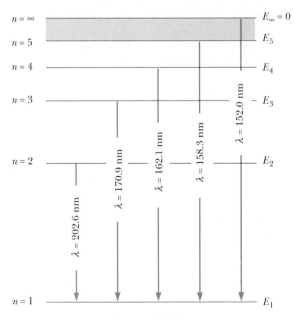

Figure P40.72

ries in the one-electron atom described in part (b).
(d) Based on this observation, explain why this atom could be hydrogen.

73. The total power per unit area radiated by a black body at a temperature T is the area under the $I(\lambda, T)$-versus-λ curve, as shown in Figure 40.3. (a) Show that this power per unit area is

$$\int_0^\infty I(\lambda, T) \, d\lambda = \sigma T^4$$

where $I(\lambda, T)$ is given by Planck's radiation law and σ is a constant independent of T. This result is known as the Stefan–Boltzmann law (see Section 20.7). To carry out the integration, you should make the change of variable $x = hc/\lambda k_B T$ and use the fact that

$$\int_0^\infty \frac{x^3 \, dx}{e^x - 1} = \frac{\pi^4}{15}$$

(b) Show that the Stefan–Boltzmann constant σ has the value

$$\sigma = \frac{2\pi^5 k_B^4}{15 c^2 h^3} = 5.67 \times 10^{-8} \, \text{W/m}^2 \cdot \text{K}^4$$

74. Derive Wien's displacement law from Planck's law. Proceed as follows: In Figure 40.3 note that the wavelength at which a black body radiates with greatest intensity is the wavelength for which the graph of $I(\lambda, T)$ versus λ has a horizontal tangent. From Equation 40.3 evaluate the derivative $dI/d\lambda$. Set it equal to zero. Solve the resulting transcendental equation numerically to prove $hc/\lambda_{\max} k_B T = 4.965 \ldots$, or $\lambda_{\max} T = hc/4.965 k_B$. Evaluate the constant as precisely as possible and compare it with Wien's experimental value.

75. A photon of initial energy E_0 undergoes Compton scattering at an angle θ from a free electron (mass m_e) initially at rest. Using relativistic equations for energy and momentum conservation, derive the following relationship for the final energy E' of the scattered photon:

$$E' = E_0[1 + (E_0/m_e c^2)(1 - \cos\theta)]^{-1}$$

76. As we learned in Section 39.4, a muon has a charge of $-e$ and a mass equal to 207 times the mass of an electron. Muonic lead is formed when a lead nucleus captures a muon. According to the Bohr theory, what are the radius and energy of the ground state of muonic lead?

77. An electron initially at rest recoils from a head-on collision with a photon. Show that the kinetic energy acquired by the electron is $2hfa/(1 + 2a)$, where a is the ratio of the photon's initial energy to the rest energy of the electron.

78. The spectral distribution function $I(\lambda, T)$ for an ideal black body at absolute temperature T is shown in Figure P40.78. (a) Show that the percentage of the total power radiated per unit area in the range $0 \le \lambda \le \lambda_{\max}$

is

$$\frac{A}{A + B} = 1 - \frac{15}{\pi^4} \int_0^{4.965} \frac{x^3}{e^x - 1} \, dx$$

independent of the value of T. (b) Using numerical integration, show that this ratio is approximately $1/4$.

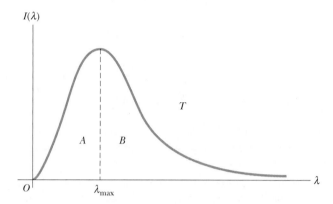

Figure P40.78

79. Show that the ratio of the Compton wavelength λ_C to the de Broglie wavelength $\lambda = h/p$ for a relativistic electron is

$$\frac{\lambda_C}{\lambda} = \left[\left(\frac{E}{m_e c^2}\right)^2 - 1\right]^{1/2}$$

where E is the total energy of the electron and m_e is its mass.

80. The neutron has a mass of 1.67×10^{-27} kg. Neutrons emitted in nuclear reactions can be slowed down via collisions with matter. They are referred to as thermal neutrons once they come into thermal equilibrium with their surroundings. The average kinetic energy $(3k_B T/2)$ of a thermal neutron is approximately 0.04 eV. Calculate the de Broglie wavelength of a neutron with a kinetic energy of 0.040 0 eV. How does it compare with the characteristic atomic spacing in a crystal? Would you expect thermal neutrons to exhibit diffraction effects when scattered by a crystal?

81. A photon with wavelength λ_0 moves toward a free electron that is moving with speed u in the same direction

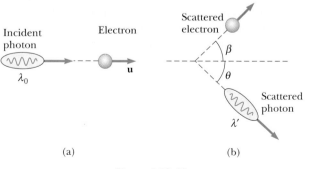

(a)

Figure P40.81

(b)

as the photon (Fig. P40.81a). If the photon scatters through an angle θ (Fig. P40.81b), show that the wavelength of the scattered photon is

$$\lambda' = \lambda_0 \left(\frac{1 - (u/c)\cos\theta}{1 - u/c} \right) + \frac{h}{m_e c} \sqrt{\frac{1 + u/c}{1 - u/c}} \, (1 - \cos\theta)$$

ANSWERS TO QUICK QUIZZES

40.1 (c). Ultraviolet light has the highest frequencies of the three, and hence each photon delivers more energy to a skin cell. (This explains why you can become sunburned on a cloudy day: Clouds block visible light but not much ultraviolet light. You usually do not become sunburned through window glass, even though you can feel the warmth due to the Sun's infrared rays, because the glass blocks ultraviolet light.)

40.2 Comparing Equation 40.8 with the slope–intercept form of the equation for a straight line, $y = mx + b$, we see that the slope in Figure 40.9 is Planck's constant h and that the y intercept is $-\phi$, the negative of the work function. If a different metal were used, the slope would remain the same but the work function would be different. Thus, data for different metals appear as parallel lines on the graph.

40.3 Classical physics predicts that light of sufficient intensity causes photoelectron emission, independent of frequency and certainly without a cutoff frequency. Also, the greater the intensity, the greater the maximum ki-

netic energy, with some time delay in emission at low intensities. Thus, the classical expectation (which did not match experiment) yields a graph that looks like this:

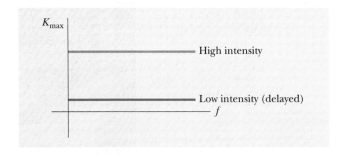

40.4 The fractional change in wavelength $\Delta\lambda / \lambda_0$ is greater (and thus easier to measure) for smaller wavelengths, and x-rays have much smaller wavelengths than visible light.

ROSE IS ROSE reprinted by permission of United Feature Syndicate, Inc.

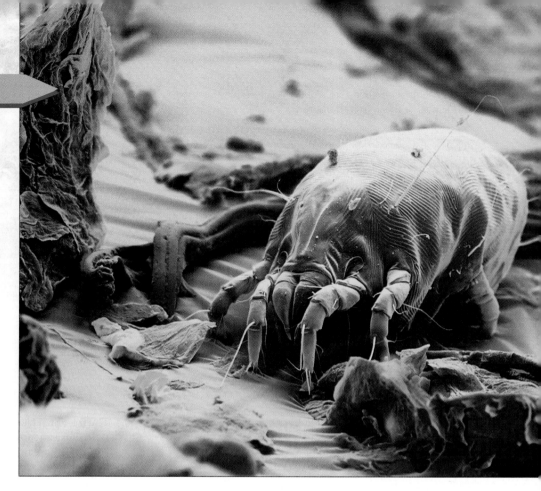
c h a p t e r

41

Quantum Mechanics

T he Bohr model of the hydrogen atom, which we presented in Chapter 40, has severe limitations. It depicts the electron as moving on the circumference of a flat circle, but scattering experiments show that the electron fills a sphere around the nucleus, with an exponentially decreasing probability of being found at greater and greater distances from the nucleus. The Bohr model does not take account of the wave motion of the electron. Bohr supposed that the minimum angular momentum of the electron was $\hbar$; in fact, it is zero. Also, the model cannot be extended to explain the absorption and emission spectra of complex atoms, nor does it predict such details as variations in spectral line intensities and the splittings observed in certain spectral lines under controlled laboratory conditions. Finally, it does not enable us to understand how atoms interact with each other and how such interactions affect the observed physical and chemical properties of matter.

In this chapter we introduce *quantum mechanics*, an extremely successful theory for explaining atomic structure. This theory, developed from 1925 to 1926 by Erwin Schrödinger, Werner Heisenberg, and others, addresses the limitations of the Bohr model and enables us to understand a host of phenomena involving atoms, molecules, nuclei, and solids. Basically, we shall be studying the equation of motion of matter waves, as well as some of the basic features of quantum mechanics and their application to simple one-dimensional systems. For example, we shall treat the problem of a particle confined to a potential well having infinitely high barriers.

41.1 ▶ THE DOUBLE-SLIT EXPERIMENT REVISITED

As we saw in Chapter 40, the concept of wave–particle duality in modern physics is very difficult to understand. One way to crystallize our ideas about this duality is to consider the diffraction of electrons passing through a double slit. This experiment shows the impossibility of *simultaneously* measuring wave and particle properties and embodies all the bizarre consequences of quantum mechanics.

Consider a beam of electrons all having the same energy and all incident on a double-slit barrier, as shown in Figure 41.1, where the slit widths are much less than the slit separation D. An electron detector is positioned far from the slits at a distance much greater than D. **If the detector detects electrons at different positions for a sufficiently long period of time, one finds an interference pattern representing the number of electrons arriving at any position along the detector line.** Such an interference pattern cannot occur if electrons behave as classical particles, and so we must infer that the electrons are behaving as waves. If the experiment is carried out at lower beam intensities over a long period of time, the interference pattern is still observed. At first, one observes only individual blips that are like photon "bullets" hitting in an apparently random pattern, but after long exposure a pattern of blips is observed. This is illustrated in the computer-simulated patterns in Figure 41.2. Note that the interference pattern becomes clearer as the number of electrons reaching the detector increases.

If a single electron produces in-phase waves as it reaches one of the slits, standard wave theory can be used to find the angular separation θ between the central probability maximum and its neighboring minimum. The minimum occurs when the path-length difference between paths A and B in Figure 41.1 is half a wavelength, or

$$D \sin \theta = \frac{\lambda}{2}$$

(a) After 28 electrons

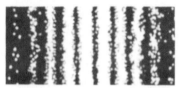

(b) After 1000 electrons

(c) After 10000 electrons

(d) Two-slit electron pattern

Figure 41.2   (a), (b), (c) Computer-simulated interference patterns for a beam of electrons incident on a double slit. *(From E. R. Huggins,* Physics I, *New York, W. A. Benjamin, 1968)*   (d) Photograph of a double-slit interference pattern produced by electrons. *(From C. Jönsson,* Zeitschrift für Physik *161:454, 1961; used with permission.)*

Wave function ψ

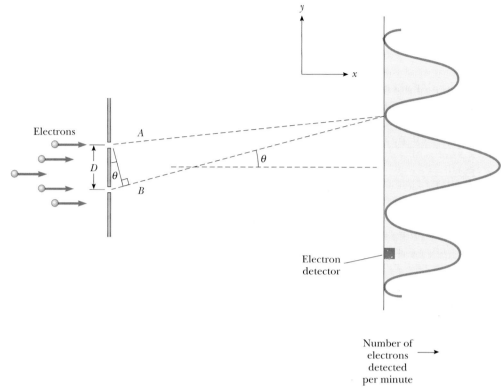

Figure 41.1   Electron diffraction. The slit separation D is much greater than the individual slit widths and much less than the distance between slits and detector.

Because the electron's de Broglie wavelength is given by $\lambda = h/p_x$, we see that, for small θ,

$$\sin\theta \approx \theta = \frac{h}{2p_x D}$$

Thus, the dual nature of electrons is clearly shown in this experiment: **Although the electrons are detected as particles at a localized spot at some instant of time, the probability of arrival at that spot is determined by the intensity of two interfering matter waves.**

In quantum mechanics, matter waves are described by the complex-valued **wave function** ψ. The absolute square $|\psi|^2 = \psi^*\psi$, where ψ^* is the complex conjugate of ψ, gives the probability of finding a particle at a given point at some instant. The wave function contains all the information that can be known about the particle.

Let us use the notion of the wave function to investigate some other unusual results from the double-slit experiment. If one slit is covered during the experiment, the result is a symmetric curve that peaks around the center of the open slit, much like the pattern formed by bullets shot through a hole in armor plate. The two overlapping blue curves in the center of Figure 41.3 are plots of electrons detected per minute with only one slit open. These curves are expressed as $|\psi_1|^2 = \psi_1^*\psi_1$ and $|\psi_2|^2 = \psi_2^*\psi_2$, where ψ_1 and ψ_2 represent the electron passing through slit 1 and slit 2, respectively.

If an experiment is performed with slit 2 blocked for the first half of the experiment and then slit 1 blocked during the remaining time, the accumulated pattern of electrons detected per minute, shown by the single blue curve on the right in Figure 41.3, is completely different from the pattern obtained with both slits open (red curve). In the single-slit curve, a maximum probability of arrival no longer occurs at $\theta = 0$. In fact, **the interference pattern has been lost, and the accumulated result is simply the sum of the individual results.** Because the electron must pass through either slit 1 or slit 2, it is just as localized and indivisible at the slits as it is when measured at the detector. Thus, the blue pattern on the right in Figure 41.3 must represent the sum of those electrons that come through slit 1, $|\psi_1|^2$, and those that come through slit 2, $|\psi_2|^2$.

When both slits are open, it is tempting to assume that the electron goes through either slit 1 or slit 2 and that the counts per minute are again given by $|\psi_1|^2 + |\psi_2|^2$. However, the experimental results, indicated by the red interference pattern in Figure 41.3, contradict this assumption. Thus, our assumption that the electron is localized and goes through only one slit when both slits are open must be wrong (a painful conclusion!). Somehow the wave property of the electron has a presence at both slits.

To find the probability of detecting the electron at a particular point at the detector when both slits are open, we may say that the electron is in a *superposition state,* given by

$$\psi = \psi_1 + \psi_2$$

Thus, the probability of detecting the electron at the detector is $|\psi_1 + \psi_2|^2$, and not $|\psi_1|^2 + |\psi_2|^2$. Because in general matter waves that start out in phase at the slits travel different distances to the detector, ψ_1 and ψ_2 have a relative phase difference ϕ at the detector. Using a phasor diagram (Fig. 41.4) to find $|\psi_1 + \psi_2|^2$ immediately yields

$$|\psi|^2 = |\psi_1 + \psi_2|^2 = |\psi_1|^2 + |\psi_2|^2 + 2|\psi_1||\psi_2|\cos\phi$$

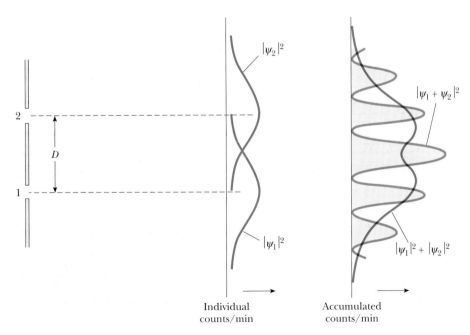

Individual
counts/min

Accumulated
counts/min

Figure 41.3 The two blue curves in the middle represent the patterns of the individual slits with the upper or lower slit closed. The single blue curve on the right represents the accumulated pattern of counts per minute when each slit is closed half the time. The red curve represents the diffraction pattern with both slits open at the same time.

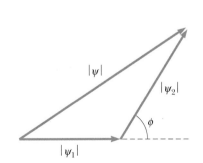

Figure 41.4 Phasor diagram to represent the addition of two complex quantities ψ_1 and ψ_2.

where $|\psi_1|^2$ is the probability of detection with slit 1 open and slit 2 closed, and $|\psi_2|^2$ is the probability of detection with slit 2 open and slit 1 closed. The term $2|\psi_1||\psi_2|\cos\phi$ in this expression is the interference term, which arises from the relative phase ϕ of the waves, in analogy with the phasor addition used in wave optics (see Chapter 37).

To interpret these results, we are forced to conclude that *an electron's wave property interacts with both slits simultaneously.* If we attempt to determine experimentally which slit the electron goes through, the simple act of measurement destroys the interference pattern. Thus, it is impossible to make such a determination. In effect, we can say only that **the electron passes through both slits!** The same arguments apply to photons.

Quick Quiz 41.1

Describe the signal from an electron detector as it is moved laterally far in front of three slits from which electrons with the same energy are being diffracted.

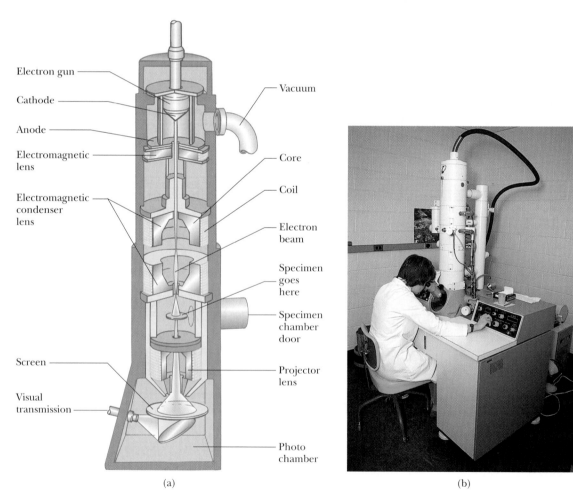

Electron gun

Cathode

Anode

Electromagnetic lens

Electromagnetic condenser lens

Screen

Visual transmission

Vacuum

Core

Coil

Electron beam

Specimen goes here

Specimen chamber door

Projector lens

Photo chamber

(a)

(b)

Figure 41.5   (a) Diagram of a transmission electron microscope for viewing a thinly sectioned sample. The "lenses" that control the electron beam are magnetic deflection coils. (b) An electron microscope. *(W. Ormerod/Visuals Unlimited)*

The Electron Microscope

A practical device that relies on the wave characteristics of electrons is the **electron microscope.** A *transmission* electron microscope, used for viewing flat, very thin samples, is shown in Figure 41.5. In many respects it is similar to an optical microscope, but the electron microscope has a much greater resolving power because it can accelerate electrons to very high kinetic energies, giving them very short wavelengths. No microscope can resolve details that are significantly smaller than the wavelength of the radiation used to illuminate the object. Typically, the wavelengths of electrons are about 100 times shorter than those of the visible light used in optical microscopes. As a result, an electron microscope with ideal lenses would be able to distinguish details about 100 times smaller than those distinguished by an optical microscope. (Radiation of the same wavelength as the electrons in an electron microscope is in the x-ray region of the spectrum.)

The electron beam in an electron microscope is controlled by electrostatic or magnetic deflection, which acts on the electrons to focus the beam to an image. Rather than examining the image through an eyepiece as in an optical microscope, the viewer looks at an image formed on a fluorescent screen. (The viewing screen must be fluorescent because otherwise the image produced would not be visible.)

A photograph taken by a *scanning* electron microscope, which operates in a somewhat different manner to reveal surface details of a three-dimensional sample, is shown at the beginning of the chapter.

41.2 ▶ THE UNCERTAINTY PRINCIPLE

If you were to measure the position and speed of a particle at any instant, you would always be faced with experimental uncertainties in your measurements. According to classical mechanics, no fundamental barrier to an ultimate refinement of the apparatus or experimental procedures exists. In other words, it is possible, in principle, to make such measurements with arbitrarily small uncertainty. Quantum theory predicts, however, that such a barrier does exist. In 1927, Werner Heisenberg (1901–1976) introduced this notion, which is now known as the **Heisenberg uncertainty principle:**

> If a measurement of position is made with precision Δx and a simultaneous measurement of linear momentum is made with precision Δp_x, then the product of the two uncertainties can never be smaller than $\hbar/2$:

$$\Delta x \, \Delta p_x \geq \frac{\hbar}{2} \qquad \text{(41.1)}$$

◀ Heisenberg uncertainty principle

where $\hbar = h/2\pi$. In other words, **it is physically impossible to measure simultaneously the exact position and exact linear momentum of a particle.** If Δx is very small, then Δp_x is large, and vice versa. Heisenberg was careful to point out that the inescapable uncertainties Δx and Δp_x do not arise from imperfections in measuring instruments. Rather, they arise from the quantum structure of matter—from effects such as the unpredictable recoil of an electron when struck by a photon or the diffraction of light or electrons passing through a small opening.

To understand the uncertainty principle, consider the following thought experiment introduced by Heisenberg: Suppose you wish to measure the position and linear momentum of an electron as accurately as possible. You might be able to do this by viewing the electron with a powerful light microscope. For you to see the electron and thus determine its location, at least one photon of light must bounce off the electron, as shown in Figure 41.6a, and then pass through the microscope into your eye, as shown in Figure 41.6b. When it strikes the electron, however, the photon transfers some unknown amount of its momentum to the electron. Thus, in the process of your locating the electron very accurately—that is, making Δx very small by using light with a short wavelength (and consequently a high momentum)—the very light that enables you to succeed changes the electron's momentum to some undeterminable extent (making Δp_x very great).

Let us analyze the collision by first noting that the incoming photon has momentum h/λ. As a result of the collision, the photon transfers part or all of its momentum along the x axis to the electron. Thus, the *uncertainty* in the electron's momentum after the collision is as great as the momentum of the incoming photon: $\Delta p_x = h/\lambda$. Furthermore, because the photon also has wave properties, we expect to be able to determine its position to within one wavelength of the light being used to view it, so $\Delta x = \lambda$. Multiplying these two uncertainties gives

$$\Delta x \, \Delta p_x = \lambda \left(\frac{h}{\lambda} \right) = h$$

This value h represents the minimum in the products of the uncertainties. Because the uncertainty can always be greater than this minimum, we have

$$\Delta x \, \Delta p_x \geq h$$

Apart from the numerical factor $1/4\pi$ introduced by Heisenberg's more precise analysis, this result agrees with Equation 41.1.

Quick Quiz 41.2

To determine the location of an electron, we can send it through a narrow slit. The narrower the slit, the more precisely we know the electron's location. Why does this fact not provide an escape from the limitations of the Heisenberg uncertainty principle?

The Heisenberg uncertainty principle enables us to better understand the dual wave–particle nature of light and matter. We have seen that the wave description of whatever entity we are studying is quite different from the particle description. Therefore, if an experiment (such as the photoelectric effect) is designed to reveal the particle character of, say, an electron, the electron's wave character becomes less apparent. If an experiment (such as diffraction from a crystal) is designed to measure the electron's wave properties, its particle character becomes less apparent.

Another uncertainty relationship sets a limit on the accuracy with which the energy of a system ΔE can be measured in a finite time interval Δt:

$$\Delta E \, \Delta t \geq \frac{\hbar}{2} \tag{41.2}$$

This relationship is plausible if a frequency measurement of any wave is considered. For example, consider measuring the frequency of a 1 000-Hz electromagnetic wave. If our frequency-measuring device has a fixed sensitivity of ± 1 cycle, in 1 s we measure a frequency of $(1\,000 \pm 1)$ cycles/1 s, but in 2 s we measure a frequency of $(2\,000 \pm 1)$ cycles/2 s. Thus, the uncertainty in frequency Δf is inversely proportional to Δt, the time interval during which the measurement is

Werner Heisenberg **German theoretical physicist (1901–1976)**
Heisenberg made many significant contributions to physics, including his famous uncertainty principle, for which he received a Nobel Prize in 1932; the development of an abstract model of quantum mechanics called matrix mechanics; the prediction of two forms of molecular hydrogen; and theoretical models of the nucleus.
(Courtesy of the University of Hamburg)

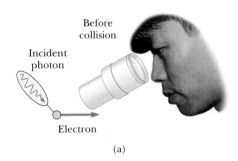

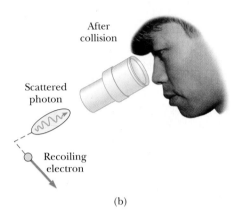

Figure 41.6 A thought experiment for viewing an electron with a powerful optical microscope. (a) The electron is moving to the right before colliding with the photon. (b) The electron recoils (its momentum changes) as a result of the collision with the photon.

made. This relationship may be stated as

$$\Delta f \Delta t \approx 1$$

Because all quantum systems are wave-like and can be described by the relationship $E = hf$, we may substitute $\Delta f = \Delta E / h$ into the preceding expression to obtain

$$\Delta E \, \Delta t \approx h$$

in basic agreement with Equation 41.2, apart from the factor of $1/4\pi$.

We conclude this section with examples of the types of calculations that can be performed with the uncertainty principle. These back-of-the-envelope calculations are surprising for their simplicity and for their essential description of quantum systems of which the details are unknown.

CONCEPTUAL EXAMPLE 41.1 ▶ Is the Bohr Model Realistic?

According to the Bohr model of the hydrogen atom, the electron in the ground state moves in a circular orbit of radius 0.529×10^{-10} m. In view of the Heisenberg uncertainty principle, is this model realistic?

Solution According to the uncertainty principle, the product $\Delta p_r \Delta r \geq \hbar/2$, where Δp_r is the uncertainty in the linear momentum of the electron in the radial direction. Let us calculate this uncertainty. The model specifies the radius of the circular orbit very precisely. When we quote the radius to three significant digits, we imply that the uncertainty in the radial position is at most $\Delta r \approx 0.000\,5 \times 10^{-10}$ m. The corresponding uncertainty in momentum of the electron in the radial direction is at least

$$\Delta p_r \approx \frac{\hbar}{2\Delta r} = \frac{1.05 \times 10^{-34}\,\text{J}\cdot\text{s}}{2(0.000\,5 \times 10^{-10}\,\text{m})} \approx 1 \times 10^{-21}\,\text{kg}\cdot\text{m/s}$$

The corresponding uncertainty in the radial speed of the electron (using a nonrelativistic calculation) is

$$\Delta v_r \approx \frac{\Delta p_r}{m_e} = \frac{1 \times 10^{-21}\,\text{kg}\cdot\text{m/s}}{9.1 \times 10^{-31}\,\text{kg}} \approx 1 \times 10^{9}\,\text{m/s}$$

A relativistic calculation would also give a large uncertainty in the speed. Because the uncertainty of the radial speed is on the order of ten times the speed of light, we must conclude that the Bohr model is not a reasonable description of the hydrogen atom!

EXAMPLE 41.2   Locating an Electron

The speed of an electron is measured to be 5.00×10^3 m/s to an accuracy of 0.003 00%. Find the minimum uncertainty in determining the position of this electron.

Solution   The momentum of the electron is

$$p_x = mv = (9.11 \times 10^{-31} \text{ kg})(5.00 \times 10^3 \text{ m/s})$$
$$= 4.56 \times 10^{-27} \text{ kg} \cdot \text{m/s}$$

The uncertainty in p_x is 0.003 00% of this value:

$$\Delta p_x = (0.000\ 030\ 0)(4.56 \times 10^{-27} \text{ kg} \cdot \text{m/s})$$
$$= 1.37 \times 10^{-31} \text{ kg} \cdot \text{m/s}$$

We can now calculate the minimum uncertainty in position by using this value of Δp_x and Equation 41.1:

$$\Delta x \, \Delta p_x \geq \frac{\hbar}{2}$$

$$\Delta x \geq \frac{\hbar}{2\Delta p_x} = \frac{1.05 \times 10^{-34} \text{ J} \cdot \text{s}}{2(1.37 \times 10^{-31} \text{ kg} \cdot \text{m/s})} = \boxed{0.383 \text{ mm}}$$

EXAMPLE 41.3   The Width of Spectral Lines

Although an excited atom can radiate at any time from $t = 0$ to $t = \infty$, the average time after excitation at which a group of atoms radiates is called the **lifetime** τ. (a) If $\tau = 1.0 \times 10^{-8}$ s, use the uncertainty principle to compute the line width Δf produced by this finite lifetime.

Solution   We use $\Delta E \, \Delta t \geq \hbar/2$, where $\Delta E = h \, \Delta f$ and $\Delta t = 1.0 \times 10^{-8}$ s is the average time available to measure the excited state. Thus, the minimum value of Δf is

$$\Delta f = \frac{1}{4\pi(1.0 \times 10^{-8} \text{ s})} = \boxed{8.0 \times 10^6 \text{ Hz}}$$

Note that ΔE is the uncertainty in the energy of the excited atom. It is also the uncertainty in the energy of the photon emitted by an atom in this state. (Note that in Bohr's theory, spectral lines should have vanishingly small line widths because the energy levels are precise.)

(b) If the wavelength of the spectral line involved in this process is 500 nm, what is the fractional broadening $\Delta f/f$?

Solution   First we find the frequency f of this line:

$$f = \frac{c}{\lambda} = \frac{3.00 \times 10^8 \text{ m/s}}{500 \times 10^{-9} \text{ m}} = 6.00 \times 10^{14} \text{ Hz}$$

Hence,

$$\frac{\Delta f}{f} = \frac{8.0 \times 10^6 \text{ Hz}}{6.00 \times 10^{14} \text{ Hz}} = \boxed{1.3 \times 10^{-8}}$$

This narrow natural line width can be seen with a sensitive interferometer. Usually, however, temperature and pressure effects overshadow the natural line width and broaden the line through mechanisms associated with the Doppler effect and collisions.

41.3   **PROBABILITY DENSITY**

Chapters 34, 37, and 40 revealed various aspects of light, and we can give a profound summary of the nature of light as follows: A photon is a quantum particle that has zero mass and transports energy and momentum as it moves as a wave of electric and magnetic fields. Its equation of motion is the wave equation for electromagnetic waves:

$$\frac{\partial^2 E}{\partial x^2} = \mu_0 \epsilon_0 \frac{\partial^2 E}{\partial t^2}$$

for the electric field, and a similar equation for the magnetic field. The intensity of the wave is proportional to the square of the electric field and is measured as the rate of photon bombardment at a detector.

     Our purpose in the current chapter is to give an analogous account of any material particle (one having nonzero mass). As noted in Section 41.1, the probability of finding a matter particle at a given point at some instant of time is given by $|\psi|^2$, the absolute square of a complex-valued wave function ψ. This wave function contains all the information that can be known about the particle. This interpretation of matter waves was first suggested by Max Born (1882–1970) in 1928. In 1926

Erwin Schrödinger (1887–1961) proposed a wave equation that describes how matter waves change in space and time. (The analogous propagation of electromagnetic waves is governed by Maxwell's equations.) The *Schrödinger equation* represents a key element in the theory of quantum mechanics.

A question arises quite naturally from the statement that matter has both a wave nature and a particle nature: If we are describing a particle, what do we conceive is waving? In the cases of waves on strings, water waves, and sound waves, the wave is represented by some quantity that varies with time and position. In a similar manner, the wave function ψ for matter waves depends on both the positions of all the particles in a system and on time, and therefore is often written $\psi(x, y, z, t)$. If ψ is known for a particle, then the particular properties of that particle can be described. In fact, the fundamental problem of quantum mechanics is this: Given the wave function at some instant, find the wave function at some later time t.

In Section 40.7 we found that the de Broglie equation relates the momentum of a particle to its wavelength through the relationship $p = h/\lambda$. If a free particle has a precisely known momentum, its wave function is a sinusoidal wave of wavelength $\lambda = h/p$, and the particle has equal probability of being at any point along the x axis. The wave function for such a free particle moving along the x axis can be written as

$$\psi(x) = A \sin\left(\frac{2\pi x}{\lambda}\right) = A \sin(kx) \qquad \textbf{(41.3)}$$

where $k = 2\pi/\lambda$ is the angular wave number and A is a constant amplitude. As we mentioned earlier, the wave function is generally a function of both position and time. Equation 41.3 represents the part of the wave function that depends on position only. For this reason, we can view $\psi(x)$ as a snapshot of the wave function at a given instant, as shown in Figure 41.7a. The wave function for a particle whose wavelength is not precisely defined is shown in Figure 41.7b. Because the wavelength is not precisely defined, it follows that the linear momentum is known only approximately. That is, if the momentum of the particle were measured, the result would have any value over some range, determined by the spread in wavelength. The greater the uncertainty in the momentum, the more the particle is localized. This is reflected in an increased probability density at the location of the particle.

Although we cannot measure ψ, we saw in Section 41.1 that we can measure $|\psi|^2$, a quantity that describes the probability of finding the particle at a particular location and at a certain time. To be more specific, if ψ represents a single particle, then $|\psi(x)|^2$—called the **probability density**—is the probability per unit volume that the particle will be found within an infinitesimal volume containing the point x. This interpretation, first suggested by Born in 1928, can also be stated in the following manner: If dV is a small volume element surrounding some point, then the probability of finding the particle in that volume element is $|\psi|^2 \, dV$. In this chapter we deal only with one-dimensional systems, in which the particle must be located along the x axis; thus, we replace dV with dx. In this case, the probability $P(x) \, dx$ that the particle will be found in the infinitesimal interval dx around the point x is

Probability density $|\psi|^2$

$$P(x) \, dx = |\psi|^2 \, dx$$

Because the particle must be somewhere along the x axis, the sum of the probabilities over all values of x must be 1:

$$\int_{-\infty}^{\infty} |\psi|^2 \, dx = 1 \qquad \textbf{(41.4)}$$

Normalization condition on ψ

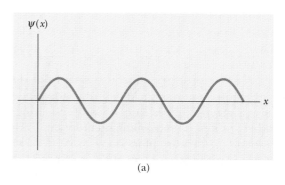

(a)

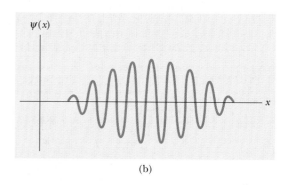

Figure 41.7   (a) Wave function for
a particle whose wavelength is precisely known. (b) Wave function for
a particle whose wavelength is not
precisely known and hence whose
momentum is known only over some
range of values.

(b)

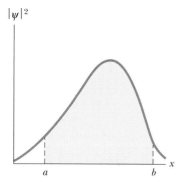

Figure 41.8   The probability of a
particle's being in the interval
$a \leq x \leq b$ is the area under the
curve from a to b.

Any wave function satisfying Equation 41.4 is said to be normalized and to fulfill
the **normalization condition.** Normalization is simply a statement that the particle exists at some point at all times. Therefore, although it is not possible to specify
the position of a particle with complete certainty, it is possible, through $|\psi|^2$, to
specify the probability of observing it at a given location. Furthermore, *the probability of finding the particle in the interval $a \leq x \leq b$ is*

$$P_{ab} = \int_a^b |\psi|^2 \, dx \qquad \textbf{(41.5)}$$

The probability P_{ab} is the area under the curve of probability density versus x between $x = a$ and $x = b$ in Figure 41.8.

Experimentally, a finite probability of finding a particle at some point and at
some instant always exists, so the value of the probability must lie between the limits 0 and 1. For example, if the probability density is 0.3 m^{-1} at some point, the
probability of finding the particle in some small interval Δx centered on that point
is 0.3 Δx.

The wave function ψ satisfies a wave equation, just as the electric field associated with an electromagnetic wave satisfies a wave equation that follows from
Maxwell's equations. The wave equation satisfied by ψ, which is the Schrödinger
equation, cannot be derived from any more fundamental laws, but ψ can be com-
puted from it. Although ψ itself cannot be measured, all measurable quantities of a
particle, such as its energy and linear momentum, can be derived from a knowledge of ψ. For example, once the wave function for a particle is known, it is possible to calculate the average position x of the particle, after many experimental trials. This average position is called the **expectation value** of x and is defined by

the equation

$$\langle x \rangle \equiv \int_{-\infty}^{\infty} x \, |\psi|^2 \, dx \qquad \textbf{(41.6)}$$

Expectation value of x

(Brackets $\langle \quad \rangle$ denote expectation values.) This expression implies that the particle is in a definite state, so the probability density is time-independent. Note that the expectation value is equivalent to the average value of x that would be obtained if we were dealing with a large number of particles in the same state. Furthermore, we can find the expectation value of any function $f(x)$ by using Equation 41.6 with x replaced by $f(x)$.

41.4 ▶ A PARTICLE IN A BOX

From a classical viewpoint, if a particle is confined to moving parallel to an x axis and to bouncing back and forth between two impenetrable walls (Fig. 41.9), its motion is easy to describe. If the speed of the particle is v, then the magnitude of its linear momentum (mv) remains constant, as does its kinetic energy. Furthermore, classical physics places no restrictions on the values of its momentum and energy. The quantum-mechanical approach to this problem is quite different and requires that we find the appropriate wave function consistent with the given conditions.

Before we address this problem, it is instructive to review the classical situation of standing waves on a stretched string (see Sections 18.2 and 18.3). If a string of length L is fixed at both ends, standing waves set up in the string must have nodes at the ends, as shown in Figure 41.10, because the wave function must vanish at the boundaries. Standing waves exist only when the length L of the string is some integral multiple of half-wavelengths. That is, we require that

$$L = n \frac{\lambda}{2}$$

or

$$\lambda = \frac{2L}{n} \qquad n = 1, 2, 3, \ldots$$

This result shows that *the wavelength for a standing wave on a string is quantized.*

As we saw in Section 18.2, each point on a standing wave oscillates with simple harmonic motion. Furthermore, all points oscillate with the same frequency, but the amplitude y of the simple harmonic motion of any particle in the medium differs from one point to the next and depends on how far a given point is from one end. We found that the position-dependent part of the wave function for a standing wave is

$$y(x) = A \sin(kx) \qquad \textbf{(41.7)}$$

where A is the maximum amplitude of the wave and $k = 2\pi/\lambda$. Because $\lambda = 2L/n$, we see that

$$k = \frac{2\pi}{\lambda} = \frac{2\pi}{2L/n} = n \frac{\pi}{L}$$

Substituting this result into Equation 41.7 gives

$$y(x) = A \sin\left(\frac{n\pi x}{L}\right)$$

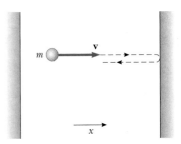

Figure 41.9 A particle of mass m and velocity **v** confined to moving parallel to the x axis and bouncing between two impenetrable walls.

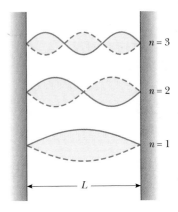

Figure 41.10 Standing waves set up in a stretched string of length L.

From this expression, we see that the wave function for a standing wave on a string meets the required boundary conditions—namely, that for all values of n, $y = 0$ at $x = 0$ and at $x = L$. The wave functions for $n = 1$, 2, and 3 are plotted in Figure 41.10.

Now let us return to the quantum-mechanical description of a particle in a box. Because the walls are impenetrable, the wave function $\psi(x) = 0$ for $x \le 0$ and for $x \ge L$, where L is now the distance between the two walls. This means that the particle can never be found outside the box. Furthermore, because the wave function must be continuous everywhere, we require that $\psi(0) = \psi(L) = 0$. Only wave functions that satisfy this condition are allowed. In analogy with standing waves on a string, the allowed wave functions for the particle in the box are sinusoidal and are given by

$$\psi(x) = A \sin\left(\frac{n\pi x}{L}\right) \qquad n = 1, 2, 3, \ldots \qquad \textbf{(41.8)}$$

where A is the maximum value of the wave function. This expression shows that, for a particle confined to a box and having a well-defined de Broglie wavelength, ψ is represented by a sinusoidal wave. The allowed wavelengths are those for which $L = n\lambda/2$. These allowed states of the system are called **stationary states** because they are standing waves.

Figure 41.11 shows plots of ψ versus x and $|\psi|^2$ versus x for $n = 1$, 2, and 3. As we shall soon see, these states correspond to the three lowest allowed energies for the particle. For $n = 1$, the probability of finding the particle is greatest at $x = L/2$—this is the *most probable position* for a particle in this state. For $n = 2$, $|\psi|^2$ is a maximum at $x = L/4$ and again at $x = 3L/4$; this means that both points are equally likely places for a particle in this state to be found.

There are also points within the box at which it is impossible to find the particle. For $n = 2$, $|\psi|^2$ is zero at the midpoint, $x = L/2$; for $n = 3$, $|\psi|^2 = 0$ at $x = L/3$ and $x = 2L/3$; and so on. But how does our particle get from one place

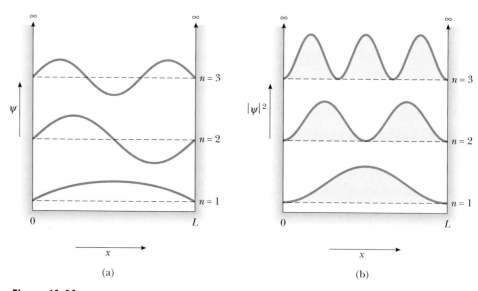

(a)   (b)

Figure 41.11   The first three allowed stationary states for a particle confined to a one-dimensional box. (a) The wave functions ψ for $n = 1$, 2, and 3. (b) The probability densities $|\psi|^2$ for $n = 1$, 2, and 3.

to another when no probability exists for its ever being at points between? This is one of the bizarre consequences of quantum mechanics—we must give up our notion that a particle moves from one point to another by occupying all intervening positions. In quantum mechanics, objects are not viewed as particles but as more complicated things having both particle *and* wave attributes.

Quick Quiz 41.3

Redraw Figure 41.11b, the probability of finding a particle at a particular location in a box, on the basis of classical mechanics instead of quantum mechanics.

Quick Quiz 41.4

(a) Make a sketch like Figure 41.11b for $n = 20$. Imagine placing within the confines of the box a detector that samples the probability of finding a particle within some narrow limits Δx. What would the detector measure as n approached infinity?

Because the wavelengths of the particle are restricted by the condition $\lambda = 2L/n$, the magnitude of the linear momentum is restricted to the values

$$p = \frac{h}{\lambda} = \frac{h}{2L/n} = \frac{nh}{2L}$$

The potential energy is constant inside the box, and it is convenient to set it at $U = 0$. Therefore, the total energy of the particle is equal to its kinetic energy. Using $p = mv$, we find that the allowed values of the energy are

$$E_n = \frac{1}{2}mv^2 = \frac{p^2}{2m} = \frac{(nh/2L)^2}{2m}$$

$$E_n = \left(\frac{h^2}{8mL^2}\right)n^2 \qquad n = 1, 2, 3, \ldots \qquad \textbf{(41.9)}$$

As we see from this expression, *the energy of the particle is quantized,* as we would expect. The lowest allowed energy corresponds to $n = 1$, for which $E_1 = h^2/8mL^2$. Because $E_n = n^2E_1$, the excited states corresponding to $n = 2, 3, 4, \ldots$ have energies given by $4E_1, 9E_1, 16E_1, \ldots$. Figure 41.12 is an energy level diagram describing the positions of the allowed states. Note that the state $n = 0$ is not allowed. This means that, according to quantum mechanics, the particle can never be at rest. The least energy the particle can have, corresponding to $n = 1$, is called the **zero-point energy.** This result clearly contradicts the classical viewpoint, in which $E = 0$ is an acceptable state. In our quantum-mechanical analysis, only nonzero positive values of E are allowed because the total energy E equals the kinetic energy and the potential energy is zero.

Energy levels are of special importance for the following reason. If the particle is electrically charged, it can emit a photon when it drops from an excited state, such as E_3, to a lower-lying state, such as E_2. It can also absorb a photon whose energy matches the difference in energy between two allowed states. For example, if the photon frequency is f, the particle jumps from state E_1 to state E_2 if $hf = E_2 - E_1$. As noted in Chapter 40, photon emission and absorption can be observed by spectroscopy, and spectral wavelengths are a direct measurement of such energy differences.

Allowed energies for a particle in a box

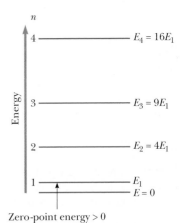

Zero-point energy > 0

Figure 41.12 Energy level diagram for a particle confined to a one-dimensional box of width L. The lowest allowed energy is $E_1 = h^2/8mL^2$.

EXAMPLE 41.4 ▶ A Bound Electron

An electron is confined between two impenetrable walls 0.200 nm apart. Determine the energy levels for the states $n = 1, 2,$ and 3.

Solution   For the state $n = 1$, Equation 41.9 gives

$$E_1 = \frac{h^2}{8mL^2} = \frac{(6.63 \times 10^{-34} \, \text{J} \cdot \text{s})^2}{8(9.11 \times 10^{-31} \, \text{kg})(2.00 \times 10^{-10} \, \text{m})^2}$$

$$= 1.51 \times 10^{-18} \, \text{J} = \boxed{9.42 \text{ eV}}$$

For $n = 2$ and $n = 3$, $E_2 = 4E_1 = 37.7 \text{ eV}$ and $E_3 = 9E_1 = 84.8 \text{ eV}$. Although this is a rather primitive model, it can be used to describe an electron trapped in a vacant crystal site.

EXAMPLE 41.5 ▶ Energy Quantization for a Macroscopic Object

A 1.00-mg object is confined to moving between two rigid walls separated by 1.00 cm. Calculate the minimum speed of the object.

Solution   The minimum speed corresponds to the state for which $n = 1$. Using Equation 41.9 with $n = 1$ gives the zero-point energy:

$$E_1 = \frac{h^2}{8mL^2} = \frac{(6.63 \times 10^{-34} \, \text{J} \cdot \text{s})^2}{8(1.00 \times 10^{-6} \, \text{kg})(1.00 \times 10^{-2} \, \text{m})^2}$$

$$= 5.49 \times 10^{-58} \, \text{J}$$

Because $E = K = \frac{1}{2}mv^2$, we have

$$\tfrac{1}{2}mv^2 = 5.49 \times 10^{-58} \, \text{J}$$

$$v = \left[\frac{2(5.49 \times 10^{-58} \, \text{J})}{1.00 \times 10^{-6} \, \text{kg}} \right]^{1/2} = \boxed{3.31 \times 10^{-26} \, \text{m/s}}$$

This speed is so small that the object can be considered to be at rest, which is what one would expect for the minimum speed of a macroscopic object.

Exercise   If the speed of the particle is 3.00 cm/s, find its energy and the value of n that corresponds to this energy.

Answer 4.50×10^{-10} J; $n = 9.05 \times 10^{23}$. (Note that for values of n this great, we would never be able to distinguish the quantized nature of the energy levels because the difference between the $n = 9.05 \times 10^{23}$ and $n + 1 = 9.05 \times 10^{23} + 1$ levels is so small.)

EXAMPLE 41.6 ▶ Model of an Atom

An atom can be viewed as several electrons moving around a positively charged nucleus, where the electrons are subject mainly to the electrical attraction of the nucleus. (This attraction is partially "screened" by the inner-core electrons and is therefore diminished.) Figure 41.13 represents the potential energy of the electron as a function of r. (a) Use the simple model of a particle in a box to *estimate* the energy (in electron volts) required to raise an electron from the state $n = 1$ to the state $n = 2$, assuming that the atom has a radius of 0.100 nm.

Solution   Using Equation 41.9 and taking the length L of the box to be 0.200 nm (the diameter of the atom) and $m = 9.11 \times 10^{-31}$ kg, we find that, as in Example 41.4,

$$E_n = \left(\frac{h^2}{8mL^2} \right) n^2$$

$$= \frac{(6.63 \times 10^{-34} \, \text{J} \cdot \text{s})^2}{8(9.11 \times 10^{-31} \, \text{kg})(2.00 \times 10^{-10} \, \text{m})^2} \, n^2$$

$$= (1.51 \times 10^{-18}) n^2 \, \text{J} = 9.42 n^2 \, \text{eV}$$

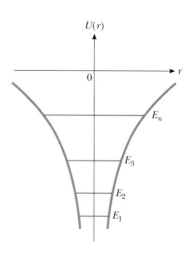

Figure 41.13   Model of potential energy versus r for an atom.

Hence, the energy difference between the states $n = 1$ and $n = 2$ is

$$\Delta E = E_2 - E_1 = 9.42(2)^2 \text{ eV} - 9.42(1)^2 \text{ eV} = \boxed{28.3 \text{ eV}}$$

(b) Calculate the wavelength of the photon that would cause this transition.

Solution   Using the fact that $\Delta E = hc/\lambda$, we obtain

$$\lambda = \frac{hc}{\Delta E} = \frac{(6.63 \times 10^{-34} \text{ J} \cdot \text{s})(3.00 \times 10^{8} \text{ m/s})}{(28.3 \text{ eV} \times 1.60 \times 10^{-19} \text{ J/eV})}$$

$$= 4.39 \times 10^{-8} \text{ m} = \boxed{43.9 \text{ nm}}$$

This wavelength is in the far ultraviolet region, and it is interesting to note that the result is roughly correct. Although this oversimplified model gives a good estimate for transitions between lowest-lying levels of the atom, the estimate becomes progressively worse for higher-energy transitions.

41.5 THE SCHRÖDINGER EQUATION

As we mentioned earlier, the wave function for de Broglie waves must satisfy an equation developed by Schrödinger. One of the methods of quantum mechanics is to determine a solution to this equation, which in turn yields the allowed wave functions and energy levels of the system under consideration. Proper manipulation of the wave functions enables calculation of all measurable features of the system.

In Section 16.9 we derived Equation 16.26, the general form of the wave equation for waves traveling along the x axis:

$$\frac{\partial^2 y}{\partial x^2} = \frac{1}{v^2} \frac{\partial^2 y}{\partial t^2} \qquad \textbf{(41.10)}$$

where v is the wave speed and the variable y depends on x and t. Matter waves are more complicated and do not obey this wave equation.

The wave function for a particle confined to one dimension is

$$\Psi(x, t) = \psi(x) \, e^{-i\omega t} \qquad \textbf{(41.11)}$$

where ω is the angular frequency of the matter wave and $\Psi(x, t)$ represents the full time-dependent and space-dependent wave function. In our investigations, we shall need to focus only on $\psi(x)$, the spatial part of the wave function. This function satisfies the equation

$$\frac{d^2 \psi}{dx^2} = -\frac{2m}{\hbar^2} (E - U)\psi \qquad \textbf{(41.12)}$$

This is the famous **Schrödinger equation** as it applies to a particle confined to moving along the x axis. Because this equation is independent of time, it is commonly referred to as the *time-independent Schrödinger equation*. (We shall not discuss the time-dependent Schrödinger equation in this text.)

In principle, if the potential energy $U(x)$ is known for the system, we can solve Equation 41.12 and obtain the wave functions and energies for the allowed states. Because U may vary with position, it may be necessary to solve the equation in pieces. In the process, the wave functions for the different regions must join smoothly at the boundaries. In the language of mathematics, we require that $\psi(x)$ be *continuous*. Furthermore, for $\psi(x)$ to obey the normalization condition (see text following Equation 41.4), we require that $\psi(x)$ approach zero as x approaches

Erwin Schrödinger **Austrian theoretical physicist (1887–1961)** Schrödinger is best known as the creator of quantum mechanics. He also produced important papers in the fields of statistical mechanics, color vision, and general relativity. Schrödinger did much to hasten the universal acceptance of quantum theory by demonstrating the mathematical equivalence between his quantum mechanics and the more abstract matrix mechanics developed by Heisenberg.

Time-independent Schrödinger equation

Required conditions for $\psi(x)$

$\pm \infty$. Finally, $\psi(x)$ must be *single-valued*, and $d\psi/dx$ must also be continuous for finite values of $U(x)$.

The task of solving the Schrödinger equation may be very difficult, depending on the form of the potential energy function. As it turns out, the Schrödinger equation has been extremely successful in explaining the behavior of atomic and nuclear systems, whereas classical physics has failed to explain this behavior. Furthermore, when quantum mechanics is applied to macroscopic objects, the results agree with classical physics, as required by the correspondence principle.

The Particle in a Box Revisited

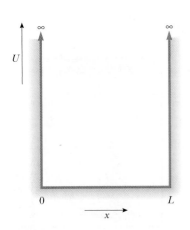

Figure 41.14   A one-dimensional box of width L and infinitely high walls.

Let us solve the Schrödinger equation for our particle in a one-dimensional box of width L (Fig. 41.14). The walls are infinitely high, corresponding to $U(x) = \infty$ for $x = 0$ and $x = L$. The potential energy is constant within the box, and it is again convenient to choose $U = 0$ as its value. Hence, in the region $0 < x < L$, we can express the Schrödinger equation in the form

$$\frac{d^2\psi}{dx^2} = -\frac{2mE}{\hbar^2}\psi = -k^2\psi \qquad \textbf{(41.13)}$$

where

$$k = \frac{\sqrt{2mE}}{\hbar}$$

Because the walls are infinitely high, the particle cannot exist outside the box. The particle is permanently bound in the box and cannot be found outside the interval $0 < x < L$. Consequently, $\psi(x)$ must be zero outside the box and at the walls. The solution of Equation 41.13 that meets the boundary conditions $\psi(x) = 0$ at $x = 0$ and $x = L$ is

$$\psi(x) = A\sin(kx) \qquad \textbf{(41.14)}$$

This can easily be verified by substitution into Equation 41.13. Note that the first boundary condition, $\psi(0) = 0$, is satisfied by Equation 41.14 because $\sin 0 = 0$. The second boundary condition, $\psi(L) = 0$, is satisfied only if kL is an integer multiple of π—that is, if $kL = n\pi$, where n is an integer. Because $k = \sqrt{2mE}/\hbar$, we see that

$$kL = \frac{\sqrt{2mE}}{\hbar}L = n\pi$$

Solving for the allowed energies E gives

$$E_n = \left(\frac{h^2}{8mL^2}\right)n^2$$

Likewise, the allowed wave functions are given by

$$\psi_n(x) = A\sin\left(\frac{n\pi x}{L}\right)$$

These results agree with those obtained in the previous section (Eqs. 41.8 and 41.9). It is left as a problem (Problem 25) to show that the normalization constant A for this solution is equal to $(2/L)^{1/2}$.

Optional Section

41.6 ▸ A PARTICLE IN A WELL OF FINITE HEIGHT

Consider a particle having a potential energy that is zero in the region $0 < x < L$—which we call a *potential well*—and has a finite value U outside this region, as in Figure 41.15. If the total energy E of the particle is less than U, classically the particle is permanently bound in the potential well. However, according to quantum mechanics, a finite probability exists that the particle can be found outside the well even if $E < U$. That is, the wave function ψ is generally nonzero outside the well—in regions I and III in Figure 41.15—so the probability density $|\psi|^2$ is also nonzero in these regions.

In region II, where $U = 0$, the allowed wave functions are again sinusoidal because they represent solutions of Equation 41.13. However, the boundary conditions no longer require that ψ be zero at the walls, as was the case with infinitely high walls.

The Schrödinger equation for regions I and III may be written

$$\frac{d^2\psi}{dx^2} = \frac{2m(U - E)}{\hbar^2}\,\psi \tag{41.15}$$

Because $U > E$, the coefficient on the right-hand side is necessarily positive. Therefore, we can express Equation 41.15 in the form

$$\frac{d^2\psi}{dx^2} = C^2\psi \tag{41.16}$$

where $C^2 = 2m(U - E)/\hbar^2$ is a positive constant in regions I and III. As you can verify by substitution, the general solution of Equation 41.16 is

$$\psi = Ae^{Cx} + Be^{-Cx}$$

where A and B are constants.

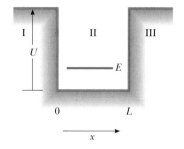

Figure 41.15 Potential-energy diagram of a well of finite height U and width L. The total energy E of the particle is less than U.

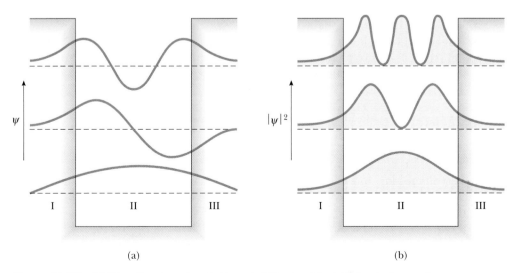

Figure 41.16 (a) Wave functions ψ and (b) probability densities $|\psi|^2$ for the lowest three energy states for a particle in a potential well of finite height.

We can use this general solution as a starting point for determining the appropriate solution for regions I and III. The function we choose for our solution must remain finite over the entire region under consideration. In region I, where $x < 0$, we must rule out the term Be^{-Cx}. In other words, we must require that $B = 0$ in region I to avoid an infinite value for ψ for large negative values of x. Likewise, in region III, where $x > L$, we must rule out the term Ae^{Cx}; this is accomplished by taking $A = 0$ in this region. This choice avoids an infinite value for ψ for large positive x values. Hence, the solutions in regions I and III are

$$\psi_{I} = Ae^{Cx} \qquad \text{for } x < 0$$

$$\psi_{III} = Be^{-Cx} \qquad \text{for } x > L$$

In region II the wave function is sinusoidal and has the general form

$$\psi_{II}(x) = F\sin(kx) + G\cos(kx)$$

where F and G are constants.

These results show that the wave functions outside the potential well (where classical physics forbids the presence of the particle) decay exponentially with distance. At great negative x values, ψ_{I} approaches zero exponentially; at great positive x values, ψ_{III} approaches zero exponentially. These functions, together with the sinusoidal solution in region II, are shown in Figure 41.16a for the first three energy states. In evaluating the complete wave function, we require that

$$\psi_{I} = \psi_{II} \qquad \text{and} \qquad \frac{d\psi_{I}}{dx} = \frac{d\psi_{II}}{dx} \qquad \text{at } x = 0$$

$$\psi_{II} = \psi_{III} \qquad \text{and} \qquad \frac{d\psi_{II}}{dx} = \frac{d\psi_{III}}{dx} \qquad \text{at } x = L$$

Figure 41.16b plots the probability densities for these states. Note that in each case the inside and outside wave functions join smoothly at the boundaries of the potential well. These boundary conditions and plots follow from the Schrödinger equation.

Figure 41.16a shows that the wave functions ψ are not equal to zero at the walls of the potential well and in the exterior regions. Therefore, the probability densities $|\psi|^2$ are nonzero at these points. The fact that ψ is nonzero at the walls increases the de Broglie wavelength in region II (compare the case of a particle in a potential well of infinite depth; see Fig. 41.11), and this, in turn, lowers the energy and linear momentum of the particle.

Optional Section

41.7 ▶ TUNNELING THROUGH A BARRIER

A very interesting and peculiar phenomenon occurs when a particle strikes a barrier of finite height and width. Consider a particle of energy E incident on a rectangular barrier of height U and width L, where $E < U$ (Fig. 41.17). Classically, the particle is reflected because it does not have sufficient energy to cross or even penetrate the barrier. Thus, regions II and III are classically *forbidden* to the particle.

According to quantum mechanics, however, **all regions are accessible to the particle, regardless of its energy,** because the amplitude of the de Broglie matter wave associated with the particle is nonzero everywhere. A typical waveform for this case, illustrated in Figure 41.17, shows the wave penetrating into the barrier and beyond. The wave functions are sinusoidal to the left (region I) and right (region III) of the barrier and join smoothly with an exponentially decaying function within the barrier (region II).

Because the probability of locating the particle is proportional to $|\psi|^2$, which is nonzero within the barrier and beyond, we conclude that the particle may be found in region III. This barrier penetration is in complete contradiction to classical physics. The possibility of finding the particle on the far side of the barrier is called **tunneling** or **barrier penetration.** A detailed analysis shows that if tunneling takes place, the barrier must be sufficiently narrow that the time of passage Δt is very short. Then the uncertainty in energy $\Delta E \leq \hbar/2\Delta t$ is so great that we cannot actually attribute negative kinetic energy to the tunneling particle.

Hard as it is to believe, there is a finite (although *extremely* small) possibility that a marble placed inside a shoebox will suddenly appear outside the box (see Problem 47)! However, you would probably have to wait through several life spans of the Universe to observe this.

The probability of tunneling can be described with a transmission coefficient T and a reflection coefficient R. The transmission coefficient is the probability that the particle passes through the barrier, and the reflection coefficient is the probability that the particle is reflected by the barrier. Because the incident particle is either reflected or transmitted, we must require that $T + R = 1$. An approximate expression for T when $T \ll 1$ (a very high or very wide barrier) is

$$T \approx e^{-2CL} \qquad \textbf{(41.17)}$$

where

$$C = \frac{\sqrt{2m(U - E)}}{\hbar} \qquad \textbf{(41.18)}$$

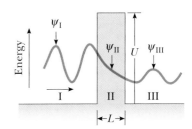

Figure 41.17 Wave function ψ for a particle incident from the left on a barrier of height U. The wave function is sinusoidal in regions I and III but exponentially decaying in region II.

EXAMPLE 41.7 ▶ Transmission Coefficient for an Electron

A 30-eV electron is incident on a barrier whose cross-section is a square of height 40 eV. What is the probability that the electron will tunnel through the barrier if its thickness is (a) 1.0 nm? (b) 0.10 nm?

Solution (a) In this situation, the quantity $U - E$ has the value

$$U - E = (40 \text{ eV} - 30 \text{ eV}) = 10 \text{ eV} = 1.6 \times 10^{-18} \text{ J}$$

Using Equation 41.18, we find that

$$2CL = 2 \frac{\sqrt{2(9.11 \times 10^{-31} \text{ kg})(1.6 \times 10^{-18} \text{ J})}}{1.054 \times 10^{-34} \text{ J} \cdot \text{s}}$$
$$\times (1.0 \times 10^{-9} \text{ m}) = 32.4$$

Thus, the probability of tunneling through the barrier is

$$T \approx e^{-2CL} = e^{-32.4} = \boxed{8.5 \times 10^{-15}}$$

The electron has about 1 chance in 10^{14} of tunneling through the 1.0-nm thickness.

(b) For $L = 0.10$ nm, $2CL = 3.24$, and

$$T \approx e^{-2CL} = e^{-3.24} = \boxed{0.039}$$

Now the electron has a high probability (a 4% chance) of penetrating the barrier. Thus, reducing the thickness of the barrier by only one order of magnitude increases the probability of tunneling by about 12 orders of magnitude!

Some Applications of Tunneling

As we have seen, tunneling is a quantum phenomenon, a manifestation of the wave nature of matter. Many examples for which tunneling is very important exist in nature, on the atomic and nuclear scales.

- **Tunnel diode**   The tunnel diode is a semiconductor device consisting of two oppositely charged regions separated by a very narrow electrically neutral region. The electric current (basically, the rate of tunneling) can be controlled over a wide range by varying the potential difference across the charged regions, which is equivalent to changing the height of the barrier.

- **Josephson junction**   The Josephson junction consists of two superconductors separated by a thin insulating oxide layer, 1 to 2 nm thick. Under appropriate conditions, electrons in the superconductors travel as pairs and tunnel from one superconductor to the other through the oxide layer. Several effects have been observed in this type of junction. For example, a direct current is observed across the junction *in the absence of electric or magnetic fields.* The current is proportional to sin ϕ, where ϕ is the phase difference between the wave functions in the two superconductors. When a potential difference ΔV is applied across the junction, the current oscillates with a frequency $f = 2e\,\Delta V/h$, where e is the charge on the electron.

- **Alpha decay**   One form of radioactive decay is the emission of alpha particles (the nuclei of helium atoms) by unstable heavy nuclei. To escape from the nucleus, an alpha particle must penetrate a barrier that arises from the combination of the attractive nuclear force and the electrical repulsion between the positively charged alpha particle and the rest of the (positively charged) nucleus. Occasionally an alpha particle tunnels through the barrier, which explains the basic mechanism for this type of decay and the great variations in the mean lifetimes of various radioactive nuclei.

- **Solar energy**   According to classical physics, positively charged hydrogen ions in the Sun cannot overcome their mutual repulsion and penetrate the barrier caused by electrical repulsion. Quantum-mechanically, however, the ions are able to tunnel through the barrier and fuse together to form helium. This is the basic reaction that powers the Sun and, indirectly, almost everything else in the Solar System.

Figure 41.18   A quantum corral consisting of a ring of 48 iron atoms located on a copper surface. The diameter of the ring is 143 nm, and the photograph was obtained using a low-temperature scanning tunneling microscope. Corrals and other structures can confine electron waves. The study of such structures will play an important role in determining the future of small electronic devices. *(IBM Corporation Research Division)*

- **Quantum traps**   Scientists are beginning to experiment with *quantum dots* that trap a single electron and *quantum corrals* made up of a small number of atoms, as shown in Figure 41.18. Such tiny traps may eventually be put to use in electronic devices.
- **Scanning tunneling microscopes,** discussed in Section 41.8.

Optional Section

41.8 ▶ THE SCANNING TUNNELING MICROSCOPE[1]

One of the basic phenomena of quantum mechanics—tunneling—is at the heart of a very practical device, the scanning tunneling microscope (STM), which enables us to obtain highly detailed images of surfaces at resolutions comparable to the size of a *single atom.* Figures 41.18 (iron atoms on copper) and 41.19 (the surface of a piece of graphite) show what the STM can do. Note the high quality of the images and the discernible rings of carbon atoms in Figure 41.19. What makes this image so remarkable is that its resolution is about 0.2 nm. For an optical microscope, the resolution is limited by the wavelength of the light used to make the image. Thus, an optical microscope has a resolution no better than 200 nm, about half the wavelength of visible light, and so could never show the detail displayed in Figure 41.19. An ideal electron microscope (Section 41.1) could have a resolution of 0.2 nm by using electron waves of this wavelength, given by the de Broglie formula $\lambda = h/p$. The linear momentum p of an electron required to give this wavelength is 10 000 eV/c, corresponding to an electron speed of 2% of the speed of light. Electrons traveling at this speed would penetrate into the interior of the graphite in Figure 41.19 and thus could not give us information about individual surface atoms.

The STM achieves its very fine resolution by using the basic idea shown in Figure 41.20. An electrically conducting probe with a very sharp tip is brought near the surface to be studied. The empty space between tip and surface represents the "barrier" we have been discussing, and the tip and surface are the two walls of the "potential well." Because electrons obey quantum rules rather than Newtonian rules, they can "tunnel" across the barrier of empty space. If a voltage is applied between surface and tip, electrons in the atoms of the surface material can be made to tunnel preferentially from surface to tip to produce a tunneling current. In this way the tip samples the distribution of electrons just above the surface.

Because of the nature of tunneling, the STM is very sensitive to the distance z from tip to surface—in other words, to the thickness of the barrier (see Example 41.7). The reason is that in the empty space between tip and surface, the electron wave function falls off exponentially (see Fig. 41.17, region II) with a decay length of order 0.1 nm; that is, the wave function decreases by $1/e$ over that distance. For distances $z > 1$ nm (that is, beyond a few atomic diameters), essentially no tunneling takes place. This exponential behavior causes the current of electrons tunneling from surface to tip to depend very strongly on z. This sensitivity is the basis of the operation of the STM: By monitoring the tunneling current as the tip is scanned over the surface, scientists obtain a sensitive measure of the topography of the electron distribution on the surface. The result of this scan is used to make images like that in Figure 41.19. In this way the STM can measure the height of surface features to within 0.001 nm, approximately 1/100 of an atomic diameter!

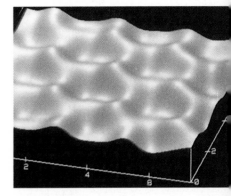

Figure 41.19 The surface of graphite as "viewed" with a scanning tunneling microscope. This type of microscope enables scientists to see details with a lateral resolution of about 0.2 nm and a vertical resolution of 0.001 nm. The contours seen here represent the ring-like arrangement of individual carbon atoms on the crystal surface.

---

[1] This section was written by Roger A. Freedman and Paul K. Hansma, University of California—Santa Barbara.

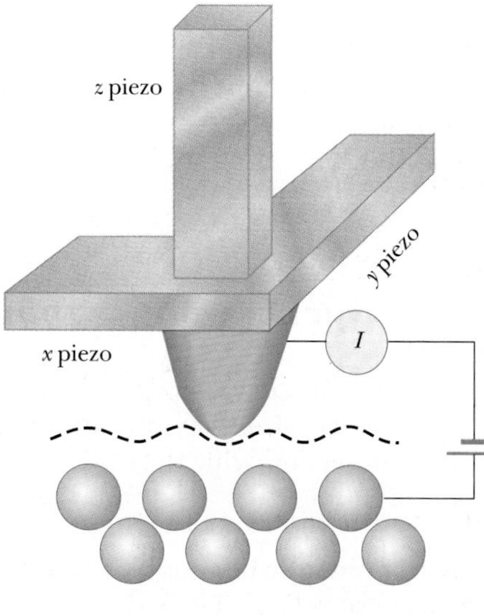

Figure 41.20 Schematic view of an STM. The tip, shown as a rounded cone, is mounted on a piezoelectric *xyz* scanner. A scan of the tip over the sample can reveal contours of the surface down to the atomic level. An STM image is composed of a series of scans displaced laterally from one another. *(Based on a drawing from P. K. Hansma, V. B. Elings, O. Marti, and C. Bracker,* Science *242:209, 1988. Copyright 1988 by the AAAS.)*

You can see just how sensitive the STM is by examining Figure 41.19. Of the six carbon atoms in each ring, three appear lower than the other three. In fact, all six atoms are at the same level, but all have slightly different electron distributions. The three atoms that appear lower are bonded to other carbon atoms directly beneath them in the underlying atomic layer; as a result, their electron distributions, which are responsible for the bonding, extend downward beneath the surface. The atoms in the surface layer that appear higher do not lie directly over subsurface atoms and hence are not bonded to any underlying atoms. For these higher-appearing atoms, the electron distribution extends upward into the space above the surface. This extra electron density is what makes these electrons appear higher in Figure 41.19, because what the STM maps is the topography of the electron distribution.

The STM has, however, one serious limitation: It depends on the electrical conductivity of the sample and the tip. Unfortunately, most materials are not electrically conductive at their surfaces. Even metals, which are usually excellent electrical conductors, are covered with nonconductive oxides. A newer microscope, the atomic force microscope (AFM), overcomes this limitation. It measures the electric force acting between a tip and the sample, rather than an electric current. This force, which typically is a result of the exclusion principle, depends very strongly on the tip–sample separation distance, just as the electron tunneling current does in an STM. Thus, the AFM has comparable sensitivity for measuring topography and has become widely used for technological applications.

Perhaps the most remarkable thing about the STM is that its operation is based on a quantum-mechanical phenomenon—tunneling—that was well understood in the 1920s, even though the first STM was not built until the 1980s. What other applications of quantum mechanics may yet be waiting to be discovered?

Optional Section

41.9 THE SIMPLE HARMONIC OSCILLATOR

Finally, let us consider the problem of a particle that is subject to a linear restoring force $F = -kx$, where x is the magnitude of the displacement of the particle from equilibrium ($x = 0$) and k is the force constant. (This is an important situation to understand because the forces between atoms in a solid can be approximated by this type of interaction.) The classical motion of a particle subject to such a force is simple harmonic motion, which was discussed in Chapter 13. The potential energy of the system is, from Equation 13.21,

$$U = \tfrac{1}{2}kx^2 = \tfrac{1}{2}m\omega^2x^2$$

where the angular frequency of vibration is $\omega = \sqrt{k/m}$. Classically, if the particle is displaced from its equilibrium position and released, it oscillates between the points $x = -A$ and $x = A$, where A is the amplitude of the motion. Furthermore, its total energy E is, from Equation 13.22,

$$E = K + U = \tfrac{1}{2}kA^2 = \tfrac{1}{2}m\omega^2A^2$$

In the classical model, any value of E is allowed, including $E = 0$, which is the total energy when the particle is at rest at $x = 0$.

The Schrödinger equation for this problem is obtained by substituting $U = \tfrac{1}{2}m\omega^2x^2$ into Equation 41.12:

$$\frac{d^2\psi}{dx^2} = -\left[\left(\frac{2mE}{\hbar^2}\right) - \left(\frac{m\omega}{\hbar}\right)^2 x^2\right]\psi \qquad \textbf{(41.19)}$$

The mathematical technique for solving this equation is beyond the level of this text. However, it is instructive to guess at a solution. We take as our guess the following wave function:

$$\psi = Be^{-Cx^2} \qquad \textbf{(41.20)}$$

Substituting this function into Equation 41.19, we find that it is a satisfactory solution to the Schrödinger equation, provided that

$$C = \frac{m\omega}{2\hbar} \qquad \text{and} \qquad E = \tfrac{1}{2}\hbar\omega$$

It turns out that the solution we have guessed corresponds to the ground state of the system, which has an energy $\tfrac{1}{2}\hbar\omega$, the zero-point energy of the system. Because $C = m\omega/2\hbar$, it follows from Equation 41.20 that the wave function for this state is

$$\psi = Be^{-(m\omega/2\hbar)x^2} \qquad \textbf{(41.21)}$$

This is only one solution to Equation 41.19. The remaining solutions that describe the excited states are more complicated, but all solutions include the exponential factor e^{-Cx^2}.

The energy levels of a harmonic oscillator are quantized, as we would expect when we use quantum mechanics to analyze the situation. The energy of the state for which the quantum number is n is

$$E_n = (n + \tfrac{1}{2})\hbar\omega \qquad n = 0, 1, 2, \ldots$$

The state $n = 0$ corresponds to the ground state, where $E_0 = \tfrac{1}{2}\hbar\omega$; the state $n = 1$ corresponds to the first excited state, where $E_1 = \tfrac{3}{2}\hbar\omega$; and so on. The energy level diagram for this system is shown in Figure 41.21. Note that the separations between adjacent levels are equal and are given by

$$\Delta E = \hbar\omega \qquad \textbf{(41.22)}$$

The red curves in Figure 41.22 indicate the probability densities $|\psi|^2$ for the first three states of a simple harmonic oscillator. The blue curves represent the classical probability densities that correspond to the same energy and are provided

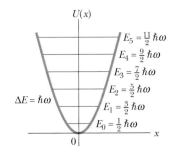

Figure 41.21 Energy level diagram for a simple harmonic oscillator. The levels are equally spaced, with separation $\hbar\omega$. The zero-point energy is $E_0 = \tfrac{1}{2}\hbar\omega$.

Wave function for the ground state of a simple harmonic oscillator

Allowed energies for a simple harmonic oscillator

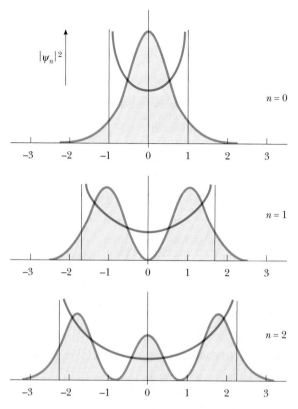

Figure 41.22   The red curves represent probability densities $|\psi|^2$ for the first three states of a simple harmonic oscillator. The blue curves represent classical probability densities corresponding to the same energies.   *(From C.W. Sherwin,* Introduction to Quantum Mechanics, *New York, Holt, Rinehart and Winston, 1959. Used with permission.)*

for comparison. Note that as n increases, the agreement between the classical and quantum-mechanical results improves, as expected.

Quick Quiz 41.5

(a) Why do the classical probability density curves in Figure 41.22 bend up at the ends? (b) What do you expect the quantum-mechanical probability density curves to look like at very large values of n?

SUMMARY

The **Heisenberg uncertainty principle** states that if a measurement of position is made with precision Δx and a simultaneous measurement of linear momentum is made with precision Δp_x, the product of the two uncertainties can never be less than $\hbar/2$:

$$\Delta x \, \Delta p_x \geq \frac{\hbar}{2} \tag{41.1}$$

In quantum mechanics, de Broglie matter waves are represented by a wave function $\psi(x, y, z, t)$. The probability per unit volume (or probability density) that a particle will be found at a point is $|\psi|^2$. If the particle is confined to moving along the x axis, then the probability that it is located in an interval dx is $|\psi|^2\, dx$. Furthermore, the sum of all these probabilities over all values of x must be 1:

$$\int_{-\infty}^{\infty} |\psi|^2\, dx = 1 \qquad \textbf{(41.4)}$$

This is called the **normalization condition.** The measured position x of the particle, averaged over many trials, is called the **expectation value** of x and is defined by

$$\langle x \rangle \equiv \int_{-\infty}^{\infty} x\,|\psi|^2\, dx \qquad \textbf{(41.6)}$$

If a particle of mass m is confined to moving in a one-dimensional box of width L whose walls are impenetrable, we require that ψ be zero at the walls and outside the box. The allowed wave functions for the particle are given by

$$\psi(x) = A \sin\left(\frac{n\pi x}{L}\right) \qquad n = 1, 2, 3, \ldots \qquad \textbf{(41.8)}$$

where A is the maximum value of ψ. The particle has a well-defined wavelength λ with values such that $L = n\lambda/2$. These allowed states are called **stationary states** of the system. The energies of a particle in a box are quantized and are given by

$$E_n = \left(\frac{h^2}{8mL^2}\right) n^2 \qquad n = 1, 2, 3, \ldots \qquad \textbf{(41.9)}$$

The wave function must satisfy the **Schrödinger equation.** The time-independent Schrödinger equation for a particle confined to moving along the x axis is

$$\frac{d^2\psi}{dx^2} = -\frac{2m}{\hbar^2}(E - U)\psi \qquad \textbf{(41.12)}$$

where E is the total energy of the system and U is the potential energy.

The approach of quantum mechanics is to solve Equation 41.12 for ψ and E, given the potential energy $U(x)$ for the system. In doing so, we must place restrictions on $\psi(x)$: (1) $\psi(x)$ must be continuous, (2) $\psi(x)$ must approach zero as x approaches $\pm\infty$, (3) $\psi(x)$ must be single-valued, and (4) $d\psi/dx$ must be continuous for all finite values of $U(x)$.

QUESTIONS

1. Is an electron a particle or a wave? Support your answer by citing some experimental results.
2. An electron and a proton are accelerated from rest through the same potential difference. Which particle has the greater wavelength?
3. If matter has a wave nature, why is this wave-like characteristic not observable in our daily experiences?
4. In what way does Bohr's model of the hydrogen atom violate the uncertainty principle?
5. Why is it impossible to measure simultaneously, with infinite accuracy, the position and speed of a particle?
6. In describing electrons passing through a slit and arriving at a screen, the physicist Richard Feynman said that "electrons arrive in lumps, like particles, but the probability of arrival of these lumps is determined as the intensity of the waves would be. It is in this sense that the electron behaves sometimes like a particle and sometimes like a wave." Elaborate on this point in your own words. (For a further discussion of this point, see R. Feynman, *The Character of Physical Law,* Cambridge, MA, MIT Press, 1980, Chapter 6.)
7. For a particle in a box, the probability density at certain points is zero, as seen in Figure 41.11b. Does this imply that the particle cannot move across these points? Explain.
8. Discuss the relationship between zero-point energy and the uncertainty principle.

9. As a particle of energy E is reflected from a potential barrier of height U, where $E < U$, how does the amplitude of the reflected wave change as the barrier height is reduced?

10. A philosopher once said that "it is necessary for the very existence of science that the same conditions always produce the same results." In view of what has been discussed in this chapter, present an argument showing that this statement is false. How might the statement be reworded to make it true?

11. In quantum mechanics it is possible for the energy E of a particle to be less than the potential energy, but classically this is not possible. Explain.

12. Consider two square wells of the same width, one with finite walls and the other with infinite walls. Compare the energy and momentum of a particle trapped in the finite well with the energy and momentum of an identical particle in the infinite well.

13. Why is it impossible for the lowest energy state of a harmonic oscillator to be zero?

14. Why is an electron microscope more suitable than an optical microscope for "seeing" objects less than 1 μm in size?

15. What is the Schrödinger equation? How is it useful in describing atomic phenomena?

PROBLEMS

1, 2, 3 = straightforward, intermediate, challenging   ☐ = full solution available in the *Student Solutions Manual and Study Guide*
WEB = solution posted at **http://www.saunderscollege.com/physics/**   🖳 = Computer useful in solving problem   🖱 = Interactive Physics
☐ = paired numerical/symbolic problems

Section 41.1   The Double-Slit Experiment Revisited

WEB 1. Neutrons traveling at 0.400 m/s are directed through a double slit having a 1.00-mm separation. An array of detectors is placed 10.0 m from the slit. (a) What is the de Broglie wavelength of the neutrons? (b) How far off axis is the first zero-intensity point on the detector array? (c) When a neutron reaches a detector, can we say which slit the neutron passed through? Explain.

2. A modified oscilloscope is used to perform an electron interference experiment. Electrons are incident on a pair of narrow slits 0.060 0 μm apart. The bright bands in the interference pattern are separated by 0.400 mm on a screen 20.0 cm from the slits. Determine the potential difference through which the electrons were accelerated to give this pattern.

3. (a) Show that the wavelength of a neutron is

$$\lambda = \frac{2.86 \times 10^{-11}}{\sqrt{K_n}} \text{ m}$$

where K_n is the kinetic energy of the neutron in electron volts. (b) What is the wavelength of a 1.00-keV neutron?

4. The distance between adjacent atoms in crystals is on the order of 0.1 nm. The use of electrons in diffraction studies of crystals requires that the de Broglie wavelength of the electrons be of the same order as the distance between atoms of the crystals. What is the order of magnitude of the minimum required energy (in electron volts) of electrons to be used for this purpose?

5. The resolving power of a microscope depends on the wavelength used. If one wished to "see" an atom, a resolution of approximately 1.00×10^{-11} m would be required. (a) If electrons are used (in an electron microscope), what minimum kinetic energy is required for the electrons? (b) If photons are used, what minimum photon energy is needed to obtain the required resolution?

6. Electrons are accelerated through 40 000 V in an electron microscope. What, ideally, is the smallest observable distance between objects seen with this microscope?

7. A beam of electrons with a kinetic energy of 1.00 MeV is incident normally on an array of atoms in rows separated by 0.250 nm. If the array acts like a flat diffraction grating, in what direction can we expect the electrons in the fifth order to be moving?

Section 41.2   The Uncertainty Principle

8. Suppose Fuzzy, a quantum-mechanical duck, lives in a world in which $h = 2\pi$ J·s. Fuzzy has a mass of 2.00 kg and initially is known to be within a pond 1.00 m wide. (a) What is the minimum uncertainty in his speed? (b) Assuming that this uncertainty in speed prevails for 5.00 s, determine the uncertainty in position after that time.

WEB 9. An electron ($m_e = 9.11 \times 10^{-31}$ kg) and a bullet ($m = 0.020\ 0$ kg) each have a speed of 500 m/s, accurate to within 0.010 0%. Within what limits could we determine the positions of the objects?

10. An air rifle is used to shoot 1.00-g particles at 100 m/s through a hole of diameter 2.00 mm. How far from the rifle must an observer be to see the beam spread by 1.00 cm because of the uncertainty principle? Compare this answer with the diameter of the visible Universe (2×10^{26} m).

11. A light source is used to determine the location of an electron in an atom to a precision of 0.050 0 nm. What is the minimum possible uncertainty in the speed of the electron?

12. Use the uncertainty principle to show that if an electron were confined inside an atomic nucleus of diameter 2×10^{-15} m, it would have to be moving relativistically,

while a proton confined to the same nucleus could be moving nonrelativistically.

13. A woman on a ladder drops small pellets toward a point target on the floor. (a) Show that, according to the uncertainty principle, the average miss distance must be at least

$$\Delta x_f = (2/\hbar m)^{1/2}(2/Hg)^{1/4}$$

where H is the initial height of each pellet above the floor and m is the mass of each pellet. Assume that the spread in impact points is given by $\Delta x_f = \Delta x_i + (\Delta v_x)t$. (b) If $H = 2.00$ m and $m = 0.500$ g, what is Δx_f?

Section 41.3   Probability Density

14. The wave function for a particle is

$$\psi(x) = \sqrt{\frac{a}{\pi(x^2 + a^2)}}$$

for $a > 0$ and $-\infty < x < +\infty$. Determine the probability that the particle is located somewhere between $x = -a$ and $x = +a$.

WEB 15. A free electron has a wave function

$$\psi(x) = A\sin(5.00 \times 10^{10}\,x)$$

where x is in meters. Find (a) the de Broglie wavelength, (b) the linear momentum, and (c) the kinetic energy in electron volts.

Section 41.4   A Particle in a Box

16. An electron that has an energy of approximately 6 eV moves between rigid walls 1.00 nm apart. Find (a) the quantum number n for the energy state that the electron occupies, and (b) the precise energy of the electron.

WEB 17. An electron is contained in a one-dimensional box of width 0.100 nm. (a) Draw an energy-level diagram for the electron for levels up to $n = 4$. (b) Find the wavelengths of all photons that can be emitted by the electron in making transitions from the $n = 4$ state to the $n = 1$ state (by all spontaneous paths).

18. An electron is confined to a one-dimensional region in which its ground-state ($n = 1$) energy is 2.00 eV. (a) What is the width of the region? (b) How much energy is required to promote the electron to its first excited state?

19. A ruby laser emits 694.3-nm light. Assuming that this light is due to a transition of an electron in a box from the $n = 2$ state to the $n = 1$ state, find the width of the box.

20. A laser emits light of wavelength λ. Assuming that this light is due to a transition of an electron in a box from the $n = 2$ state to the $n = 1$ state, find the width of the box.

21. The nuclear potential energy that binds protons and neutrons in a nucleus is often approximated by a square

well. Imagine a proton confined in an infinitely high square well of width 10.0 fm, a typical nuclear diameter. Calculate the wavelength and energy associated with the photon emitted when the proton moves from the $n = 2$ state to the ground state. In what region of the electromagnetic spectrum does this wavelength belong?

22. An alpha particle in a nucleus can be modeled as a particle moving in a box of width 1.00×10^{-14} m (the approximate diameter of a nucleus). Using this model, estimate the energy and momentum of an alpha particle in its lowest energy state ($m_\alpha = 6.64 \times 10^{-27}$ kg).

23. Use the particle-in-a-box model to calculate the first three energy levels of a neutron trapped in a nucleus of diameter 20.0 fm. Do the energy-level differences have a realistic order of magnitude?

24. A particle in an infinitely deep square well has a wave function that is given by

$$\psi_2(x) = \sqrt{\frac{2}{L}}\sin\left(\frac{2\pi x}{L}\right)$$

for $0 \le x \le L$ and is zero otherwise. (a) Determine the expectation value of x. (b) Determine the probability of finding the particle near $L/2$, by calculating the probability that the particle lies in the range $0.490L \le x \le 0.510L$. (c) Determine the probability of finding the particle near $L/4$, by calculating the probability that the particle lies in the range $0.240L \le x \le 0.260L$. (d) Argue that no contradiction exists between the result of part (a) and the results of parts (b) and (c).

25. The wave function for a particle confined to moving in a one-dimensional box is

$$\psi(x) = A\sin\left(\frac{n\pi x}{L}\right)$$

Use the normalization condition on ψ to show that

$$A = \sqrt{\frac{2}{L}}$$

Hint: Because the box width is L, the wave function is zero for $x < 0$ and for $x > L$, so the normalization condition (Eq. 41.4) reduces to

$$\int_0^L |\psi|^2\, dx = 1$$

26. The wave function of an electron is

$$\psi_2(x) = \sqrt{\frac{2}{L}}\sin\left(\frac{2\pi x}{L}\right)$$

Calculate the probability of finding the electron between $x = 0$ and $x = L/4$.

27. An electron in an infinitely deep square well has a wave function that is given by

$$\psi_2(x) = \sqrt{\frac{2}{L}}\sin\left(\frac{2\pi x}{L}\right)$$

for $0 \leq x \leq L$ and is zero otherwise. What are the most probable positions of the electron?

28. A particle in an infinite square well has a wave function that is given by

$$\psi_1(x) = \sqrt{\frac{2}{L}} \sin\left(\frac{\pi x}{L}\right)$$

for $0 \leq x \leq L$ and is zero otherwise. (a) Determine the probability of finding the particle between $x = 0$ and $x = L/3$. (b) Use the result of this calculation and symmetry arguments to find the probability of finding the particle between $x = L/3$ and $x = 2L/3$. Do not reevaluate the integral. (c) Compare the result of part (a) with the classical probability.

29. A proton is confined to moving in a one-dimensional box of width 0.200 nm. (a) Find the lowest possible energy of the proton. (b) What is the lowest possible energy of an electron confined to the same box? (c) How do you account for the great difference in your results for (a) and (b)?

30. Consider a particle moving in a one-dimensional box for which the walls are at $x = -L/2$ and $x = L/2$. (a) Write the wave functions and probability densities for $n = 1$, $n = 2$, and $n = 3$. (b) Sketch the wave functions and probability densities. (*Hint:* Make an analogy with the case of a particle in a box for which the walls are at $x = 0$ and $x = L$.)

Section 41.5   The Schrödinger Equation

31. Show that the wave function $\psi = Ae^{i(kx-\omega t)}$ is a solution to the Schrödinger equation (Eq. 41.12), where $k = 2\pi/\lambda$ and $U = 0$.

32. The wave function of a particle is given by

$$\psi(x) = A\cos(kx) + B\sin(kx)$$

where A, B, and k are constants. Show that ψ is a solution of the Schrödinger equation (Eq. 41.12), assuming that the particle is free ($U = 0$), and find the corresponding energy E of the particle.

33. A particle of mass m moves in a potential well of width $2L$. Its potential energy is infinite for $x < -L$ and for $x > +L$. Inside the region $-L < x < L$, its potential energy is given by

$$U(x) = \frac{-\hbar^2 x^2}{mL^2(L^2 - x^2)}$$

In addition, the particle is in a stationary state described by the wave function $\psi(x) = A(1 - x^2/L^2)$ for $-L < x < +L$, and by $\psi(x) = 0$ elsewhere. (a) Determine the energy of the particle in terms of $\hbar$, m, and L. (*Hint:* Use the Schrödinger equation, Eq. 41.12.) (b) Show that $A = (15/16L)^{1/2}$. (c) Determine the probability that the particle is located between $x = -L/3$ and $x = +L/3$.

34. In a region of space, a particle with zero total energy has a wave function

$$\psi(x) = Axe^{-x^2/L^2}$$

(a) Find the potential energy U as a function of x. (b) Make a sketch of $U(x)$ versus x.

(Optional)
Section 41.6   A Particle in a Well of Finite Height

35. Suppose a particle is trapped in its ground state in a box that has infinitely high walls (see Fig. 41.11). Now suppose the left-hand wall is suddenly lowered to a finite height and width. (a) Qualitatively sketch the wave function for the particle a short time later. (b) If the box has a width L, what is the wavelength of the wave that penetrates the left-hand wall?

36. Sketch the wave function $\psi(x)$ and the probability density $|\psi(x)|^2$ for the $n = 4$ state of a particle in a finite potential well. (See Fig. 41.16.)

(Optional)
Section 41.7   Tunneling Through a Barrier

37. An electron with kinetic energy $E = 5.00$ eV is incident on a barrier with thickness $L = 0.200$ nm and height $U = 10.0$ eV (Fig. P41.37). What is the probability that the electron (a) will tunnel through the barrier? (b) will be reflected?

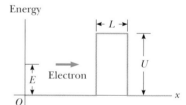

Figure P41.37   Problems 37 and 38.

38. An electron having total energy $E = 4.50$ eV approaches a rectangular energy barrier with $U = 5.00$ eV and $L = 950$ pm, as in Figure P41.37. Classically, the electron cannot pass through the barrier because $E < U$. However, quantum-mechanically a finite probability of tunneling exists. Calculate this probability, which is the transmission coefficient.

39. In Problem 38, by how much would the width L of the potential barrier have to be increased for the chance of an incident 4.50-eV electron tunneling through the barrier to be one in a million?

(Optional)
Section 41.8   The Scanning Tunneling Microscope

40. A scanning tunneling microscope (STM) can precisely determine the depths of surface features because the

current through its tip is very sensitive to differences in the width of the gap between the tip and the sample surface. Assume that in this direction the electron wave function falls off exponentially with a decay length of 0.100 nm—that is, with $C = 10.0/\text{nm}$. Determine the ratio of the current when the STM tip is 0.500 nm above a surface feature to the current when the tip is 0.515 nm above the surface.

41. The design criterion for a typical scanning tunneling microscope specifies that it must be able to detect, on the sample below its tip, surface features that differ in height by only 0.002 00 nm. What percentage change in electron transmission must the electronics of the STM be able to detect, to achieve this resolution? Assume that the electron transmission coefficient is e^{-2CL}, with $C = 10.0/\text{nm}$.

(Optional)
Section 41.9   The Simple Harmonic Oscillator

42. Show that Equation 41.21 is a solution of Equation 41.19 with energy $E = \frac{1}{2}\hbar\omega$.

43. A one-dimensional harmonic oscillator wave function is

$$\psi = Axe^{-bx^2}$$

(a) Show that ψ satisfies Equation 41.19. (b) Find b and the total energy E. (c) Is this a ground state or a first excited state?

44. A quantum simple harmonic oscillator consists of an electron bound by a restoring force proportional to its displacement from a certain equilibrium point. The proportionality constant is 8.99 N/m. What is the longest wavelength of light that can excite the oscillator?

45. (a) Normalize the wave function for the ground state of a simple harmonic oscillator. That is, apply Equation 41.4 to Equation 41.21 and find the required value for the coefficient B, in terms of m, ω, and constants. (b) Determine the probability of finding the oscillator in a narrow interval $-\delta/2 < x < \delta/2$ around its equilibrium position.

46. The total energy of a particle moving with simple harmonic motion along the x axis is

$$E = \frac{p_x^2}{2m} + \frac{kx^2}{2}$$

where p_x is the momentum of the particle and k is the spring constant. (a) Using the uncertainty principle, show that this expression can also be written

$$E \geq \frac{p_x^2}{2m} + \frac{k\hbar^2}{8p_x^2}$$

(b) Show that the minimum energy of the harmonic oscillator is

$$E_{\text{min}} = K + U = \frac{1}{4}\hbar\sqrt{\frac{k}{m}} + \frac{\hbar\omega}{4} = \frac{\hbar\omega}{2}$$

ADDITIONAL PROBLEMS

47. Keeping a constant speed of 0.8 m/s, a marble rolls back and forth across a shoebox. Make an order-of-magnitude estimate of the probability of its escaping through the wall of the box by quantum tunneling. State the quantities you take as data and the values you measure or estimate for them.

48. A particle of mass 2.00×10^{-28} kg is confined to a one-dimensional box of width 1.00×10^{-10} m. For $n = 1$, what are (a) the particle's wavelength, (b) its momentum, and (c) its ground-state energy?

WEB 49. An electron is represented by the time-independent wave function

$$\psi(x) = \begin{cases} Ae^{-\alpha x} & \text{for } x > 0 \\ Ae^{+\alpha x} & \text{for } x < 0 \end{cases}$$

(a) Sketch the wave function as a function of x.
(b) Sketch the probability that the electron is found between x and $x + dx$. (c) Argue that this can be a physically reasonable wave function. (d) Normalize the wave function. (e) Determine the probability of finding the electron somewhere in the range

$$x_1 = -\frac{1}{2\alpha} \qquad \text{to} \qquad x_2 = \frac{1}{2\alpha}$$

50. Particles incident from the left are confronted with a step in potential energy as shown in Figure P41.50. The step has a height U, and the particles have energy $E > U$. Classically, we would expect all of the particles to continue on, although with reduced speed. According to quantum mechanics, a fraction of the particles are reflected at the barrier. (a) Prove that the reflection coefficient R for this case is

$$R = \frac{(k_1 - k_2)^2}{(k_1 + k_2)^2}$$

where $k_1 = 2\pi/\lambda_1$ and $k_2 = 2\pi/\lambda_2$ are the angular wavenumbers for the incident and transmitted particles. Proceed as follows: Show that the wave function $\psi_1 = A\cos k_1 x + B\cos(-k_1 x)$ satisfies the Schrödinger equation in region 1, where $x < 0$. Here $A\cos k_1 x$ represents the incident beam, and $B\cos(-k_1 x)$ represents the reflected particles. Show that $\psi_2 = C\cos k_2 x$ satisfies the Schrödinger equation in region 2, for $x > 0$.

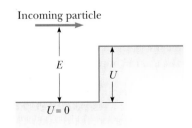

Figure P41.50   Problems 50 and 51.

Impose the boundary conditions $\psi_1 = \psi_2$ and $d\psi_1/dx = d\psi_2/dx$ at $x = 0$, to find the relationship between B and A. Then evaluate $R = B^2/A^2$. (b) A particle that has kinetic energy $E = 7.00$ eV is incident from a region where the potential energy is zero onto one in which $U = 5.00$ eV. Find its probability of being reflected and its probability of being transmitted.

51. Particles incident from the left are confronted with a step in potential energy as shown in Figure P41.50. The step has a height U, and the particles have energy $E = 2U$. Classically, all the particles would pass into the region of higher potential energy at the right. However, according to quantum mechanics, a fraction of the particles are reflected at the barrier. Use the result of Problem 50 to determine the fraction of the incident particles that are reflected. (This situation is analogous to the partial reflection and transmission of light striking an interface between two different media.)

52. An electron is trapped at a defect in a crystal. (A defect is an imperfection in the otherwise orderly arrangement of atoms.) The defect may be modeled as a one-dimensional, rigid-walled box of width 1.00 nm. (a) Sketch the wave functions and probability densities for the $n = 1$ and $n = 2$ states. (b) For the $n = 1$ state, calculate the probability of finding the electron between $x_1 = 0.150$ nm and $x_2 = 0.350$ nm, where $x = 0$ is the left side of the box. (c) Repeat part (b) for the $n = 2$ state. (d) Calculate the energies, in electron volts, of the $n = 1$ and $n = 2$ states. *Hint:* For parts (b) and (c), use Equation 41.5 and note that

$$\int \sin^2 ax\, dx = \tfrac{1}{2}x - \frac{1}{4a}\sin 2ax$$

53. Johnny Jumper's favorite trick is to step out of his 16th-story window and fall 50.0 m into a pool. A news reporter takes a picture of 75.0-kg Johnny just before he makes a splash, using an exposure time of 5.00 ms. Find (a) Johnny's de Broglie wavelength at this moment, (b) the uncertainty of his kinetic energy measurement during such a period of time, and (c) the percent error caused by such an uncertainty.

54. A π^0 meson is an unstable particle produced in high-energy particle collisions. Its rest energy is about 135 MeV, and it exists for an average lifetime of only 8.70×10^{-17} s before decaying into two gamma rays. Using the uncertainty principle, estimate the fractional uncertainty $\Delta m/m$ in its mass determination.

55. An atom in an excited state 1.80 eV above the ground state remains in that excited state 2.00 μs before moving to the ground state. Find the (a) frequency and (b) wavelength of the emitted photon. (c) Find the approximate uncertainty in energy of the photon.

56. An atom in an excited state E above the ground state remains in that excited state for a time T before moving to the ground state. Find the (a) frequency and (b) wavelength of the emitted photon. (c) Find the approximate uncertainty in energy of the photon.

57. For a particle described by a wave function $\psi(x)$, the expectation value of a physical quantity $f(x)$ associated with the particle is defined by

$$\langle f(x) \rangle \equiv \int_{-\infty}^{\infty} f(x)|\psi|^2\, dx$$

For a particle in a one-dimensional box extending from $x = 0$ to $x = L$, show that

$$\langle x^2 \rangle = \frac{L^2}{3} - \frac{L^2}{2n^2\pi^2}$$

58. A particle is described by the wave function

$$\psi(x) = \begin{cases} A\cos\left(\dfrac{2\pi x}{L}\right) & \text{for } -\dfrac{L}{4} \le x \le \dfrac{L}{4} \\ 0 & \text{for other values of } x \end{cases}$$

(a) Determine the normalization constant A. (b) What is the probability that the particle will be found between $x = 0$ and $x = L/8$ if its position is measured? (*Hint:* Use Eq. 41.5.)

59. A particle has a wave function

$$\psi(x) = \begin{cases} \sqrt{\dfrac{2}{a}}\, e^{-x/a} & \text{for } x > 0 \\ 0 & \text{for } x < 0 \end{cases}$$

(a) Find and sketch the probability density. (b) Find the probability that the particle will be at any point where $x < 0$. (c) Show that ψ is normalized, and then find the probability that the particle will be found between $x = 0$ and $x = a$.

60. A certain electron microscope accelerates electrons to an energy of 65.0 keV. (a) Find the wavelength of these electrons. (b) If one can resolve two points separated by at least 50.0 wavelengths, what is the smallest separation (or the minimum-sized object) that can be resolved with this microscope?

61. An electron of momentum p is at a distance r from a stationary proton. The electron has kinetic energy $K = p^2/2m_e$, potential energy $U = -k_e e^2/r$, and total energy $E = K + U$. If the electron is bound to the proton to form a hydrogen atom, its average position is at the proton, but the uncertainty in its position is approximately equal to the radius r of its orbit. The electron's average vector momentum is zero, but its average squared momentum is approximately equal to the squared uncertainty in its momentum, as given by the uncertainty principle. Treating the atom as a one-dimensional system, (a) estimate the uncertainty in the electron's momentum in terms of r. (b) Estimate the electron's kinetic, potential, and total energies in terms of r. (c) The actual value of r is the one that *minimizes the total energy*, resulting in a stable atom. Find that value of r and the resulting total energy. Compare your answer with the predictions of the Bohr theory.

62. A particle of mass m is placed in a one-dimensional box of width L. Assume that the box is so small that the par-

ticle's motion is *relativistic,* so $E = p^2/2m$ is not valid. (a) Derive an expression for the energy levels of the particle. (b) If the particle is an electron in a box of width $L = 1.00 \times 10^{-12}$ m, find its lowest possible kinetic energy. By what percentage is the nonrelativistic formula in error? (*Hint:* See Eq. 39.26.)

63. Consider a "crystal" consisting of two nuclei and two electrons as shown in Figure P41.63. (a) Taking into account all the pairs of interactions, find the potential energy of the system as a function of d. (b) Assuming that the electrons are restricted to a one-dimensional box of width $3d$, find the minimum kinetic energy of the two electrons. (c) Find the value of d for which the total energy is a minimum. (d) Compare this value of d with the spacing of atoms in lithium, which has a density of 0.530 g/cm^3 and an atomic mass of 7 u. (This type of calculation can be used to estimate the densities of crystals and certain stars.)

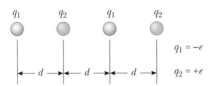

Figure P41.63

64. *The simple harmonic oscillator in an excited state.* The wave function

$$\psi(x) = Bxe^{-(m\omega/2\hbar)x^2}$$

is also a solution to the simple-harmonic-oscillator problem. (a) Find the energy of this state. (b) At what position are you least likely to find the particle? (c) At what positions are you most likely to find the particle? (d) Determine the value of B required to normalize the wave function. (e) Determine the classical probability of finding the particle in an interval of small width δ centered at the position $x = 2(\hbar/m\omega)^{1/2}$. (f) What is the actual probability of finding the particle in this interval?

65. *Normalization of wave functions:* (a) Find the normalization constant A for a wave function made up of the two lowest states of a particle in a box:

$$\psi(x) = A\left[\sin\left(\frac{\pi x}{L}\right) + 4 \sin\left(\frac{2\pi x}{L}\right)\right]$$

(b) A particle is described in the space $-a \le x \le a$ by the wave function

$$\psi(x) = A \cos\left(\frac{\pi x}{2a}\right) + B \sin\left(\frac{\pi x}{a}\right)$$

Determine the relationship between the values of A and B that are required for normalization. (*Hint:* Use the identity $\sin 2\theta = 2 \sin\theta \cos\theta$.)

66. A two-slit electron diffraction experiment is done with slits of *unequal* widths. The number of electrons reach-

ing the screen per second when only slit 1 is open is 25.0 times the number of electrons reaching the screen per second when only slit 2 is open. When both slits are open, an interference pattern results in which the destructive interference is not complete. Find the ratio of the probability that an electron will arrive at an interference maximum to the probability that an electron will arrive at an adjacent interference minimum. (*Hint:* Use the superposition principle.)

67. **Review Problem.**   Consider an extension of Young's double-slit experiment, performed with photons. Think of Figure 41.3 as a top view, with the reader looking down on the apparatus. The viewing screen can be a large flat array of *charge-coupled detectors.* Each cell in the array registers individual photons with high efficiency, so we can see where individual photons strike the screen in real time. We cover slit 1 with a polarizer with its transmission axis horizontal, and slit 2 with a polarizer with vertical transmission axis. Any one photon is either absorbed by a polarizing filter or allowed to pass through. The photons that come through a polarizer have their electric field oscillating in the plane defined by their direction of motion and the filter axis. Now we place another large square sheet of polarizing material just in front of the screen. For experimental trial 1, we make the transmission axis of this third polarizer horizontal. This choice, in effect, blocks slit 2. After many photons have been sent through the apparatus, their distribution on the viewing screen is shown by the blue curve $|\psi_1|^2$ in Figure 41.3. For trial 2, we turn the polarizer at the screen to make its transmission axis vertical. Then the screen receives photons only by way of slit 2, and their distribution is shown as $|\psi_2|^2$. For trial 3, we temporarily remove the third sheet of polarizing material. Then the interference pattern shown by the red curve $|\psi_1 + \psi_2|^2$ appears. (a) Is the light that is arriving at the screen to form the interference pattern polarized? Explain your answer. (b) Next, in trial 4 we replace the large square of polarizing material in front of the screen and set its transmission axis to 45°, halfway between horizontal and vertical. What appears on the screen? (c) Suppose we repeat all of trials 1 through 4 with very low light intensity, so that only one photon is present in the apparatus at a time. What are the results now? (d) We go back to high light intensity for convenience, and in trial 5 make the large square of polarizer turn slowly and steadily about a rotation axis through its center and perpendicular to its area. What appears on the screen? (e) At last, we go back to very low light intensity and replace the large square sheet of polarizing plastic with a flat layer of liquid crystal, to which we can apply an electric field in either a horizontal or a vertical direction. With the applied field we can very rapidly switch the liquid crystal so that it transmits only photons with horizontal electric field, acts as a polarizer with a vertical transmission axis, or transmits all photons with high efficiency. We keep track of photons as they are emitted individually by the source. For each photon we

wait until it has passed through the pair of slits. Then we quickly choose the setting of the liquid crystal, and make that photon encounter a horizontal polarizer, a vertical polarizer, or no polarizer before it arrives at the detector array. We can alternate among the conditions

we set up earlier in trials 1, 2, and 3. We keep track of our settings of the liquid crystal and sort out how photons behave under the different conditions, to end up with full sets of data for all three of those trials. What are the results?

ANSWERS TO QUICK QUIZZES

41.1 The diffraction pattern looks like the pattern for light waves going through three slits, which is shown in Figure 37.13.

41.2 If the slit is wide, as in part (a) of the figure at the bottom of the page, we cannot precisely know an individual electron's horizontal position within the electron beam. If we squeeze the beam through a narrower slit, as in part (b) of the figure, we do decrease Δx, but the increasing effects of diffraction mean that we have increased the uncertainty in p_x. (Depending on its momentum, an electron can appear anywhere in a wide horizontal area of the viewing screen.)

41.3 Classically, we expect the particle to bounce back and forth between the two walls at constant speed. Thus, we are as likely to find it at the left side of the box as at the middle, the right side, or anywhere else inside the box. Our graph of probability density versus x would therefore be a horizontal line, with a total area under the line of unity.

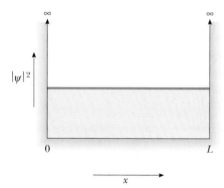

41.4 Figure 41.11b drawn for $n = 20$ would have 20 peaks quite close together. As the value of n increases, the peaks become progressively closer to one another, and

the detector is more likely to detect several peaks and valleys at once. As $n \to \infty$, the detector reports the average value of many cycles of the oscillating wave function. It gives this same average value anywhere inside the box, matching the classical answer given in Quick Quiz 41.3.

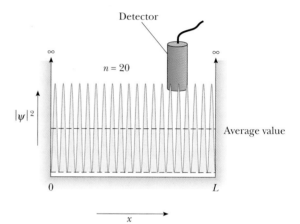

41.5 (a) The upward-curving parts of the classical graphs represent higher probability density values, which means the harmonic oscillator spends more time near the maximum displacement points. A block oscillating at the end of a vertical spring, for example, moves more slowly when the spring is near its fully stretched and fully compressed positions. Thus, a person taking a quick glance is more likely to see the block near one of these two points of maximum excursion from equilibrium. (b) As n increases, if we average over the unresolvable peaks and valleys in the probability density, the quantum-mechanical predictions become progressively closer to classical predictions; for great enough n, the two curves are indistinguishable from each other.

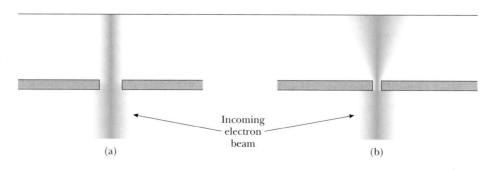

(a)   Incoming electron beam   (b)

PUZZLER

This supermarket scanner uses light from a laser to identify products being purchased. The word *laser* is an acronym for *l*ight *a*mplification by *st*imulated *e*mission of *r*adiation. How does a laser work, and what gives special properties to the light emitted by these devices? *(Paul Shambroom/Photo Researchers, Inc.)*

c h a p t e r

42

Atomic Physics

In Chapter 41 we introduced some of the basic concepts and techniques used in quantum mechanics, along with their applications to various one-dimensional systems. This chapter applies quantum mechanics to the real world of atomic structure, and a large portion of the chapter is quantum mechanics applied to the study of the hydrogen atom. Understanding the hydrogen atom, the simplest atomic system, is important for several reasons:

- The hydrogen atom is the only atomic system that can be solved exactly.
- Much of what is learned about the hydrogen atom, with its single electron, can be extended to such single-electron ions as He^+ and Li^{2+}.
- The hydrogen atom is an ideal system for performing precise tests of theory against experiment and for improving our overall understanding of atomic structure.
- The quantum numbers that are used to characterize the allowed states of hydrogen can also be used to describe the allowed states of more complex atoms, and such description enables us to understand the periodic table of the elements. This understanding is one of the greatest triumphs of quantum mechanics.
- The basic ideas about atomic structure must be well understood before we attempt to deal with the complexities of molecular structures and the electronic structure of solids.

The full mathematical solution of the Schrödinger equation applied to the hydrogen atom gives a complete and beautiful description of the atom's properties. However, because the mathematical procedures that are involved are beyond the scope of this text, the details are omitted. The solutions for some states of hydrogen are discussed, together with the quantum numbers used to characterize various allowed stationary states. We also discuss the physical significance of the quantum numbers and the effect of a magnetic field on certain quantum states.

A new physical idea, the *exclusion principle*, is presented in this chapter. This principle is extremely important for understanding the properties of multielectron atoms and the arrangement of elements in the periodic table. In fact, the implications of the exclusion principle are almost as far-reaching as those of the Schrödinger equation.

Finally, we apply our knowledge of atomic structure to describe the mechanisms involved in the production of x-rays and in the operation of a laser.

Joseph John Thomson
English physicist (1856–1940)
The recipient of a Nobel Prize in 1906, Thomson is usually considered the discoverer of the electron. He opened up the field of subatomic particle physics with his extensive work on the deflection of cathode rays (electrons) in an electric field.   *(Stock Montage, Inc.)*

42.1   EARLY MODELS OF THE ATOM

The model of the atom in the days of Newton was a tiny, hard, indestructible sphere. Although this model provided a good basis for the kinetic theory of gases, new models had to be devised when experiments revealed the electrical nature of atoms. J. J. Thomson suggested a model that describes the atom as a volume of positive charge with electrons embedded throughout the volume, much like the seeds in a watermelon or raisins in thick pudding (Fig. 42.1).

In 1911, Ernest Rutherford (1871–1937) and his students Hans Geiger and Ernest Marsden performed a critical experiment that showed that Thomson's model could not be correct. In this experiment, a beam of positively charged alpha particles (helium nuclei) was projected into a thin metallic foil, such as the target in Figure 42.2a. Most of the particles passed through the foil as if it were empty space. But some of the results of the experiment were astounding: Many of the particles deflected from their original direction of travel were scattered through *large* angles. Some particles were even deflected backward, reversing their

Electron

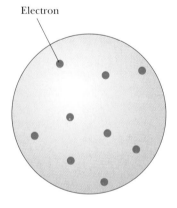

Figure 42.1   Thomson's model of the atom: negatively charged electrons in a volume of continuous positive charge.

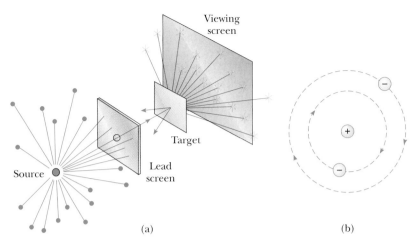

Figure 42.2 (a) Rutherford's technique for observing the scattering of alpha particles from a thin foil target. The source is a naturally occurring radioactive substance, such as radium. (b) Rutherford's planetary model of the atom.

direction of travel! When Geiger informed Rutherford that some alpha particles were scattered backward, Rutherford wrote, "It was quite the most incredible event that has ever happened to me in my life. It was almost as incredible as if you fired a 15-inch shell at a piece of tissue paper and it came back and hit you."

Such large deflections were not expected on the basis of Thomson's model. According to that model, the positive charge of an atom in the foil is spread out over such a great volume (the entire atom) that a positively charged alpha particle would never come close enough to a positive charge strong enough to cause any large-angle deflections. Rutherford explained his astonishing results by developing a new atomic model, one that assumed that the positive charge in the atom was concentrated in a region that was small relative to the size of the atom. He called this concentration of positive charge the **nucleus** of the atom. Any electrons belonging to the atom were assumed to be in the relatively large volume outside the nucleus. To explain why these electrons were not pulled into the nucleus, Rutherford modeled them as moving in orbits around the positively charged nucleus in the same manner as the planets orbit the Sun (Fig. 42.2b).

Two basic difficulties exist with Rutherford's planetary model. As we saw in Chapter 40, an atom emits certain characteristic frequencies of electromagnetic radiation and no others; the Rutherford model cannot explain this phenomenon. A second difficulty is that Rutherford's electrons are undergoing a centripetal acceleration. According to Maxwell's theory of electromagnetism, centripetally accelerated charges revolving with frequency f should radiate electromagnetic waves of frequency f. Unfortunately, this classical model leads to disaster when applied to the atom. As the electron radiates energy, the radius of its orbit steadily decreases and its frequency of revolution increases. This would lead to an ever-increasing frequency of emitted radiation and an ultimate collapse of the atom as the electron plunges into the nucleus (Fig. 42.3).

Now the stage was set for Bohr. To circumvent the erroneous deductions of electrons falling into the nucleus and a continuous emission from atoms, Bohr postulated that classical radiation theory did not hold for atomic-sized systems. By applying Planck's ideas of quantized energy levels to orbiting atomic electrons, he overcame the problem of a classical electron that continuously loses energy. Thus, Bohr postulated that atoms are generally confined to stable, nonradiating energy

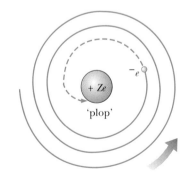

Figure 42.3 The classical model of the nuclear atom.

levels, each level representing a stationary state (see Section 40.5). Furthermore, he applied Einstein's concept of the photon to arrive at an expression for the frequency of the light that is emitted when an electron jumps from one stationary state to another.

One of the first indications that the Bohr theory needed to be modified arose when improved spectroscopic techniques were used to examine the spectral lines of hydrogen. It was found that many of the lines in the Balmer and other series were not single lines at all. Instead, each was a group of lines spaced very close together. An additional difficulty arose when it was observed that, in some situations, certain single spectral lines were split into three closely spaced lines when the atoms were placed in a strong magnetic field. Efforts to explain these deviations from the Bohr model led to improvements in the theory. One of the changes introduced into the original theory was the postulate that the electron could spin on its axis. In addition, Arnold Sommerfeld (1868–1951) improved the Bohr theory by introducing the theory of relativity into the analysis of the electron's motion.

42.2 ▷ THE HYDROGEN ATOM REVISITED

In Chapter 40 we described how the Bohr model views the electron as a particle orbiting around the nucleus in nonradiating, quantized energy levels. The de Broglie model assigns a wavelike nature to the electron, a model that allows some deeper understanding of the hydrogen atom. However, that model does not address all objections to the Bohr model and introduces some of its own difficulties. Fortunately, those difficulties are removed when the methods of quantum mechanics are used to describe atoms.

The potential energy function for the hydrogen atom is

$$U(r) = -k_e \frac{e^2}{r} \tag{42.1}$$

where $k_e = 8.99 \times 10^9 \,\text{N} \cdot \text{m}^2/\text{C}^2$ is the Coulomb constant, and r is the radial distance from the proton (situated at $r = 0$) to the electron. Figure 42.4 is a plot of this function versus r/a_0, where a_0 is the Bohr radius, 0.052 9 nm (see Eq. 40.24).

The formal procedure for solving the problem of the hydrogen atom is to substitute $U(r)$ into the Schrödinger equation and find appropriate solutions to the equation, as we did for the particle in a box in Chapter 41. The present problem is more complicated, however, because it is three-dimensional and because U depends on the radial coordinate r. We do not attempt to carry out these solutions. Rather, we simply describe their properties and some of their implications with regard to atomic structure.

According to quantum mechanics, the energies of the allowed states for the hydrogen atom are

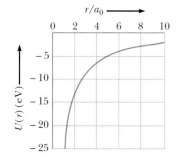

Figure 42.4 Potential energy $U(r)$ versus the ratio r/a_0 for the hydrogen atom. The constant a_0 is the Bohr radius, and r is the electron–proton separation distance.

Allowed energies for the hydrogen atom

$$E_n = -\left(\frac{k_e e^2}{2a_0}\right)\frac{1}{n^2} = -\frac{13.606}{n^2}\,\text{eV} \qquad n = 1, 2, 3, \ldots \tag{42.2}$$

This result is in exact agreement with that obtained in the Bohr theory.

In this solution to the Schrödinger equation, three quantum numbers all having integer values are needed for each stationary state, corresponding to three independent degrees of freedom for the electron: **principal quantum number** n, **orbital quantum number** ℓ, and **orbital magnetic quantum number** m_ℓ. A

TABLE 42.1   Three Quantum Numbers for the Hydrogen Atom

Quantum Number	Name	Allowed Values	Number of Allowed States
n	Principal quantum number	$1, 2, 3, \ldots$	Any number
ℓ	Orbital quantum number	$0, 1, 2, \ldots, n-1$	n
m_ℓ	Orbital magnetic quantum number	$-\ell, -\ell+1, \ldots,$ $0, \ldots, \ell-1, \ell$	$2\ell + 1$

fourth quantum number, resulting from a relativistic treatment of the hydrogen atom, is discussed in Section 42.3.

Certain important relationships exist among these three quantum numbers, as well as certain restrictions on their values:

> The values of n can range from 1 to ∞.
> The values of ℓ can range from 0 to $n-1$.
> The values of m_ℓ can range from $-\ell$ to ℓ.

◁ Restrictions on the values of quantum numbers

For example, if $n = 1$, only $\ell = 0$ and $m_\ell = 0$ are permitted. If $n = 2$, ℓ may be 0 or 1; if $\ell = 0$, then $m_\ell = 0$; but if $\ell = 1$, then m_ℓ may be 1, 0, or -1. Table 42.1 summarizes the rules for determining the allowed values of ℓ and m_ℓ for a given n.

For historical reasons, **all states having the same principal quantum number are said to form a shell.** Shells are identified by the letters K, L, M, $\ldots$, which designate the states for which $n = 1, 2, 3, \ldots$. Likewise, **all states having the same values of n and ℓ are said to form a subshell.** The letters[1] s, p, d, f, g, h, $\ldots$ are used to designate the subshells for which $\ell = 0, 1, 2, 3, \ldots$. For example, the state designated by $3p$ has the quantum numbers $n = 3$ and $\ell = 1$, and the $2s$ state has the quantum numbers $n = 2$ and $\ell = 0$. These notations are summarized in Table 42.2.

States that violate the rules given in Table 42.1 do not exist. For instance, the $2d$ state, which would have $n = 2$ and $\ell = 2$, cannot exist because the highest allowed value of ℓ is $n - 1$, which in this case is 1. Thus, for $n = 2$, $2s$ and $2p$ are allowed states but $2d$, $2f$, $\ldots$ are not. For $n = 3$, the allowed subshells are $3s$, $3p$, and $3d$.

TABLE 42.2   Atomic Shell and Subshell Notations

n	Shell Symbol	ℓ	Subshell Symbol
1	K	0	s
2	L	1	p
3	M	2	d
4	N	3	f
5	O	4	g
6	P	5	h
$\ldots$		$\ldots$	

---

[1] The first four of these letters come from early classifications of spectral lines: sharp, principal, diffuse, and fundamental. The remaining letters are in alphabetical order.

EXAMPLE 42.1   The $n = 2$ Level of Hydrogen

For a hydrogen atom, determine the number of allowed states corresponding to the principal quantum number $n = 2$, and calculate the energies of these states.

Solution   When $n = 2$, ℓ can be 0 or 1. If $\ell = 0$, the only value that m_ℓ can have is 0; for $\ell = 1$, m_ℓ can be -1, 0, or 1. Hence, we have one state, designated as the 2s state, that is associated with the quantum numbers $n = 2$, $\ell = 0$, and $m_\ell = 0$, and three states, designated as 2p states, for which the quantum numbers are $n = 2$, $\ell = 1$, $m_\ell = -1$; $n = 2$, $\ell = 1$, $m_\ell = 0$; and $n = 2$, $\ell = 1$, $m_\ell = 1$.

Because all four of these states have the same principal quantum number $n = 2$, they all have the same energy, according to Equation 42.2:

$$E_2 = -\frac{13.606 \text{ eV}}{2^2} = \boxed{-3.401 \text{ eV}}$$

Exercise   How many possible states exist for the $n = 3$ level of hydrogen? For the $n = 4$ level?

Answer   9; 16.

42.3   THE SPIN MAGNETIC QUANTUM NUMBER

In Example 42.1, we found four quantum states corresponding to $n = 2$. In reality, however, eight such states occur. The additional four states can be explained by requiring a fourth quantum number for each state: **spin magnetic quantum number** m_s.

The need for this new quantum number came about because of an unusual feature that was noted in the spectra of certain gases, such as sodium vapor. Close examination of one prominent line in the emission spectrum of sodium reveals that the line is, in fact, two closely spaced lines called a *doublet*. The wavelengths of these lines occur in the yellow region of the electromagnetic spectrum at 589.0 nm and 589.6 nm. In 1925, when this doublet was first observed, atomic theory could not explain it. To resolve this dilemma, Samuel Goudsmit (1902–1978) and George Uhlenbeck (1900–1988), following a suggestion made by the Austrian physicist Wolfgang Pauli (1900–1958), proposed the spin quantum number.

To describe this new quantum number, it is convenient (but technically incorrect) to think of the electron as spinning about its axis as it orbits the nucleus, as described in Section 30.8. Only two directions exist for the electron spin, as illustrated in Figure 42.5. If the direction of spin is as shown in Figure 42.5a, the electron is said to have *spin up*. If the direction of spin is reversed as in Figure 42.5b, the electron is said to have *spin down*. In the presence of an external magnetic field, the energy of the electron is slightly different for the two spin directions, and this energy difference accounts for the sodium doublet. The quantum numbers associated with the spin of the electron are $m_s = \frac{1}{2}$ for the spin-up state and $m_s = -\frac{1}{2}$ for the spin-down state.

The classical description of electron spin—as resulting from a spinning electron—is incorrect because quantum mechanics tells us that a rotational degree of freedom would require too many quantum numbers, and more recent theory indicates that the electron is a point particle, without spatial extent. Thus, the electron cannot be considered to be spinning as pictured in Figure 42.5. Despite this conceptual difficulty, all experimental evidence supports the idea that an electron does have some intrinsic property that can be described by the spin magnetic quantum number. Sommerfeld and Paul Dirac (1902–1984) showed that this fourth quantum number originates in the relativistic properties of the electron.

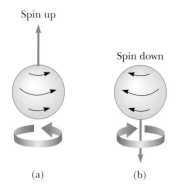

Figure 42.5   The spin of an electron can be either (a) up or (b) down relative to an external magnetic field.

EXAMPLE 42.2 ▶ **Putting Some Spin on Hydrogen**

For a hydrogen atom, determine the quantum numbers associated with the possible states that correspond to the principal quantum number $n = 2$.

Solution With the addition of the spin quantum number, we have the possibilities given in the accompanying table.

Exercise Show that for $n = 3$, there are 18 possible states.

n	ℓ	m_ℓ	m_s	Subshell	Shell	Number of Electrons in Subshell
2	0	0	$\frac{1}{2}$	$2s$	L	2
2	0	0	$-\frac{1}{2}$			
2	1	1	$\frac{1}{2}$	$2p$	L	6
2	1	1	$-\frac{1}{2}$			
2	1	0	$\frac{1}{2}$			
2	1	0	$-\frac{1}{2}$			
2	1	-1	$\frac{1}{2}$			
2	1	-1	$-\frac{1}{2}$			

42.4 ▶ THE WAVE FUNCTIONS FOR HYDROGEN

If we neglect electron spin for the present, the potential energy of the hydrogen atom depends only on the radial distance r between nucleus and electron. We therefore expect that some of the allowed states for this atom can be represented by wave functions that depend only on r. This indeed is the case. The simplest wave function for hydrogen is the one that describes the $1s$ state and is designated $\psi_{1s}(r)$:

$$\psi_{1s}(r) = \frac{1}{\sqrt{\pi a_0^3}} e^{-r/a_0} \quad \textbf{(42.3)}$$

Wave function for hydrogen in its ground state

where a_0 is the Bohr radius. Note that ψ_{1s} approaches zero as r approaches ∞ and is normalized as presented (see Eq. 41.4). Furthermore, because ψ_{1s} depends only on r, it is *spherically symmetric*. This, in fact, is true for all s states.

Recall that the probability of finding the electron in any region is equal to an integral of the probability density $|\psi|^2$ over the region. The probability density for the $1s$ state is

$$|\psi_{1s}|^2 = \left(\frac{1}{\pi a_0^3}\right) e^{-2r/a_0} \quad \textbf{(42.4)}$$

and the actual probability of finding the electron in a volume element dV is $|\psi|^2 \, dV$. It is convenient to define the *radial probability density function* $P(r)$ as the probability per unit radial length of finding the electron in a spherical shell of radius r and thickness dr. Thus, $P(r) \, dr$ is the probability of finding the electron in this shell. The volume dV of such an infinitesimally thin shell equals its surface area $4\pi r^2$ multiplied by the shell thickness dr (Fig. 42.6), so we can write this probability as

$$P(r) \, dr = |\psi|^2 \, dV = |\psi|^2 \, 4\pi r^2 \, dr$$

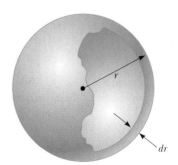

Figure 42.6 A spherical shell of radius r and thickness dr has a volume equal to $4\pi r^2 \, dr$.

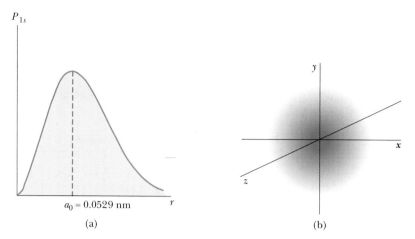

Figure 42.7 (a) The probability of finding the electron as a function of distance from the nucleus for the hydrogen atom in the 1s (ground) state. Note that the probability has its maximum value when r equals the Bohr radius a_0. (b) The spherical electronic charge distribution for the hydrogen atom in its 1s state.

Thus, the radial probability density function is

$$P(r) = 4\pi r^2 |\psi|^2 \qquad \textbf{(42.5)}$$

Substituting Equation 42.4 into Equation 42.5 gives the radial probability density function for the hydrogen atom in its ground state:

Radial probability density for the 1s state of hydrogen

$$P_{1s}(r) = \left(\frac{4r^2}{a_0{}^3}\right) e^{-2r/a_0} \qquad \textbf{(42.6)}$$

A plot of the function $P_{1s}(r)$ versus r is presented in Figure 42.7a. The peak of the curve corresponds to the most probable value of r for this particular state. The spherical symmetry of the radial probability density function is shown in Figure 42.7b.

Quick Quiz 42.1

Sketch a cross-section of the three-dimensional charge distribution shown in Figure 42.7b, imagining the sphere as being "sliced" in the xy plane.

EXAMPLE 42.3   The Ground State of Hydrogen

Calculate the most probable value of r for an electron in the ground state of the hydrogen atom.

Solution The most probable value of r corresponds to the peak of the plot of $P(r)$ versus r. Because the slope of the curve at this point is zero, we can evaluate the most probable value of r by setting $dP/dr = 0$ and solving for r. Using Equation 42.6, we obtain

$$\frac{dP}{dr} = \frac{d}{dr}\left[\left(\frac{4r^2}{a_0{}^3}\right) e^{-2r/a_0}\right] = 0$$

Carrying out the derivative operation and simplifying the expression, we obtain

$$e^{-2r/a_0}\frac{d}{dr}(r^2) + r^2\frac{d}{dr}(e^{-2r/a_0}) = 0$$

$$2re^{-2r/a_0} + r^2(-2/a_0)e^{-2r/a_0} = 0$$

$$(1)\qquad 2r[1 - (r/a_0)]e^{-2r/a_0} = 0$$

This expression is satisfied if

$$1 - \frac{r}{a_0} = 0$$

$$r = \boxed{a_0}$$

The most probable value of r is the Bohr radius! This result and Equation 42.2 are interesting connections between the Bohr theory and the more sophisticated quantum theory.

Equation (1) is also satisfied at $r = 0$. This is a point of *minimum* probability, which is equal to zero, as seen in Figure 42.7a.

EXAMPLE 42.4 ▶ **Probabilities for the Electron in Hydrogen**

Calculate the probability that the electron in the ground state of hydrogen will be found outside the first Bohr radius.

Solution The probability is found by integrating the radial probability density function for this state $P_{1s}(r)$ from the Bohr radius a_0 to ∞. Using Equation 42.6, we obtain

$$P = \int_{a_0}^{\infty} P_{1s}(r) \, dr = \frac{4}{a_0{}^3} \int_{a_0}^{\infty} r^2 e^{-2r/a_0} \, dr$$

We can put the integral in dimensionless form by changing variables from r to $z = 2r/a_0$. Noting that $z = 2$ when $r = a_0$ and that $dr = (a_0/2) \, dz$, we obtain

$$P = \frac{1}{2} \int_{2}^{\infty} z^2 e^{-z} \, dz = -\frac{1}{2} (z^2 + 2z + 2) e^{-z} \Big|_{2}^{\infty}$$

$$P = 5e^{-2} = \boxed{0.677 \text{ or } 67.7\%}$$

Example 42.3 shows that, for the ground state of hydrogen, the most probable value of r equals the Bohr radius a_0. It turns out that the average value of r for the ground state of hydrogen is $\frac{3}{2}a_0$, which is 50% greater than the most probable value (see Problem 49). The reason the average value is so much greater is the asymmetry in the radial probability density function (Fig. 42.7a), which has more area to the right of the peak. According to quantum mechanics, the atom has no sharply defined boundary. Therefore, the probability distribution for the electron can be viewed as a diffuse region of space, commonly referred to as an *electron cloud*.

The next-simplest wave function for the hydrogen atom is the one corresponding to the $2s$ state ($n = 2$, $\ell = 0$). The normalized wave function for this state is

$$\psi_{2s}(r) = \frac{1}{4\sqrt{2\pi}} \left(\frac{1}{a_0}\right)^{3/2} \left(2 - \frac{r}{a_0}\right) e^{-r/2a_0} \qquad \text{(42.7)}$$

Wave function for hydrogen in the $2s$ state

Again we see that ψ_{2s} depends only on r and is spherically symmetric. The energy corresponding to this state is $E_2 = -(13.606/4)$ eV $= -3.401$ eV. This energy level represents the first excited state of hydrogen. Plots of the radial probability density function for this state and several others are shown in Figure 42.8. The plot for the $2s$ state has two peaks. In this case, the most probable value corresponds to that value of r that has the highest value of $P (\approx 5a_0)$. An electron in the $2s$ state would be much farther from the nucleus (on the average) than an electron in the $1s$ state. The average value of r is even greater for the $3d$, $3p$, and $4d$ states.

As we have mentioned, all s states have spherically symmetric wave functions. The other states are not spherically symmetric. For example, the three wave functions corresponding to the states for which $n = 2$, $\ell = 1$ ($m_\ell = 1$, 0, or -1) can be expressed as appropriate linear combinations of the three p states. Although quantum mechanics limits our knowledge of angular momentum to the projection along any one axis at a time, these p states may be described mathematically as linear combinations of mutually perpendicular functions p_x, p_y, and p_z, as repre-

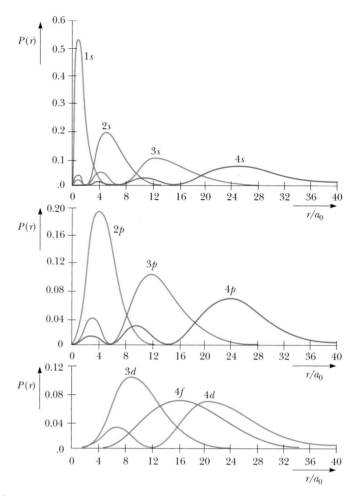

Figure 42.8 The radial probability density function versus r/a_0 for several states of the hydrogen atom. *(From E. U. Condon and G. H. Shortley,* The Theory of Atomic Spectra, *Cambridge, England, Cambridge University Press, 1953. Used with permission.)*

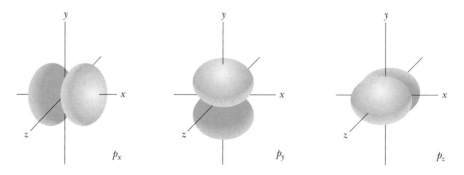

Figure 42.9 Angular dependence of the charge distribution for an electron in a p state. The three charge distributions p_x, p_y, and p_z have the same structure and differ only in their orientations in space.

sented in Figure 42.9, where only the angular dependence of these functions is shown. Note that the three clouds have identical structure but differ in their orientations with respect to the x, y, and z axes. The nonspherical wave functions for these states are

$$\psi_{2p_x} = xF(r)$$
$$\psi_{2p_y} = yF(r) \tag{42.8}$$
$$\psi_{2p_z} = zF(r)$$

Wave functions for the $2p$ state

where $F(r)$ is some exponential function of r. Wave functions with a highly directional character such as these are convenient for describing chemical bonding, the formation of molecules, and chemical properties.

42.5 ▷ THE OTHER QUANTUM NUMBERS

The energy of a particular state in the hydrogen atom depends on the principal quantum number n. Now let us see what the other three quantum numbers contribute to our atomic model.

The Orbital Quantum Number ℓ

If a particle moves in a circle of radius r, the magnitude of its angular momentum relative to the center of the circle is $L = mvr$. The direction of $\mathbf{L}$ is perpendicular to the plane of the circle and is given by a right-hand rule.[2] According to classical physics, L can have any value. However, the Bohr model of hydrogen postulates that the magnitude of the angular momentum of the electron is restricted to multiples of $\hbar$; that is, $mvr = n\hbar$. This model must be modified because it predicts (incorrectly) that the ground state of hydrogen ($n = 1$) has one unit of angular momentum. Furthermore, if L is taken to be zero in the Bohr model, we are forced to accept a picture of the electron as a particle oscillating along a straight line through the nucleus, a physically unacceptable situation.

These difficulties are resolved with the quantum-mechanical model of the atom. According to quantum mechanics, an atom in a state whose principal quantum number is n can take on the following *discrete* values of the magnitude of the orbital angular momentum:

$$L = \sqrt{\ell(\ell + 1)}\,\hbar \qquad \ell = 0, 1, 2, \ldots, n - 1 \tag{42.9}$$

Allowed values of L

Because ℓ is restricted to the values $\ell = 0, 1, 2, \ldots, n - 1$, we see that $L = 0$ (corresponding to $\ell = 0$) is an acceptable value of the magnitude of the angular momentum. The fact that L can be zero in this model serves to point out the inherent difficulties in any attempt to describe results based on quantum mechanics in terms of a purely particle-like (classical) model. In the quantum-mechanical interpretation, the electron cloud for the $L = 0$ state is spherically symmetric and has no fundamental axis of revolution.

[2] See Sections 11.3 and 11.4 for details on angular momentum and a review of this material.

EXAMPLE 42.5   **Calculating *L* for a *p* State**

Calculate the magnitude of the orbital angular momentum of an electron in a *p* state of hydrogen.

Solution Because we know that $\hbar = 1.054 \times 10^{-34}$ J·s, we can use Equation 42.9 to calculate *L*. With $\ell = 1$ for a *p* state, we have

$$L = \sqrt{1(1+1)}\,\hbar = \sqrt{2}\,\hbar = \boxed{1.49 \times 10^{-34}\ \text{J·s}}$$

This number is extremely small relative to, say, the orbital angular momentum of the Earth orbiting the Sun, which is approximately 2.7×10^{40} J·s. The quantum number that describes *L* for macroscopic objects, such as the Earth, is so large that the separation between adjacent states cannot be measured. Once again, the correspondence principle is upheld.

The Orbital Magnetic Quantum Number m_ℓ

Because angular momentum is a vector, its direction must be specified. Recall from Chapter 30 that an orbiting electron can be considered an effective current loop with a corresponding magnetic moment. Such a moment placed in a magnetic field **B** interacts with the field. Suppose a weak magnetic field applied along the *z* axis defines a direction in space. According to quantum mechanics, L^2 and L_z (the projection of **L** along the *z* axis) can have only discrete values. The orbital magnetic quantum number m_ℓ specifies the allowed values of the *z* component of the orbital angular momentum according to the expression

Allowed values of L_z

$$L_z = m_\ell \hbar \tag{42.10}$$

Space quantization

The quantization of the direction of **L** with respect to an external magnetic field is often referred to as **space quantization.**

Let us look at the possible orientations of **L** for a given value of ℓ. Recall that m_ℓ can have values ranging from $-\ell$ to ℓ. If $\ell = 0$, then $m_\ell = 0$ and $L_z = 0$. If $\ell = 1$, then the possible values of m_ℓ are -1, 0, and 1; hence, L_z may be $-\hbar$, 0, or $\hbar$. If $\ell = 2$, then m_ℓ can be -2, -1, 0, 1, or 2, corresponding to L_z values of $-2\hbar$, $-\hbar$, 0, $\hbar$, $2\hbar$, and so on.

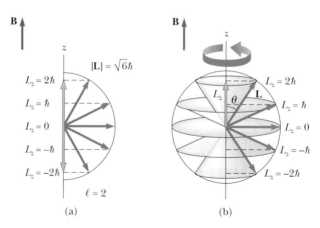

Figure 42.10 (a) The allowed projections of the orbital angular momentum **L** for the case $\ell = 2$. (b) The orbital angular momentum vector **L** lies on the surface of a cone and precesses about the *z* axis when a magnetic field **B** is applied in the *z* direction.

Figure 42.10a shows a vector model that describes space quantization for $\ell = 2$. Note that **L** can never be aligned parallel or antiparallel to **B** because L_z must be less than the total angular momentum L. For L_z to be zero, **L** must be perpendicular to **B**. From a three-dimensional viewpoint, **L** must lie on the surface of a cone that makes an angle θ with the z axis, as shown in Figure 42.10b. From the figure, we see that θ is also quantized and that its values are specified through the relationship

$$\cos \theta = \frac{L_z}{|\mathbf{L}|} = \frac{m_\ell}{\sqrt{\ell(\ell + 1)}} \qquad \textbf{(42.11)}$$

 θ is quantized

Note that m_ℓ is never greater than ℓ, and therefore θ can never be zero. (Classically, θ can have any value.)

Because of the uncertainty principle, **L** does not point in a specific direction. We can imagine it to trace out a cone in space. If **L** were known exactly, then all three components L_x, L_y, and L_z would be specified. For the moment, let us assume that this is the case and suppose that the electron moves in the xy plane, so **L** is in the z direction and the z component of its linear momentum $p_z = 0$. This means that p_z is precisely known, which is in violation of the uncertainty principle, $\Delta p_z \, \Delta z \geq \hbar/2$. In reality, only the magnitude of **L** and one component (say, L_z) can have definite values. In other words, quantum mechanics allows us to specify L and L_z but not L_x and L_y. Because the direction of **L** is constantly changing as we imagine it to precess about the z axis, the average values of L_x and L_y are zero and L_z maintains a fixed value of $m_\ell \hbar$.

The additional energy levels provided by the orbital magnetic quantum number explain the *Zeeman effect*, in which spectral lines are observed to split when a magnetic field is present, as shown in Figure 42.11.

QuickLab

Spin a top or a gyroscope rapidly, and watch as its axis of rotation slowly precesses about a vertical line passing through the point of support. This models the precession of the angular-momentum vector as shown in Figure 42.10b. In the photograph, Wolfgang Pauli and Niels Bohr are seeing this for themselves. *(Courtesy of AIP Niels Bohr Library, Margarethe Bohr Collection)*

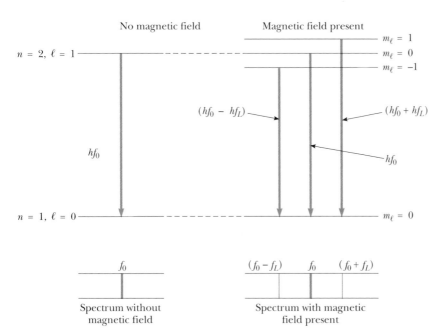

Figure 42.11 Split energy levels for the ground and first excited states of a hydrogen atom immersed in a magnetic field **B**. An atom in one of the excited states decays to the ground state with the emission of a photon, giving rise to emission lines at f_0, $f_0 + f_L$, and $f_0 - f_L$. This is the Zeeman effect. When **B** = 0, only the line at f_0 is observed.

EXAMPLE 42.6   **Space Quantization for Hydrogen**

Consider the hydrogen atom in the $\ell = 3$ state. Calculate the magnitude of **L** and the allowed values of L_z and θ.

Solution   We can calculate the magnitude of the orbital angular momentum using Equation 42.9:

$$L = \sqrt{\ell(\ell + 1)}\,\hbar = \sqrt{3(3 + 1)}\,\hbar = \boxed{2\sqrt{3}\,\hbar}$$

The allowed values of L_z can be calculated using $L_z = m_\ell \hbar$ with $m_\ell = -3, -2, -1, 0, 1, 2,$ and 3:

$$\boxed{L_z = -3\hbar, -2\hbar, -\hbar, 0, \hbar, 2\hbar, \text{ and } 3\hbar}$$

Finally, we calculate the allowed values of θ using Equation 42.11:

$$\cos \theta = \frac{m_\ell}{2\sqrt{3}}$$

Substituting the allowed values of m_ℓ gives

$$\cos \theta = \pm 0.866, \pm 0.577, \pm 0.289, \text{ and } 0$$

$$\boxed{\theta = 30.0°, 54.8°, 73.2°, 90.0°, 107°, 125°, \text{ and } 150°}$$

Quick Quiz 42.2

Make two drawings like Figure 42.10a, one for $\ell = 1$ and the other for $\ell = 3$.

The Spin Magnetic Quantum Number m_s

In 1921, Otto Stern (1888–1969) and Walter Gerlach (1889–1979) performed an experiment that demonstrated space quantization. However, their results were not in quantitative agreement with the theory that existed at that time. In their experiment, a beam of silver atoms sent through a nonuniform magnetic field was split into two components (Fig. 42.12). They repeated the experiment using other atoms, and in each case the beam split into two or more components. The classical argument is as follows: If the z direction is chosen to be the direction of the maximum nonuniformity of **B**, the net magnetic force on the atoms is along the z axis and is proportional to the component of the magnetic moment $\boldsymbol{\mu}$ of the atom in the z direction. (See Quick Quiz 29.4 for a review of the cause of this.) Classically, $\boldsymbol{\mu}$ can have any orientation, so the deflected beam should be spread out continuously. According to quantum mechanics, however, the deflected beam has several

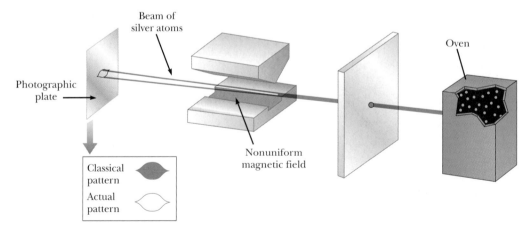

Figure 42.12   The apparatus used by Stern and Gerlach to verify space quantization. A beam of silver atoms is split in two by a nonuniform magnetic field.

components, and the number of components determines the number of possible values of μ_z. Hence, because the Stern–Gerlach experiment showed split beams, space quantization was at least qualitatively verified.

For the moment, let us assume that the magnetic moment $\boldsymbol{\mu}$ of the atom is due to the orbital angular momentum. Because μ_z is proportional to m_ℓ, the number of possible values of μ_z is $2\ell + 1$. Furthermore, because ℓ is an integer, the number of values of μ_z is always odd. This prediction is clearly not consistent with Stern and Gerlach's observation of two components (an *even* number) in the deflected beam of silver atoms. Hence, we are forced to conclude that either quantum mechanics is incorrect or the model is in need of refinement.

In 1927, Phipps and Taylor repeated the Stern–Gerlach experiment using a beam of hydrogen atoms. That experiment was important because it dealt with an atom containing a single electron in its ground state, for which the quantum theory makes reliable predictions. Recall that $\ell = 0$ for hydrogen in its ground state, so $m_\ell = 0$. Hence, we would not expect the beam to be deflected by the field because the magnetic moment $\boldsymbol{\mu}$ of the atom is zero. However, the beam in the Phipps–Taylor experiment was again split into two components. On the basis of that result, we can come to only one conclusion: Something other than the orbital motion is contributing to the magnetic moment.

As we learned earlier, Goudsmit and Uhlenbeck had proposed that the electron has an intrinsic angular momentum apart from its orbital angular momentum. From a classical viewpoint, this intrinsic angular momentum is attributed to the spinning of the charged electron about its own axis and hence is called electron spin.[3] In other words, the total angular momentum of the electron in a particular electronic state contains both an orbital contribution $\mathbf{L}$ and a spin contribution $\mathbf{S}$. The Phipps–Taylor result confirmed the hypothesis of Goudsmit and Uhlenbeck.

Quick Quiz 42.3

Explain why classical theory predicts the result labeled "classical pattern" in Figure 42.12.

In 1929, Dirac used the relativistic form of the total energy of a system to solve the relativistic wave equation for the electron in a potential well. His analysis confirmed the fundamental nature of electron spin. (Spin, like mass and charge, is an intrinsic property of a particle, independent of its surroundings.) Furthermore, the analysis showed that electron spin can be described by a single quantum number s, whose value can be only $\frac{1}{2}$. The spin angular momentum of the electron *never changes*. This notion contradicts classical laws, which dictate that a rotating charge slows down in the presence of an applied magnetic field because of the Faraday emf that accompanies the changing field. Furthermore, if the electron were viewed as a spinning ball of charge subject to classical laws, parts of it near its surface would be rotating with speeds exceeding the speed of light. Thus, the classical picture must not be pressed too far; ultimately, the spinning electron is a quantum entity defying any simple classical description.

The magnitude of the **spin angular momentum S** for the electron is

$$S = \sqrt{s(s+1)}\,\hbar = \frac{\sqrt{3}}{2}\hbar \tag{42.12}$$

Spin angular momentum of an electron

[3] Physicists often use the word *spin* when referring to spin angular momentum. For example, it is common to make the statement "The electron has a spin of $\frac{1}{2}$."

Like orbital angular momentum **L**, spin angular momentum **S** is quantized in space, as described in Figure 42.13. It can have two orientations relative to an external magnetic field, specified by the spin magnetic quantum number $m_s = \pm\frac{1}{2}$. The z component of spin angular momentum is

$$S_z = m_s\hbar = \pm\tfrac{1}{2}\hbar \qquad (42.13)$$

The two values $\pm\hbar/2$ for S_z correspond to the two possible orientations for **S** shown in Figure 42.13. The value $m_s = +\frac{1}{2}$ refers to the spin-up case, and the value $m_s = -\frac{1}{2}$ refers to the spin-down case.

The spin magnetic moment $\boldsymbol{\mu}_{\text{spin}}$ of the electron is related to its spin angular momentum **S** by the expression

$$\boldsymbol{\mu}_{\text{spin}} = -\frac{e}{m_e}\,\mathbf{S} \qquad (42.14)$$

where e is the electronic charge and m_e is the mass of the electron. Because $S_z = \pm\frac{1}{2}\hbar$, the z component of the spin magnetic moment can have the values

$$\mu_{\text{spin}, z} = \pm\frac{e\hbar}{2m_e} \qquad (42.15)$$

As we learned in Section 30.8, the quantity $e\hbar/2m_e$ is the Bohr magneton $\mu_{\text{B}} = 9.27 \times 10^{-24}\,\text{J/T}$. Note that the ratio of magnetic moment to angular momentum is twice as great for spin angular momentum (Eq. 42.14) as it is for orbital angular momentum (Eq. 30.25). The factor of 2 is explained in a relativistic treatment first carried out by Dirac.

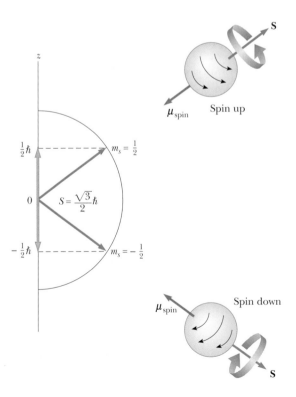

Figure 42.13 Spin angular momentum **S** exhibits space quantization. This figure shows the two allowed orientations of the spin angular momentum vector **S** and the spin magnetic moment $\boldsymbol{\mu}_{\text{spin}}$ for a spin $\frac{1}{2}$ particle, such as the electron.

Today physicists explain the Stern–Gerlach experiment as follows. The observed magnetic moments for both silver and hydrogen are due to spin angular momentum only, with no contribution from orbital angular momentum. A single-electron atom such as hydrogen has its electron spin quantized in the magnetic field in such a way that the z component of spin angular momentum is either $\frac{1}{2}\hbar$ or $-\frac{1}{2}\hbar$, corresponding to $m_s = \pm\frac{1}{2}$. Electrons with spin $+\frac{1}{2}$ are deflected downward, and those with spin $-\frac{1}{2}$ are deflected upward.

The Stern–Gerlach experiment provided two important results. First, it verified the concept of space quantization. Second, it showed that spin angular momentum exists—even though this property was not recognized until four years after the experiments were performed.

42.6 THE EXCLUSION PRINCIPLE AND THE PERIODIC TABLE

Earlier we found that the state of a hydrogen atom is specified by four quantum numbers: n, ℓ, m_ℓ, and m_s. As it turns out, the state of an electron in any other atom may also be specified by this same set of quantum numbers. In fact, these four quantum numbers can be used to describe all the electronic states of an atom regardless of the number of electrons in its structure.

An obvious question that arises here is: "How many electrons can be in a particular quantum state?" Pauli answered this important question in 1925, in a statement known as the **exclusion principle:**

> No two electrons in the same atom can ever be in the same quantum state; therefore, no two electrons in the same atom can have the same set of quantum numbers.

Exclusion principle

If this principle were not valid, every electron in an atom would end up in the lowest possible energy state of the atom, and the chemical behavior of the elements would be grossly modified. Nature as we know it would not exist!

In reality, we can view the electronic structure of complex atoms as a succession of filled levels increasing in energy. As a general rule, the order of filling of an atom's subshells is as follows. Once a subshell is filled, the next electron goes into the lowest-energy vacant subshell. We can understand this behavior by recognizing that if the atom were not in the lowest energy state available to it, it would radiate energy until it reached this state.

Before we discuss the electronic configuration of various elements, it is convenient to define an *orbital* as the state of an electron characterized by the quantum numbers n, ℓ, and m_ℓ. From the exclusion principle, we see that **only two electrons can be present in any orbital.** One of these electrons has a spin magnetic quantum number $m_s = +\frac{1}{2}$, and the other has $m_s = -\frac{1}{2}$. Because each orbital is limited to two electrons, the number of electrons that can occupy the various shells is also limited.

Table 42.3 shows the number of allowed quantum states for an atom for which $n = 3$. The arrows pointing upward indicate an atom in which the electron is described by $m_s = \frac{1}{2}$, and those pointing downward indicate that $m_s = -\frac{1}{2}$. The $n = 1$ shell can accommodate only two electrons because $m_\ell = 0$ means that only one orbital is allowed. (The three quantum numbers describing this orbital are $n = 1$, $\ell = 0$, and $m_\ell = 0$.) The $n = 2$ shell has two subshells, one for $\ell = 0$ and

TABLE 42.3   Allowed Quantum States for an Atom Having $n = 3$

n	1	2				3								
ℓ	0	0	1			0	1			2				
m_ℓ	0	0	1	0	-1	0	1	0	-1	2	1	0	-1	-2
m_s	↑↓	↑↓	↑↓	↑↓	↑↓	↑↓	↑↓	↑↓	↑↓	↑↓	↑↓	↑↓	↑↓	↑↓

Wolfgang Pauli   Austrian theoretical physicist (1900–1958)   An extremely talented theoretician who made important contributions in many areas of modern physics, Pauli gained public recognition at the age of 21 with a masterful review article on relativity that is still considered one of the finest and most comprehensive introductions to the subject. His other major contributions were the discovery of the exclusion principle, the explanation of the connection between particle spin and statistics, and theories of relativistic quantum electrodynamics, the neutrino hypothesis, and the hypothesis of nuclear spin.
(CERN, courtesy of AIP Emilio Segre Visual Archive)

one for $\ell = 1$. The $\ell = 0$ subshell is limited to two electrons because $m_\ell = 0$. The $\ell = 1$ subshell has three allowed orbitals, corresponding to $m_\ell = 1$, 0, and -1. Because each orbital can accommodate two electrons, the $\ell = 1$ subshell can hold six electrons. Thus, the $n = 2$ shell can contain eight electrons. The $n = 3$ shell has three subshells ($\ell = 0, 1, 2$) and nine orbitals and can accommodate up to 18 electrons. In general, each shell can accommodate up to $2n^2$ electrons.

The exclusion principle can be illustrated by an examination of the electronic arrangement in a few of the lighter atoms. First, recall from Section 1.2 that the atomic number Z of any element is the number of protons in the nucleus of an atom of that element. Hydrogen ($Z = 1$) has only one electron—which, in the ground state of the atom, can be described by either of two sets of quantum numbers: 1, 0, 0, $\frac{1}{2}$ or 1, 0, 0, $-\frac{1}{2}$. This electronic configuration is often written $1s^1$. The notation $1s$ refers to a state for which $n = 1$ and $\ell = 0$, and the superscript indicates that one electron is present in the s subshell.

Neutral helium ($Z = 2$) has two electrons. In the ground state, their quantum numbers are 1, 0, 0, $\frac{1}{2}$, and 1, 0, 0, $-\frac{1}{2}$. No other possible combinations of quantum numbers exist for this level, and we say that the K shell is filled. This electronic configuration is written $1s^2$.

Neutral lithium ($Z = 3$) has three electrons. In the ground state, two of these are in the $1s$ subshell. The third is in the $2s$ subshell because this subshell is slightly lower in energy than the $2p$ subshell.[4] Hence, the electronic configuration for lithium is $1s^2 2s^1$.

The electronic configurations of lithium and the next several elements are provided in Figure 42.14. The electronic configuration of beryllium ($Z = 4$), with its four electrons, is $1s^2 2s^2$, and boron ($Z = 5$) has a configuration of $1s^2 2s^2 2p^1$. The $2p$ electron in boron may be described by any of six equally probable sets of quantum numbers. In Figure 42.14, we show this electron in the leftmost $2p$ box with spin up, but it is equally likely to be in any $2p$ box with spin either up or down.

Carbon ($Z = 6$) has six electrons, giving rise to a question concerning how to assign the two $2p$ electrons. Do they go into the same orbital with paired spins (↑ ↓), or do they occupy different orbitals with unpaired spins (↑ ↑)? Experimental data show that the most stable configuration (that is, the one that is energetically preferred) is the latter, in which the spins are unpaired. Hence, the two $2p$ electrons in carbon and the three $2p$ electrons in nitrogen ($Z = 7$) have unpaired spins, as Figure 42.14 shows. The general rule that governs such situations, called **Hund's rule,** states that

Hund's rule

when an atom has orbitals of equal energy, the order in which they are filled by electrons is such that a maximum number of electrons have unpaired spins.

[4] To a first approximation, energy depends only on the quantum number n, as we have discussed. Because of the effect of the electronic charge shielding the nuclear charge in multielectron atoms, however, energy depends on ℓ also. We shall discuss these shielding effects in Section 42.7.

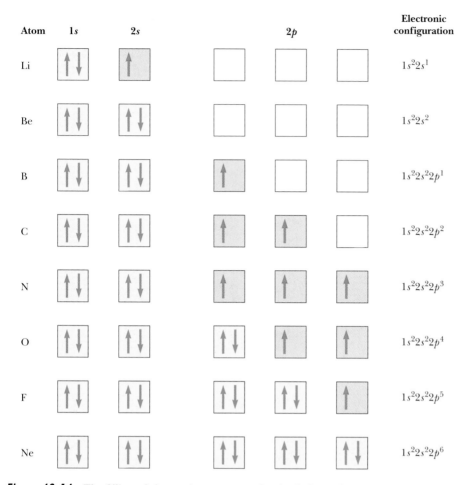

Figure 42.14 The filling of electronic states must obey both the exclusion principle and Hund's rule.

Some exceptions to this rule occur in elements having subshells that are close to being filled or half-filled.

A complete list of electronic configurations is provided in Table 42.4. In 1871, without any understanding of quantum mechanics that we now have, the Russian chemist Dmitri Mendeleev (1834–1907) made an early attempt at finding some order among the elements. He was trying to organize the elements for the table of contents of a book he was writing. He arranged the atoms in a table similar to that shown in Appendix C, according to their atomic masses and chemical similarities. The first table Mendeleev proposed contained many blank spaces, and he boldly stated that the gaps were there only because the elements had not yet been discovered. By noting the columns in which some missing elements should be located, he was able to make rough predictions about their chemical properties. Within 20 years of this announcement, most of these elements were indeed discovered.

The elements in the **periodic table** are arranged so that all those in a column have similar chemical properties. For example, consider the elements in the last column, which are all gases at room temperature: He (helium), Ne (neon), Ar (argon), Kr (krypton), Xe (xenon), and Rn (radon). The outstanding characteristic of all these elements is that they do not normally take part in chemical reactions—

TABLE 42.4 Electronic Configuration of the Elements

Atomic Number Z	Symbol	Ground-State Configuration	Ionization Energy (eV)
1	H	$1s^1$	13.595
2	He	$1s^2$	24.581
3	Li	[He] $2s^1$	5.39
4	Be	$2s^2$	9.320
5	B	$2s^2 2p^1$	8.296
6	C	$2s^2 2p^2$	11.256
7	N	$2s^2 2p^3$	14.545
8	O	$2s^2 2p^4$	13.614
9	F	$2s^2 2p^5$	17.418
10	Ne	$2s^2 2p^6$	21.559
11	Na	[Ne] $3s^1$	5.138
12	Mg	$3s^2$	7.644
13	Al	$3s^2 3p^1$	5.984
14	Si	$3s^2 3p^2$	8.149
15	P	$3s^2 3p^3$	10.484
16	S	$3s^2 3p^4$	10.357
17	Cl	$3s^2 3p^5$	13.01
18	Ar	$3s^2 3p^6$	15.755
19	K	[Ar] $4s^1$	4.339
20	Ca	$4s^2$	6.111
21	Sc	$3d^1 4s^2$	6.54
22	Ti	$3d^2 4s^2$	6.83
23	V	$3d^3 4s^2$	6.74
24	Cr	$3d^5 4s^1$	6.76
25	Mn	$3d^5 4s^2$	7.432
26	Fe	$3d^6 4s^2$	7.87
27	Co	$3d^7 4s^2$	7.86
28	Ni	$3d^8 4s^2$	7.633
29	Cu	$3d^{10} 4s^1$	7.724
30	Zn	$3d^{10} 4s^2$	9.391
31	Ga	$3d^{10} 4s^2 4p^1$	6.00
32	Ge	$3d^{10} 4s^2 4p^2$	7.88
33	As	$3d^{10} 4s^2 4p^3$	9.81
34	Se	$3d^{10} 4s^2 4p^4$	9.75
35	Br	$3d^{10} 4s^2 4p^5$	11.84
36	Kr	$3d^{10} 4s^2 4p^6$	13.996
37	Rb	[Kr] $5s^1$	4.176
38	Sr	$5s^2$	5.692
39	Y	$4d^1 5s^2$	6.377
40	Zr	$4d^2 5s^2$	
41	Nb	$4d^4 5s^1$	6.881
42	Mo	$4d^5 5s^1$	7.10
43	Tc	$4d^5 5s^2$	7.228
44	Ru	$4d^7 5s^1$	7.365

Note: The bracket notation is used as a shorthand method to avoid repetition in indicating inner-shell electrons. Thus, [He] represents $1s^2$, [Ne] represents $1s^2 2s^2 2p^6$, [Ar] represents $1s^2 2s^2 2p^6 3s^2 3p^6$, and so on.

TABLE 42.4   *Continued*

Atomic Number Z	Symbol	Ground-State Configuration	Ionization Energy (eV)
45	Rh	$4d^85s^1$	7.461
46	Pd	$4d^{10}$	8.33
47	Ag	$4d^{10}5s^1$	7.574
48	Cd	$4d^{10}5s^2$	8.991
49	In	$4d^{10}5s^25p^1$	
50	Sn	$4d^{10}5s^25p^2$	7.342
51	Sb	$4d^{10}5s^25p^3$	8.639
52	Te	$4d^{10}5s^25p^4$	9.01
53	I	$4d^{10}5s^25p^5$	10.454
54	Xe	$4d^{10}5s^25p^6$	12.127
55	Cs	[Xe] $6s^1$	3.893
56	Ba	$6s^2$	5.210
57	La	$5d^16s^2$	5.61
58	Ce	$4f^15d^16s^2$	6.54
59	Pr	$4f^36s^2$	5.48
60	Nd	$4f^46s^2$	5.51
61	Pm	$4f^56s^2$	
62	Fm	$4f^66s^2$	5.6
63	Eu	$4f^76s^2$	5.67
64	Gd	$4f^75d^16s^2$	6.16
65	Tb	$4f^96s^2$	6.74
66	Dy	$4f^{10}6s^2$	
67	Ho	$4f^{11}6s^2$	
68	Er	$4f^{12}6s^2$	
69	Tm	$4f^{13}6s^2$	
70	Yb	$4f^{14}6s^2$	6.22
71	Lu	$4f^{14}5d^16s^2$	6.15
72	Hf	$4f^{14}5d^26s^2$	7.0
73	Ta	$4f^{14}5d^36s^2$	7.88
74	W	$4f^{14}5d^46s^2$	7.98
75	Re	$4f^{14}5d^56s^2$	7.87
76	Os	$4f^{14}5d^66s^2$	8.7
77	Ir	$4f^{14}5d^76s^2$	9.2
78	Pt	$4f^{14}5d^86s^2$	8.88
79	Au	[Xe, $4f^{14}5d^{10}$] $6s^1$	9.22
80	Hg	$6s^2$	10.434
81	Tl	$6s^26p^1$	6.106
82	Pb	$6s^26p^2$	7.415
83	Bi	$6s^26p^3$	7.287
84	Po	$6s^26p^4$	8.43
85	At	$6s^26p^5$	
86	Rn	$6s^26p^6$	10.745
87	Fr	[Rn] $7s^1$	
88	Ra	$7s^2$	5.277
89	Ac	$6d^17s^2$	6.9
90	Th	$6d^27s^2$	
91	Pa	$5f^26d^17s^2$	

continued

TABLE 42.4   *Continued*

Atomic Number Z	Symbol	Ground-State Configuration	Ionization Energy (eV)
92	U	$5f^36d^17s^2$	4.0
93	Np	$5f^46d^17s^2$	
94	Pu	$5f^67s^2$	
95	Am	$5f^77s^2$	
96	Cm	$5f^76d^17s^2$	
97	Bk	$5f^86d^17s^2$	
98	Cf	$5f^{10}6d^07s^2$	
99	Es	$5f^{11}6d^07s^2$	
100	Fm	$5f^{12}6d^07s^2$	
101	Md	$5f^{13}6d^07s^2$	
102	No	$5f^{14}6d^07s^2$	
103	Lr	$5f^{14}6d^17s^2$	
104	Rf	$5f^{14}6d^27s^2$	

that is, they do not join with other atoms to form molecules. They are therefore called *inert gases.*

We can partially understand this behavior by looking at the electronic configurations in Table 42.4. The chemical behavior of an element depends on the outermost shell that contains electrons. Shells inside the outermost one are filled and do not contribute to chemical behavior. The electronic configuration for helium is $1s^2$—the $n = 1$ shell (which is the outermost shell because it is the only shell) is filled. Additionally, the energy of the atom in this configuration is considerably lower than the energy for the configuration in which an electron is in the next available level, the $2s$ subshell. Next, look at the electronic configuration for neon, $1s^22s^22p^6$. Again, the outermost shell ($n = 2$ in this case) is filled, and a wide gap in energy occurs between the filled $2p$ subshell and the next available one, the $3s$ subshell. Argon has the configuration $1s^22s^22p^63s^23p^6$. Here, it is only the $3p$ subshell that is filled, but again a wide gap in energy occurs between the filled $3p$ subshell and the next available one, the $3d$ subshell. We could continue this procedure through all the inert gases; the pattern remains the same. An inert gas is formed when either a shell or a subshell is filled and a large gap in energy occurs between the filled shell or subshell and the next highest available one.

42.7 ▶ ATOMIC SPECTRA

In Chapter 40 we briefly discussed the origin of the visible spectral lines for the hydrogen atom and for hydrogen-like ions. Recall that an atom emits electromagnetic radiation if the atom in an excited state makes a transition to a lower energy state. The set of wavelengths that is observed when a specific atom undergoes such processes is called the **emission spectrum** for that atom. Likewise, atoms having electrons in the ground-state configuration can absorb electromagnetic radiation at specific wavelengths, giving rise to an **absorption spectrum.** Such spectra can be used to identify elements.

The energy level diagram for hydrogen is shown in Figure 42.15. The various diagonal lines represent allowed transitions between stationary states. Whenever an atom makes a transition from a higher energy state to a lower one, a photon of

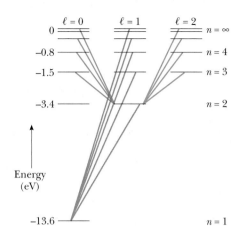

Figure 42.15 Some allowed electronic transitions for hydrogen, represented by the colored lines. These transitions must obey the selection rule $\Delta\ell = \pm 1$.

light is emitted. The frequency of this photon is $f = \Delta E/h$, where ΔE is the energy difference between the two states and h is Planck's constant. The **selection rules** for the *allowed transitions* are

$$\Delta\ell = \pm 1 \qquad \text{and} \qquad \Delta m_\ell = 0, \pm 1 \qquad \textbf{(42.16)}$$

▸ Selection rules for allowed atomic transitions

Transitions that do not obey these selection rules are said to be *forbidden*. (Such transitions can occur, but their probability is low relative to the probability of allowed transitions.)

Because the orbital angular momentum of an atom changes when a photon is emitted or absorbed (that is, as a result of a transition between states) and because angular momentum must be conserved, we conclude that **the photon involved in the process must carry angular momentum.** In fact, the photon has an angular momentum equivalent to that of a particle having a spin of 1. Hence, a photon has energy, linear momentum, and angular momentum.

▸ The photon carries angular momentum

Recall from Equation 40.32 that the allowed energies for one-electron atoms, such as hydrogen and He$^+$, are

$$E_n = -\frac{k_e e^2}{2a_0}\left(\frac{Z^2}{n^2}\right) = -\frac{13.6Z^2}{n^2}\ \text{eV} \qquad \textbf{(42.17)}$$

▸ Allowed energies for one-electron atoms

For multielectron atoms, the positive nuclear charge Ze is largely shielded by the negative charge of the inner-shell electrons. Hence, the outer electrons interact with a net charge that is much smaller than the nuclear charge. The expression for the allowed energies for multielectron atoms has the same form as Equation 42.17 with Z replaced by an effective atomic number Z_{eff}:

$$E_n = -\frac{13.6Z_{\text{eff}}^2}{n^2}\ \text{eV} \qquad \textbf{(42.18)}$$

▸ Allowed energies for multielectron atoms

where Z_{eff} depends on n and ℓ.

It is interesting to plot ionization energy (see Section 40.5) versus atomic number Z, as in Figure 42.16. Note the pattern of $\Delta Z = 2, 8, 8, 18, 18, 32$ for the various peaks. This pattern follows from the exclusion principle and helps explain why the elements repeat their chemical properties in groups. For example, the peaks at $Z = 2, 10, 18$, and 36 correspond to the elements helium, neon, argon, and krypton, which all have filled outermost shells. These elements have relatively high ionization energies and similar chemical behavior.

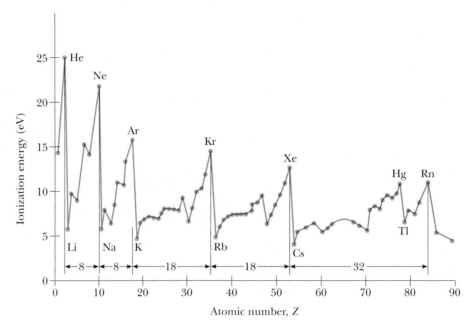

Figure 42.16   Ionization energy of the elements versus atomic number Z. *(Adapted from J. Orear, Physics, New York, Macmillan, 1979.)*

X-Ray Spectra

X-rays are emitted when high-energy electrons or any other charged particles bombard a metal target. The x-ray spectrum typically consists of a broad continuous band containing a series of sharp lines, as shown in Figure 42.17. The continuous spectrum is the result of collisions between incoming electrons and atoms in the target. The kinetic energy lost by the electrons during the collisions emerges as the energy ($E = hf$) of the x-ray photons radiated from the target. The sharp lines superimposed on the continuous spectrum are known as **characteristic x-rays** because they are characteristic of the target material. They were discovered in 1908, but their origin remained unexplained until the details of atomic structure, particularly the shell structure of the atom, were discovered.

Characteristic x-ray emission occurs when a bombarding electron that collides with a target atom has sufficient energy to remove an inner-shell electron from the atom. The vacancy created in the shell is filled when an electron from a higher level drops down into it. This transition is accompanied by the emission of a photon whose energy equals the difference in energy between the two levels. Typically, the energy of such transitions is greater than 10 000 eV, and the emitted photons have wavelengths in the range of 0.001 nm to 0.1 nm, in the x-ray region of the electromagnetic spectrum.

Let us assume that the incoming electron has dislodged an atomic electron from the innermost shell—the K shell. If the vacancy is filled by an electron dropping from the next higher shell—the L shell—the photon emitted has an energy corresponding to the K_α characteristic x-ray line on the curve of Figure 42.17. If the vacancy is filled by an electron dropping from the M shell, the K_β line in Figure 42.17 is produced.

Other characteristic x-ray lines are formed when electrons drop from upper levels to vacancies other than those in the K shell. For example, L lines are produced when vacancies in the L shell are filled by electrons dropping from higher

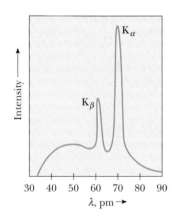

Figure 42.17   The x-ray spectrum of a metal target consists of a broad continuous spectrum containing a number of sharp lines; the lines are due to *characteristic x-rays*. The data shown were obtained when 37-keV electrons bombarded a molybdenum target.

shells. An L_α line is produced as an electron drops from the M shell to the L shell, and an L_β line is produced by a transition from the N shell to the L shell.

Although multielectron atoms cannot be analyzed exactly with either the Bohr model or the Schrödinger equation, we can apply our knowledge of Gauss's law from Chapter 24 to make some surprisingly accurate estimates of expected x-ray energies and wavelengths. Consider an atom of atomic number Z in which one of the two electrons in the K shell has been ejected. Imagine that we draw a gaussian sphere just inside the most probable radius of the L electrons. The electric field at the position of the L electrons is a combination of the fields created by the nucleus, the single K electron, the other L electrons, and the outer electrons. The wave functions of the outer electrons are such that the electrons have a very high probability of being farther from the nucleus than the L electrons are. Thus, they are much more likely to be outside the gaussian surface than inside and, on the average, do not contribute significantly to the electric field at the position of the L electrons. The effective charge inside the gaussian surface is the positive nuclear charge and one negative charge due to the single K electron. If we ignore the interactions between L electrons, a single L electron behaves as if it experiences an electric field due to a charge $(Z - 1)e$ enclosed by the gaussian surface. The nuclear charge is shielded by the electron in the K shell such that Z_{eff} in Equation 42.18 is $Z - 1$. For higher-level shells, the nuclear charge is shielded by electrons in all of the inner shells.

We can now use Equation 42.18 to estimate the energy associated with an electron in the L shell:

$$E_L = -(Z - 1)^2 \frac{13.6}{2^2} \text{ eV}$$

After the atom makes the transition, there are two electrons in the K shell. Using a similar argument for a gaussian surface drawn just inside the most probable radius for the single K electron, we can argue that the energy associated with one of these electrons is approximately that of a one-electron atom with the nuclear charge reduced by the negative charge of the other electron. Thus,

$$E_K = -(Z - 1)^2 (13.6 \text{ eV}) \qquad \textbf{(42.19)}$$

As Example 42.7 shows, the energy of the atom with an electron in an M shell can be estimated in a similar fashion. Taking the energy difference between the initial and final levels, we can then calculate the energy and wavelength of the emitted photon.

Quick Quiz 42.4

Note in Figure 42.17 that the continuous spectrum stops abruptly at the cutoff wavelength of about 34 pm. Why does a cutoff wavelength occur?

In 1914, Henry G. J. Moseley (1887–1915) plotted the Z values for a number of elements versus $\sqrt{1/\lambda}$, where λ is the wavelength of the K_α line of each element. He found that the plot is a straight line, as in Figure 42.18. This is consistent with rough calculations of the energy levels given by Equation 42.19. From this plot, Moseley determined the Z values of elements that had not yet been discovered and produced a periodic table in excellent agreement with the known chemical properties of the elements. Until that experiment, atomic numbers had been merely placeholders for the elements that appeared in the periodic table, the elements being ordered according to mass.

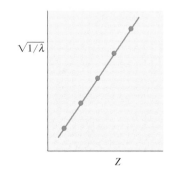

Figure 42.18 A Moseley plot of $\sqrt{1/\lambda}$ versus Z, where λ is the wavelength of the K_α x-ray line of the element of atomic number Z.

EXAMPLE 42.7   **The Energy of an X-Ray**

Determine the energy of the characteristic x-ray emitted from a tungsten (chemical symbol W) target when an electron drops from the M shell ($n = 3$) to a vacancy in the K shell ($n = 1$).

Solution   The atomic number of tungsten is $Z = 74$. Using Equation 42.19, we see that the energy associated with the electron in the K shell is approximately

$$E_K = -(74 - 1)^2(13.6 \text{ eV}) = -72\,500 \text{ eV}$$

An electron in the M shell is subject to an effective nuclear charge that depends on the number of electrons in the $n = 1$ and $n = 2$ states because these electrons shield the M electrons from the nucleus. Because there are eight electrons in the $n = 2$ state and one remaining in the $n = 1$ state, roughly nine electrons shield M electrons from the nucleus, so $Z_{\text{eff}} = Z - 9$. Hence, the energy associated with an electron in the M shell is, from Equation 42.18,

$$E_M = -\frac{13.6 Z_{\text{eff}}^2}{3^2} \text{ eV} = -\frac{13.6(Z - 9)^2}{3^2} \text{ eV}$$

$$= -\frac{(13.6)(74 - 9)^2}{9} \text{ eV} = -6\,380 \text{ eV}$$

Therefore, the emitted x-ray has an energy equal to $E_M - E_K = -6\,380 \text{ eV} - (-72\,500 \text{ eV}) = 66\,100 \text{ eV}$. Despite the approximations we have made in developing Equations 42.18 and 42.19 and the estimation of the effective nuclear charge, this result is in excellent agreement with measurements made on x-rays from tungsten targets.

Exercise   Calculate the wavelength of the emitted x-ray for this transition.

Answer   0.018 8 nm.

42.8   ATOMIC TRANSITIONS

We have seen that an atom absorbs and emits electromagnetic radiation only at frequencies that correspond to the energy separation between allowed states. Let us now look at the details of these processes. Consider an atom having the allowed energy levels labeled E_1, E_2, E_3, . . . in Figure 42.19. When radiation is incident on the atom, only those photons whose energy hf matches the energy separation ΔE between two energy levels can be absorbed by the atom. Figure 42.20 is a schematic diagram representing this process, which is called **stimulated absorption** because the photon stimulates the atom to make the upward transition. At ordinary temperatures, most of the atoms in a sample are in the ground state. If a vessel containing many atoms of a gaseous element is illuminated with radiation of all possible photon frequencies (that is, a continuous spectrum), only those photons having energy $E_2 - E_1$, $E_3 - E_1$, $E_4 - E_1$, $E_3 - E_2$, $E_4 - E_2$, and so on are absorbed by the atoms. As a result of this absorption, some of the atoms are raised to allowed higher energy levels, which, as we learned in Section 40.5, are called **excited states.**

Once an atom is in an excited state, some probability exists that the excited atom will jump back to a lower energy level and emit a photon in the process, as in Figure 42.21. This process is known as **spontaneous emission** because it happens randomly, without requiring an event to trigger the transition. Typically, an atom remains in an excited state for only about 10^{-8} s.

When an atom in an excited state returns to the ground state via two or more intermediate steps, the photons emitted during the process are lower in energy than the original photon absorbed by the atom. This process is called *fluorescence.* In a fluorescent light tube, electrons leaving a filament at the end of the tube collide with atoms of mercury vapor present in the tube, causing the mercury atoms to be elevated into excited states. As these atoms make transitions to lower states, they emit ultraviolet photons that strike a coating on the inner surface of the tube. The coating absorbs the photons and emits visible light by means of fluorescence.

Figure 42.19 Energy level diagram of an atom having various allowed states. The lowest-energy state E_1 is the ground state. All others are excited states.

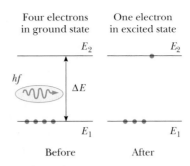

Figure 42.20 Stimulated absorption of a photon. The dots represent electrons. One electron is transferred from the ground state to the excited state when the atom absorbs a photon of energy $hf = E_2 - E_1$.

Phosphorescent materials glow because of a similar process, but the excited atoms may remain in an excited state for periods ranging from a few seconds to several hours. Eventually, the atoms drop to the ground state and while doing so emit visible light. For this reason, phosphorescent materials emit light long after being placed in the dark.

Quick Quiz 42.5

Make a drawing similar to Figure 42.21 for fluorescence.

In addition to spontaneous emission, **stimulated emission** occurs. Suppose an atom is in an excited state E_2, as in Figure 42.22. If the excited state is a *metastable state*—that is, if its lifetime is much longer than the typical 10^{-8} s life-

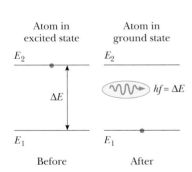

Figure 42.21 Spontaneous emission of a photon by an atom that is initially in the excited state E_2. When the atom falls to the ground state, it emits a photon of energy $hf = E_2 - E_1$.

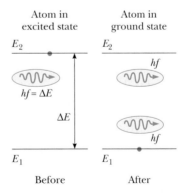

Figure 42.22 Stimulated emission of a photon by an incoming photon of energy hf. Initially, the atom is in the excited state. The incoming photon stimulates the atom to emit a second photon of energy $hf = E_2 - E_1$.

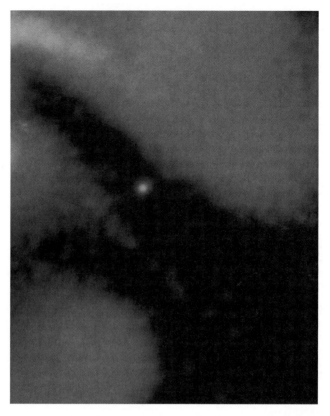

Figure 42.23 A single barium ion (the small dot in the center) glows because it is stimulated by a laser beam (not shown). The surrounding structure is the electromagnetic trap that holds the ion in place. *(Courtesy of David Wineland, National Institute of Standards and Technology)*

time of excited states—then the time interval until spontaneous emission occurs will be relatively long. Let us imagine that in that interval a photon of energy $hf = E_2 - E_1$ is incident on the atom. One possibility is that the photon energy will be sufficient for the photon to ionize the atom. Another possibility is that the interaction between the incoming photon and the atom will cause the atom to return to the ground state and thereby emit a second photon with energy $hf = E_2 - E_1$. In this process the incident photon is not absorbed; thus, after the stimulated emission, two photons with identical energy exist—the incident photon and the emitted photon. The two are in phase—an important consideration in lasers, which we shall discuss in the next section.

In the mid-1980s it became possible to electromagnetically "trap" a single ion (Fig. 42.23) and stimulate it to emit light. This procedure directly confirmed the existence of discrete energy levels in atoms.

Optional Section

42.9 ▸ LASERS AND HOLOGRAPHY

We have described how an incident photon can cause atomic energy transitions either upward (stimulated absorption) or downward (stimulated emission). The two processes are equally probable. When light is incident on a collection of atoms, a

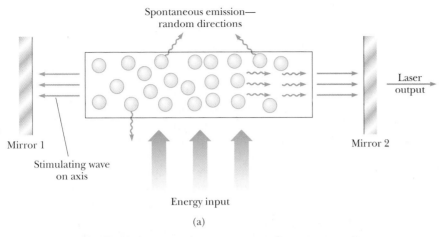

Spontaneous emission—
random directions

Laser
output

Mirror 1

Stimulating wave
on axis

Mirror 2

Energy input

(a)

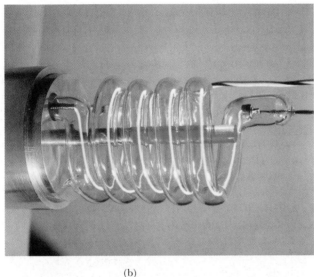

(b)

Figure 42.24 (a) Schematic diagram of a laser design. The tube contains the atoms that are the active medium. An external source of energy (for example, an optical or electrical device) "pumps" the atoms to excited states. The parallel end mirrors confine the photons to the tube, but mirror 2 is slightly transparent. (b) Photograph of the first ruby laser, showing the flash lamp surrounding the ruby rod. *(b, Courtesy of HRL Laboratories LLC, Malibu, CA)*

net absorption of energy usually occurs because, when the system is in thermal equilibrium, many more atoms are in the ground state than in excited states. However, if the situation can be inverted so that more atoms are in an excited state than in the ground state, a net emission of photons can result. Such a condition is called **population inversion.**

This, in fact, is the fundamental principle involved in the operation of a **laser**—an acronym for *l*ight *a*mplification by *s*timulated *e*mission of *r*adiation. The amplification corresponds to a buildup of photons in the system as the result of a chain reaction of events. The following three conditions must be satisfied to achieve laser action:

- The system must be in a state of population inversion.
- The excited state of the system must be a metastable state. When this condition is met, stimulated emission occurs before spontaneous emission.

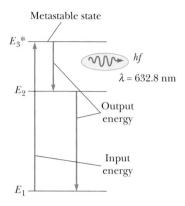

Figure 42.25 Energy level diagram for a neon atom in a helium–neon laser. The atom emits 632.8-nm photons through stimulated emission in the transition E_3^*–E_2. This is the source of coherent light in the laser.

- The emitted photons must be confined in the system long enough to stimulate further emission from other excited atoms. This confinement is achieved through the use of reflecting mirrors at the ends of the system. One end is made totally reflecting, and the other is slightly transparent to allow part of the laser beam to escape.

In a helium–neon gas laser, a mixture of helium and neon is confined in a sealed glass tube (Fig. 42.24a). An oscillator connected to the tube causes electrons to sweep through it, colliding with the gas atoms and raising them to excited states. As Figure 42.25 shows, neon atoms are excited to state E_3^* through this process (the asterisk * indicates a metastable state) and also as a result of collisions with excited helium atoms. Stimulated emission occurs as the neon atoms make a transition to state E_2 and neighboring excited atoms are stimulated. This results in the production of coherent light at a wavelength of 632.8 nm.

Applications

Since the development of the first laser in 1960 (shown in Fig. 42.24b), tremendous growth has occurred in laser technology. Lasers that cover wavelengths in the infrared, visible, and ultraviolet regions are now available. Applications include surgical "welding" of detached retinas, precision surveying and length measurement, precision cutting of metals and other materials (such as the fabric in Figure 42.26), and telephone communication along optical fibers. These and other applications are possible because of the unique characteristics of laser light. In addition to being highly monochromatic, laser light is also highly directional and can be sharply focused to produce regions of extremely intense light energy (with energy densities 10^{12} times that in the flame of a typical cutting torch).

Lasers are used in precision long-range distance measurement (range finding). In recent years it has become important, for astronomical and geophysical

Figure 42.26 This robot carrying laser scissors, which can cut up to 50 layers of fabric at a time, is one of the many applications of laser technology. *(Philippe Plailly/SPL/Photo Researchers, Inc.)*

purposes, to measure as precisely as possible the distances from various points on the surface of the Earth to a point on the Moon's surface. To facilitate this, the Apollo astronauts set up a 0.5-m square of reflector prisms on the Moon, which enables laser pulses directed from an Earth station to be retroreflected to the same station (see Fig. 35.8a). Using the known speed of light and the measured round-trip travel time of a 1-ns pulse, the Earth–Moon distance, 380 000 km, can be determined to a precision of better than 10 cm.

Medical applications use the fact that various laser wavelengths can be absorbed in specific biological tissues. For example, certain laser procedures have greatly reduced blindness in glaucoma and diabetes patients. Glaucoma is a widespread eye condition characterized by a high fluid pressure in the eye, a condition that can lead to destruction of the optic nerve. A simple laser operation (iridectomy) can "burn" open a tiny hole in a clogged membrane, relieving the destructive pressure. A serious side effect of diabetes is neovascularization, the proliferation of weak blood vessels, which often leak blood. When this occurs in the retina, vision deteriorates (diabetic retinopathy) and finally is destroyed. It is now possible to direct the green light from an argon ion laser through the clear eye lens and eye fluid, focus on the retina edges, and photocoagulate the leaky vessels. Even people who have only minor vision defects such as nearsightedness are benefiting from the use of lasers to reshape the cornea, changing its focal length and reducing the need for eyeglasses.

Laser surgery is now an everyday occurrence at hospitals around the world. Infrared light at 10 μm from a carbon dioxide laser can cut through muscle tissue, primarily by vaporizing the water contained in cellular material. Laser power of about 100 W is required in this technique. The advantage of the "laser knife" over conventional methods is that laser radiation cuts tissue and coagulates blood at the same time, leading to a substantial reduction in blood loss. In addition, the technique virtually eliminates cell migration, an important consideration when tumors are being removed.

A laser beam can be trapped in fine glass-fiber light guides (endoscopes) by means of total internal reflection. The light fibers can thus be introduced through natural orifices, conducted around internal organs, and directed to specific interior body locations, eliminating the need for invasive surgery. For example, bleeding in the gastrointestinal tract can be optically cauterized by fiber-optic endoscopes inserted through the mouth.

In biological and medical research, it is often important to isolate and collect unusual cells for study and growth. A laser cell separator exploits the fact that specific cells can be tagged with fluorescent dyes. All cells are then dropped from a tiny charged nozzle and laser-scanned for the dye tag. If triggered by the correct light-emitting tag, a small voltage applied to parallel plates deflects the falling electrically charged cell into a collection beaker. This is an efficient method for extracting the proverbial needle from the haystack.

One of the most unusual and interesting applications of the laser is in the production of three-dimensional images in a process called **holography.** Figure 42.27 shows how a hologram is made. Light from the laser is split into two parts by a half-silvered mirror at B. After passing through lens L_1, which diverges the light rays, one part of the beam reflects off the object to be photographed and strikes an ordinary photographic film. The other part of the beam is diverged by lens L_2, reflects from mirrors M_1 and M_2, and finally strikes the film. The two beams overlap on the film to form an extremely complicated interference pattern. Such an interference pattern can be produced only if the phase relationship of the waves in the two beams is constant throughout the exposure of the film. This condition is met by illuminating the scene with coherent laser radiation. The hologram records not

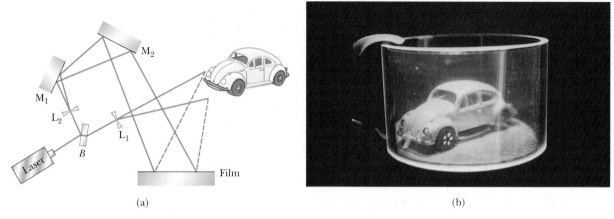

Figure 42.27   (a) Arrangement for producing a hologram. (b) Photograph of a hologram that uses a cylindrical film. *(Courtesy of Central Scientific Company.)*

only the intensity of the light scattered from the object (as in a conventional photograph) but the phase difference between the beam reflected from the mirrors and the beam scattered from the object. Because of this phase difference, the interference pattern that is formed produces an image having a full three-dimensional perspective.

A hologram is best viewed by allowing coherent light to pass through the developed film as one looks back along the direction from which the beam comes. The light passing through the hologram is diffracted. It emerges in a form identical to the light that left the object while the hologram was being recorded; as a result, viewing the hologram is almost like looking through a window at the object. Figure 42.27b is a photograph of a hologram that was made using a cylindrical film. One sees a three-dimensional image in which the perspective changes as the viewer's head moves.

SUMMARY

Quantum mechanics can be applied to the hydrogen atom by the use of the potential energy function $U(r) = -k_e e^2/r$ in the Schrödinger equation. The solution to this equation yields wave functions for allowed states and allowed energies:

$$E_n = -\left(\frac{k_e e^2}{2a_0}\right)\frac{1}{n^2} = -\frac{13.606}{n^2}\ \text{eV} \qquad n = 1, 2, 3, \ldots \qquad \textbf{(42.2)}$$

where n is the **principal quantum number.** The allowed wave functions depend on three quantum numbers: n, ℓ, and m_ℓ, where ℓ is the **orbital quantum number** and m_ℓ is the **orbital magnetic quantum number.** The restrictions on the quantum numbers are

$$n = 1, 2, 3, \ldots$$

$$\ell = 0, 1, 2, \ldots, n - 1$$

$$m_\ell = -\ell, -\ell + 1, \ldots, \ell - 1, \ell$$

All states having the same principal quantum number n form a **shell,** identified by the letters K, L, M, . . . (corresponding to $n = 1, 2, 3, \ldots$). All states having the same values of n and ℓ form a **subshell,** designated by the letters s, p, d, f, . . . (corresponding to $\ell = 0, 1, 2, 3, \ldots$).

To completely describe a quantum state, it is necessary to include a fourth quantum number m_s, called the **spin magnetic quantum number.** This quantum number can have only two values, $\pm\frac{1}{2}$.

An atom in a state characterized by a specific value of n can have the following values of L, the magnitude of the atom's orbital angular momentum $\mathbf{L}$:

$$L = \sqrt{\ell\,(\ell+1)}\,\hbar \qquad \ell = 0, 1, 2, \ldots, n-1 \qquad \textbf{(42.9)}$$

The allowed values of the projection of $\mathbf{L}$ along the z axis are

$$L_z = m_\ell \hbar \qquad\qquad \textbf{(42.10)}$$

Only discrete values of L_z are allowed, as determined by the restrictions on m_ℓ. This quantization of L_z is referred to as **space quantization.**

The electron has an intrinsic angular momentum called the **spin angular momentum.** That is, the total angular momentum of an electron in an atom has two contributions, one arising from the spin of the electron ($\mathbf{S}$) and one arising from the orbital motion of the electron ($\mathbf{L}$). Electron spin can be described by a single quantum number $s = \frac{1}{2}$. The magnitude of the spin angular momentum is

$$S = \frac{\sqrt{3}}{2}\,\hbar \qquad\qquad \textbf{(42.12)}$$

and the z component of $\mathbf{S}$ is

$$S_z = m_s\hbar = \pm\tfrac{1}{2}\hbar \qquad\qquad \textbf{(42.13)}$$

That is, the spin angular momentum is also quantized in space, as specified by the spin magnetic quantum number $m_s = \pm\frac{1}{2}$.

The magnetic moment $\boldsymbol{\mu}_{\text{spin}}$ associated with the spin angular momentum of an electron is

$$\boldsymbol{\mu}_{\text{spin}} = -\frac{e}{m_e}\,\mathbf{S} \qquad\qquad \textbf{(42.14)}$$

The z component of $\boldsymbol{\mu}_{\text{spin}}$ can have the values

$$\mu_{\text{spin},\,z} = \pm\frac{e\hbar}{2m_e} \qquad\qquad \textbf{(42.15)}$$

The **exclusion principle** states that **no two electrons in an atom can be in the same quantum state.** In other words, no two electrons can have the same set of quantum numbers n, ℓ, m_ℓ, and m_s. Using this principle, the electronic configurations of the elements can be determined. This serves as a basis for understanding atomic structure and the chemical properties of the elements.

The allowed electronic transitions between any two levels in an atom are governed by the **selection rules**

$$\Delta\ell = \pm 1 \qquad \text{and} \qquad \Delta m_\ell = 0, \pm 1 \qquad \textbf{(42.16)}$$

The x-ray spectrum of a metal target consists of a set of sharp characteristic lines superimposed on a broad continuous spectrum. **Characteristic x-rays** are emitted by atoms when an electron undergoes a transition from an outer shell to a vacancy in an inner shell.

Atomic transitions can be described with three processes: **stimulated absorption,** in which an incoming photon raises the atom to a higher energy state; **spontaneous emission,** in which the atom makes a transition to a lower energy state, emitting a photon; and **stimulated emission,** in which an incident photon causes an excited atom to make a downward transition, emitting a photon identical to the incident one.

QUESTIONS

1. Why are three quantum numbers needed to describe the state of a one-electron atom (neglecting spin)?
2. Compare the Bohr theory and the Schrödinger treatment of the hydrogen atom. Comment on the total energy and the orbital angular momentum.
3. Why is the direction of the orbital angular momentum of an electron opposite that of its magnetic moment?
4. Why is a nonuniform magnetic field used in the Stern–Gerlach experiment?
5. Could the Stern–Gerlach experiment be performed with ions rather than neutral atoms? Explain.
6. Describe some experiments that support the conclusion that the spin magnetic quantum number for electrons can have only the values $\pm\frac{1}{2}$.
7. Discuss some of the consequences of the exclusion principle.
8. Why do lithium, potassium, and sodium exhibit similar chemical properties?
9. Explain why a photon must have a spin of 1.
10. An energy of about 21 eV is required to excite an electron in a helium atom from the 1s state to the 2s state.

The same transition for the He$^+$ ion requires approximately twice as much energy. Explain.
11. Does the intensity of light from a laser fall off as $1/r^2$?
12. The absorption or emission spectrum of a gas consists of lines that broaden as the density of gas molecules increases. Why do you suppose this occurs?
13. How is it possible that electrons, whose positions are described by a probability distribution around a nucleus, can exist in states of *definite* energy (e.g., 1s, 2p, 3d, . . .)?
14. It is easy to understand how two electrons (one spin up, one spin down) can fill the 1s shell for a helium atom. How is it possible that eight more electrons can fit into the 2s, 2p level to complete the $1s^2 2s^2 2p^6$ shell for a neon atom?
15. In 1914, Henry Moseley discovered how to define the atomic number of an element from its characteristic x-ray spectrum. How was this possible? (*Hint:* See Figs. 42.17 and 42.18.)
16. What are the advantages of using monochromatic light to view a holographic image?
17. Why is stimulated emission so important in the operation of a laser?

PROBLEMS

1, 2, 3 = straightforward, intermediate, challenging   ☐ = full solution available in the *Student Solutions Manual and Study Guide*
WEB = solution posted at **http://www.saunderscollege.com/physics/**   🖥 = Computer useful in solving problem   📱 = Interactive Physics
☐ = paired numerical/symbolic problems

Section 42.1   Early Models of the Atom

1. In the Rutherford scattering experiment, 4.00-MeV alpha particles (^{4}He nuclei containing 2 protons and 2 neutrons) scatter off gold nuclei (containing 79 protons and 118 neutrons). If an alpha particle makes a direct head-on collision with the gold nucleus and scatters backward at 180°, determine (a) the distance of closest approach of the alpha particle to the gold nucleus and (b) the maximum force exerted on the alpha particle. Assume that the gold nucleus remains fixed throughout the entire process.
2. In the Rutherford scattering experiment, alpha particles of energy E (^{4}He nuclei containing 2 protons and 2 neutrons) scatter off a target whose atoms have an atomic number Z. If an alpha particle makes a direct head-on collision with a target nucleus and scatters backward at 180°, determine (a) the distance of closest approach of the alpha particle to the target nucleus and (b) the maximum force exerted on the alpha particle. Assume that the target nucleus remains fixed throughout the entire process.

Section 42.2   The Hydrogen Atom Revisited

3. A photon with energy 2.28 eV is barely capable of causing a photoelectric effect when it strikes a sodium plate. Suppose that the photon is instead absorbed by hydrogen. Find (a) the minimum n for a hydrogen atom that can be ionized by such a photon and (b) the speed of the released electron far from the nucleus.
4. The Balmer series for the hydrogen atom corresponds to electronic transitions that terminate in the state with quantum number $n = 2$, as shown in Figure 40.18. (a) Consider the photon of longest wavelength; determine its energy and wavelength. (b) Consider the spectral line of shortest wavelength; find its photon energy and wavelength.
5. A general expression for the energy levels of one-electron atoms and ions is

$$E_n = -\frac{\mu k_e^2 q_1^2 q_2^2}{2\hbar^2 n^2}$$

where k_e is the Coulomb constant, q_1 and q_2 are the charges of the two particles, and μ is the reduced mass, given by $\mu = m_1 m_2/(m_1 + m_2)$. In Problem 4 we found that the wavelength for the $n = 3$ to $n = 2$ transition of the hydrogen atom is 656.3 nm (visible red light). What are the wavelengths for this same transition in (a) positronium, which consists of an electron and a positron, and (b) singly ionized helium? (*Note:* A positron is a positively charged electron.)

6. Ordinary hydrogen gas is a mixture of two kinds of atoms (isotopes) containing either one- or two-particle nuclei. These isotopes are hydrogen-1 with a proton nucleus and deuterium with a deuteron nucleus. (A deuteron is one proton and one neutron bound together.) Hydrogen-1 and deuterium have identical chemical properties but can be separated via an ultra-centrifuge or other methods. Their emission spectra show lines of the same colors at very slightly different wavelengths. (a) Use the equation given in Problem 5 to show that the difference in wavelength between the hydrogen and deuterium spectral lines associated with a particular electronic transition is given by

$$\lambda_H - \lambda_D = (1 - \mu_H/\mu_D)\lambda_H$$

(b) Evaluate the wavelength difference for the H_α line of hydrogen, with wavelength 656.3 nm, emitted by an atom making a transition from an $n = 3$ state to an $n = 2$ state.

Section 42.3   The Spin Magnetic Quantum Number

7. List the possible sets of quantum numbers for electrons in (a) the $3d$ subshell and (b) the $3p$ subshell.

Section 42.4   The Wave Functions for Hydrogen

8. Plot the wave function $\psi_{1s}(r)$ (see Eq. 42.3) and the radial probability density function $P_{1s}(r)$ (see Eq. 42.6) for hydrogen. Let r range from 0 to $1.5a_0$, where a_0 is the Bohr radius.

9. The ground-state wave function for the electron in a hydrogen atom is

$$\psi_{1s}(r) = \frac{1}{\sqrt{\pi a_0^3}} e^{-r/a_0}$$

where r is the radial coordinate of the electron and a_0 is the Bohr radius. (a) Show that the wave function as given is normalized. (b) Find the probability of locating the electron between $r_1 = a_0/2$ and $r_2 = 3a_0/2$.

10. The wave function for an electron in the $2p$ state of hydrogen is described by the expression

$$\psi_{2p}(r) = \frac{1}{\sqrt{3}(2a_0)^{3/2}} \frac{r}{a_0} e^{-r/2a_0}$$

What is the most likely distance from the nucleus to find an electron in the $2p$ state? (See Fig. 42.8.)

WEB 11. Show that the $1s$ wave function for an electron in hydrogen,

$$\psi_{1s}(r) = \frac{1}{\sqrt{\pi a_0^3}} e^{-r/a_0}$$

satisfies the radially symmetric Schrödinger equation,

$$-\frac{\hbar^2}{2m_e}\left(\frac{d^2\psi}{dr^2} + \frac{2}{r}\frac{d\psi}{dr}\right) - \frac{k_e e^2}{r}\psi = E\psi$$

12. During a particular period of time, an electron in the ground state of a hydrogen atom is "observed" 1 000 times at a distance $a_0/2$ from the nucleus. How many times is this electron observed at a distance $2a_0$ from the nucleus during this period of "observation"?

Section 42.5   The Other Quantum Numbers

13. Calculate the angular momentum for an electron in (a) the $4d$ state and (b) the $6f$ state.

14. If an electron has an orbital angular momentum of $4.714 \times 10^{-34}\,\text{J}\cdot\text{s}$, what is the orbital quantum number for the state of the electron?

15. A hydrogen atom is in its fifth excited state. The atom emits a 1 090-nm wavelength photon. Determine the maximum possible orbital angular momentum of the electron after emission.

16. Find all possible values of L, L_z, and θ for an electron in a $3d$ state of hydrogen.

WEB 17. How many sets of quantum numbers are possible for an electron for which (a) $n = 1$, (b) $n = 2$, (c) $n = 3$, (d) $n = 4$, and (e) $n = 5$? Check your results to show that they agree with the general rule that the number of sets of quantum numbers is equal to $2n^2$.

18. The z component of the electron's spin magnetic moment is given by the Bohr magneton, $\mu_B = e\hbar/2m_e$. Show that the Bohr magneton has the numerical value $9.27 \times 10^{-24}\,\text{J/T} = 5.79 \times 10^{-5}\,\text{eV/T}$.

19. (a) Find the mass density of a proton, picturing it as a solid sphere of radius 1.00×10^{-15} m. (b) Consider a classical model of an electron as a solid sphere with the same density as the proton. Find its radius. (c) If this electron possesses spin angular momentum $I\omega = \hbar/2$ because of classical rotation about the z axis, determine the speed of a point on the equator of the electron, and (d) compare this speed to the speed of light.

20. All objects, large and small, behave quantum-mechanically. (a) Estimate the quantum number ℓ for the Earth in its orbit about the Sun. (b) What energy change (in joules) would occur if the Earth made a transition to an adjacent allowed state?

21. Like the electron, the nucleus of an atom has spin angular momentum and a corresponding magnetic moment. The z component of the spin magnetic moment for a nucleus is characterized by the *nuclear magneton* $\mu_n = e\hbar/2m_p$, where m_p is the proton mass. (a) Calculate the value of μ_n in joules per tesla and in electron volts per tesla. (b) Determine the ratio μ_n/μ_B, and comment on your result.

22. An electron is in the N shell. Determine the maximum value of the z component of its angular momentum.

23. The ρ-meson has a charge of $-e$, a spin quantum number of 1, and a mass 1 507 times that of the electron.

Imagine that the electrons in atoms are replaced by ρ-mesons, and list the possible sets of quantum numbers for ρ-mesons in the $3d$ subshell.

Section 42.6   The Exclusion Principle and the Periodic Table

24. (a) Write out the electronic configuration for the ground state of oxygen ($Z = 8$). (b) Write out the values for the set of quantum numbers n, ℓ, m_ℓ, and m_s for each electron in oxygen.

25. Going down the periodic table, which subshell is filled first, the $3d$ or the $4s$ subshell? Which electronic configuration has a lower energy: $[\text{Ar}]3d^4 4s^2$ or $[\text{Ar}]3d^5 4s^1$? Which has the greater number of unpaired spins? Identify this element, and discuss Hund's rule in this case. (*Note:* The notation [Ar] represents the filled configuration for argon.)

26. Two electrons in the same atom both have $n = 3$ and $\ell = 1$. List the quantum numbers for the possible states of the atom. (b) How many states would be possible if the exclusion principle were inoperative?

27. Consider an atom in its ground state, with its outer electrons completely filling the M shell. (a) Identify the atom. (b) List the number of electrons in each subshell.

28. For a neutral atom of element 110, what would be the probable electronic configuration?

WEB **29.** (a) Scanning through Table 42.4 in order of increasing atomic number, note that the electrons fill the subshells in such a way that the subshells with the lowest values of $n + \ell$ are filled first. If two subshells have the same value of $n + \ell$, the one with the lower value of n is filled first. Using these two rules, write the order in which the subshells are filled through $n + \ell = 7$. (b) Predict the chemical valence for the elements that have atomic numbers 15, 47, and 86, and compare your predictions with the actual valences.

30. Devise a table similar to that shown in Figure 42.14 for atoms containing 11 through 19 electrons. Use Hund's rule and educated guesswork.

Section 42.7   Atomic Spectra

31. (a) Determine the possible values of the quantum numbers ℓ and m_ℓ for the He$^+$ ion in the state corresponding to $n = 3$. (b) What is the energy of this state?

32. If you wish to produce 10.0-nm x-rays in the laboratory, what is the minimum voltage you must use in accelerating the electrons?

33. A tungsten target is struck by electrons that have been accelerated from rest through a 40.0-kV potential difference. Find the shortest wavelength of the radiation emitted.

34. In x-ray production, electrons are accelerated through a high voltage ΔV and then decelerated by striking a target. Show that the shortest-wavelength x-ray that can be produced is

$$\lambda_{\text{min}} = \frac{1\,240\text{ nm} \cdot \text{V}}{\Delta V}$$

35. Use the method illustrated in Example 42.7 to calculate the wavelength of the x-ray emitted from a molybdenum target ($Z = 42$) when an electron moves from the L shell ($n = 2$) to the K shell ($n = 1$).

36. The wavelength of characteristic x-rays corresponding to the K$_\beta$ line is 0.152 nm. Determine the material in the target.

37. Electrons are shot into a bismuth target, and x-rays are emitted. Determine (a) the M-to L-shell transitional energy for Bi and (b) the wavelength of the x-ray emitted when an electron falls from the M shell into the L shell.

38. The K series of the discrete spectrum of tungsten contains wavelengths of 0.018 5 nm, 0.020 9 nm, and 0.021 5 nm. The K-shell ionization energy is 69.5 keV. Determine the ionization energies of the L, M, and N shells. Sketch the transitions.

39. When the outermost electron of an alkali atom is excited, it is found that states with the same n but different ℓ have slightly different energies because they penetrate into the central core to different degrees. Low-ℓ orbitals penetrate more, while higher-ℓ orbitals penetrate less. The wavelengths of absorption lines are given approximately by the equation

$$1/\lambda_{n\ell \to n'\ell'} = R_H[(n - \delta_\ell)^{-2} - (n' - \delta_{\ell'})^{-2}]$$

Observe that it is like Equation 40.29, which describes hydrogen, but with the principal quantum numbers n replaced by effective quantum numbers. Here δ_ℓ is the "quantum defect" associated with orbital quantum number ℓ. The value of δ_ℓ is independent of n. For sodium (Na), $\delta_0 = 1.35$. The longest wavelength for an absorption transition carrying Na from its ground state to a state with higher principal quantum number is 330 nm. (a) For what other value of ℓ can you determine the quantum defect, and (b) what is the value of that defect?

Section 42.8   Atomic Transitions

40. The familiar yellow light from a sodium-vapor street lamp results from the $3p \to 3s$ transition in ^{11}Na. Evaluate the wavelength of this light given that the energy difference $E_{3p} - E_{3s} = 2.10$ eV.

41. Assume that a great number n of identical atoms are in a first excited state. The rate dn/dt at which this population will de-excite is $dn/dt = -Pn$, where P is the quantum transition probability rate. The transition rate is given, in turn, by $P = A + u_f B$, where A is the Einstein coefficient for spontaneous emission and B is the Einstein coefficient for stimulated emission due to the presence of photons with energy density u_f per unit frequency. Einstein showed that these coefficients are related by $A = 16\pi^2 \hbar B/\lambda^3$. For an atomic transition of wavelength 645 nm, what must be the energy density of photons for stimulated emission to be as important as spontaneous emission?

(Optional)

Section 42.9   Lasers and Holography

42. The carbon dioxide laser is one of the most powerful developed. The energy difference between the two laser levels is 0.117 eV. Determine the frequency and wavelength of the radiation emitted by this laser. In what portion of the electromagnetic spectrum is this radiation?

WEB **43.** A ruby laser delivers a 10.0-ns pulse of 1.00 MW average power. If the photons have a wavelength of 694.3 nm, how many are contained in the pulse?

44. An important characteristic of a laser is its gain G, specifying the relative enhancement of light-beam intensity over the length L of the laser. When $G = 1.05$, a 5% increase in intensity occurs as the light makes one pass through the laser. The gain is given by

$$G = e^{\sigma(n_u - n_\ell)L}$$

In this equation σ is the atomic-absorption cross-section for the lasing transition, with units of length squared. It is related to the quantum transition probability. The variables n_u and n_ℓ are the number densities (units of length^{-3}) of active atoms in the upper and lower energy states of the laser transition. If $L = 0.500$ m and $\sigma = 1.00 \times 10^{-18}$ m^2 for a particular laser, what number density inversion $n_u - n_\ell$ must be maintained to have a gain of 1.05?

45. The number N of atoms in a particular state is called the population of that state. This number depends on the energy of that state and the temperature. In thermal equilibrium the population of atoms in a state of energy E_n is given by a Boltzmann distribution expression

$$N = N_g e^{-(E_n - E_g)/k_B T}$$

where T is the absolute temperature and N_g is the population of the ground state, of energy E_g. (a) Before the power is switched on, the neon atoms in a laser are in thermal equilibrium at 27.0°C. Find the equilibrium ratio of the populations of the states $E_3{}^*$ and E_2 shown in Figure 42.25. (b) Find the equilibrium ratio at 4.00 K of the populations of the two states in a ruby laser that can produce a light beam of wavelength 694.3 nm.

46. Lasers operate by a clever artificial production of a "population inversion" between the upper and lower atomic energy states involved in the lasing transition. This means that more atoms occur with electrons in the upper excited state than in the lower one. Consider the ruby laser transition at 694.3 nm. Assume that 2% more atoms occur in the upper state than in the lower. For simplicity, assume that both levels have only one quantum state associated with them. (a) To demonstrate how unnatural such a situation is, find the temperature for which the Boltzmann distribution describes a 2.00% population inversion. (b) Why does such a situation not occur naturally?

ADDITIONAL PROBLEMS

47. An Nd:YAG laser used in eye surgery emits a 3.00-mJ pulse in 1.00 ns, focused to a spot 30.0 μm in diameter on the retina. (a) Find (in SI units) the power per unit area at the retina. (This quantity is called the irradiance.) (b) What energy is delivered to an area of molecular size, taken as a circular area 0.600 nm in diameter?

48. **Review Problem.**   (a) How much energy is required to cause an electron in hydrogen to move from the $n = 1$ state to the $n = 2$ state? (b) Suppose the electrons gain this energy through collisions among hydrogen atoms at a high temperature. At what temperature would the average atomic kinetic energy $3k_B T/2$, where k_B is the Boltzmann constant, be great enough to excite the electrons?

49. Show that the average value of r for the $1s$ state of hydrogen has the value $3a_0/2$. (*Hint:* Use Eq. 42.6.)

50. Find the average (expectation) value of $1/r$ in the $1s$ state of hydrogen. It is given by

$$\langle 1/r \rangle = \int_{\text{all space}} |\psi|^2 (1/r)\, dV = \int_0^\infty P(r)\,(1/r)\, dr$$

Is the result equal to the inverse of the average value of r?

51. Suppose a hydrogen atom is in the $2s$ state. Taking $r = a_0$, calculate values for (a) $\psi_{2s}(a_0)$, (b) $|\psi_{2s}(a_0)|^2$, and (c) $P_{2s}(a_0)$. (*Hint:* Use Eq. 42.7.)

52. As noted in a previous chapter, the muon is an elementary particle with the charge of an electron but a mass 207 times greater than that of an electron. Muonium is an "atom" composed of a muon and a proton. Using the formula for the energy levels of hydrogen-like atoms given in Problem 42.5, find the ionization energy of ground-state muonium.

53. A pulsed ruby laser emits light at 694.3 nm. For a 14.0-ps pulse containing 3.00 J of energy, find (a) the physical length of the pulse as it travels through space and (b) the number of photons in it. (c) If the beam has a circular cross-section of 0.600 cm diameter, find the number of photons per cubic millimeter.

54. A pulsed laser emits light of wavelength λ. For a pulse of duration t having energy E, find (a) the physical length of the pulse as it travels through space and (b) the number of photons in it. (c) If the beam has a circular cross section of diameter d, find the number of photons per unit volume.

55. (a) Show that the most probable radial position for an electron in the $2s$ state of hydrogen is $r = 5.236a_0$. (b) Show that the wave function given by Equation 42.7 is normalized.

56. The force on a magnetic moment μ_z in a nonuniform magnetic field B_z is given by $F_z = \mu_z(dB_z/dz)$. If a beam of silver atoms travels a horizontal distance of 1.00 m through such a field and each atom has a speed of 100 m/s, how strong must be the field gradient dB_z/dz to deflect the beam 1.00 mm?

57. An electron in chromium moves from the $n = 2$ state to the $n = 1$ state without emitting a photon. Instead, the excess energy is transferred to an outer electron (one in the $n = 4$ state), which is then ejected by the atom. (This is called an Auger [pronounced 'ohjay'] process, and the ejected electron is referred to as an Auger electron.) Use the Bohr theory to find the kinetic energy of the Auger electron.

58. Suppose the ionization energy of an atom is 4.10 eV. In the spectrum of this same atom, we observe emission lines with wavelengths 310 nm, 400 nm, and 1 377.8 nm. Use this information to construct the energy-level diagram with the fewest levels. Assume that the higher levels are closer together.

59. All atoms have the same size, to an order of magnitude. (a) To show this, estimate the diameters for aluminum (with molar mass = 27.0 g/mol and density 2.70 g/cm³) and uranium (with molar mass = 238 g/mol and density 18.9 g/cm³). (b) What do the results imply about the wave functions for inner-shell electrons as we progress to higher and higher atomic mass atoms? (*Hint:* The molar volume is approximately $D^3 N_A$, where D is the atomic diameter and N_A is Avogadro's number.)

60. In interstellar space, atomic hydrogen produces the sharp spectral line called the 21-cm radiation, which astronomers find most helpful in detecting clouds of hydrogen between stars. This radiation is useful because interstellar dust that obscures visible wavelengths is transparent to these radio wavelengths. The radiation is not generated by an electronic transition between energy states characterized by n. Instead, in the ground state ($n = 1$), the electron and proton spins may be parallel or antiparallel, with a resultant slight difference in these energy states. (a) Which condition has the higher energy? (b) More precisely, the line has wavelength 21.11 cm. What is the energy difference between the states? (c) The average lifetime in the excited state is about 10^7 yr. Calculate the associated uncertainty in energy of the excited energy level.

61. For hydrogen in the $1s$ state, what is the probability of finding the electron farther than $2.50 a_0$ from the nucleus?

62. For the ground state of hydrogen, what is the probability of finding the electron closer to the nucleus than the Bohr radius?

WEB **63.** According to classical physics, a charge e moving with an acceleration a radiates at a rate

$$\frac{dE}{dt} = -\frac{1}{6\pi\epsilon_0}\frac{e^2 a^2}{c^3}$$

(a) Show that an electron in a classical hydrogen atom (see Fig. 42.3) spirals into the nucleus at a rate

$$\frac{dr}{dt} = -\frac{e^4}{12\pi^2\epsilon_0^2 r^2 m_e^2 c^3}$$

(b) Find the time it takes the electron to reach $r = 0$, starting from $r_0 = 2.00 \times 10^{-10}$ m.

64. In a lithium atom, the electron cloud of the outer electron overlaps with the electron clouds of the two K-shell electrons. A detailed calculation of the effective charge exerting an electric force on another electron may be made using quantum mechanics. For the case of the lithium atom, the effective charge on each inner electron is $-0.85e$. Use this value to find (a) the effective charge on the nucleus as "seen" by the outer valence electron and (b) the ionization energy (compare this with 5.4 eV).

65. **Review Problem.**   In the technique known as electron spin resonance (ESR), a sample containing unpaired electrons is placed in a magnetic field. Consider the simplest situation, in which only one electron is present and therefore only two energy states are possible, corresponding to $m_s = \pm\frac{1}{2}$. In ESR, the absorption of a photon causes the electron's spin magnetic moment to flip from a lower energy state to a higher energy state. (The lower energy state corresponds to the case in which the magnetic moment $\boldsymbol{\mu}_{spin}$ is aligned with the magnetic field, and the higher energy state corresponds to the case in which $\boldsymbol{\mu}_{spin}$ is aligned opposite the field.) What photon frequency is required to excite an ESR transition in a 0.350-T magnetic field?

66. Figure P42.66 shows the energy-level diagrams of He and Ne. An electrical voltage excites the He atom from its ground state to its excited state of 20.61 eV. The excited He atom collides with an Ne atom in its ground state and excites this atom to the state at 20.66 eV. Lasing action takes place for electronic transitions from E_3^* to E_2 in the Ne atoms. Show that the wavelength of the red He–Ne laser light is approximately 633 nm.

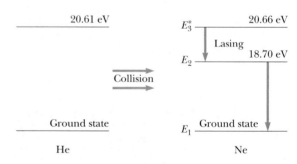

Figure P42.66

67. A dimensionless number that often appears in atomic physics is the fine-structure constant $\alpha = k_e e^2/\hbar c$, where k_e is the Coulomb constant. (a) Obtain a numerical value for $1/\alpha$. (b) In scattering experiments, the electron size is taken to be the classical electron radius, $r_e = k_e e^2/m_e c^2$. In terms of α, what is the ratio of the

Compton wavelength (see Section 40.3), $\lambda_C = h/m_e c$, to the classical electron radius? (c) In terms of α, what is the ratio of the Bohr radius, a_0, to the Compton wavelength? (d) In terms of α, what is the ratio of the Rydberg wavelength, $1/R_H$, to the Bohr radius (see Section 40.5)?

68. Show that the wave function for an electron in the 2s-state in hydrogen

$$\psi_{2s}(r) = \frac{1}{4\sqrt{2\pi}} \left(\frac{1}{a_0}\right)^{3/2} \left(2 - \frac{r}{a_0}\right) e^{-r/2a_0}$$

satisfies the radially symmetric Schrödinger equation given in Problem 11.

69. A collimated light beam of frequency f passes in the x direction through a sample of a transparent substance with index of refraction n. The length of its path is L. The frequency is tuned to be resonant with a transition between two atomic levels in the substance. The beam can induce stimulated emission from atoms in the upper state, and can be absorbed by atoms in the lower state. The beam intensity is therefore a function $I(x)$ of position. The number of transitions per unit time and per area that the beam will induce over a small distance dx in the material is equal to $BNI(x)n\,dx/c$, where B is Einstein's coefficient for the transition (related to the transition probability) and N is the number density (i.e., population density) of atoms of the initial state in the transition. The same equation holds for both stimulated emission and absorption. Show that if I_0 is the intensity of the beam before it enters the material, the intensity of the beam at the other end is

$$I(L) = I_0 e^{-\alpha L}$$

where $\alpha = hfB\,\Delta Nn/c$, and where ΔN is the difference in number densities between lower and upper states. (*Hint:* Intensity is (energy/time)/area, and photons have energy.)

70. **Review Problem.** The 1997 Nobel Prize in physics was awarded to Steven Chu, Claude Cohen-Tannoudji, and William Phillips for "the development of methods to cool and trap atoms with laser light." One part of their work was the production of a beam of atoms (mass $\sim 10^{-25}$ kg) that move at a speed on the order of 1 km/s, similar to the speed of molecules in air at room temperature. An intense laser light beam tuned to a visible atomic transition (assume 500 nm) is then directed straight into the atomic beam. That is, the atomic beam and light beam are counterpropagating. An atom in the ground state immediately absorbs a photon. Total momentum is conserved in the absorption process. After a lifetime on the order of 10^{-8} s, the excited atom radiates by spontaneous emission. It has an equal probability of emitting a photon in any direction. Thus, the average "recoil" of the atom is zero over many absorption and emission cycles. (a) Estimate the average deceleration of the atomic beam. (b) What is the order of magnitude of the distance over which the atoms in the beam will be brought to a halt?

Answers to Quick Quizzes

42.1 The three-dimensional charge distribution shown in Figure 42.7b is not uniform — it peaks at the Bohr radius. Figure 42.7a represents the probability of finding the electron as a function of distance from the center of the nucleus. Because that probability is a function of r but not of x or y individually, the chance of finding the electron in the xy plane is a maximum at any point for which $r = a_0$. For $r < a_0$, the probability drops rapidly — indicating that we are unlikely to find the electron very close to or inside the nucleus. As r becomes very great, the probability again approaches zero, meaning that the bound electron does not have a significant probability of being far away from the nucleus either. Imagine that you are looking down the z axis of Figure 42.7b, toward the xy plane. The peak area, where we are most likely to find the electron, would appear darkest, and the areas of lower probability would be lighter.

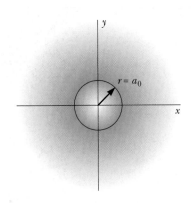

42.2 The $\ell = 3$ drawing is a graphical representation of the results of Example 42.6. Similar calculations yield the magnitude and direction of the angular momentum for the $\ell = 1$ case.

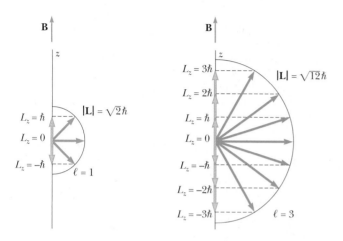

42.3 Classical physics puts no constraints on the angular momentum **L** of the atoms in the beam. Hence the magnetic moment **μ** for an atom can point in any direction. Each atom interacts differently with the nonuniform magnetic field and is deflected accordingly. This ran-dom variation in deflection results in a smoothly distributed exposure pattern on the photographic plate, like the classical pattern in Figure 42.12. Of course, this turns out not to be the case experimentally.

42.4 The bombarding electrons have an energy of 37 keV. The cutoff wavelength corresponds to one of these electrons losing all of its kinetic energy in a single collision and having that energy be emitted from the target as a single photon. We can calculate the wavelength of this photon from Equations 16.14 and 40.6: $\lambda = c/f = hc/E = 34$ pm. Wavelengths shorter than this can appear in the continuous spectrum only if the energy of the bombarding electrons is increased.

42.5

c h a p t e r

43

Molecules and Solids

The beautiful symmetry and regularity of crystalline solids have both stimulated and allowed rapid progress in the field of solid state physics in the 20th century. The most random atomic arrangement, that of a gas, was well understood in the 1800s, as we discussed in Chapter 21. In the 1900s great progress was first made in accounting for the properties of the most regular atomic arrangements, those of crystalline solids. Quite recently, our understanding of liquids and amorphous (irregular) solids has advanced. The recent interest in the physics of low-cost amorphous materials has been driven by their use in such devices as solar cells, memory elements, and fiber optic waveguides.

In this chapter, we study the aggregates of atoms known as molecules. We begin by describing the bonding mechanisms in molecules, the various modes of molecular excitation, and the radiation emitted or absorbed by molecules. We then take the next logical step and show how molecules combine to form solids. Then, by examining their electronic distributions, we explain the differences between insulating, conducting, semiconducting, and superconducting materials. The chapter also includes discussions of semiconducting junctions and several semiconductor devices, and it concludes with further treatment of superconductors.

43.1 MOLECULAR BONDS

The energy of a stable molecule is less than the total energy of the separated atoms. The bonding mechanisms in a molecule are primarily due to electric forces between atoms (or ions). When two atoms are separated by an infinite distance, the electric force between them is zero, as is the electric potential energy of the system they constitute. As the atoms are brought closer together, both attractive and repulsive forces act. At very large separation distances, the dominant forces are attractive. At small separation distances, electrostatic forces and the exclusion principle result in a repulsive force, as we shall discuss shortly.

The potential energy of a system of atoms can be positive or negative, depending on the distance between the constituent atoms. As we saw in Example 8.11, the total potential energy of a system of two atoms can be approximated by an expres-

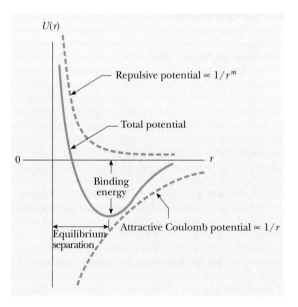

Figure 43.1 Total energy as a function of internuclear separation distance for a system of two atoms.

sion of the form

$$U = -\frac{A}{r^n} + \frac{B}{r^m} \tag{43.1}$$

where r is the internuclear separation distance, A and B are parameters associated with the attractive and repulsive forces, and n and m are small integers. Total potential energy versus internuclear separation distance for such a two-atom system is graphed in Figure 43.1. At large separation distances, the slope of the curve is positive, corresponding to a net attractive force. When the atoms are close together, the slope is negative, indicating a net repulsive force. At the equilibrium separation distance, the attractive and repulsive forces just balance, the potential energy has its minimum value, and the slope of the curve is zero.

A complete description of the binding mechanisms in molecules is highly complex because binding involves the mutual interactions of many particles. In this section, therefore, we discuss only some simplified models: ionic bonding, covalent bonding, van der Waals bonding, and hydrogen bonding.

Ionic Bonding

When two atoms combine in such a way that one atom gives one or more of its outer electrons to the other atom, the bond formed is called an **ionic bond.** Ionic bonds are fundamentally caused by the Coulomb attraction between oppositely charged ions.

A familiar example of an ionically bonded solid is sodium chloride, NaCl, which is common table salt. Sodium, which has the electronic configuration $1s^2 2s^2 2p^6 3s$, is ionized relatively easily, giving up its $3s$ electron to form a Na^+ ion. The energy required to ionize the atom to form Na^+ is 5.1 eV. Chlorine, which has the electronic configuration $1s^2 2s^2 2p^5$, is one electron short of the filled-shell structure of argon. Because filled-shell configurations are energetically more favorable than unfilled-shell configurations, the Cl^- ion is more stable than the neutral Cl atom. The energy released when an atom takes on an electron is called the **electron affinity** of the atom. For chlorine, the electron affinity is 3.7 eV. Therefore, the energy required to form Na^+ and Cl^- from isolated atoms is $5.1 - 3.7 = 1.4$ eV.

Total energy versus internuclear separation distance for NaCl is graphed in Figure 43.2. The total energy has a minimum value of -4.2 eV at the equilibrium separation distance, which is about 0.24 nm. This means that the energy required to break the Na^+–Cl^- bond and form neutral sodium and chlorine atoms, called the **dissociation energy,** is 4.2 eV.

When the two ions are brought to within 0.24 nm of each other, their filled outer shells overlap, and this results in a repulsion between the shells. This repulsion is partly electrostatic in origin and partly the result of the exclusion principle. Because all electrons must obey the exclusion principle, some of them in the overlapping shells are forced into higher energy states, and the system energy increases, as if a repulsive force existed between them.

Quick Quiz 43.1

Figure 43.2 shows the total energy versus the internuclear separation distance for Na^+ and Cl^- ions. Once the ions are more than 0.24 nm apart, the energy increases but not without limit. What is the maximum value of the energy for $r > 0.24$ nm?

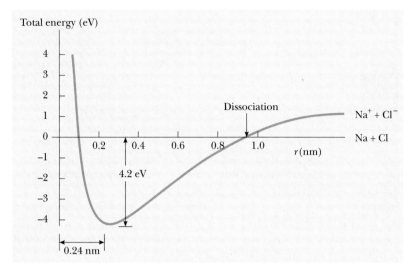

Figure 43.2   Total energy versus internuclear separation distance for Na^+ and Cl^- ions. The energy required to separate the NaCl molecule into neutral atoms of Na and Cl is 4.2 eV.

Covalent Bonding

A **covalent bond** between two atoms is one in which electrons supplied by either one or both atoms are shared. Many diatomic molecules, such as H_2, F_2, and CO, owe their stability to covalent bonds. In the case of the H_2 molecule, the two electrons are equally shared between the nuclei and occupy what is called a *molecular orbital*. The electron density is great in the region between the two nuclei, with the electrons acting as the "glue" holding the nuclei together.

The molecular orbital formed from the *s* orbitals of the two hydrogen atoms in H_2 is shown in Figure 43.3. Because of the exclusion principle, the two electrons in the ground state of H_2 must have antiparallel spins. Also because of the exclusion principle, if a third H atom is brought near the H_2 molecule, the third electron would have to occupy a higher energy level, which is an energetically unfavorable situation. Hence, the H_3 molecule is not stable and does not form.

Stable molecules more complex than H_2, such as H_2O, CO_2, and CH_4, also contain covalent bonds. Consider methane, CH_4, a typical organic molecule shown schematically in the electron-sharing diagram of Figure 43.4a. In this case, one covalent bond is formed between the carbon atom and each hydrogen atom, resulting in a total of four C–H covalent bonds. The geometrical arrangement of the four bonds is shown in Figure 43.4b. The four hydrogen nuclei are at the corners of a regular tetrahedron, and the carbon nucleus is at the center.

Because the outermost molecular orbitals of covalent molecules are full, the interactions between such molecules are quite weak. In fact, many covalent molecules form gases or liquids rather than solids.

Van der Waals Bonding

If two molecules are some distance apart, they are attracted to each other by weak electrostatic forces called **van der Waals forces.** Likewise, atoms that do not form ionic or covalent bonds are attracted to each other by van der Waals forces. For this reason, at sufficiently low temperatures at which thermal excitations are negli-

Figure 43.3   Covalent bond formed by the two 1*s* electrons of the H_2 molecule. The depth of blue color at any location is proportional to the probability of finding an electron in that location.

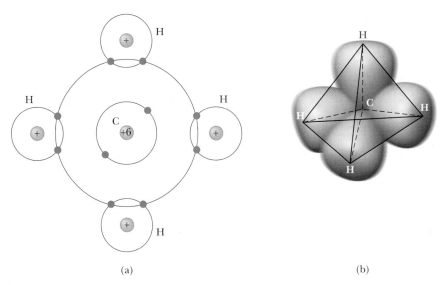

(a)   (b)

Figure 43.4   (a) A highly schematic representation of the four covalent bonds in the CH_4 molecule. (b) The spatial arrangement of these four bonds. The carbon atom is at the center of a tetrahedron having hydrogen atoms at its corners. The electron density is greatest between the nuclei.

gible, gases first condense to liquids and then solidify (with the exception of helium, which does not solidify at atmospheric pressure).

There are three types of van der Waals forces. The first type, called the *dipole–dipole force,* is an interaction between two molecules each having a permanent electric dipole moment—for example, polar molecules such as H_2O have permanent electric dipole moments and attract other polar molecules (Fig. 43.5). In effect, one molecule interacts with the electric field produced by another molecule.

The second type, the *dipole–induced dipole force,* results when a polar molecule having a permanent electric dipole moment induces a dipole moment in a nonpolar molecule.

The third type is called the *dispersion force,* an attractive force that occurs between two nonpolar molecules. In this case, the interaction results from the fact that, although the average dipole moment of a nonpolar molecule is zero, the average of the square of the dipole moment is nonzero because of charge fluctuations. Consequently, two nonpolar molecules near each other tend to be correlated so as to produce an attractive van der Waals force.

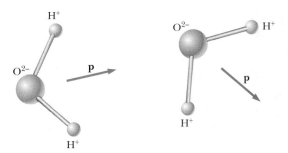

Figure 43.5   Water molecules have a permanent electric dipole moment **p**. The molecules attract each other because the electric field produced by one molecule interacts with and orients the dipole moment of a nearby molecule.

Hydrogen Bonding

Because hydrogen has only one electron, it is expected to form a covalent bond with only one other atom. A hydrogen atom in a given molecule also can form a second type of bond, however, bonding with an atom in another molecule via a **hydrogen bond.** Let us use the water molecule H_2O as an example. In the two covalent bonds in this molecule, the electrons from the hydrogen atoms are more likely to be found near the oxygen atom than near the hydrogen atoms. This leaves essentially bare protons at the positions of the hydrogen atoms. This unshielded positive charge can be attracted to the negative end of another polar molecule. Because the proton is unshielded by electrons, the negative end of the other molecule can come very close to the proton to form a bond that is strong enough to form a solid crystalline structure, such as that of ice. The bonds *within* a water molecule are covalent, but the bonds *between* water molecules in ice are hydrogen bonds. Because hydrogen bonds are relatively weak, ice melts at the low temperature of 0°C.

The hydrogen bond has a binding energy of about 0.1 eV. Although it is relatively weak compared with other chemical bonds, hydrogen bonding is the mechanism responsible for the linking of biological molecules and polymers. For example, in the case of the famous DNA (deoxyribonucleic acid) molecule, which has a double-helix structure (Fig. 43.6), hydrogen bonds formed by the sharing of a proton between two atoms create linkages between the turns of the helix.

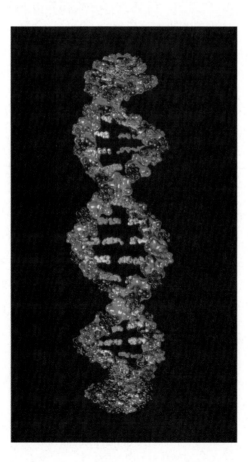

Figure 43.6   DNA molecules are held together by hydrogen bonds. *(Douglas Struthers/Tony Stone Images)*

APPLICATION ▶ **Measuring Molecular Bonding Forces with an Atomic Force Microscope (Example 8.11 Revisited)**

As noted in Section 41.8, an atomic force microscope (AFM) uses a very sharp tip mounted on a cantilever in close proximity to a surface in order to image surface topography with nanometer resolution. The AFM is similar to the scanning tunneling microscope, except here the tip interacts with the surface to measure force instead of a tunneling current. A variation of the AFM technique allows the measurement of bonding forces between atoms or between linkages in biological molecules or other macromolecules, as illustrated in Figure 43.7. In this figure, the molecule is bonded to a surface that can move vertically with nanometer precision. The force exerted on the tip where another part of the molecule is anchored causes the cantilever to bend. The cantilever can be approximated as a simple spring; thus its deflection is proportional to the force exerted on it (Hooke's law). The AFM can measure forces as minute as piconewtons. In an experiment to measure bonding forces, a specially prepared tip is carefully brought in contact with a surface coated with the molecules of interest. Several molecules might attach to the tip, but careful continued extension (by moving the surface downward) leaves attached to the tip only the longest molecule bridging the surface–tip distance. The force needed to break the next weakest bond is measured by recording cantilever deflection as the tip is retracted.

Consider the force required to break a single covalent bond in a molecule. We can greatly simplify this problem by considering bond breakage for a potential energy function with which we are familiar: the van der Waals interaction described earlier in this section. The potential energy function stated in general form in Equation 43.1 takes the form of the Lennard–Jones equation cited in Example 8.11:

$$U(r) = 4\epsilon\left[\left(\frac{\sigma}{r}\right)^{12} - \left(\frac{\sigma}{r}\right)^{6}\right]$$

This function is plotted in Figure 43.8 for argon, an inert gas that interacts by the van der Waals force, with experimentally determined parameters $\sigma = 0.340$ nm and $\epsilon = 0.0104$ eV. The shape of this potential energy function is generic for many types of bonds (compare with the plot in Fig. 43.2). The position of the energy minimum represents the equilibrium distance of the bond. Repulsive interactions between inner-shell electrons cause a large increase in energy if the atoms are moved closer together, and the interaction energy approaches zero when the atoms are sufficiently far apart. (Note that for very small deviations from the equilibrium distance, the first term in a Taylor series expansion is quadratic, which is the form of the potential energy function for a harmonic oscillator.)

We calculate the equilibrium internuclear distance by finding the position of the minimum in $U(r)$, that is, where $dU(r)/dr = 0$:

$$\frac{dU(r)}{dr} = -\frac{4\epsilon}{\sigma}\left[12\left(\frac{\sigma}{r}\right)^{13} - 6\left(\frac{\sigma}{r}\right)^{7}\right] = 0$$

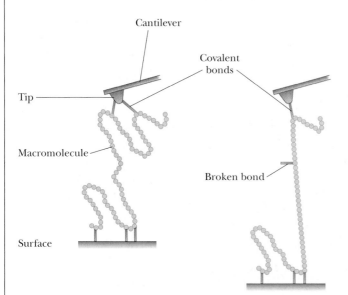

Figure 43.7 Measuring bonding forces with an AFM tip. A single macromolecular chain is covalently attached to the tip at two sites. Increased stretching breaks one bond, producing a sudden change in the force versus separation curve as the separation is increased. *(Adapted from M. Grandbois, M. Beyer, M. Reif, H. Clausen-Schaumann, H. Goub, "How Strong Is a Covalent Bond?" Science 283: 1727–1730, 1999.)*

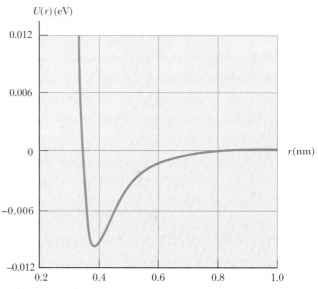

Figure 43.8 Potential energy versus internuclear separation distance for two argon atoms bonded by the van der Waals interaction.

This is equivalent to the position of zero net force. Solving for the equilibrium position r_{eq}, we find it at

$$r_{eq} = 2^{1/6}\,\sigma$$

which for argon is 0.382 nm.

When force is applied by an AFM, the bond ruptures at the point at which the force is a maximum. Because the force is $F = -dU/dr$, the maximum attractive force occurs at $d^2U/dr^2 = 0$. We calculate

$$\frac{d^2 U(r)}{dr^2} = \frac{4\epsilon}{\sigma^2}\left[156\left(\frac{\sigma}{r}\right)^{14} - 42\left(\frac{\sigma}{r}\right)^{8}\right] = 0$$

and find that $r_{rupture} = 0.423$ nm, which corresponds to a force of 7.33×10^{-2} eV/nm = 11.7 pN. A cantilever with a spring constant of 0.12 N/m would deflect about 0.1 nm at this force. Using similar techniques, Grandbois and coworkers[1] found that the covalent silicon–carbon bond breaks with a force of about 2 nN and that the sulfur–gold bond breaks with a force of about 1.4 nN.

A complete analysis of an AFM experiment would also include the potential energy of the bending cantilever.[2] However, our simplified analysis of bond rupture for a model system suggests how AFM techniques can be applied to more complex and important macromolecular systems.

43.2 ▷ THE ENERGY AND SPECTRA OF MOLECULES

As in the case of atoms, we can study the structure and properties of molecules by examining the radiation they emit or absorb. Before we describe these processes, it is important to understand the various ways of exciting a molecule.

Consider a single molecule in the gaseous phase. The energy of the molecule can be divided into four categories: (1) electronic energy, due to the interactions between the molecule's electrons and nuclei; (2) translational energy, due to the motion of the molecule's center of mass through space; (3) rotational energy, due to the rotation of the molecule about its center of mass; and (4) vibrational energy, due to the vibration of the molecule's constituent atoms:

Total energy of a molecule

$$E = E_{el} + E_{trans} + E_{rot} + E_{vib}$$

The electronic energy of a molecule is very complex because it involves the interaction of many charged particles, but various techniques have been developed to approximate its values. Because the translational energy is unrelated to internal structure, this molecular energy is unimportant in interpreting molecular spectra.

Rotational Motion of Molecules

Let us consider the rotation of a molecule around its center of mass, confining our discussion to the diatomic case (Fig. 43.9a) but noting that the same ideas can be extended to polyatomic molecules. A diatomic molecule aligned along an x axis has only two rotational degrees of freedom, corresponding to rotations around the y and z axes. If ω is the angular frequency of rotation around one of these axes, the rotational kinetic energy of the molecule about that axis can be expressed in the form

$$E_{rot} = \tfrac{1}{2}I\omega^2 \tag{43.2}$$

where I is the moment of inertia of the molecule, given by

Moment of inertia for a diatomic molecule

$$I = \left(\frac{m_1 m_2}{m_1 + m_2}\right) r^2 = \mu r^2 \tag{43.3}$$

---

[1] M. Grandbois, M. Beyer, M. Rief, H. Clausen–Schaumann, and H. Gaub, "How Strong Is a Covalent Bond?" *Science* 283:1727–1730, 1999.

[2] For details, see B. Shapiro and H. Qian, "A Quantitative Analysis of Single Protein–Ligand Complex Separation with the Atomic Force Microscope," *Biophys. Chem.* 67:211–219, 1997.

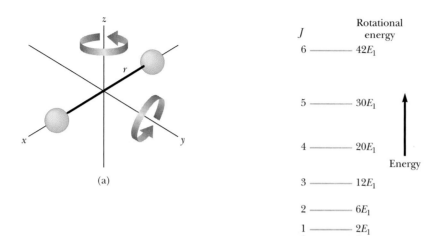

$$E_1 = \frac{\hbar^2}{2I}$$

(a)

(b)

Figure 43.9 (a) A diatomic molecule oriented along the x axis has two rotational degrees of freedom, corresponding to rotation around the y and z axes. (b) Allowed rotational energies of a diatomic molecule calculated with Equation 43.6.

where m_1 and m_2 are the masses of the atoms that form the molecule, r is the atomic separation, and μ is the **reduced mass** of the molecule:

$$\mu = \frac{m_1 m_2}{m_1 + m_2} \qquad \textbf{(43.4)}$$

Reduced mass of a molecule

The magnitude of the angular momentum of the molecule is $I\omega$, which classically can have any value. Quantum mechanics, however, restricts angular momentum to the values

$$I\omega = \sqrt{J(J+1)}\,\hbar \qquad J = 0, 1, 2, \ldots \qquad \textbf{(43.5)}$$

Allowed values of rotational angular momentum

where J is an integer called the **rotational quantum number.** Substituting Equation 43.5 into Equation 43.2, we obtain an expression for the allowed values of the rotational kinetic energy:

$$E_{\text{rot}} = \tfrac{1}{2} I\omega^2 = \frac{1}{2I}(I\omega)^2 = \frac{\left(\sqrt{J(J+1)}\,\hbar\right)^2}{2I}$$

$$E_{\text{rot}} = \frac{\hbar^2}{2I} J(J+1) \qquad J = 0, 1, 2, \ldots \qquad \textbf{(43.6)}$$

Allowed values of rotational energy

Thus, we see that **the rotational energy of the molecule is quantized and depends on its moment of inertia.** The allowed rotational energies of a diatomic molecule are plotted in Figure 43.9b.

For most molecules, the transitions between adjacent rotational energy levels result in radiation that lies in the microwave range of frequencies ($f \sim 10^{11}$ Hz). When a molecule absorbs a microwave photon, the molecule jumps from a lower rotational energy level to a higher one. The allowed rotational transitions of linear

TABLE 43.1	**Several Rotational Transitions of the CO Molecule**	
Rotational Transition	**Wavelength of Absorbed Photon (m)**	**Frequency of Absorbed Photon (Hz)**
$J = 0 \rightarrow J = 1$	2.60×10^{-3}	1.15×10^{11}
$J = 1 \rightarrow J = 2$	1.30×10^{-3}	2.30×10^{11}
$J = 2 \rightarrow J = 3$	8.77×10^{-4}	3.46×10^{11}
$J = 3 \rightarrow J = 4$	6.50×10^{-4}	4.61×10^{11}

From G. M. Barrows, *The Structure of Molecules,* New York, W. A. Benjamin, 1963.

molecules are regulated by the selection rule $\Delta J = \pm 1$. That is, an absorption line in the microwave spectrum of a linear molecule corresponds to an energy separation equal to $E_J - E_{J-1}$. From Equation 43.6, we see that the allowed transitions are given by the condition

$$\Delta E = E_J - E_{J-1} = \frac{\hbar^2}{2I} [J(J + 1) - (J - 1)J]$$

> Separation between adjacent rotational levels

$$\Delta E = \frac{\hbar^2}{I} J = \frac{h^2}{4\pi^2 I} J \qquad \text{(43.7)}$$

where J is the rotational quantum number of the higher energy state. Because $\Delta E = hf$, where f is the frequency of the absorbed photon, we see that the allowed frequency for the transition $J = 0$ to $J = 1$ is $f_1 = h/4\pi^2 I$. The frequency corresponding to the $J = 1$ to $J = 2$ transition is $2f_1$, and so on. These predictions are in excellent agreement with the observed frequencies.

The wavelengths and frequencies for the microwave absorption spectrum of the carbon monoxide molecule are given in Table 43.1. From these data, we can evaluate the moment of inertia and bond length of the molecule.

EXAMPLE 43.1   Rotation of the CO Molecule

The $J = 0$ to $J = 1$ rotational transition of the CO molecule occurs at 1.15×10^{11} Hz. (a) Use this information to calculate the moment of inertia of the molecule.

Solution   From Equation 43.7, we see that the energy difference between the $J = 0$ and $J = 1$ rotational levels is $h^2/4\pi^2 I$. Equating this ΔE value to the energy of the absorbed photon, we have

$$\Delta E = \frac{h^2}{4\pi^2 I} = hf$$

Solving for I gives

$$I = \frac{h}{4\pi^2 f} = \frac{6.626 \times 10^{-34}\,\text{J} \cdot \text{s}}{4\pi^2 (1.15 \times 10^{11}\,\text{s}^{-1})}$$

$$= \boxed{1.46 \times 10^{-46}\,\text{kg} \cdot \text{m}^2}$$

(b) Calculate the bond length of the molecule.

Solution   We can use Equation 43.3 to calculate the bond length, but we first need to know the value for the reduced mass μ of the CO molecule:

$$\mu = \frac{m_1 m_2}{m_1 + m_2} = \frac{(12\,\text{u})(16\,\text{u})}{12\,\text{u} + 16\,\text{u}} = 6.86\,\text{u}$$

$$= (6.86\,\text{u})\left(1.66 \times 10^{-27}\,\frac{\text{kg}}{\text{u}}\right) = 1.14 \times 10^{-26}\,\text{kg}$$

where we have used the fact that $1\,\text{u} = 1.66 \times 10^{-27}$ kg.

Substituting this value and the result of part (a) into Equation 43.3 and solving for r, we obtain

$$r = \sqrt{\frac{I}{\mu}} = \sqrt{\frac{1.46 \times 10^{-46}\,\text{kg} \cdot \text{m}^2}{1.14 \times 10^{-26}\,\text{kg}}}$$

$$= 1.13 \times 10^{-10}\,\text{m} = \boxed{0.113\,\text{nm}}$$

Vibrational Motion of Molecules

A molecule is a flexible structure in which the atoms are bonded together by what can be considered "effective springs" (see Fig. 13.11). If disturbed, the molecule can vibrate and acquire vibrational energy. This vibrational motion and corresponding vibrational energy can be altered if the molecule is exposed to electromagnetic waves of the proper frequency.

Consider the diatomic molecule shown in Figure 43.10a. Its effective spring has a force constant k. A plot of potential energy versus atomic separation for such a molecule is sketched in Figure 43.10b, where r_0 is the equilibrium atomic separation. According to classical mechanics, the frequency of vibration for this system is

$$f = \frac{1}{2\pi}\sqrt{\frac{k}{\mu}} \qquad \textbf{(43.8)}$$

where again μ is the reduced mass given by Equation 43.4.

As we expect, the quantum mechanical solution to this system shows that the energy is quantized, with allowed energies

$$E_{\text{vib}} = (v + \tfrac{1}{2})hf \qquad v = 0, 1, 2, \ldots \qquad \textbf{(43.9)}$$

where v is an integer called the **vibrational quantum number.** If the system is in the lowest vibrational state, for which $v = 0$, its zero-point energy is $\frac{1}{2}hf$. The accompanying vibration—the *zero-point motion*—is always present, even if the molecule is not excited. In the first excited state, $v = 1$ and the vibrational energy is $\frac{3}{2}hf$, and so on.

Substituting Equation 43.8 into Equation 43.9, we obtain the following expression for the vibrational energy:

$$E_{\text{vib}} = (v + \tfrac{1}{2})\frac{h}{2\pi}\sqrt{\frac{k}{\mu}} \qquad v = 0, 1, 2, \ldots \qquad \textbf{(43.10)}$$

◀ Allowed values of vibrational energy

The selection rule for the allowed vibrational transitions is $\Delta v = \pm 1$. From Equation 43.10, we see that the energy difference between any two successive vibra-

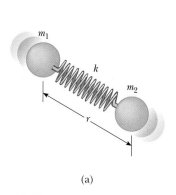

(a)

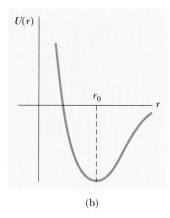

(b)

Figure 43.10 (a) Effective-spring model of a diatomic molecule. The vibration is along the molecular axis. (b) Plot of the potential energy of a diatomic molecule versus atomic separation distance, where r_0 is the equilibrium separation distance of the atoms.

TABLE 43.2   **Photon Frequency and Effective-Spring Force Constant for $v = 0$ to $v = 1$ Transition in Some Diatomic Molecules**

Molecule	Photon Frequency (Hz)	Force Constant k (N/m)
HF	8.72×10^{13}	970
HCl	8.66×10^{13}	480
HBr	7.68×10^{13}	410
HI	6.69×10^{13}	320
CO	6.42×10^{13}	1 850
NO	5.63×10^{13}	1 530

From G. M. Barrows, *The Structure of Molecules,* New York, W. A. Benjamin, 1963. The k values were calculated from Equation 43.11.

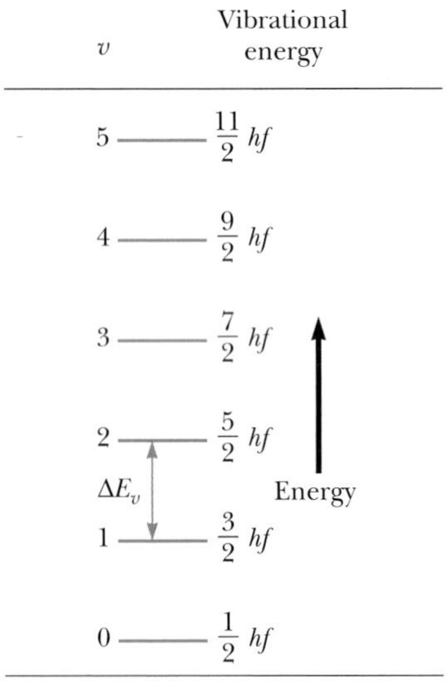

	Vibrational
v	energy
5	$\frac{11}{2} hf$
4	$\frac{9}{2} hf$
3	$\frac{7}{2} hf$
2	$\frac{5}{2} hf$
ΔE_v	Energy
1	$\frac{3}{2} hf$
0	$\frac{1}{2} hf$

Figure 43.11 Allowed vibrational energies of a diatomic molecule, where f is the frequency of vibration of the molecule, given by Equation 43.8. The spacings between adjacent vibrational levels are equal if the molecule behaves as a harmonic oscillator.

tional levels is

$$\Delta E_{\text{vib}} = \frac{h}{2\pi}\sqrt{\frac{k}{\mu}} = hf \tag{43.11}$$

The vibrational energies of a diatomic molecule are plotted in Figure 43.11. At ordinary temperatures, most molecules have vibrational energies corresponding to the $v = 0$ state because the spacing between vibrational states is much greater than $k_B T$. The molecules are not thermally excited into higher states.

Transitions between vibrational levels are caused by absorption in the infrared region of the spectrum. That is, a molecule jumps from a lower to a higher vibrational energy level by absorbing a photon having a frequency in the infrared range. The photon frequencies corresponding to the $v = 0$ to $v = 1$ transition are listed in Table 43.2 for several diatomic molecules, together with the force constants of the effective springs holding the molecules together. The latter values were calculated by using Equation 43.11. The "stiffness" of a bond can be measured by the size of the effective force constant.

EXAMPLE 43.2   **Vibration of the CO Molecule**

The frequency of the photon that causes the $v = 0$ to $v = 1$ transition in the CO molecule is 6.42×10^{13} Hz. (a) Calculate the force constant k for this molecule.

Solution We can use Equation 43.11 and the value $\mu = 1.14 \times 10^{-26}$ kg we calculated in Example 43.1b:

$$\frac{h}{2\pi}\sqrt{\frac{k}{\mu}} = hf$$

$$k = 4\pi^2 \mu f^2$$
$$= 4\pi^2 (1.14 \times 10^{-26}\ \text{kg})(6.42 \times 10^{13}\ \text{s}^{-1})^2$$
$$= \boxed{1.85 \times 10^3\ \text{N/m}}$$

(b) What is the maximum amplitude of vibration for this molecule in the $v = 0$ vibrational state?

Solution The maximum potential energy stored in the molecule is $\frac{1}{2}kA^2$, where A is the amplitude of vibration. Equating this maximum energy to the vibrational energy given by Equation 43.10 with $v = 0$, we have

$$\frac{1}{2}kA^2 = \frac{h}{4\pi}\sqrt{\frac{k}{\mu}}$$

Substituting the value $k = 1.85 \times 10^3$ N/m and the value for μ from part (a), we obtain

$$A^2 = \frac{h}{2\pi k}\sqrt{\frac{k}{\mu}}$$
$$= \frac{6.626 \times 10^{-34}\ \text{J·s}}{2\pi(1.85 \times 10^3\ \text{N/m})}\sqrt{\frac{1.85 \times 10^3\ \text{N/m}}{1.14 \times 10^{-26}\ \text{kg}}}$$
$$= 2.30 \times 10^{-23}\ \text{m}^2$$

Thus,

$$A = 4.79 \times 10^{-12} \text{ m} = \boxed{4.79 \times 10^{-3} \text{ nm}}$$

Comparing this result with the bond length of 0.113 nm we

calculated in Example 43.1b, we see that the amplitude of vibration is about 4% of the bond length. Thus, we see that infrared spectroscopy provides useful information on the elastic properties (bond strengths) of molecules.

Molecular Spectra

In general, a molecule vibrates and rotates simultaneously. To a first approximation, these motions are independent of each other, and so the total energy of the molecule is the sum of Equations 43.6 and 43.9:

$$E = (v + \tfrac{1}{2})hf + \frac{\hbar^2}{2I}J(J + 1) \qquad \textbf{(43.12)}$$

The energy levels of any molecule can be calculated from this expression, and each level is indexed by the two quantum numbers J and v. From these calculations, an energy level diagram like the one shown in Figure 43.12a can be constructed. For each allowed value of the vibrational quantum number v, there is a complete set of rotational levels corresponding to $J = 0, 1, 2, \ldots$. Note that the energy separation between successive rotational levels is much smaller than the separation between successive vibrational levels. As noted earlier, most molecules

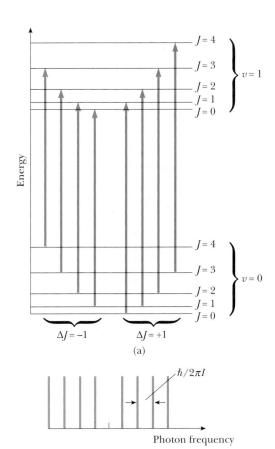

(a)

(b)

Figure 43.12 (a) Absorptive transitions between the $v = 0$ and $v = 1$ vibrational states of a diatomic molecule. The transitions obey the selection rule $\Delta J = \pm 1$ and fall into two sequences, those for $\Delta J = +1$ and those for $\Delta J = -1$. The transition energies are given by Equation 43.12. (b) Expected lines in the absorption spectrum of a molecule. The lines to the right of the center mark correspond to transitions in which J changes by $+1$; the lines to the left of the center mark correspond to transitions for which J changes by -1. These same lines appear in the emission spectrum.

at ordinary temperatures are in the $v = 0$ vibrational state; these molecules can be in various rotational states, as Figure 43.12a shows.

When a molecule absorbs an infrared photon, the vibrational quantum number v increases by one unit while the rotational quantum number J either increases or decreases by one unit, as can be seen in Figure 43.12. Thus, the molecular absorption spectrum consists of two groups of lines: one group to the right of center and satisfying the selection rules $\Delta J = 1$ and $\Delta v = 1$, and the other group to the left of center and satisfying[3] the selection rules $\Delta J = -1$ and $\Delta v = 1$.

The energies of the absorbed photons can be calculated from Equation 43.12:

$$\Delta E = hf + \frac{\hbar^2}{I}(J+1) \qquad J = 0, 1, 2, \ldots (\Delta J = +1) \qquad \textbf{(43.13)}$$

$$\Delta E = hf - \frac{\hbar^2}{I}J \qquad J = 1, 2, 3, \ldots (\Delta J = -1) \qquad \textbf{(43.14)}$$

where now J is the rotational quantum number of the *initial* state. Equation 43.13 generates the series of equally spaced lines *higher* than the frequency f, whereas Equation 43.14 generates the series *lower* than this frequency. Adjacent lines are separated in frequency by the fundamental unit $\hbar/2\pi I$; we can see this by substituting hf for ΔE in Equation 43.7 and setting $J = 1$. Figure 43.12b shows the expected frequencies in the absorption spectrum of the molecule; these same frequencies appear in the emission spectrum.

The absorption spectrum of the HCl molecule shown in Figure 43.13 follows this pattern very well and reinforces our model. However, one peculiarity is apparent: Each line is split into a doublet. This doubling occurs because two chlorine isotopes (see Section 1.2) were present in the sample used to obtain this spectrum. Because the isotopes have different masses, the two HCl molecules have different values of I.

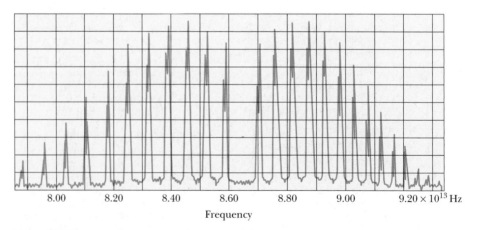

Figure 43.13 Absorption spectrum of the HCl molecule. Each line is split into a doublet because the sample contained two chlorine isotopes that had different masses and therefore different moments of inertia.

---

[3] The selection rule $\Delta J = \pm 1$ implies that the photon causing the transition is a spin-1 particle having a spin quantum number $s = 1$. Hence, this selection rule describes conservation of angular momentum for the system that consists of the molecule and the photon.

Using Figure 43.13, estimate the moment of inertia of an HCl molecule.

For CO_2 molecules, most of the absorption lines are in the infrared portion of the spectrum. Thus, visible light from the Sun is not absorbed by atmospheric CO_2 but instead strikes the Earth's surface, warming it. In turn, the Earth emits infrared radiation. These IR waves are absorbed by the CO_2 in the air instead of radiating out into space. Thus, atmospheric CO_2 acts like a one-way valve for energy from the Sun. The burning of fossil fuels can add more CO_2 to the atmosphere. Many scientists fear that substantial climatic changes might result from an enhanced "greenhouse effect."

43.3 BONDING IN SOLIDS

A crystalline solid consists of a large number of atoms arranged in a regular array, forming a periodic (in other words, repeating) structure. Two of the bonding mechanisms described in Section 43.1—ionic and covalent—are appropriate to use in describing bonds in solids. For example, the ions in the NaCl crystal are ionically bonded, as already noted, and the carbon atoms in the crystal that we call diamond form covalent bonds with one another. The metallic bond described at the end of this section is responsible for the cohesion of copper, silver, sodium, and other solid metals.

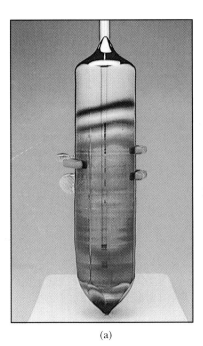

(a)

(b)

Crystalline solids. (a) A cylinder of nearly pure crystalline silicon (Si), approximately 25 cm long. Such crystals are cut into wafers and processed to make various semiconductor devices. (b) Although this crystal is called a Herkimer "diamond," it is natural quartz (SiO_2), one of the most common minerals on the Earth. Quartz crystals are used to make special camera lenses and prisms and in certain electronic applications. *(Charles D. Winters)*

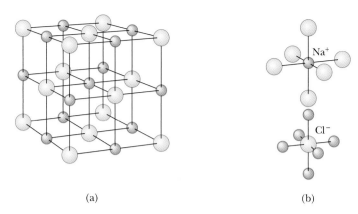

(a)   (b)

Figure 43.14   (a) Crystalline structure of NaCl. (b) Each positive sodium ion (red spheres) is surrounded by six negative chloride ions (blue spheres), and each chloride ion is surrounded by six sodium ions.

Ionic Solids

Many crystals are formed by ionic bonding, in which the dominant interaction between ions is the Coulomb interaction. Consider the NaCl crystal in Figure 43.14. Each Na^+ ion has six nearest-neighbor Cl^- ions, and each Cl^- ion has six nearest-neighbor Na^+ ions. Each Na^+ ion is attracted to its six Cl^- neighbors. The corresponding attractive potential energy is $-6k_e e^2/r$, where k_e is the Coulomb constant and r is the separation distance between each Na^+ and Cl^-. In addition, there are 12 Na^+ ions at a distance of $\sqrt{2}\,r$ from the Na^+, and these 12 positive ions exert weaker repulsive forces on the central Na^+. Furthermore, beyond these 12 Na^+ ions are more Cl^- ions that exert an attractive force, and so on. The net effect of all these interactions is a resultant negative electric potential energy

$$U_{\text{attractive}} = -\alpha k_e \frac{e^2}{r} \qquad \textbf{(43.15)}$$

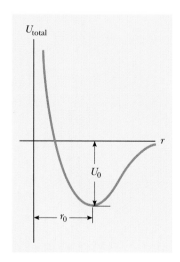

Figure 43.15   Total potential energy versus ion separation distance for an ionic solid, where U_0 is the ionic cohesive energy and r_0 is the equilibrium separation distance between ions.

where α is a pure number known as the **Madelung constant.** The value of α depends only on crystalline structure. For example, $\alpha = 1.747\,6$ for the NaCl structure. When the constituent ions of a crystal are brought close together, a repulsive force exists because of electrostatic forces and the exclusion principle, as discussed in Section 43.1. The potential energy term B/r^m in Equation 43.1 accounts for this. Therefore, the total potential energy is

$$U_{\text{total}} = -\alpha k_e \frac{e^2}{r} + \frac{B}{r^m} \qquad \textbf{(43.16)}$$

where m in this expression is some small integer.

A plot of total potential energy versus ion separation distance is shown in Figure 43.15. The potential energy has its minimum value U_0 at the equilibrium separation, when $r = r_0$. It is left as a problem for you (Problem 47) to show that

$$U_0 = -\alpha k_e \frac{e^2}{r_0}\left(1 - \frac{1}{m}\right) \qquad \textbf{(43.17)}$$

This minimum energy U_0 is called the **ionic cohesive energy** of the solid, and its absolute value represents the energy required to separate the solid into a collection of isolated positive and negative ions. Its value for NaCl is -7.84 eV per ion pair.

To calculate the **atomic cohesive energy,** which is the binding energy relative to the energy of the neutral atoms, we must add 5.14 eV to the ionic cohesive energy value to account for the transition from Na^+ to Na, and we must subtract 3.61 eV to account for the conversion of Cl^- to Cl. Thus, the atomic cohesive energy of NaCl is

$$-7.84 \text{ eV} + 5.14 \text{ eV} - 3.61 \text{ eV} = -6.31 \text{ eV}$$

Ionic crystals have the following general properties:

- They form relatively stable, hard crystals. (The melting point of NaCl is 801°C.)
- They are poor electrical conductors because they contain no free electrons.
- They have high vaporization temperatures.
- They are transparent to visible radiation but absorb strongly in the infrared region. No visible light is absorbed because the shells formed by the electrons in ionic solids are so tightly bound that visible radiation does not possess sufficient energy to promote electrons to the next allowed shell. Infrared radiation is absorbed strongly because the vibrations of the ions have a natural resonant frequency in the low-energy infrared region.
- Many are quite soluble in polar liquids, such as water. The polar solvent molecules exert an attractive electric force on the charged ions, which breaks the ionic bonds and dissolves the solid.

Covalent Solids

Solid carbon, in the form of diamond, is a crystal whose atoms are covalently bonded. Because atomic carbon has the electronic configuration $1s^2 2s^2 2p^2$, it is four electrons short of filling its $n = 2$ shell, which can accommodate eight electrons. Hence, two carbon atoms have a strong attraction for each other, with a cohesive energy of 7.37 eV.

In the diamond structure, each carbon atom is covalently bonded to four other carbon atoms located at four corners of a cube, as shown in Figure 43.16a.

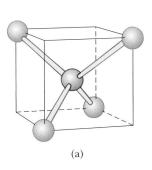

(a)

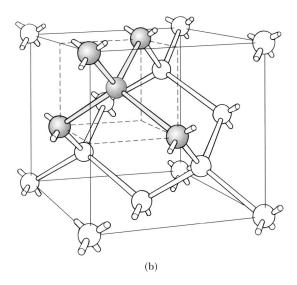

(b)

Figure 43.16 (a) Each carbon atom in a diamond crystal is covalently bonded to four other carbon atoms and forms a tetrahedral structure. (b) The crystal structure of diamond, showing the tetrahedral bond arrangement.

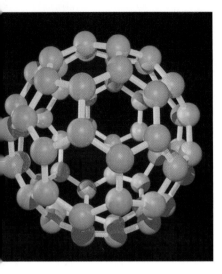

Figure 43.17 Computer rendering of a "buckyball," short for the molecule buckminsterfullerene. These nearly spherical molecular structures that look like soccer balls were named for R. Buckminster Fuller, inventor of the geodesic dome. This form of carbon, C_{60}, was discovered by astrophysicists while investigating the carbon gas that exists between stars. Scientists are actively studying the properties and potential uses of buckminsterfullerene and related molecules. *(Charles D. Winters)*

To form such a configuration of bonds, one $2s$ electron of each atom must be promoted to the $2p$ subshell so that the electronic configuration becomes $1s^2 2s^1 2p^3$, which corresponds to a half-filled p subshell. The promotion of this electron requires an energy input of about 4 eV.

The crystalline structure of diamond is shown in Figure 43.16b. Note that each carbon atom forms covalent bonds with four nearest-neighbor atoms. The basic structure of diamond is called tetrahedral (each carbon atom is at the center of a regular tetrahedron), and the angle between the bonds is 109.5°. Such other crystals as silicon and germanium have a similar structure.

When carbon atoms form a large hollow structure, the compound is called **buckminsterfullerene** after the famous architect who invented the geodesic dome. The unique shape of this molecule (Fig. 43.17) provides a "cage" to hold other atoms or molecules. Related structures, called "buckytubes" because of their long, narrow cylindrical arrangements of carbon atoms, may provide the basis for extremely strong, yet lightweight materials.

The atomic cohesive energies of some covalent solids are given in Table 43.3. The large energies account for the hardness of covalent solids. Diamond is particularly hard and has an extremely high melting point (about 4 000 K). Covalently bonded solids are usually very hard, have high bond energies and high melting points, and are good electrical insulators.

Metallic Solids

Metallic bonds are generally weaker than ionic or covalent bonds. The outer electrons in the atoms of a metal are relatively free to move throughout the material, and the number of such mobile electrons in a metal is large. The metallic structure can be viewed as a "sea" or a "gas" of nearly free electrons surrounding a lattice of positive ions (Fig. 43.18). The binding mechanism in a metal is the attractive force between the positive ions and the electron gas. Metals have a cohesive energy in the range of 1 to 3 eV, which is less than the cohesive energies of ionic or covalent solids.

Light interacts strongly with the free electrons in metals. Hence, visible light is absorbed and re-emitted quite close to the surface of a metal, which accounts for the shiny nature of metal surfaces. (Compare this to the transparency of the glass shown in the photograph at the beginning of this chapter. Visible light does not interact strongly with the electrons of the glass.) In addition to the high electrical conductivity of metals produced by the free electrons, the nondirectional nature

Metal ion
Electron gas

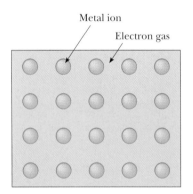

Figure 43.18 Highly schematic diagram of a metal. The blue area represents the electron gas, and the red circles represent the positive metal ions.

TABLE 43.3	Cohesive Energies of Some Covalent Solids
Solid	**Cohesive Energy (eV)**
C (diamond)	7.37
Si	4.63
Ge	3.85
InAs	5.70
SiC	6.15
ZnS	6.32
CuCl	9.24

of the metallic bond allows many different types of metal atoms to be dissolved in a host metal in varying amounts. The resulting *solid solutions,* or *alloys,* may be designed to have particular properties, such as tensile strength, ductility, electrical and thermal conductivity, and resistance to corrosion. Such properties are usually controllable and in many cases predictable.

43.4 ▶ BAND THEORY OF SOLIDS

If two identical atoms are very far apart, they do not interact and their electronic energy levels can be considered to be those of isolated atoms. Suppose that the two atoms are sodium, each having a lone $3s$ electron that has a specific, well-defined energy. As the two sodium atoms are brought closer together, their outer orbits begin to overlap. When the interaction between them is sufficiently strong, two $3s$ energy levels are formed,[4] as shown in Figure 43.19a.

When a large number of atoms are brought together to form a solid, a similar phenomenon occurs. As the atoms are brought close together, the various isolated-atom energy levels split. This splitting in levels for six atoms in close proximity is shown in Figure 43.19b. In this case, there are six energy levels corresponding to six different combinations of isolated-atom wave functions. Because the range of energy values into which overlapping levels split is not a function of the number of atoms being combined, the six-atom energy levels are more closely spaced than the two-atom levels.

If we extend this argument to the large number of atoms found in solids (of the order of 10^{23} atoms per cm^3), we obtain a large number of levels so closely spaced that they may be regarded as a continuous **band** of energy levels, as shown in Figure 43.19c. In the case of sodium, it is customary to refer to the continuous distribution of allowed energy levels as the $3s$ band because the band originates from the $3s$ levels of the individual sodium atoms.

In general, a crystalline solid has a large number of allowed energy bands that arise from the various atomic energy levels. Figure 43.20 shows the allowed energy bands of sodium. Note that energy gaps, called *forbidden energy bands,* occur between the allowed bands.

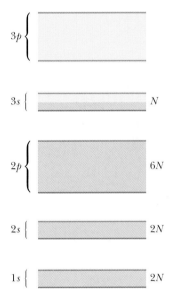

Figure 43.20 Energy bands of sodium. Note the energy gaps (white regions) between the allowed bands; electrons cannot occupy states that lie in these forbidden gaps. Blue represents energy bands occupied by the 11 sodium electrons when the atom is in its ground state. Gold represents energy bands that are empty.

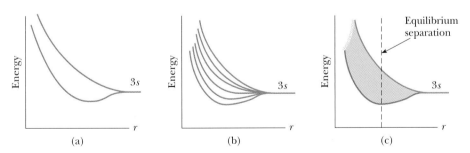

Figure 43.19 (a) Splitting of the $3s$ levels when two sodium atoms are brought together. (b) Splitting of the $3s$ levels when six sodium atoms are brought together. (c) Formation of a $3s$ band when a large number of sodium atoms are assembled to form a solid.

---

[4] There is a split in energy levels because of the two ways the wave functions of the atoms can combine. If ψ_1 and ψ_2 are the two wave functions, the $\psi_1 + \psi_2$ state results in a high electron density between the two atomic cores, whereas the $\psi_1 - \psi_2$ state has zero probability of finding the electron between the two cores. Because the electron is more tightly bound (has lower energy) when it is between the atoms, the $\psi_1 + \psi_2$ state has slightly lower energy than the $\psi_1 - \psi_2$ state. Thus, there are two slightly separated energy levels for the electron when the two atoms are brought close together.

For N atoms combined in any solid, each energy band contains N energy levels. The $1s$, $2s$, and $2p$ bands of sodium are each full, as indicated by the blue-shaded areas in Figure 43.20. An energy level in which the orbital angular momentum is ℓ can hold $2(2\ell + 1)$ electrons. The factor 2 arises from the two possible electron spin orientations, and the factor $2\ell + 1$ corresponds to the number of possible orientations of the orbital angular momentum. The capacity of each band for a system of N atoms is $2(2\ell + 1)N$ electrons. Hence, the $1s$ and $2s$ bands each contain $2N$ electrons ($\ell = 0$), and the $2p$ band contains $6N$ electrons ($\ell = 1$). Because sodium has only one $3s$ electron and there are a total of N atoms in the solid, the $3s$ band contains only N electrons and is only half full, as indicated by the half-blue, half-gold coloring in Figure 43.20. The $3p$ band, which is above the $3s$ band, is completely empty (all gold in the figure).

In Section 43.6, we shall discuss how the band theory allows us to understand the behavior of conductors, insulators, and semiconductors.

43.5 ▸ FREE-ELECTRON THEORY OF METALS

In Section 27.3 we described a classical theory of electrical conduction in metals that led to Ohm's law. According to this theory, a metal is modeled as a classical gas of conduction electrons moving through a fixed lattice of ion cores. Although this theory predicts the correct functional form of Ohm's law, it does not predict the correct values of electrical and thermal conductivities.

In this section, we discuss the free-electron theory of metals, which remedies the shortcomings of the classical model by taking into account the wave nature of the electrons. In this model, one imagines that the outer-shell electrons are free to move through the metal but are trapped within a cavity formed by the surface of the metal.

Statistical physics can be applied to a collection of particles in an effort to relate microscopic properties to macroscopic properties. In the case of electrons, it is necessary to use *quantum statistics*, with the requirement that each state of the system can be occupied by only one electron. Each state is specified by a set of quantum numbers. The probability that a particular state having energy E is occupied by one of the electrons in a solid is given by

Fermi–Dirac distribution function

$$f(E) = \frac{1}{e^{(E - E_F)/k_B T} + 1} \tag{43.18}$$

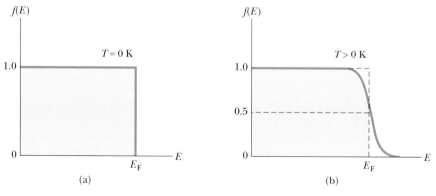

Figure 43.21   Plot of the Fermi–Dirac distribution function $f(E)$ versus energy at (a) $T = 0$ K and (b) $T > 0$ K. The energy E_F is the Fermi energy.

where $f(E)$ is called the **Fermi–Dirac distribution function** and E_F is called the **Fermi energy.** This expression describes how energy is distributed among the electrons. A plot of $f(E)$ versus E at $T = 0$ K is shown in Figure 43.21a. Note that $f(E) = 1$ for $E < E_F$ and $f(E) = 0$ for $E > E_F$. That is, at 0 K all states having energies less than the Fermi energy are occupied, and all states having energies greater than the Fermi energy are vacant. A plot of $f(E)$ versus E at some temperature $T > 0$ K is shown in Figure 43.21b. This curve shows that as T increases, the distribution rounds off slightly, with states between E and $E - k_B T$ losing population and states between E and $E + k_B T$ gaining population. The Fermi energy E_F also depends on temperature, but the dependence is weak in metals.

Quick Quiz 43.3

In Figure 43.21b, what is the physical meaning of the curved part of the plot near E_F?

Quick Quiz 43.4

Where is the Fermi energy level in Figure 43.20?

In Section 41.4 we found that if a particle of mass m is confined to move in a one-dimensional box of length L, the allowed states have quantized energy levels:

$$E_n = \frac{h^2}{8mL^2} n^2 = \frac{\hbar^2 \pi^2}{2mL^2} n^2 \qquad n = 1, 2, 3, \ldots$$

The wave functions for these allowed states are standing waves given by $\psi = A \sin(n\pi x / L)$, which satisfy the boundary condition $\psi = 0$ at $x = 0$ and $x = L$.

Now imagine a piece of metal in the shape of a solid cube of sides L and volume L^3, and let us focus on one electron that is free to move anywhere in this volume. In this model, we require that $\psi(x, y, z) = 0$ at the boundaries. This requirement results in solutions that are standing waves in three dimensions. It can be shown (see Problem 30) that the energy for such an electron is

$$E = \frac{\hbar^2 \pi^2}{2m_e L^2} (n_x^2 + n_y^2 + n_z^2) \tag{43.19}$$

where m_e is the mass of the electron and n_x, n_y, and n_z are quantum numbers. Again, the energy levels are quantized, and each is characterized by this set of three quantum numbers (one for each degree of freedom) and the spin quantum number m_s. For example, the ground state, corresponding to $n_x = n_y = n_z = 1$, has an energy equal to $3\hbar^2 \pi^2 / 2m_e L^2$.

If the quantum numbers are treated as continuous variables, the number of allowed states per unit volume that have energies between E and $E + dE$ is

$$g(E) dE = CE^{1/2} dE \tag{43.20}$$

where

$$C = \frac{8\sqrt{2}\, \pi m_e^{3/2}}{h^3} \tag{43.21}$$

The function $g(E) = CE^{1/2}$ is called the **density-of-states function.**

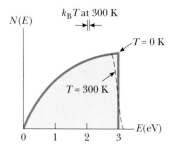

Figure 43.22 Plot of electronic distribution versus energy in a metal at $T = 0$ K (curved and vertical rust lines) and $T = 300$ K (curved rust line and dashed black line). The Fermi energy E_F is 3 eV.

In a metal in thermal equilibrium, the number of electrons per unit volume that have energy between E and $E + dE$ is equal to the product $f(E) g(E) \, dE$:

$$N(E) \, dE = C \frac{E^{1/2} dE}{e^{(E-E_F)/k_B T} + 1} \tag{43.22}$$

A plot of $N(E)$ versus E for two temperatures is given in Figure 43.22.

If n_e is the total number of electrons per unit volume, we require that

$$n_e = \int_0^\infty N(E) \, dE = C \int_0^\infty \frac{E^{1/2} dE}{e^{(E-E_F)/k_B T} + 1} \tag{43.23}$$

We can use this condition to calculate the Fermi energy. At $T = 0$ K, the Fermi distribution function $f(E) = 1$ for $E < E_F$ and $f(E) = 0$ for $E > E_F$. Therefore, at $T = 0$ K, Equation 43.23 becomes

$$n_e = C \int_0^{E_F} E^{1/2} dE = \tfrac{2}{3} C E_F^{3/2} \tag{43.24}$$

Substituting Equation 43.21 into Equation 43.24 gives for the Fermi energy at 0 K

$$E_F(0) = \frac{h^2}{2m_e} \left(\frac{3n_e}{8\pi} \right)^{2/3} \tag{43.25}$$

According to this result, E_F shows a gradual increase with increasing electron concentration. This is expected because the electrons fill the available energy states, two electrons per state, in accordance with the exclusion principle, up to the Fermi energy.

The order of magnitude of the Fermi energy for metals is about 5 eV. Representative values for various metals are given in Table 43.4, together with values for the Fermi speed v_F of the electrons, which is the speed of the electrons when their energy is equal to the Fermi energy. The Fermi speed is defined by the relationship

$$\tfrac{1}{2} m_e v_F^2 \equiv E_F \tag{43.26}$$

Table 43.4 also lists Fermi temperatures T_F, defined by the relationship

$$k_B T_F \equiv E_F \tag{43.27}$$

It is left as a problem for you (Problem 29) to show that the average energy of a free electron in a metal at 0 K is

$$E_{av} = \tfrac{3}{5} E_F \tag{43.28}$$

In summary, we can consider a metal to be a system comprising a very large number of energy levels available to the free electrons. These electrons fill these levels in accordance with the exclusion principle, beginning with $E = 0$ and ending with E_F. At $T = 0$ K, all levels below the Fermi energy are filled and all levels above the Fermi energy are empty. Although the levels are discrete, they are so close together that the electrons have an almost continuous distribution of energy. At 300 K, a very small fraction of the free electrons are excited above the Fermi energy.

TABLE 43.4 Calculated Values of Various Parameters for Metals at 300 K Based on the Free-Electron Theory

Metal	Electron Concentration (m^{-3})	Fermi Energy (eV)	Fermi Speed (m/s)	Fermi Temperature (K)
Li	4.70×10^{28}	4.72	1.29×10^6	5.48×10^4
Na	2.65×10^{28}	3.23	1.07×10^6	3.75×10^4
K	1.40×10^{28}	2.12	0.86×10^6	2.46×10^4
Cu	8.49×10^{28}	7.05	1.57×10^6	8.12×10^4
Ag	5.85×10^{28}	5.48	1.39×10^6	6.36×10^4
Au	5.90×10^{28}	5.53	1.39×10^6	6.41×10^4

EXAMPLE 43.3   The Fermi Energy of Gold

Each atom of gold (Au) contributes one free electron to the metal. Compute (a) the Fermi energy, (b) the Fermi speed, and (c) the Fermi temperature for gold.

Solution   (a) The concentration of free electrons in gold is 5.90×10^{28} m^{-3} (see Table 43.4). Substitution of this value into Equation 43.25 gives

$$E_F(0) = \frac{h^2}{2m_e}\left(\frac{3n_e}{8\pi}\right)^{2/3}$$

$$= \frac{(6.626 \times 10^{-34}\,\text{J·s})^2}{2(9.11 \times 10^{-31}\,\text{kg})}\left(\frac{3 \times 5.90 \times 10^{28}\,\text{m}^{-3}}{8\pi}\right)^{2/3}$$

$$= 8.85 \times 10^{-19}\,\text{J} = \boxed{5.53\,\text{eV}}$$

(b) The Fermi speed is defined by Equation 43.26, $\frac{1}{2}m_e v_F^2 = E_F$. Solving for v_F gives

$$v_F = \left(\frac{2E_F}{m_e}\right)^{1/2} = \left(\frac{2 \times 8.85 \times 10^{-19}\,\text{J}}{9.11 \times 10^{-31}\,\text{kg}}\right)^{1/2}$$

$$= \boxed{1.39 \times 10^6\,\text{m/s}}$$

(c) The Fermi temperature is defined by Equation 43.27:

$$T_F = \frac{E_F}{k_B} = \frac{8.85 \times 10^{-19}\,\text{J}}{1.38 \times 10^{-23}\,\text{J/K}} = \boxed{6.41 \times 10^4\,\text{K}}$$

Thus, the temperature of a gas of classical particles would have to be raised to approximately 64 000 K to have an average energy per particle equal to the Fermi energy at 0 K!

43.6 ELECTRICAL CONDUCTION IN METALS, INSULATORS, AND SEMICONDUCTORS

Good electrical conductors contain a high density of charge carriers, and the density of charge carriers in insulators is nearly zero. Semiconductors are a class of technologically important materials in which charge-carrier densities are intermediate between those of insulators and those of conductors. In this section, we discuss the mechanisms of conduction in these three classes of materials. The enormous variation in electrical conductivity of these materials can be explained in terms of energy bands.

Metals

If a material is to conduct electricity, the charge carriers in the material must be free to move in response to an applied electric field. Let us consider the electrons in a metal as the charge carriers we shall investigate. The motion of the electrons in response to an electric field represents an increase in energy corresponding to the additional kinetic energy of the moving electrons. Thus, to respond to an electric field, electrons must move upward to a higher energy state on an energy level diagram. For them to be able to do this, energy states must be available above the filled states in the band. If a band is completely filled with electrons, no such states are available, and the electrons cannot respond to the electric field by moving.

In Section 43.4, we described the energy-band configuration for the ground state of metallic sodium. We can obtain a better understanding of how metals act as electrical conductors by considering a half-filled band, such as the $3s$ band of sodium.

Figure 43.23 shows a half-filled band in a metal at $T = 0$ K, where the blue represents levels filled with electrons. Because electrons obey Fermi–Dirac statistics, all levels below the Fermi energy are filled with electrons, and all levels above the Fermi energy are empty. The Fermi energy lies in the middle of the band, as was discussed for sodium in Quick Quiz 43.4. At temperatures slightly greater than 0 K, some electrons are thermally excited to levels above E_F, but overall there is little change from the 0 K case. However, **if a potential difference is applied to the metal, electrons having energies near the Fermi energy require only a small amount of additional energy from the applied field to reach nearby empty energy states above the Fermi energy.** Thus, electrons in a metal experiencing only a small applied field are free to move because there are many empty levels available close to the occupied energy levels. We conclude from this high degree of electron mobility that metals are excellent electrical conductors.

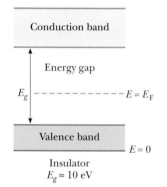

E = E_F

E = 0

Metal

Figure 43.23   Half-filled band of a metal, an electrical conductor. At $T = 0$ K, the Fermi energy lies in the middle of the band.

Insulators

Now consider the two outermost energy bands of a material, where the lower band is filled with electrons and the higher band is empty at 0 K (Fig. 43.24). It is common to refer to the energy separation between the outermost filled band and the adjacent empty band as the **energy gap** E_g of the material. The energy gap for an insulator is large (≈ 10 eV). The lower, filled band is called the **valence band,** and the upper, empty band is the **conduction band.** The Fermi energy lies somewhere in the energy gap, as shown in Figure 43.24. At 300 K (room temperature), $k_B T = 0.025$ eV, which is much smaller than the energy gap in an insulator. At such temperatures, the Fermi–Dirac distribution predicts very few electrons thermally excited into the conduction band. Thus, although an insulator has many vacant states in its conduction band that can accept electrons, so few electrons occupy these states that the overall electrical conductivity is very small, resulting in a high resistivity for insulators.

Conduction band

Energy gap

E_g - - - - - - - - - - - - $E = E_F$

Valence band

$E = 0$

Insulator
$E_g \approx 10$ eV

Figure 43.24   An electrical insulator at $T = 0$ K has a filled valence band and an empty conduction band. The Fermi level lies somewhere between these bands in the region known as the energy gap.

Semiconductors

Semiconductors have the band structure of an insulator and an energy gap on the order of 1 eV. Table 43.5 shows the energy gaps for some representative materials. At $T = 0$ K, all electrons in these materials are in the valence band, and no energy is available to excite them across the energy gap. Thus, semiconductors are poor conductors at very low temperatures. At ordinary temperatures, however, the situation is quite different. For example, the conductivity of silicon at room temperature is about 1.6×10^{-3} $(\Omega \cdot m)^{-1}$.

The band structure of a semiconductor is shown in Figure 43.25. Because the Fermi level is located near the middle of the gap for a semiconductor and because

TABLE 43.5	Energy Gap Values for Some Semiconductors*	
	E_g (eV)	
Crystal	0 K	300 K
Si	1.17	1.14
Ge	0.744	0.67
InP	1.42	1.35
GaP	2.32	2.26
GaAs	1.52	1.43
CdS	2.582	2.42
CdTe	1.607	1.45
ZnO	3.436	3.2
ZnS	3.91	3.6

*From C. Kittel, *Introduction to Solid State Physics*, 5th ed., New York, John Wiley & Sons, 1976.

E_g is small, appreciable numbers of electrons are thermally excited from the valence band to the conduction band. There are many empty levels in the conduction band; therefore, a small applied potential difference can easily raise the energy of the electrons in the conduction band, resulting in a moderate current. Because electrons' being thermally excited across the narrow gap is more probable at higher temperatures, the conductivity of semiconductors increases rapidly with temperature. This contrasts sharply with the conductivity of metals, which decreases slowly with temperature, as described at the end of Section 27.3.

Charge carriers in a semiconductor can be negative or positive, or both. When an electron moves from the valence band into the conduction band, it leaves behind a vacant site, called a **hole,** in the otherwise filled valence band. This hole (electron-deficient site) appears as a positive charge $+e$ and acts as a charge carrier in the sense that a free electron from a nearby site can transfer into the hole. Whenever an electron does so, it creates a new hole at the site it abandoned. Thus, the net effect can be viewed as the hole migrating through the material in the direction opposite the direction of electron movement.

In a pure crystal containing only one element or one compound, there are equal numbers of conduction electrons and holes. Such combinations of charges are called **electron-hole pairs,** and a pure semiconductor that contains such pairs is called an **intrinsic semiconductor** (Fig. 43.26). In the presence of an external electric field, the holes move in the direction of the field, and the conduction electrons move in the direction opposite the field.

Doped Semiconductors

When impurities are added to a semiconductor, both the band structure of the semiconductor and its resistivity are modified. The process of adding impurities, called **doping,** is important in making devices and semiconductors having well-defined regions of different conductivities. For example, when an atom containing five outer-shell electrons, such as arsenic, is added to a semiconductor, four of the electrons form covalent bonds with atoms of the semiconductor and one is left

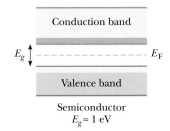

Figure 43.25 Band structure of a semiconductor at ordinary temperatures ($T \approx 300$ K). The energy gap is much smaller than in an insulator, and many electrons occupy states in the conduction band.

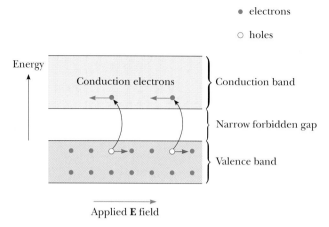

Figure 43.26   Movement of charges (holes and electrons) in an intrinsic semiconductor. The electrons move in the direction opposite the direction of the external electric field, and the holes move in the direction of the field.

over (Fig. 43.27a). This extra electron is nearly free of its parent atom and has an energy level that lies in the energy gap, just below the conduction band (Fig. 43.27b). Such a pentavalent atom in effect donates an electron to the structure and hence is referred to as a **donor atom.** Because the spacing between the energy level of the electron of the donor atom and the bottom of the conduction band is very small (typically, about 0.05 eV), only a small amount of thermal excitation is needed to cause this electron to move into the conduction band. (Recall that the average energy of an electron at room temperature is about $k_B T \approx 0.025$ eV). Semiconductors doped with donor atoms are called **n-type semiconductors** because the majority of charge carriers are electrons, which are **n**egatively charged.

If a semiconductor is doped with atoms containing three outer-shell electrons, such as indium and aluminum, the three form covalent bonds with neighboring

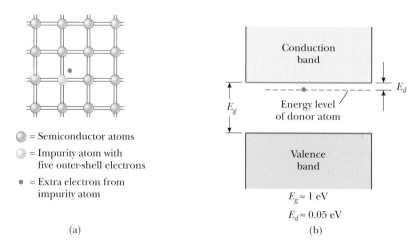

◉ = Semiconductor atoms

◯ = Impurity atom with five outer-shell electrons

• = Extra electron from impurity atom

(a)

$E_g \approx 1$ eV
$E_d \approx 0.05$ eV

(b)

Figure 43.27   (a) Two-dimensional representation of a semiconductor (gray) containing an impurity atom (yellow-orange) that has five outer-shell electrons. Each double line represents a covalent bond. (b) Energy-band diagram for a semiconductor in which the nearly free electron of the impurity atom lies in the forbidden gap, just below the bottom of the conduction band.

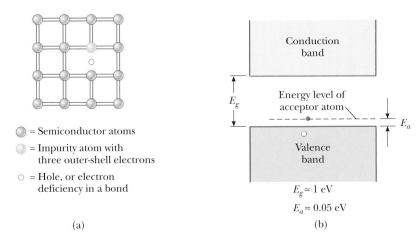

Figure 43.28 (a) Two-dimensional representation of a semiconductor (gray) containing an impurity atom (yellow-orange) having three outer-shell electrons. (b) Energy-band diagram for a semiconductor in which the hole resulting from the trivalent impurity atom lies in the forbidden gap, just above the top of the valence band.

semiconductor atoms, leaving an electron deficiency—a hole—where the fourth bond would be if an impurity-atom electron were available to form it (Fig. 43.28a). The energy level of this hole lies in the energy gap, just above the valence band, as shown in Figure 43.28b. An electron from the valence band has enough energy at room temperature to fill this impurity level, leaving behind a hole in the valence band. Because a trivalent atom in effect accepts an electron from the valence band, such impurities are referred to as **acceptor atoms.** A semiconductor doped with trivalent (acceptor) impurities is known as a **p-type semiconductor** because the majority of charge carriers are **p**ositively charged holes.

When conduction in a semiconductor is the result of acceptor or donor impurities, the material is called an **extrinsic semiconductor.** The typical range of doping densities for extrinsic semiconductors is 10^{13} to 10^{19} cm^{-3}, whereas the electron density in a typical semiconductor is roughly 10^{21} cm^{-3}.

Optional Section

43.7 SEMICONDUCTOR DEVICES

The $p-n$ Junction

Now let us consider what happens when a p-type semiconductor is joined to an n-type semiconductor to form a $p-n$ junction. The junction consists of the three distinct regions shown in Figure 43.29a: a p region, a depletion region, and an n region.

The depletion region, which extends several micrometers to either side of the center of the junction, may be visualized as arising when the two halves of the junction are brought together. The mobile n-side donor electrons nearest the junction (deep-blue area in Fig. 43.29a) diffuse to the p side, leaving behind immobile positive ions. At the same time, holes from the p side nearest the junction diffuse to the n side and leave behind a region (brown area in Fig. 43.29a) of fixed negative ions.

The depletion region contains an internal electric field (arising from the charges of the fixed ions) on the order of 10^4 to 10^6 V/cm (see Fig. 43.29b). This

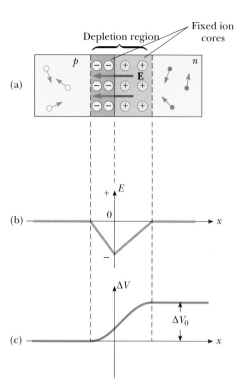

Figure 43.29 (a) Physical arrangement of a $p-n$ junction. (b) Internal electric field versus x for the $p-n$ junction. (c) Internal electric potential difference ΔV versus x for the $p-n$ junction. The potential difference ΔV_0 represents the potential difference across the junction in the absence of an applied electric field.

field sweeps mobile charges out of the depletion region. Thus, the depletion region is so named because it is depleted of mobile charge carriers. This internal electric field creates an internal potential difference ΔV_0 that prevents further diffusion of holes and electrons across the junction and thereby ensures zero current through the junction when no potential difference is applied.

Perhaps the most notable feature of the $p-n$ junction is its ability to pass current in only one direction. Such action is easiest to understand in terms of the potential difference graph shown in Figure 43.29c. If a voltage ΔV is applied to the junction such that the p side is connected to the positive terminal of a voltage source, as shown in Figure 43.30a, the internal potential difference ΔV_0 across the junction is decreased; the decrease results in a current that increases exponentially with increasing forward voltage, or *forward bias*. For *reverse bias* (where the n side of the junction is connected to the positive terminal of a voltage source), the internal potential difference ΔV_0 increases with increasing reverse bias; the increase results in a very small reverse current that quickly reaches a saturation value I_0. The current–voltage relationship for an ideal diode is

$$I = I_0(e^{e\Delta V/k_B T} - 1) \qquad \textbf{(43.29)}$$

where the first e is the base of the natural logarithm, the second e represents the magnitude of the electron charge, k_B is Boltzmann's constant, and T is the temperature in kelvins. Figure 43.30b shows an $I-\Delta V$ plot characteristic of a real $p-n$ junction.

Quick Quiz 43.5

Does the $p-n$ junction described in Figure 43.30 obey Ohm's law?

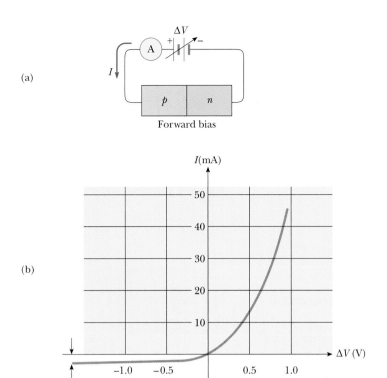

Figure 43.30 (a) Schematic of a p–n junction under forward bias. (b) The characteristic curve for a real p–n junction.

Light-Emitting and Light-Absorbing Diodes

Light emission and absorption in semiconductors is similar to light emission and absorption by gaseous atoms, except that in the discussion of semiconductors discrete atomic energy levels must be replaced by bands. As shown in Figure 43.31a, an electron excited electrically into the conduction band can easily recombine with a hole (especially if the electron is injected into a p region). As this recombination takes place, a photon of energy E_g is emitted. Light-emitting diodes (LEDs) and semiconductor lasers are common examples of devices using this phenomenon.

- electrons
- ○ holes

(a) Light emission

(b) Light absorption

Figure 43.31 (a) Light emission from a semiconductor. (b) Light absorption by a semiconductor.

QuickLab

Aim a television remote control at the lens of a camcorder and record for a few seconds while pressing buttons on the remote. When you play the tape, you'll see the "invisible" flashing of the remote. The semiconductor device in the camcorder that senses the light entering the lens is sensitive to the infrared pulses from the remote control.

The *Sojourner*, seen here cruising on the surface of Mars in 1997, used photovoltaic solar cells to convert sunlight to electricity. *(Courtesy of NASA)*

Conversely, an electron in the valence band may absorb a photon of light and be promoted to the conduction band, leaving a hole behind (Fig. 43.31b). One device that operates on this principle is the photovoltaic solar cell.

EXAMPLE 43.4   Where's the Remote?

Estimate the band gap of the semiconductor in the infrared LED of a typical television remote control.

Solution   In Chapter 34 we learned that the wavelength of infrared light ranges from 700 nm to 1 mm. Let us pick a number that is easy to work with, such as 1 000 nm. (This is

not a bad estimate. Remotes typically operate in the range of 880 to 950 nm.)

The energy of a photon is given by $E = hc/\lambda$, and so the energy of the photons from the remote control is about 2.0×10^{-19} J = 1.2 eV. This corresponds to an energy gap E_g of approximately 1.2 eV in the LED's semiconductor.

To learn more about TV remotes and hundreds of other everyday devices, visit **www.howstuffworks.com**

The Junction Transistor

The invention of the transistor by John Bardeen (1908–1991), Walter Brattain (1902–1987), and William Shockley (1910–1989) in 1948 totally revolutionized the world of electronics. For this work, these three men shared a Nobel Prize in 1956. By 1960, the transistor had replaced the vacuum tube in many electronic applications. The advent of the transistor created a multitrillion dollar industry that produces such popular devices as pocket radios, hand-held calculators, computers, television receivers, and electronic games.

One form of the junction transistor consists of a semiconducting material in which a very narrow *n* region is sandwiched between two *p* regions. This configuration is called a ***pnp* transistor.** Another configuration is the ***npn* transistor,** which consists of a *p* region sandwiched between two *n* regions. Because the operation of the two transistors is essentially the same, we describe only the *pnp* transistor. The structure of the *pnp* transistor, together with its circuit symbol, is shown in Figure 43.32. The outer regions are called the **emitter** and the **collector,** and the narrow central region is called the **base.** The configuration contains two junctions: the emitter–base interface and the collector–base interface.

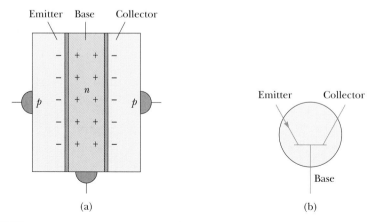

Figure 43.32   (a) The *pnp* transistor consists of an *n* region (base) sandwiched between two *p* regions (emitter and collector). (b) Circuit symbol for the *pnp* transistor.

Suppose that a voltage is applied to the transistor such that the emitter is at a higher electric potential than the collector. (This is accomplished with the battery labeled ΔV_{ec} in Fig. 43.33.) If we think of the transistor as two p–n juctions back to back, we see that the emitter–base junction is forward-biased and that the base–collector junction is reverse-biased. The emitter is heavily doped relative to the base, and as a result nearly all the current consists of holes moving across the emitter–base junction. Most of these holes do not recombine with electrons in the base because it is very narrow. Instead, they are accelerated across the reverse-biased base-collector junction, producing the emitter current I_e shown in Figure 43.33.

Although only a small percentage of holes recombine in the base, those that do limit the emitter current to a small value because positive charge carriers accumulating in the base prevent holes from flowing in. To prevent this current limitation, some of the positive charge on the base must be drawn off; this is accomplished by connecting the base to the battery labeled ΔV_{eb}, as shown in Figure 43.33. Those positive charges that are not swept across the base–collector junction leave the base through this added pathway. **This base current I_b is very small, but a small change in it can significantly change the collector current I_c.** If the transistor is properly biased, the collector (output) current is directly proportional to the base (input) current, and the transistor acts as a current amplifier. This condition may be written

$$I_c = \beta I_b$$

where β, the *current gain* factor, is typically in the range from 10 to 100.

The transistor may be used to amplify a small signal. A small voltage to be amplified is placed in series with the battery ΔV_{eb}. The input signal produces a small variation in the base current, resulting in a large change in the collector current and hence a large change in the voltage across the output resistor.

The Integrated Circuit

Invented independently by Jack Kilby (b. 1923) at Texas Instruments in late 1958 and by Robert Noyce (b. 1927) at Fairchild Camera and Instrument in early 1959, the integrated circuit has been justly called "the most remarkable technology ever to hit mankind." Kilby's first device is shown in Figure 43.34. Integrated circuits have indeed started a "second industrial revolution" and are found at the heart of

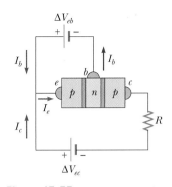

Figure 43.33 A bias voltage ΔV_{eb} applied to the base as shown produces a small base current I_b that is used to control the collector current I_c in a *pnp* transistor.

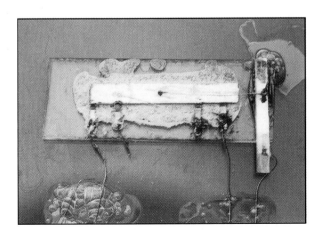

Figure 43.34 Jack Kilby's first integrated circuit, tested on September 12, 1958. *(Courtesy of Texas Instruments, Inc.)*

Figure 43.35 Integrated circuits continue to shrink in size and price while simultaneously growing in capability. *(Courtesy of Intel Corporation)*

web

For more information, visit
www.intel.com

computers, watches, cameras, automobiles, aircraft, robots, space vehicles, and all sorts of communication and switching networks.

In simplest terms, an **integrated circuit** is a collection of interconnected transistors, diodes, resistors, and capacitors fabricated on a single piece of silicon known as a *chip*. State-of-the-art chips easily contain several million components within a 1-cm^2 area (Fig. 43.35), with the number of components per square inch having doubled every year since the integrated circuit was invented. Figure 43.36 illustrates the dramatic advances made in chip technology in the past 30 years.

Integrated circuits were invented partly to solve the interconnection problem spawned by the transistor. In the era of vacuum tubes, power and size considera-

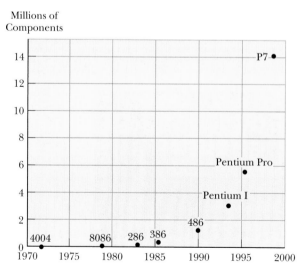

Figure 43.36 This plot illustrates dramatic advances in chip technology: The number of components fitting on a single computer chip versus year of manufacture.

tions of individual components set modest limits on the number of components that could be interconnected in a given circuit. With the advent of the tiny, low-power, highly reliable transistor, design limits on the number of components disappeared and were replaced by the problem of wiring together hundreds of thousands of components. The magnitude of this problem can be appreciated when we consider that second-generation computers (consisting of discrete transistors rather than integrated circuits) contained several hundred thousand components requiring more than a million joints that had to be hand-soldered and tested.

In addition to solving the interconnection problem, integrated circuits possess the advantages of miniaturization and fast response, two attributes critical for high-speed computers. The fast response results from the miniaturization and close packing of components, because the response time of a circuit depends on the time it takes for electrical signals traveling at about 0.3 m/ns to pass from one component to another. This time is reduced by the closely packed components.

Optional Section

43.8 SUPERCONDUCTIVITY

We learned in Section 27.5 that there is a class of metals and compounds known as **superconductors** whose electrical resistance decreases to virtually zero below a certain temperature T_c called the *critical temperature* (Table 43.6). Let us now look at these amazing materials in greater detail, using what we have just learned about the properties of solids to help us understand the behavior of superconductors.

Let us start by examining the Meissner effect, described in Section 30.8 as the exclusion of magnetic flux from the interior of superconductors. Simple arguments based on the laws of electricity and magnetism can be used to show that the magnetic field inside a superconductor cannot change with time. According to Equation 27.8, $R = \Delta V/I$, and because the potential difference ΔV across a conductor is proportional to the electric field inside the conductor, we see that the electric field is proportional to the resistance of the conductor. Thus, because $R = 0$ for a superconductor at or below its critical temperature, *the electric field in its interior must be zero.* Now recall that Faraday's law of induction can be expressed in the form shown in Equation 31.9:

$$\oint \mathbf{E} \cdot d\mathbf{s} = -\frac{d\Phi_B}{dt} \qquad \textbf{(43.30)}$$

That is, the line integral of the electric field around any closed loop is equal to the negative rate of change in the magnetic flux Φ_B through the loop. Because $\mathbf{E}$ is zero everywhere inside the superconductor, the integral over any closed path inside the superconductor is zero. Hence, $d\Phi_B/dt = 0$; this tells us that **the magnetic flux in the superconductor cannot change.** From this information, we can conclude that $B \, (= \Phi_B/A)$ must remain constant inside the superconductor.

Before 1933, it was assumed that superconductivity was a manifestation of perfect conductivity. If a perfect conductor is cooled below its critical temperature in the presence of an applied magnetic field, the field should be trapped in the interior of the conductor even after the external field is removed. In addition, the final state of the perfect conductor should depend on which occurs first, the application of the field or the cooling to below T_c. If the field is applied after the material has been cooled, the field should be expelled from the superconductor. If the field is applied before the material is cooled, the field should not be ex-

For a more detailed discussion on the field of superconductivity, visit the website for the text
www.saunderscollege.com/physics/

TABLE 43.6
Critical Temperatures for Various Superconductors

Material	T_c (K)
Zn	0.88
Al	1.19
Sn	3.72
Hg	4.15
Pb	7.18
Nb	9.46
Nb_3Sn	18.05
Nb_3Ge	23.2
$YBa_2Cu_3O_7$	92
Bi-Sr-Ca-Cu-O	105
Tl-Ba-Ca-Cu-O	125

pelled once the material has been cooled. In 1933, however, W. Hans Meissner and Robert Ochsenfeld discovered that, when a metal becomes superconducting in the presence of a weak magnetic field, the field is expelled. Thus, the same final state **B** = 0 is achieved whether the field is applied before or after the material is cooled below its critical temperature.

The Meissner effect is illustrated in Figure 43.37 for a superconducting material in the shape of a long cylinder. Note that the field penetrates the cylinder when its temperature is greater than T_c (Fig. 43.37a). As the temperature is lowered to below T_c, however, the field lines are spontaneously expelled from the interior of the superconductor (Fig. 43.37b). Thus, a superconductor is more than a perfect conductor (resistivity $\rho = 0$); it is also a perfect diamagnet (**B** = 0). The property that **B** = 0 in the interior of a superconductor is as fundamental as the property of zero resistance. If the magnitude of the applied magnetic field exceeds a critical value B_c, defined as the value of B that destroys a material's superconducting properties, the field again penetrates the sample.

Because a superconductor is a perfect diamagnet having a negative magnetic susceptibility, it repels a permanent magnet. In fact, one can perform a demonstration of the Meissner effect by floating a small permanent magnet above a superconductor and achieving magnetic levitation, as seen in Figure 30.34.

You should recall from our study of electricity that a good conductor expels static electric fields by moving charges to its surface. In effect, the surface charges produce an electric field that exactly cancels the externally applied field inside the conductor. In a similar manner, a superconductor expels magnetic fields by forming surface currents. To see why this happens, consider again the superconductor shown in Figure 43.37. Let us assume that the sample is initially at a temperature $T > T_c$, as illustrated in Figure 43.37a, so that the magnetic field penetrates the cylinder. As the cylinder is cooled to a temperature $T < T_c$, the field is expelled, as shown in Figure 43.37b. Surface currents induced on the superconductor's surface produce a magnetic field that exactly cancels the externally applied field inside the superconductor. As you would expect, the surface currents disappear when the external magnetic field is removed.

An important development in physics that elicited much excitement in the scientific community was the discovery of high-temperature copper oxide–based superconductors. The excitement began with a 1986 publication by J. Georg Bed-

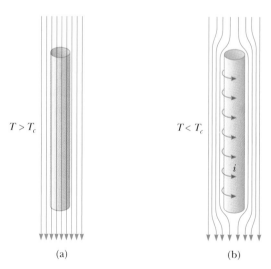

$T > T_c$

$T < T_c$

i

(a)   (b)

Figure 43.37   A superconductor in the form of a long cylinder in the presence of an external magnetic field. (a) At temperatures above T_c, the field lines penetrate the cylinder because it is in its normal state. (b) When the cylinder is cooled to $T < T_c$ and becomes superconducting, magnetic flux is excluded from its interior by the induction of surface currents.

norz (b. 1950) and K. Alex Müller (b. 1927), scientists at the IBM Zurich Research Laboratory in Switzerland. In their paper,[5] Bednorz and Müller reported strong evidence for superconductivity at 30 K in an oxide of barium, lanthanum, and copper. They were awarded the Nobel Prize for physics in 1987 for their remarkable discovery. Shortly thereafter, a new family of compounds was open for investigation, and research activity in the field of superconductivity proceeded vigorously. In early 1987, groups at the University of Alabama at Huntsville and the University of Houston announced superconductivity at about 92 K in an oxide of yttrium, barium, and copper ($YBa_2Cu_3O_7$). Later that year, teams of scientists from Japan and the United States reported superconductivity at 105 K in an oxide of bismuth, strontium, calcium, and copper. More recently, scientists have reported superconductivity at temperatures as high as 150 K in an oxide containing mercury (see Fig. 27.13). Today, one cannot rule out the possibility of room-temperature superconductivity, and the mechanisms responsible for the behavior of high-temperature superconductors are still under investigation. The search for novel superconducting materials continues both for scientific reasons and because practical applications become more probable and widespread as the critical temperature is raised.

SUMMARY

Two or more atoms combine to form molecules because of a net attractive force between the atoms. The mechanisms responsible for molecular bonding can be classified as follows:

- **Ionic bonds** form primarily because of the Coulomb attraction between oppositely charged ions. Sodium chloride (NaCl) is one example.
- **Covalent bonds** form when the constituent atoms of a molecule share electrons. For example, the two electrons of the H_2 molecule are equally shared between the two nuclei.
- **Van der Waals bonds** are weak electrostatic bonds between atoms that do not form ionic or covalent bonds. These bonds are responsible for the condensation of inert gas atoms and nonpolar molecules into the liquid phase.
- **Hydrogen bonds** form between the center of positive charge in a polar molecule that includes one or more hydrogen atoms and the center of negative charge in another polar molecule.

The energy of a gas molecule consists of contributions from the electronic energy in the bonds and from the translational, rotational, and vibrational motions of the molecule.

The allowed values of the rotational energy of a diatomic molecule are

$$E_{\text{rot}} = \frac{\hbar^2}{2I} J(J+1) \qquad J = 0, 1, 2, \ldots \qquad \textbf{(43.6)}$$

where I is the moment of inertia of the molecule and J is an integer called the **rotational quantum number.** The selection rule for transitions between rotational states is given by $\Delta J = \pm 1$.

The allowed values of the vibrational energy of a diatomic molecule are

$$E_{\text{vib}} = (v + \tfrac{1}{2}) \frac{h}{2\pi} \sqrt{\frac{k}{\mu}} \qquad v = 0, 1, 2, \ldots \qquad \textbf{(43.10)}$$

where v is the **vibrational quantum number,** k is the force constant of the "effec-

[5] J. G. Bednorz and K. A. Müller, *Z. Phys. B* 64:189, 1986.

tive spring" bonding the molecule, and μ is the **reduced mass** of the molecule. The selection rule for allowed vibrational transitions is $\Delta v = \pm 1$, and the energy difference between any two adjacent levels is the same regardless of which two levels are involved.

Bonding mechanisms in solids can be classified in a manner similar to the schemes for molecules. For example, the Na^+ and Cl^- ions in NaCl form **ionic bonds,** while the carbon atoms in diamond form **covalent bonds.** The **metallic bond** is characterized by a net attractive force between positive ion cores and the mobile free electrons of a metal.

In a crystalline solid, the energy levels of the system form a set of **bands.** Electrons occupy the lowest-energy states, with no more than one electron per state. Energy gaps are present between the bands of allowed states.

In the **free-electron theory of metals,** the free electrons fill the quantized levels in accordance with the Pauli exclusion principle. The number of states per unit volume available to the conduction electrons having energies between E and $E + dE$ is

$$N(E)\,dE = C\frac{E^{1/2}dE}{e^{(E-E_{\mathrm{F}})/k_{\mathrm{B}}T} + 1} \qquad \textbf{(43.22)}$$

where C is a constant and E_{F} is the **Fermi energy.** At $T = 0$ K, all levels below E_{F} are filled, all levels above E_{F} are empty, and

$$E_{\mathrm{F}}(0) = \frac{h^2}{2m_e}\left(\frac{3n_e}{8\pi}\right)^{2/3} \qquad \textbf{(43.25)}$$

where n_e is the total number of conduction electrons per unit volume. Only those electrons having energies near E_{F} can contribute to the electrical conductivity of the metal.

A **semiconductor** is a material having an energy gap of approximately 1 eV and a valence band that is filled at $T = 0$ K. Because of the small energy gap, a significant number of electrons can be thermally excited from the valence band into the conduction band. The band structures and electrical properties of a semiconductor can be modified by the addition of either donor atoms containing five outer-shell electrons (such as arsenic) or acceptor atoms containing three outer-shell electrons (such as indium). A semiconductor **doped** with donor impurity atoms is called an ***n*-type semiconductor,** and one doped with acceptor impurity atoms is called a ***p*-type semiconductor.** The energy levels of these impurity atoms fall within the energy gap of the material.

QUESTIONS

1. Discuss the three major forms of excitation of a molecule (other than translational motion) and the relative energies associated with these three forms.
2. Explain the role of the Pauli exclusion principle in describing the electrical properties of metals.
3. Discuss the properties of a material that determine whether it is a good electrical insulator or a good conductor.
4. Table 43.5 shows that the energy gaps for semiconductors decrease with increasing temperature. What do you suppose accounts for this behavior?

5. The resistivity of metals increases with increasing temperature, whereas the resistivity of an intrinsic semiconductor decreases with increasing temperature. Explain.
6. Discuss the differences in the band structures of metals, insulators, and semiconductors. How does the band-structure model enable you to better understand the electrical properties of these materials?
7. Discuss models for the different types of bonds that form stable molecules.
8. Discuss the electrical, physical, and optical properties of

ionically bonded solids. Compare your expectations with tabulated properties for such solids.

9. Discuss the electrical and physical properties of covalently bonded solids. Compare your expectations with tabulated properties for such solids.

10. Discuss the electrical and physical properties of metals.

11. When a photon is absorbed by a semiconductor, an electron–hole pair is created. Give a physical explanation of this statement, using the energy-band model as the basis for your description.

12. Pentavalent atoms such as arsenic are donor atoms in a semiconductor such as silicon, while trivalent atoms such as indium are acceptors. Inspect the periodic table in Appendix C, and determine what other elements might make good donors or acceptors.

13. What are the essential assumptions made in the free-electron theory of metals? How does the energy-band model differ from the free-electron theory in describing the properties of metals?

14. How do the vibrational and rotational levels of heavy hydrogen (D_2) molecules compare with those of H_2 molecules?

15. Which is easier to excite in a diatomic molecule, rotational or vibrational motion?

16. The energy of visible light ranges between 1.8 and 3.1 eV. Does this explain why silicon, with an energy gap of 1.1 eV (see Table 43.5), appears opaque, whereas diamond, with an energy gap of 5.5 eV, appears transparent?

17. Why is a *pnp* or *npn* sandwich (whose central region is very thin) essential to transistor operation?

18. How can the analysis of the rotational spectrum of a molecule lead to an estimate of the size of that molecule?

PROBLEMS

1, 2, 3 = straightforward, intermediate, challenging ⬜ = full solution available in the *Student Solutions Manual and Study Guide*
WEB = solution posted at **http://www.saunderscollege.com/physics/** 🖥 = Computer useful in solving problem 🔲 = Interactive Physics
⬜ = paired numerical/symbolic problems

Section 43.1 Molecular Bonds

WEB 1. **Review Problem.** A K^+ ion and a Cl^- ion are separated by a distance of 5.00×10^{-10} m. Assuming that the two ions act like point charges, determine (a) the force each ion exerts on the other and (b) the potential energy of attraction in electron volts.

2. Potassium chloride is an ionically bonded molecule, sold as a salt substitute for use in a low-sodium diet. The electron affinity of chlorine is 3.6 eV. An energy input of 0.7 eV is required to form separate K^+ and Cl^- ions from separate K and Cl atoms. What is the ionization energy of K?

3. One description of the potential energy of a diatomic molecule is given by the Lennard–Jones potential,

$$U = \frac{A}{r^{12}} - \frac{B}{r^6}$$

where A and B are constants. Find, in terms of A and B, (a) the value r_0 at which the energy is a minimum and (b) the energy E required to break up a diatomic molecule. (c) Evaluate r_0 in meters and E in electron volts for the H_2 molecule. In your calculations, use the values $A = 0.124 \times 10^{-120}$ eV·m^{12} and $B = 1.488 \times 10^{-60}$ eV·m^6. (*Note:* Although this potential is widely used for modeling, it is known to have serious defects. For example, its behavior at both small and large values of *r* is greatly in error.)

4. A van der Waals dispersion force between helium atoms produces a very shallow potential well, with a depth on the order of 1 meV. At about what temperature would you expect helium to condense?

Section 43.2 The Energy and Spectra of Molecules

5. The cesium iodide (CsI) molecule has an atomic separation of 0.127 nm. (a) Determine the energy of the lowest excited rotational state and the frequency of the photon absorbed in the $J = 0$ to $J = 1$ transition. (b) What would be the fractional change in this frequency if the estimate of the atomic separation is off by 10%?

6. The CO molecule makes a transition from the $J = 1$ to the $J = 2$ rotational state when it absorbs a photon of frequency 2.30×10^{11} Hz. Find the moment of inertia of this molecule from these data.

WEB 7. A HCl molecule is excited to its first rotational-energy level, corresponding to $J = 1$. If the distance between its nuclei is 0.127 5 nm, what is the angular speed of the molecule about its center of mass?

8. A diatomic molecule consists of two atoms having masses m_1 and m_2 and separated by a distance r. Show that the moment of inertia about an axis through the center of mass of the molecule is given by Equation 43.3, $I = \mu r^2$.

9. (a) Calculate the moment of inertia of a NaCl molecule about its center of mass. The atoms are separated by a distance $r = 0.28$ nm. (b) Calculate the wavelength of radiation emitted if a NaCl molecule undergoes a transition from the $J = 2$ state to the $J = 1$ state.

10. The rotational spectrum of the HCl molecule contains lines with wavelengths of 0.060 4, 0.069 0, 0.080 4, 0.096 4, and 0.120 4 mm. What is the moment of inertia of the molecule?

WEB 11. If the effective force constant of a vibrating HCl molecule is $k = 480$ N/m, find the energy difference between the ground state and the first excited vibrational level.

12. Use the data in Table 43.2 to calculate the minimum amplitude of vibration for (a) the HI molecule and (b) the HF molecule. Which has the weaker bond?

13. The nuclei of the O_2 molecule are separated by 1.2×10^{-10} m. The mass of each oxygen atom in the molecule is 2.66×10^{-26} kg. (a) Determine the rotational energies of an oxygen molecule in electron volts for the levels corresponding to $J = 0$, 1, and 2. (b) The effective force constant k between the atoms in the oxygen molecule is 1 177 N/m. Determine the vibrational energies (in electron volts) corresponding to $v = 0$, 1, and 2.

14. Figure P43.14 is a model of a benzene molecule. All atoms lie in a plane, and the carbon atoms form a regular hexagon, as do the hydrogen atoms. The carbon atoms are 0.110 nm apart center to center. Determine the allowed energies of rotation about an axis perpendicular to the plane of the paper through the center point O. Hydrogen and carbon atoms have masses of 1.67×10^{-27} kg and 1.99×10^{-26} kg, respectively.

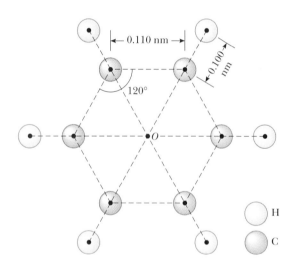

Figure P43.14

15. If the CO molecule were rigid, the rotational transition into what J state would absorb the same wavelength photon as the 0 to 1 vibrational transition? (Use information given in Table 43.2.)

16. Photons of what frequencies can be spontaneously emitted by CO molecules in the state with $v = 1$ and $J = 0$?

17. Most of the mass of an atom is in its nucleus. Model the mass distribution in a diatomic molecule as two spheres, each of radius 2.00×10^{-15} m and mass 1.00×10^{-26} kg, located at points along the x axis as shown in Figure 43.9 and separated by 2.00×10^{-10} m. Rotation about the axis joining the nuclei in the diatomic molecule is ordinarily ignored because the first

excited state would have an energy that is too high to access. To see why, calculate the ratio of the energy of the first excited state for rotation about the x axis, to the energy of the first excited state for rotation about the y axis.

Section 43.3   Bonding in Solids

18. Use a magnifying glass to look at the table salt that comes out of a salt shaker. Compare what you see with Figure 43.14(a). The distance between a sodium ion and a nearest-neighbor chloride ion is 0.261 nm. Make an order-of-magnitude estimate of the number N of atoms in a typical grain of salt. Suppose that you had a number of grains of salt equal to this number N. What would be the volume of this quantity of salt?

19. Use Equation 43.17 to calculate the ionic cohesive energy for NaCl. Take $\alpha = 1.747\ 6$, $r_0 = 0.281$ nm, and $m = 8$.

20. The distance between the K^+ and Cl^- ions in a KCl crystal is 0.314 nm. Calculate the distances from one K^+ ion to its nearest-neighbor K^+ ions, to its second-nearest neighbor K^+ ions, and to its third-nearest neighbor K^+ ions.

21. Consider a one-dimensional chain of alternating positive and negative ions. Show that the potential energy associated with an ion in this hypothetical crystal is

$$U(r) = -k_e\, \alpha\, \frac{e^2}{r}$$

where the Madelung constant is $\alpha = 2 \ln 2$ and r is the interionic spacing. [*Hint:* Use the series expansion for $\ln(1 + x)$.]

Section 43.4   Band Theory of Solids
Section 43.5   Free-Electron Theory of Metals

22. Show that Equation 43.25 can be expressed as $E_F = (3.65 \times 10^{-19})n^{2/3}$ eV where E_F is in electron volts when n is in electrons per cubic meter.

23. The Fermi energy for silver is 5.48 eV. Silver has a density of 10.6×10^3 kg/m^3 and an atomic mass of 108. Use this information to show that silver has one free electron per atom.

24. (a) Find the typical speed of a conduction electron in copper, taking its kinetic energy as equal to the Fermi energy, 7.05 eV. (b) How does this compare with a drift speed of 0.1 mm/s?

25. Sodium is a monovalent metal having a density of 0.971 g/cm^3 and molar mass of 23.0 g/mol. Use this information to calculate (a) the density of charge carriers, (b) the Fermi energy, and (c) the Fermi speed for sodium.

26. When solid silver starts to melt, what is the approximate fraction of the conduction electrons that are thermally excited above the Fermi level?

WEB 27. Calculate the energy of a conduction electron in silver at 800 K if the probability of finding an electron in that state is 0.950. The Fermi energy is 5.48 eV at this temperature.

28. Consider a cube of gold 1.00 mm on an edge. Calculate the approximate number of conduction electrons in this cube whose energies lie in the range 4.000 to 4.025 eV.

29. Show that the average kinetic energy of a conduction electron in a metal at 0 K is $E_{av} = \frac{3}{5} E_F$. (*Hint:* In general, the average kinetic energy is

$$E_{av} = \frac{1}{n_e} \int EN(E)\, dE$$

where n_e is the density of electrons, $N(E)\, dE$ is given by Equation 43.22, and the integral is over all possible values of the energy.)

30. **Review Problem.** An electron moves in a three-dimensional box of edge length L and volume L^3. If the wave function of the particle is $\psi = A \sin(k_x x) \sin(k_y y) \sin(k_z z)$, show that its energy is given by Equation 43.19:

$$E = \frac{\hbar^2 \pi^2}{2 m_e L^2} (n_x^2 + n_y^2 + n_z^2)$$

where the quantum numbers (n_x, n_y, n_z) are integers ≥ 1. (*Hint:* The Schrödinger equation in three dimensions may be written

$$\frac{\partial^2 \psi}{\partial x^2} + \frac{\partial^2 \psi}{\partial y^2} + \frac{\partial^2 \psi}{\partial z^2} = \frac{\hbar^2}{2m_e} (U - E) \psi$$

To confine the electron inside the box, take $U = 0$ inside and $U = \infty$ outside.)

31. (a) Consider a system of electrons confined to a three-dimensional box. Calculate the ratio of the number of allowed energy levels at 8.50 eV to the number at 7.00 eV. (b) Copper has a Fermi energy of 7.0 eV at 300 K. Calculate the ratio of the number of occupied levels at an energy of 8.50 eV to the number at the Fermi energy. Compare your answer with that obtained in part (a).

Section 43.6 Electrical Conduction in Metals, Insulators, and Semiconductors

32. The energy gap for silicon at 300 K is 1.14 eV. (a) Find the lowest frequency photon that will promote an electron from the valence band to the conduction band. (b) What is the wavelength of this photon?

33. Light from a hydrogen discharge tube is incident on a CdS crystal. Which spectral lines from the Balmer series are absorbed and which are transmitted?

34. A light-emitting diode (LED) made of the semiconductor GaAsP emits red light ($\lambda = 650$ nm). Determine the energy band gap E_g in the semiconductor.

WEB 35. Most solar radiation has a wavelength of 1 μm or less. What energy gap should the material in a solar cell have to absorb this radiation? Is silicon appropriate (see Table 43.5)?

36. Assume you are to build a scientific instrument that is thermally isolated from its surroundings, but such that you can use an external laser to raise the temperature of a target inside it. (It might be a calorimeter, but these design criteria could apply to other devices as well.) Since you know that diamond is transparent and a good thermal insulator, you decide to use a diamond window in the apparatus. Diamond has an energy gap of 5.5 eV between its valence and conduction bands. What is the shortest laser wavelength you can use to warm the sample inside?

(Optional)
Section 43.7 Semiconductor Devices

Note: Problem 71 in Chapter 27 also applies to this section.

37. For what value of the bias voltage ΔV in Equation 43.29 does (a) $I = 9.00 I_0$? (b) $I = -0.900 I_0$? Assume $T = 300$ K.

38. The diode shown in Figure 43.30 is connected in series with a battery and a 150-Ω resistor. What battery emf is required for a current of 25.0 mA?

39. You put a diode in a microelectronic circuit to protect the system in case an untrained person installs the battery backwards. In the correct forward-bias situation, the current is 200 mA with a potential difference of 100 mV across the diode at room temperature (300 K). If the battery were reversed, what would be the magnitude of the current through the diode?

(Optional)
Section 43.8 Superconductivity

Note: Problem 26 in Chapter 30 and Problems 76 through 79 in Chapter 32 also apply to this section.

40. A thin rod of superconducting material 2.50 cm long is placed into a 0.540-T magnetic field with its cylindrical axis along the magnetic field lines. (a) Sketch the directions of the applied field and the induced surface current. (b) Find the magnitude of the surface current on the curved surface of the rod.

41. Determine the current generated in a superconducting ring of niobium metal 2.00 cm in diameter if a 0.020 0-T magnetic field in a direction perpendicular to the ring is suddenly decreased to zero. The inductance of the ring is 3.10×10^{-8} H.

42. *A convincing demonstration of zero resistance.* A direct and relatively simple demonstration of zero dc resistance can be carried out using the four-point probe method. The probe shown in Figure P43.42 consists of a disk of $YBa_2Cu_3O_7$ (a high-T_c superconductor) to which four wires are attached by indium solder or some other suitable contact material. Current is maintained through the sample by applying a dc voltage between points a and b, and it is measured with a dc ammeter. The current can be varied with the variable resistance R. The potential difference ΔV_{cd} between c and d is measured

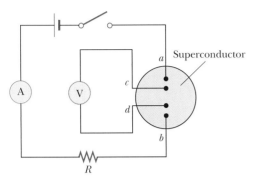

Figure P43.42 Circuit diagram used in the four-point probe measurement of the dc resistance of a sample. A dc digital ammeter is used to measure the current, and the potential difference between c and d is measured with a dc digital voltmeter. Note that there is no voltage source in the inner loop circuit where ΔV_{cd} is measured.

with a digital voltmeter. When the probe is immersed in liquid nitrogen, the sample quickly cools to 77 K, below the critical temperature of the material, 92 K. The current remains approximately constant, but ΔV_{cd} *drops abruptly to zero*. (a) Explain this observation on the basis of what you know about superconductors. (b) The data in Table P43.42 represent actual values of ΔV_{cd} for different values of I taken on the sample at room temperature. Make an I–ΔV plot of the data, and determine whether the sample behaves in a linear manner. From the data obtain a value for the dc resistance of the sample at room temperature. (c) At room temperature it is found that $\Delta V_{cd} = 2.234$ mV for $I = 100.3$ mA, but after

TABLE P43.42	Current versus Potential Difference ΔV_{cd} Measured in a Bulk Ceramic Sample of $YBa_2Cu_3O_7$ at Room Temperature[a]
I (mA)	ΔV_{cd} (mV)
57.8	1.356
61.5	1.441
68.3	1.602
76.8	1.802
87.5	2.053
102.2	2.398
123.7	2.904
155	3.61

[a]The current was supplied by a 6-V battery in series with a variable resistor R. The values of R ranged from 10 to 100 Ω. The data are from the author's (RAS) laboratory.

the sample is cooled to 77 K, $\Delta V_{cd} = 0$ and $I = 98.1$ mA. What do you think might cause the slight decrease in current?

ADDITIONAL PROBLEMS

43. As you will learn in Chapter 44, carbon-14 (^{14}C) is an isotope of carbon. It has the same chemical electronic structure of the much more abundant isotope carbon-12 (^{12}C), but it has different nuclear properties. Its mass is 14 u, greater because it has two extra neutrons in its nucleus. Assume that the CO molecular potential is the same for both isotopes of carbon and that the tables and examples in Section 43.2 refer to carbon monoxide with carbon-12 atoms. (a) What is the vibrational frequency of [14]CO? (b) What is the moment of inertia of [14]CO? (c) What wavelengths of light can be absorbed by [14]CO in the ($v = 0$, $J = 10$) state that will cause it to end up in the $v = 1$ level?

44. The effective spring constant associated with bonding in the N_2 molecule is 2 297 N/m. The nitrogen atoms each have a mass of 2.32×10^{-26} kg, and their nuclei are 0.120 nm apart. Assume that the molecule is rigid and in the ground vibrational state. Calculate the J value of the rotational state that has the same energy as the first excited vibrational state.

45. The hydrogen molecule comes apart (dissociates) when it is excited internally by 4.5 eV. Assuming that this molecule behaves like a harmonic oscillator having classical angular frequency $\omega = 8.28 \times 10^{14}$ rad/s, find the highest vibrational quantum number for a state below the 4.5-eV dissociation energy.

46. Under pressure, liquid helium can solidify as each atom bonds with four others, and each bond has an average energy of 1.74×10^{-23} J. Find the latent heat of fusion for helium in joules per gram. (The molar mass of He is 4 g/mol.)

47. Show that the ionic cohesive energy of an ionically bonded solid is given by Equation 43.17. (*Hint:* Start with Equation 43.16, and note that $dU/dr = 0$ at $r = r_0$.)

48. The dissociation energy of ground-state molecular hydrogen is 4.48 eV, whereas it only takes 3.96 eV to dissociate it when it starts in the first excited vibrational state with $J = 0$. Using this information, determine the depth of the H_2 molecular potential-energy function.

49. A particle moves in one-dimensional motion in a region where its potential energy is

$$U(x) = \frac{A}{x^3} - \frac{B}{x}$$

where $A = 0.150$ eV·nm^3 and $B = 3.68$ eV·nm. The general shape of this function is shown in Figure 43.15, where x replaces r. (a) Find the static equilibrium position x_0 of the particle. (b) Determine the depth U_0 of this potential well. (c) In moving along the x axis, what maximum force toward the negative x direction does the particle experience?

50. A particle of mass m moves in one-dimensional motion in a region where its potential energy is

$$U(x) = \frac{A}{x^3} - \frac{B}{x}$$

where A and B are constants with appropriate units. The general shape of this function is shown in Figure 43.15, where x replaces r. (a) Find the static equilibrium position x_0 of the particle in terms of m, A, and B. (b) Determine the depth U_0 of this potential well. (c) In moving along the x axis, what maximum force toward the negative x direction does the particle experience?

51. As an alternative to Equation 43.1, another useful model for the potential energy of a diatomic molecule is the Morse potential

$$U(r) = B[e^{-a(r-r_0)} - 1]^2$$

where B, a, and r_0 are parameters used to adjust the shape of the potential and its depth. (a) What is the equilibrium separation of the nuclei? (b) What is the depth of the potential well, that is, the difference in energy between the potential's minimum value and its asymptote as r approaches infinity? (c) If μ is the reduced mass of the system of two nuclei, what is the vibrational frequency of the diatomic molecule in its ground state? (Assume that the potential is nearly parabolic about the well minimum.) (d) What amount of energy needs to be supplied to the ground-state molecule to separate the two nuclei to infinity?

52. The Fermi–Dirac distribution function can be written as

$$f(E) = \frac{1}{e^{(E-E_F)/k_B T} + 1} = \frac{1}{e^{(E/E_F-1)T_F/T} + 1}$$

Write a spreadsheet to calculate and plot $f(E)$ versus E/E_F at a fixed temperature T. Examine the curves obtained for $T = 0.1T_F$, $0.2T_F$, and $0.5T_F$, where $T_F = E_F/k_B$.

53. The Madelung constant may be found by summing an infinite alternating series of terms giving the electrostatic potential energy between an Na^+ ion and its 6 nearest Cl^- neighbors, its 12 next-nearest Na^+ neighbors, and so on (Fig. 43.14a). (a) From this expression, show that the first three terms of the series yield $\alpha = 2.13$ for the NaCl structure. (b) Does this series converge rapidly? Calculate the fourth term as a check.

ANSWERS TO QUICK QUIZZES

43.1. This maximum value is the energy needed to move the sodium and chlorine ions infinitely far apart. This is sometimes called the *activation energy* and, as noted earlier, is 1.4 eV.

43.2. The spacing between adjacent peaks is approximately 0.08×10^{13} Hz. Because these lines are separated in frequency by $\hbar/2\pi I$, the moment of inertia is 1.05×10^{-34} J·s/(2π) $(0.08 \times 10^{13}$ Hz) = 2.1×10^{-47} kg·m^2, which is not much different from the value for the CO molecule calculated in Example 43.1.

43.3. At any temperature above absolute zero, internal energy $k_B T$ (≈ 0.025 eV near room temperature ≈ 300 K) is available, and this energy causes some of the electrons to have energies greater than E_F. The Fermi–Dirac distribution function $f(E)$ gives the probability of finding an electron in a particular energy level. In Figure 43.21b, that probability is not quite 1.0 for electrons having initial energies slightly less than E_F because those electrons can absorb some of the available internal energy and now have energies greater than E_F. This results in the nonzero value of $f(E)$ for energies slightly greater than E_F.

43.4. At the blue–gold boundary in the 3s band. Some electrons in the 3s blue area have enough energy to move into the 3s gold area. Thus, the horizontal boundary in Figure 43.20 and the curved part of Figure 43.21b represent the same thing.

43.5. No. If Ohm's law were obeyed, the current I would be directly proportional to the potential difference ΔV across the device (see Eq. 27.8, $I = \Delta V/R$). Instead, the curve in Figure 43.30 has a slope that varies with ΔV.

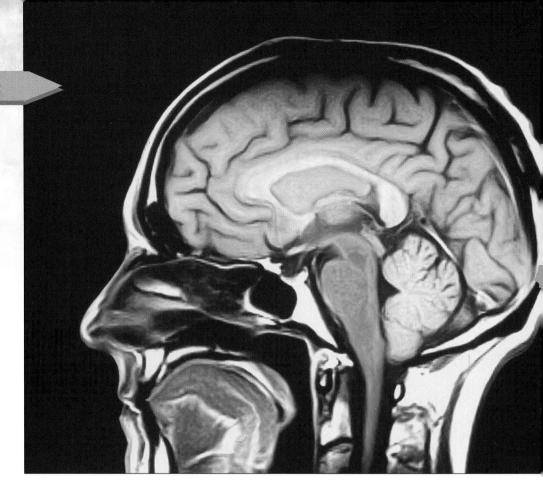

c h a p t e r

44

Nuclear Structure

Chapter Outline

In 1896, the year that marks the birth of nuclear physics, the French physicist Henri Becquerel (1852–1908) discovered radioactivity in uranium compounds. A great deal of research followed this discovery as scientists attempted to understand the nature of the radiation emitted by radioactive nuclei. Pioneering work by Rutherford showed that the emitted radiation was of three types: alpha, beta, and gamma rays, classified according to the nature of their electric charge and their ability to penetrate matter and ionize air. Later experiments showed that alpha rays are helium nuclei, beta rays are electrons, and gamma rays are high-energy photons.

In 1911, Rutherford, Geiger, and Marsden performed the alpha particle scattering experiments described in Section 42.1. These experiments established that (a) the nucleus of an atom can be regarded as essentially a point mass and point charge and that (b) most of the atomic mass is contained in the nucleus. Subsequent studies revealed the presence of a new type of force, the short-range nuclear force, which is predominant at distances less than approximately 10^{-14} m and is zero for large distances.

Other milestones in the development of nuclear physics include

- The observation of nuclear reactions in 1930 by Cockroft and Walton using artificially accelerated nuclei
- The discovery of the neutron in 1932 by Chadwick and the conclusion that neutrons make up about half of the nucleus
- The discovery of artificial radioactivity in 1933 by Joliot and Irene Curie
- The discovery of nuclear fission in 1938 by Hahn and Strassmann
- The development of the first controlled fission reactor in 1942 by Fermi and his collaborators

In this chapter we discuss the properties and structure of the atomic nucleus. We start by describing the basic properties of nuclei, and this description is followed by a discussion of nuclear forces and binding energy, nuclear models, and the phenomenon of radioactivity. We then discuss nuclear reactions and the various processes by which nuclei decay.

SOME PROPERTIES OF NUCLEI

All nuclei are composed of two types of particles: protons and neutrons. The only exception is the ordinary hydrogen nucleus, which is a single proton. In describing the atomic nucleus, we use the following quantities:

- The **atomic number** Z, which equals the number of protons in the nucleus (the atomic number is sometimes called the *charge number*)
- The **neutron number** N, which equals the number of neutrons in the nucleus
- The **mass number** A, which equals the number of **nucleons** (neutrons plus protons) in the nucleus

In representing nuclei, it is convenient to use the symbol $^{A}_{Z}X$ to show how many protons and neutrons are present, where X represents the chemical symbol of the element. For example, $^{56}_{26}Fe$ (iron) has mass number 56 and atomic number 26; therefore, it contains 26 protons and 30 neutrons. When no confusion is likely to arise, we omit the subscript Z because the chemical symbol can always be used to determine Z.

The nuclei of all atoms of a particular element contain the same number of protons but often contain different numbers of neutrons. As noted in Section 1.2, nuclei that are related in this way are called **isotopes.**

Isotopes

> The isotopes of an element have the same Z value but different N and A values.

The natural abundance of isotopes can differ substantially. For example, $^{11}_{6}C$, $^{12}_{6}C$, $^{13}_{6}C$, and $^{14}_{6}C$ are four isotopes of carbon. The natural abundance of the $^{12}_{6}C$ isotope is approximately 98.9%, whereas that of the $^{13}_{6}C$ isotope is only about 1.1%. Some isotopes, such as $^{11}_{6}C$ and $^{14}_{6}C$, do not occur naturally but can be produced by nuclear reactions in the laboratory or by cosmic rays.

Even the simplest element, hydrogen, has isotopes: $^{1}_{1}H$, the ordinary hydrogen nucleus; $^{2}_{1}H$, deuterium; and $^{3}_{1}H$, tritium.

Charge and Mass

The proton carries a single positive charge, equal in magnitude to the charge e on the electron ($|e| = 1.6 \times 10^{-19}$ C). The neutron is electrically neutral, as its name implies. Because the neutron has no charge, it is difficult to detect.

Nuclear masses can be measured with great precision with the use of a mass spectrometer (see Section 29.5) and by the analysis of nuclear reactions. The proton is approximately 1 836 times as massive as the electron, and the masses of the proton and the neutron are almost equal. In Chapter 1, we defined the atomic mass unit u in such a way that the mass of one atom of the isotope ^{12}C is exactly 12 u, where 1 u = $1.660\ 540 \times 10^{-27}$ kg. According to this definition, the proton and neutron each have a mass of approximately 1 u, and the electron has a mass that is only a small fraction of this value:

$$\text{Mass of proton} = 1.007\ 276\ u$$

$$\text{Mass of neutron} = 1.008\ 665\ u$$

$$\text{Mass of electron} = 0.000\ 548\ 6\ u$$

One might wonder how six protons and six neutrons, each having a mass larger than 1 u, can be combined with six electrons to form a carbon-12 atom having a mass of exactly 12 u. The extra mass of the separated particles appears as binding energy when the particles are combined to form the nucleus. We shall discuss this point in more detail in Section 44.3.

TABLE 44.1   **Mass of Selected Particles in Various Units**

Particle	Mass		
	kg	**u**	**MeV/c^2**
Proton	$1.672\ 62 \times 10^{-27}$	1.007 276	938.28
Neutron	$1.674\ 93 \times 10^{-27}$	1.008 665	939.57
Electron	$9.109\ 39 \times 10^{-31}$	$5.48\ 579 \times 10^{-4}$	0.510 999
$^{1}_{1}H$ atom	$1.673\ 53 \times 10^{-27}$	1.007 825	938.783
$^{4}_{2}He$ nucleus	$6.644\ 66 \times 10^{-27}$	4.001 506	3 727.38
$^{12}_{6}C$ atom	$1.992\ 65 \times 10^{-27}$	12.000 000	11 177.9

Because the rest energy of a particle is given by $E_R = mc^2$, it is often convenient to express the atomic mass unit in terms of its *rest energy equivalent*. For one atomic mass unit, we have

$$E_R = mc^2 = (1.660\ 540 \times 10^{-27}\ \text{kg})(2.997\ 924\ 58 \times 10^8\ \text{m/s})^2$$
$$= 931.494\ \text{MeV}$$

where we have used the conversion $1\ \text{eV} = 1.602\ 177 \times 10^{-19}\ \text{J}$.

Nuclear physicists often express mass in terms of the unit MeV/c^2, where

$$1\ \text{u} \equiv 931.494\ \text{MeV}/c^2$$

The masses of several nuclei and atoms are given in Table 44.1. The masses and some other properties of selected isotopes are provided in Appendix A.3.

EXAMPLE 44.1   **The Atomic Mass Unit**

Use Avogadro's number to show that $1\ \text{u} = 1.66 \times 10^{-27}\ \text{kg}$.

Solution   From the definition of the mole given in Section 1.3, we know that exactly 12 g ($= 1$ mol) of ^{12}C contains Avogadro's number of atoms, where $N_A = 6.02 \times 10^{23}$ atoms/mol. Thus, the mass of one carbon atom is

$$\text{Mass of one } {}^{12}\text{C atom} = \frac{0.012\ \text{kg}}{6.02 \times 10^{23}\ \text{atoms}}$$
$$= 1.99 \times 10^{-26}\ \text{kg}$$

Because one atom of ^{12}C is defined to have a mass of 12.0 u, we find that

$$1\ \text{u} = \frac{1.99 \times 10^{-26}\ \text{kg}}{12.0} = 1.66 \times 10^{-27}\ \text{kg}$$

The Size and Structure of Nuclei

In Rutherford's scattering experiments, positively charged nuclei of helium atoms (alpha particles) were directed at a thin piece of metallic foil. As the alpha particles moved through the foil, they often passed near a metal nucleus. Because of the positive charge on both the incident particles and the nuclei, the particles were deflected from their straight-line paths by the Coulomb repulsive force. Some particles were even deflected straight backward! These particles apparently were moving directly toward a nucleus, on a head-on collision course.

Rutherford used conservation of energy to find an expression for the separation distance d at which an alpha particle approaching a nucleus head-on is turned around by Coulomb repulsion. In such a head-on collision, he reasoned, the kinetic energy of the incoming particle must be converted completely to electric potential energy when the particle stops at the point of closest approach and turns around (Fig. 44.1). If we equate the initial kinetic energy of the alpha particle to the electric potential energy of the system (alpha particle of mass m plus target nucleus of atomic number Z), we have

$$\tfrac{1}{2}mv^2 = k_e \frac{q_1 q_2}{r} = k_e \frac{(2e)(Ze)}{d}$$

Solving for d, the distance of closest approach, we obtain

$$d = \frac{4k_e Ze^2}{mv^2}$$

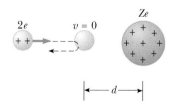

Figure 44.1   An alpha particle on a head-on collision course with a nucleus of charge Ze. Because of the Coulomb repulsion between the like charges, the alpha particle approaches to a distance d from the nucleus, called the distance of closest approach.

From this expression, Rutherford found that the alpha particles approached nuclei to within 3.2×10^{-14} m when the foil was made of gold. Thus, the radius of the gold nucleus must be less than this value. For silver atoms, the distance of closest approach was found to be 2×10^{-14} m. From the results of his scattering experiments, Rutherford concluded that the positive charge in an atom is concentrated in a small sphere, which he called the nucleus, whose radius is no greater than about 10^{-14} m.

Because such small lengths are common in nuclear physics, an often-used convenient length unit is the femtometer (fm), which is sometimes called the **fermi** and is defined as

$$1 \text{ fm} \equiv 10^{-15} \text{ m}$$

In the early 1920s it was known that the nucleus of an atom contains Z protons and has a mass nearly equivalent to that of A protons, where, on the average, $A \approx 2Z$ for lighter nuclei ($Z \le 20$) and $A > 2Z$ for heavier nuclei. To account for the nuclear mass, Rutherford proposed that each nucleus must also contain $A - Z$ neutral particles that he called neutrons. In 1932, the British physicist James Chadwick (1891–1974) discovered the neutron and was awarded the Nobel Prize for this important work.

Since the time of Rutherford's scattering experiments, a multitude of other experiments have shown that most nuclei are approximately spherical and have an average radius given by

Figure 44.2 A nucleus can be modeled as a cluster of tightly packed spheres, where each sphere is a nucleon.

Nuclear radius

$$r = r_0 A^{1/3} \tag{44.1}$$

where r_0 is a constant equal to 1.2×10^{-15} m and A is the mass number. Because the volume of a sphere is proportional to the cube of its radius, it follows from Equation 44.1 that the volume of a nucleus (assumed to be spherical) is directly proportional to A, the total number of nucleons. This proportionality suggests that *all nuclei have nearly the same density.* When nucleons combine to form a nucleus, they combine as though they were tightly packed spheres (Fig. 44.2). This fact has led to an analogy between the nucleus and a drop of liquid, in which the density of the drop is independent of its size. We shall discuss the liquid-drop model of the nucleus in Section 44.4.

EXAMPLE 44.2   The Volume and Density of a Nucleus

Find (a) an approximate expression for the mass of a nucleus of mass number A, (b) an expression for the volume of this nucleus in terms of A, and (c) a numerical value for the density of the nucleus.

Solution   (a) The mass of the proton is approximately equal to that of the neutron. Thus, if the mass of one of these particles is m, the mass of the nucleus is approximately Am.

(b) Assuming the nucleus is spherical and using Equation 44.1, we find that the volume is

$$V = \tfrac{4}{3}\pi r^3 = \tfrac{4}{3}\pi r_0^3 A$$

(c) The nuclear density is

$$\rho_n = \frac{\text{mass}}{\text{volume}} = \frac{Am}{\tfrac{4}{3}\pi r_0^3 A} = \frac{3m}{4\pi r_0^3}$$

Taking $r_0 = 1.2 \times 10^{-15}$ m and $m = 1.67 \times 10^{-27}$ kg, we find that

$$\rho_n = \frac{3(1.67 \times 10^{-27} \text{ kg})}{4\pi(1.2 \times 10^{-15} \text{ m})^3} = 2.3 \times 10^{17} \text{ kg/m}^3$$

The nuclear density is approximately 2.3×10^{14} times as great as the density of water ($\rho_{\text{water}} = 1.0 \times 10^3$ kg/m^3).

Exercise   If the Earth were compressed until it had this density, how large would it be?

Answer   A sphere of diameter 370 m!

Nuclear Stability

Because the nucleus is viewed as a closely packed collection of protons and neutrons, you might be surprised that it can exist. Because like charges (the protons) in close proximity exert very large repulsive Coulomb forces on each other, these forces should cause the nucleus to fly apart. However, nuclei are stable because of the **nuclear force,** a very-short-range (about 2 fm) attractive force that acts between all nuclear particles. The protons attract each other by means of the nuclear force, and, at the same time, they repel each other through the Coulomb force. The nuclear force also acts between pairs of neutrons and between neutrons and protons.

There are approximately 400 stable nuclei; hundreds of other nuclei have been observed, but they are unstable. A plot of neutron number N versus atomic number Z for a number of stable nuclei is given in Figure 44.3. The stable nuclei are represented by the blue dots, which lie in a narrow range called the *line of stability*. Note that light nuclei are most stable if they contain an equal number of protons and neutrons; that is, if $N = Z$. Also, note that heavy nuclei are more stable if the number of neutrons exceeds the number of protons—above $Z = 20$, the line

web

For a very detailed, "clickable" version of Figure 44.3, visit Brookhaven National Lab at

www.dne.bnl.gov/CoN/index.html

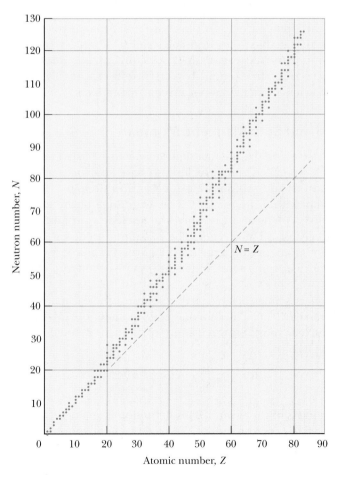

Figure 44.3 Neutron number N versus atomic number Z for the stable nuclei (blue dots). These nuclei lie in a narrow band called the line of stability. The dashed line corresponds to the condition $N = Z$.

of stability deviates upward from the line representing $N = Z$. This can be understood by recognizing that, as the number of protons increases, the strength of the Coulomb force increases, which tends to break the nucleus apart. As a result, more neutrons are needed to keep the nucleus stable because neutrons experience only the attractive nuclear force. Eventually, the repulsive Coulomb forces between protons cannot be compensated by the addition of more neutrons. This occurs when $Z = 83$, meaning that elements that contain more than 83 protons do not have stable nuclei.

Quick Quiz 44.1

The blue dots in Figure 44.3 form a sequence of vertically oriented groups. What do these groups represent?

It is interesting that most stable nuclei have an even value of A. Furthermore, only eight nuclei have odd values for both Z and N. Certain values of Z and N correspond to nuclei with unusually high stability. These values, called **magic numbers,** are

$$Z \quad \text{or} \quad N = 2, 8, 20, 28, 50, 82 \tag{44.2}$$

For example, the alpha particle (two protons and two neutrons), which has $Z = 2$ and $N = 2$, is very stable. The unusual stability of nuclei having progressively larger magic numbers suggests a shell structure of the nucleus similar to the atomic shell structure. In Section 44.4 we briefly discuss a nuclear model, the independent-particle model, that explains magic numbers.

Nuclear Spin and Spin Magnetic Moment

In Chapter 42, we discussed the fact that the electron has an intrinsic angular momentum, which we called spin. Nuclei also have spin because their component particles—neutrons and protons—each have intrinsic spin $\frac{1}{2}$, as well as orbital angular momentum within the nucleus. The magnitude of the nuclear angular momentum is $\sqrt{I(I + 1)}\,\hbar$, where I is called the **nuclear spin quantum number** and may be an integer or a half-integer. The maximum value of the z component of the spin angular momentum vector is $I\hbar$. Figure 44.4 illustrates the possible orientations of the nuclear spin vector and its projections along the z axis for the case in which $I = \frac{3}{2}$.

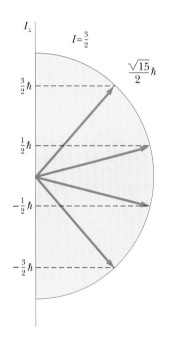

Figure 44.4 Possible orientations of the nuclear spin angular momentum vector and its projections along the z axis for the case $I = \frac{3}{2}$.

Nuclear spin has an associated, corresponding nuclear magnetic moment, similar to that of the electron. The spin magnetic moment of a nucleus is measured in terms of the **nuclear magneton** μ_n, a unit of moment defined as

Nuclear magneton

$$\mu_n \equiv \frac{e\hbar}{2m_p} = 5.05 \times 10^{-27}\,\text{J/T} \tag{44.3}$$

where m_p is the mass of the proton. This definition is analogous to that of the Bohr magneton μ_B, which corresponds to the spin magnetic moment of a free electron (see Section 42.5). Note that μ_n is smaller than μ_B ($= 9.274 \times 10^{-24}\,\text{J/T}$) by a factor of approximately 2 000 because of the large difference between the proton mass and the electron mass.

The magnetic moment of a free proton is $2.792\,8\mu_n$. Unfortunately, there is no general theory of nuclear magnetism that explains this value. Surprisingly, the uncharged neutron also has a magnetic moment, which has a value of $-1.913\ 5\mu_n$. The minus sign indicates that this moment is opposite the spin angular momentum of the neutron. The existence of a magnetic moment for the neutron suggests that it is not fundamental but rather has an underlying structure. We shall explore such structure in Chapter 46.

Quick Quiz 44.2

Which do you expect not to vary substantially between different isotopes of an element? (a) atomic mass, (b) nuclear spin magnetic moment, (c) chemical properties?

44.2 NUCLEAR MAGNETIC RESONANCE AND MAGNETIC RESONANCE IMAGING

Nuclear magnetic moments, as well as electronic magnetic moments, precess when placed in an external magnetic field. The frequency at which they precess, called the **Larmor precessional frequency** ω_p, is directly proportional to the magnitude of the magnetic field. This is described schematically in Figure 44.5a, in which the external magnetic field is along the z axis. For example, the Larmor frequency of a proton in a 1-T magnetic field is 42.577 MHz. The potential energy of a magnetic dipole moment $\boldsymbol{\mu}$ in an external magnetic field **B** is given by $-\boldsymbol{\mu}\cdot\mathbf{B}$. When the magnetic moment $\boldsymbol{\mu}$ is lined up with the field as closely as quantum physics allows, the potential energy of the dipole moment in the field has its minimum value E_{min}. When the projection of $\boldsymbol{\mu}$ is as antiparallel to the field as possible, the potential energy has its maximum value E_{max}. In general, there are other energy states between these values that correspond to the quantized directions of the magnetic moment with respect to the field. For a nucleus with spin $\frac{1}{2}$, there are only two allowed states, with energies E_{min} and E_{max}. These two energy states are shown in Figure 44.5b.

Larmor precessional frequency

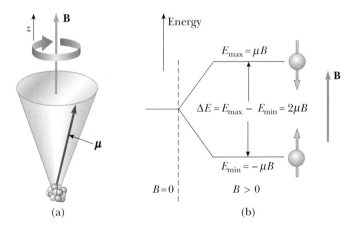

(a)　　　　　(b)

Figure 44.5 (a) When a nucleus is placed in an external magnetic field **B**, the nuclear spin magnetic moment precesses about the magnetic field with a frequency proportional to the magnitude of the field. (b) A nucleus with spin $\frac{1}{2}$ can occupy one of two energy states when placed in an external magnetic field. The lower energy state E_{min} corresponds to the case where the spin is aligned with the field as much as possible, and the higher energy state E_{max} corresponds to the case where the spin is opposite the field as much as possible.

Nuclear magnetic resonance

web

For more information on MRI and many other topics dealing with physics, visit **http://physics.miningco.com/msubtext. htm**

It is possible to observe transitions between these two spin states using a technique called **NMR,** for **nuclear magnetic resonance.** A constant magnetic field (**B** in Fig. 44.5a) is introduced to define a z axis. A second, weaker, oscillating magnetic field is then applied perpendicular to **B**. When the frequency of the oscillating field is adjusted to match the Larmor precessional frequency of the nuclei in the sample, a torque that acts on the precessing magnetic moments causes them to "flip" between the two spin states shown in Figure 44.5b. These transitions result in a net absorption of energy by the nuclei, an absorption that can be detected electronically.

A diagram of the apparatus used in nuclear magnetic resonance is illustrated in Figure 44.6. The energy absorbed by the nuclei is supplied by the generator producing the oscillating magnetic field. Nuclear magnetic resonance and a related technique called *electron spin resonance* are extremely important methods for studying nuclear and atomic systems and the ways in which these systems interact with their surroundings.

A widely used medical diagnostic technique called **MRI,** for **magnetic resonance imaging,** is based on nuclear magnetic resonance. Because nearly two thirds of the atoms in the human body are hydrogen (which gives a strong signal), MRI works exceptionally well for viewing internal tissues. The patient is placed inside a large solenoid that supplies a time-constant magnetic field whose magnitude varies spatially across the body. Because of the variation in the field, protons in different parts of the body precess at different frequencies, so the resonance signal can be used to provide information about the positions of the protons. A computer is used to analyze the position information to provide data for constructing a final image. An MRI scan showing incredible detail in internal body structure is shown in Figure 44.7.

The main advantage of MRI over other imaging techniques is that it causes minimal cellular damage. The photons associated with the radio-frequency signals used in MRI have energies of only about 10^{-7} eV. Because molecular bond strengths are much larger (approximately 1 eV), the radio-frequency radiation causes little cellular damage. In comparison, x-rays have energies ranging from 10^4 to 10^6 eV and can cause considerable cellular damage. Thus, despite some individ-

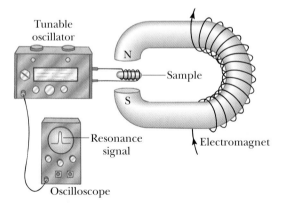

Figure 44.6   Experimental arrangement for nuclear magnetic resonance. The radio-frequency magnetic field created by the coil surrounding the sample and provided by the variable-frequency oscillator must be perpendicular to the constant magnetic field created by the electromagnet. When the nuclei in the sample meet the resonance condition, the nuclei absorb energy from the radio-frequency field of the coil, and this absorption changes the characteristics of the circuit in which the coil is included. Most modern NMR spectrometers use superconducting magnets at fixed field strengths and operate at frequencies of approximately 200 MHz.

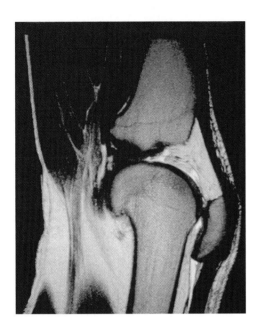

Figure 44.7 An MRI scan of a knee.
(Susie Leavines/Photo Researchers, Inc.)

 uals' fears of the word *nuclear* associated with MRI, the radio-frequency radiation involved is overwhelmingly safer than the x-rays that these individuals might accept more readily! A disadvantage of MRI is that the equipment required to conduct the procedure is quite expensive.

The magnetic field produced by the solenoid is sufficient to lift a car, and the radio signal is about the same magnitude as that from a small commercial broadcasting station!

44.3 BINDING ENERGY AND NUCLEAR FORCES

The total mass of a nucleus is always less than the sum of the masses of its individual nucleons. Because mass is a measure of energy, **the total energy of the bound system (the nucleus) is less than the combined energy of the separated nucleons.** As we learned in Example 39.14, this difference in energy is called the **binding energy** of the nucleus and can be thought of as the energy that must be added to a nucleus to break it apart into its components. Therefore, in order to separate a nucleus into protons and neutrons, energy must be delivered to the system.

Conservation of energy and the Einstein mass–energy equivalence relationship show that the binding energy E_b of any nucleus of mass M_A is

$$E_b \, (\text{MeV}) = (Z m_p + N m_n - M_A) \times 931.494 \, \text{MeV/u} \qquad \textbf{(44.4)}$$

◁ Binding energy of a nucleus

where m_p is the mass of the proton, m_n is the mass of the neutron, and the masses are all expressed in atomic mass units. In practice, it is often more convenient to use the mass of neutral atoms (nuclear mass plus mass of electrons) in computing binding energy because mass spectrometers generally measure atomic masses.[1]

---

[1] It is possible to use atomic masses rather than nuclear masses because electron masses cancel in the calculations.

A plot of binding energy per nucleon E_b/A as a function of mass number A for various stable nuclei is shown in Figure 44.8. Note that the curve in Figure 44.8 peaks in the vicinity of $A = 60$. That is, nuclei having mass numbers either greater or less than 60 are not as strongly bound as those near the middle of the periodic table. The higher values of binding energy per nucleon near $A = 60$ imply that energy is released when a heavy nucleus splits, or *fissions*, into two lighter nuclei. Energy is released in fission because the nucleons in each product nucleus are more tightly bound to one another than are the nucleons in the original nucleus. The important process of fission and a second important process of *fusion*, in which energy is released as light nuclei combine, are considered in detail in Chapter 45.

Another important feature of Figure 44.8 is that the binding energy per nucleon is approximately constant at around 8 MeV per nucleon for all nuclei with $A > 50$. For these nuclei, the nuclear forces are said to be *saturated*, meaning that, in the closely packed structure shown in Figure 44.2, a particular nucleon can form attractive bonds with only a limited number of other nucleons.

Quick Quiz 44.3

Figure 44.8 shows that, above about $A = 50$, an approximately constant amount of energy is necessary to remove a nucleon from the nucleus. Compare this with Figure 42.16, which shows the widely varying amounts of energy necessary to remove an electron from an atom. Why is there such a difference between the two graphs?

Figure 44.8 provides insight into fundamental questions about the origin of the chemical elements. In the early life of the Universe, there were only hydrogen and helium. Clouds of cosmic gas and dust coalesced under gravitational forces to form stars. As a star ages, it produces heavier elements from the lighter elements contained within it, beginning by fusing hydrogen atoms to form helium. This process continues as the star becomes older, generating atoms having larger and larger atomic numbers, up through the isotope of iron having $A = 56$, which is at the peak of the curve shown in Figure 44.8.

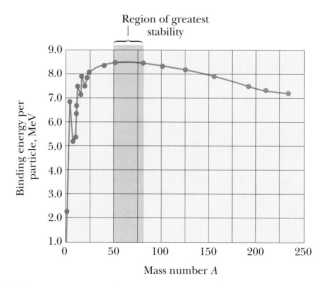

Figure 44.8   Binding energy per nucleon versus mass number for nuclei that lie along the line of stability in Figure 44.3.

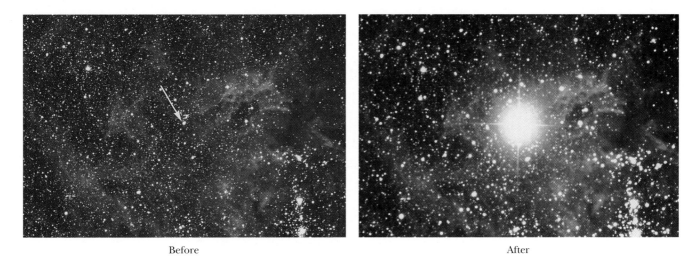

Before   After

Before and after images of the 1987A supernova in the Large Magellanic Cloud. The arrow in the "before" image points to the star that exploded. This was the brightest supernova seen within the past several hundred years. *(Anglo-Australian Telescope Board)*

It takes additional energy to create elements with mass numbers larger than 56 because of their lower binding energies per nucleon. This energy comes from the supernova explosion that occurs at the end of some large stars' lives. Thus, all the heavy atoms in your body were produced from the explosions of ancient stars. You are literally made of stardust!

The general features of the nuclear force responsible for the binding energy of nuclei have been revealed in a wide variety of experiments and are as follows:

- The nuclear force is attractive and is the strongest force in nature.
- It is a short-range force that falls to zero when the separation between nucleons exceeds several fermis. This limited range is evidenced by scattering experiments and shown in the neutron–proton potential energy plot of Figure 44.9a, obtained by scattering neutrons from a target containing hydrogen. The potential energy well is 40 to 50 MeV deep and contains a strong repulsive component that prevents the separation distance between nucleons from being less than about 0.4 fm.
- The magnitude of the nuclear force depends on the relative spin orientations of the nucleons.
- Scattering experiments and other indirect evidence show that the nuclear force is independent of the charge of the interacting nucleons. For this reason, high-speed electrons can be used to probe the properties of nuclei. The charge independence also means that the only difference between neutron–proton (n–p) and proton–proton (p–p) interactions is that the p–p potential energy is a superposition of nuclear and Coulomb interactions, as shown in Figure 44.9b. At separation distances less than 2 fm, p–p and n–p potential energies are nearly identical, but for distances greater than this, the p–p potential energy is positive, with a maximum of about 1 MeV at 4 fm.

Quick Quiz 44.4

Can either curve in Figure 44.9 be vertical at some point?

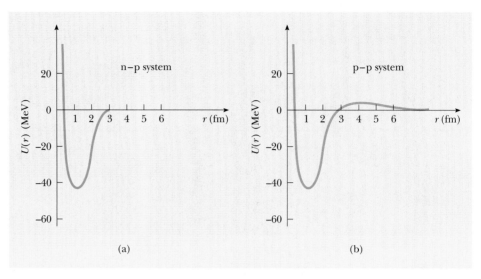

Figure 44.9   (a) Potential energy versus separation distance for a neutron–proton system. (b) Potential energy versus separation distance for a proton–proton system. The difference in the two curves is due to the large Coulomb repulsion in the case of the proton–proton interaction.

 NUCLEAR MODELS

Although the details of the nuclear force are still not well understood, several nuclear models have been proposed, and these are useful in understanding general features of nuclear experimental data and the mechanisms responsible for binding energy. The liquid-drop model accounts for nuclear binding energy, and the independent-particle model accounts for the existence of stable isotopes.

Liquid-Drop Model

In 1936, Bohr proposed treating nucleons like molecules in a drop of liquid. In this **liquid-drop model,** the nucleons interact strongly with one another and undergo frequent collisions as they jiggle around within the nucleus. This jiggling motion is analogous to the thermally agitated motion of molecules in a drop of liquid.

Three major effects influence the binding energy of the nucleus in the liquid-drop model:

- **The volume effect.** Figure 44.8 shows that, for $A > 50$, the binding energy per nucleon is approximately constant, indicating that the nuclear force saturates. This tells us that the binding energy of the nucleus is proportional to A and therefore proportional to the nuclear volume. The contribution to the binding energy of the entire nucleus is $C_1 A$, where C_1 is an adjustable constant.
- **The surface effect.** Because nucleons on the surface of the drop have fewer neighbors than those in the interior, surface nucleons reduce the binding energy by an amount proportional to their number. Because the number of surface nucleons is proportional to the surface area $4\pi r^2$ of the nucleus, and because $r^2 \propto A^{2/3}$ (Eq. 44.1), the surface term can be expressed as $-C_2 A^{2/3}$, where C_2 is a constant.

- **The Coulomb repulsion effect.** Each proton repels every other proton in the nucleus. The corresponding potential energy per pair of interacting protons is $k_e e^2/r$, where k_e is the Coulomb constant. The total Coulomb energy represents the work required to assemble Z protons from infinity to a sphere of volume V. This energy is proportional to the number of proton pairs $Z(Z-1)/2$ and inversely proportional to the nuclear radius. Consequently, the reduction in energy that results from the Coulomb effect is $-C_3 Z(Z-1)/A^{1/3}$.

Another small effect that lowers the binding energy is significant for nuclei having a large excess of neutrons—in other words, heavy nuclei. This effect gives rise to a binding energy term of the form $-C_4(N-Z)^2/A$.

Adding these contributions, we get as the total binding energy

$$E_b = C_1 A - C_2 A^{2/3} - C_3 \frac{Z(Z-1)}{A^{1/3}} - C_4 \frac{(N-Z)^2}{A} \qquad \textbf{(44.5)}$$

Semiempirical binding energy formula

This equation, often referred to as the **semiempirical binding energy formula,** contains four constants that are adjusted to fit the expression to experimental data. For nuclei having $A \geq 15$, the constants have the values

$$C_1 = 15.7 \text{ MeV} \qquad C_2 = 17.8 \text{ MeV}$$

$$C_3 = 0.71 \text{ MeV} \qquad C_4 = 23.6 \text{ MeV}$$

Equation 44.5, together with these constants, fits the known nuclear mass values very well. However, the liquid-drop model does not account for some finer details of nuclear structure, such as stability rules and angular momentum. On the other hand, it does provide a qualitative description of nuclear fission, shown schematically in Figure 44.10.

The Independent-Particle Model

In our second model of the nucleus, the **independent-particle model,** often called the *shell model,* each nucleon is assumed to exist in a shell, similar to an atomic shell for an electron. The nucleons exist in quantized energy states, and there are few collisions between nucleons. Obviously, the assumptions of this model differ greatly from those made in the liquid-drop model.

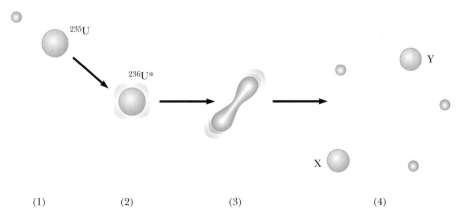

| (1) | (2) | (3) | (4) |

Figure 44.10 Steps leading to fission according to the liquid-drop model of the nucleus.

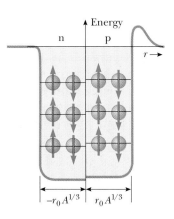

Figure 44.11 A square potential well containing 12 nucleons. The red circles represent protons, and the green circles represent neutrons. The energy levels for the protons are slightly higher than those for the neutrons because of the Coulomb potential experienced by the protons. The difference in the levels increases as Z increases. Note that only two nucleons having opposite spins can occupy a given level, as required by the exclusion principle.

Maria Goeppert-Mayer
(1906–1972) Goeppert-Mayer was born and educated in Germany. She is best known for her development of the shell model (independent-particle model) of the nucleus, published in 1950. A similar model was simultaneously developed by Hans Jensen, a German scientist. Goeppert-Mayer and Jensen were awarded the Nobel Prize in physics in 1963 for their extraordinary work in understanding the structure of the nucleus. *(Courtesy of Louise Barker/AIP Niels Bohr Library)*

The quantized states occupied by the nucleons can be described by a set of quantum numbers. Because both the proton and the neutron have spin $\frac{1}{2}$, the exclusion principle can be applied to describe the allowed states (as we did for electrons in Chapter 42). That is, each state can contain only two protons (or two neutrons) having *opposite* spins (Fig. 44.11). The protons have a set of allowed states, and these states differ from those of the neutrons because the two species move in different potential wells. The proton energy levels are higher than the neutron levels because the protons experience a superposition of Coulomb potential energy and nuclear potential energy, whereas the neutrons experience only a nuclear potential energy.

One factor influencing the observed characteristics of nuclear ground states is *nuclear spin–orbit* effects. Unlike the spin–orbit interaction between the spin of an electron and its orbital motion in an atom, an interaction that is magnetic in origin, the spin–orbit effect for nucleons in a nucleus is due to the nuclear force. It is much stronger than in the atomic case, and it has opposite sign. When these effects are taken into account, the independent-particle model is able to account for the observed magic numbers.

Finally, the independent-particle model helps us understand why nuclei containing an even number of protons and neutrons are more stable than other nuclei. (There are 160 even–even isotopes.) Any particular state is filled when it contains two protons (or two neutrons) having opposite spins. An extra proton or neutron can be added to the nucleus only at the expense of increasing the energy of the nucleus. This increase in energy leads to a nucleus that is less stable than the original nucleus. A careful inspection of the stable nuclei shows that the majority have a special stability when their nucleons combine in pairs, which results in a total angular momentum of zero. This accounts for the large number of high-stability nuclei (those having high binding energies) with the magic numbers given by Equation 44.2.

44.5 ▷ RADIOACTIVITY

In 1896, Henri Becquerel accidentally discovered that uranyl potassium sulfate crystals emit an invisible radiation that can darken a photographic plate when the plate is covered to exclude light. After a series of experiments, he concluded that the radiation emitted by the crystals was of a new type, one that requires no external stimulation and was so penetrating that it could darken protected photographic plates and ionize gases. This process of spontaneous emission of radiation by uranium was soon to be called **radioactivity.**

Subsequent experiments by other scientists showed that other substances were more powerfully radioactive. The most significant investigations of this type were conducted by Marie and Pierre Curie. After several years of careful and laborious chemical separation processes on tons of pitchblende, a radioactive ore, the Curies reported the discovery of two previously unknown elements, both radioactive. These were named polonium and radium. Subsequent experiments, including Rutherford's famous work on alpha-particle scattering, suggested that radioactivity is the result of the *decay*, or disintegration, of unstable nuclei.

Three types of radioactive decay occur in a radioactive substance: alpha (α) decay, in which the emitted particles are ^{4}He nuclei; beta (β) decay, in which the emitted particles are either electrons or positrons; and gamma (γ) decay, in which the emitted "rays" are high-energy photons. A **positron** is a particle like the electron in all respects except that the positron has a charge of $+e$ (in other words, the positron is the *antimatter twin* of the electron). The symbol e$^-$ is used to designate an electron, and e$^+$ designates a positron.

It is possible to distinguish among these three forms of radiation by using the scheme described in Figure 44.12. The radiation from a radioactive sample is directed into a region in which there is a magnetic field. The radiation beam splits into three components, two bending in opposite directions and the third experiencing no change in direction. From this simple observation, we can conclude that the radiation of the undeflected beam carries no charge (the gamma ray), the component deflected upward corresponds to positively charged particles (alpha particles), and the component deflected downward corresponds to negatively charged particles (e$^-$). If the beam includes a positron (e$^+$), it is deflected upward like the alpha particle but follows a different trajectory due to its smaller mass.

The three types of radiation have quite different penetrating powers. Alpha particles barely penetrate a sheet of paper, beta particles (electrons and positrons) can penetrate a few millimeters of aluminum, and gamma rays can penetrate several centimeters of lead.

The rate at which a particular decay process occurs in a radioactive sample is proportional to the number of radioactive nuclei present (that is, the number of nuclei that have not yet decayed). If N is the number of radioactive nuclei present

Marie Curie (1867–1934) In 1903 the Polish scientist Marie Curie shared the Nobel Prize in physics with her husband Pierre and with Becquerel for their studies of radioactive substances. In 1911 she was awarded a Nobel Prize in chemistry for the discovery of radium and polonium. She died of leukemia caused by years of exposure to radioactive substances. "I persist in believing that the ideas that then guided us are the only ones which can lead to true social progress. We cannot hope to build a better world without improving the individual. Toward this end, each of us must work toward his own highest development, accepting at the same time his share of responsibility in the general life of humanity." *(FPG International)*

The hands and numbers of this luminous watch contain minute amounts of radium mixed with a phosphorescent material. The radioactive decay of radium causes the watch to glow in the dark. *(©1990 Richard Megna/Fundamental Photographs)*

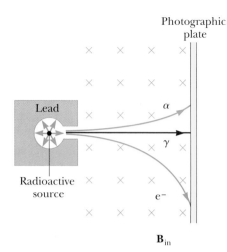

Figure 44.12 The radiation from a radioactive source can be separated into three components by using a magnetic field to deflect the charged particles. The photographic plate at the right records the events. The gamma ray is not deflected by the magnetic field.

at some instant, the rate of change of N is

$$\frac{dN}{dt} = -\lambda N \qquad \textbf{(44.6)}$$

where λ, called the **decay constant,** is the probability of decay per nucleus per second. The minus sign indicates that dN/dt is negative; that is, N decreases in time.

Equation 44.6 can be written in the form

$$\frac{dN}{N} = -\lambda \, dt$$

the solution to which is

Exponential decay

$$N = N_0 e^{-\lambda t} \qquad \textbf{(44.7)}$$

where the constant N_0 represents the number of radioactive nuclei at $t = 0$. Equation 44.7 shows that the number of radioactive nuclei in a sample decreases exponentially with time.

The **decay rate** R, which is the number of decays per second, can be obtained by differentiating Equation 44.7 with respect to time:

Half-life equation

$$R = \left| \frac{dN}{dt} \right| = N_0 \lambda e^{-\lambda t} = R_0 e^{-\lambda t} \qquad \textbf{(44.8)}$$

where $R = \lambda N$ and $R_0 = N_0 \lambda$ is the decay rate at $t = 0$. The decay rate R of a sample is often referred to as its **activity.** Note that both N and R decrease exponentially with time. The plot of N versus t shown in Figure 44.13 illustrates the exponential nature of the decay.

Another parameter useful in characterizing nuclear decay is **half-life** $T_{1/2}$:

The **half-life** of a radioactive substance is the time it takes half of a given number of radioactive nuclei to decay.

Setting $N = N_0/2$ and $t = T_{1/2}$ in Equation 44.7 gives

$$\frac{N_0}{2} = N_0 e^{-\lambda T_{1/2}}$$

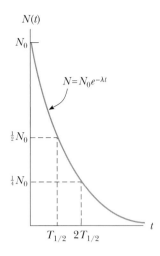

Figure 44.13 Plot of the exponential decay law for radioactive nuclei. The vertical axis represents the number of radioactive nuclei present at any time t, and the horizontal axis is time. The time $T_{1/2}$ is the half-life of the sample.

Canceling the N_0 factors and then taking the reciprocal of both sides, we obtain $e^{\lambda T_{1/2}} = 2$. Taking the natural logarithm of both sides gives

$$T_{1/2} = \frac{\ln 2}{\lambda} = \frac{0.693}{\lambda} \qquad (44.9)$$

This is a convenient expression relating half-life $T_{1/2}$ to decay constant λ. After an elapsed time equal to one half-life, there are $N_0/2$ radioactive nuclei remaining (by definition); after two half-lives, half of these have decayed and $N_0/4$ radioactive nuclei are left; after three half-lives, $N_0/8$ are left, and so on. In general, after n half-lives, the number of radioactive nuclei remaining is $N_0/2^n$.

A frequently used unit of activity is the **curie** (Ci), defined as

$$1 \text{ Ci} \equiv 3.7 \times 10^{10} \text{ decays/s}$$

The curie

This value was originally selected because it is the approximate activity of 1 g of radium. The SI unit of activity is the **becquerel** (Bq):

$$1 \text{ Bq} \equiv 1 \text{ decay/s}$$

The becquerel

Therefore, 1 Ci = 3.7×10^{10} Bq. The curie is a rather large unit, and the more frequently used activity units are the millicurie and the microcurie.

EXAMPLE 44.3 ▶ How Many Nuclei Are Left?

The isotope carbon-14, $^{14}_{6}C$, is radioactive and has a half-life of 5 730 years. If you start with a sample of 1 000 carbon-14 nuclei, how many will still be around in 22 920 years?

Solution In 5 730 years, half the sample will have decayed, leaving 500 carbon-14 nuclei remaining. In another 5 730 years (for a total elapsed time of 11 460 years), the number will be reduced to 250 nuclei. After another 5 730 years (total time 17 190 years), 125 remain. Finally, after four half-lives (22 920 years), only about 62 remain.

These numbers represent ideal circumstances. Radioactive decay is in reality an averaging process over a very large number of atoms, and the actual outcome depends on statistics. Our original sample in this example contained only 1 000 nuclei, certainly not a very large number. Thus, if we were to count the number remaining after one half-life for this small sample, it probably would not be exactly 500.

EXAMPLE 44.4 ▶ The Activity of Radium

The half-life of the radioactive nucleus radium-226, $^{226}_{88}$Ra, is 1.6×10^3 yr. (a) What is the decay constant λ of this nucleus?

Solution   We can calculate λ using Equation 44.9 and the fact that

$$T_{1/2} = (1.6 \times 10^3 \text{ yr})(3.15 \times 10^7 \text{ s/yr})$$
$$= 5.0 \times 10^{10} \text{ s}$$

Therefore,

$$\lambda = \frac{0.693}{T_{1/2}} = \frac{0.693}{5.0 \times 10^{10} \text{ s}} = \boxed{1.4 \times 10^{-11} \text{ s}^{-1}}$$

Note that this result is also the probability that any single $^{226}_{88}$Ra nucleus will decay in a time interval of 1 s.

(b) If a sample contains 3.0×10^{16} $^{226}_{88}$Ra nuclei at $t = 0$, determine its activity in curies at this time.

Solution   By definition (Eq. 44.8) R_0, the activity at $t = 0$, is λN_0, where N_0 is the number of radioactive nuclei present

at $t = 0$. With $N_0 = 3.0 \times 10^{16}$, we have

$$R_0 = \lambda N_0 = (1.4 \times 10^{-11} \text{ s}^{-1})(3.0 \times 10^{16})$$
$$= (4.2 \times 10^5 \text{ decays/s})\left(\frac{1 \text{ Ci}}{3.7 \times 10^{10} \text{ decays/s}}\right)$$
$$= \boxed{11 \ \mu\text{Ci}}$$

(c) What is the activity in becquerels after the sample is 2.0×10^3 yr old?

Solution   We use Equation 44.8 and the fact that $t = 2.0 \times 10^3$ yr $= (2.0 \times 10^3 \text{ yr})(3.15 \times 10^7 \text{ s/yr}) = 6.3 \times 10^{10}$ s:

$$R = R_0 e^{-\lambda t}$$
$$= (4.2 \times 10^5 \text{ decays/s}) e^{-(1.4 \times 10^{-11} \text{ s}^{-1})(6.3 \times 10^{10} \text{ s})}$$
$$= 1.7 \times 10^5 \text{ decays/s} = \boxed{1.7 \times 10^5 \text{ Bq}}$$

EXAMPLE 44.5 ▶ The Activity of Carbon

A radioactive sample contains 3.50 μg of pure $^{11}_{6}$C, which has a half-life of 20.4 min. (a) Determine the number of nuclei in the sample at $t = 0$.

Solution   The molar mass of $^{11}_{6}$C is approximately 11.0 g/mol, and so 11.0 g contains Avogadro's number (6.02×10^{23}) of nuclei. Therefore, 3.50 μg contains N nuclei, where

$$\frac{N}{6.02 \times 10^{23} \text{ nuclei/mol}} = \frac{3.50 \times 10^{-6} \text{ g}}{11.0 \text{ g/mol}}$$

$$N = \boxed{1.92 \times 10^{17} \text{ nuclei}}$$

(b) What is the activity in becquerels of the sample initially and after 8.00 h?

Solution   With $T_{1/2} = 20.4$ min $= 1\ 224$ s, the decay constant is

$$\lambda = \frac{0.693}{T_{1/2}} = \frac{0.693}{1\ 224 \text{ s}} = 5.66 \times 10^{-4} \text{ s}^{-1}$$

Therefore, the initial activity of the sample is

$$R_0 = \lambda N_0 = (5.66 \times 10^{-4} \text{ s}^{-1})(1.92 \times 10^{17})$$

$$= \boxed{1.09 \times 10^{14} \text{ Bq}}$$

We use Equation 44.8 to find the activity at $t = 8.00$ h $= 2.88 \times 10^4$ s:

$$R = R_0 e^{-\lambda t} = (1.09 \times 10^{14} \text{ Bq}) e^{-(5.66 \times 10^{-4} \text{ s}^{-1})(2.88 \times 10^4 \text{ s})}$$
$$= \boxed{9.09 \times 10^6 \text{ Bq}}$$

A listing of activity versus time for this situation is given in Table 44.2.

Exercise   Calculate the number of radioactive nuclei remaining after 8.00 h.

Answer 1.60×10^{10} nuclei.

TABLE 44.2   **Activity Versus Time for the Sample Described in Example 44.5**

t (h)	R (Bq)
0	1.09×10^{14}
1	1.41×10^{13}
2	1.84×10^{12}
3	2.39×10^{11}
4	3.12×10^{10}
5	4.06×10^9
6	5.28×10^8
7	6.88×10^7
8	9.09×10^6

EXAMPLE 44.6   A Radioactive Isotope of Iodine

A sample of the isotope ^{131}I, which has a half-life of 8.04 days, has an activity of 5.0 mCi at the time of shipment. Upon receipt in a medical laboratory, the activity is 4.2 mCi. How much time has elapsed between the two measurements?

Solution We use Equation 44.8 in the form

$$\frac{R}{R_0} = e^{-\lambda t}$$

Taking the natural logarithm of each side, we obtain

$$\ln\left(\frac{R}{R_0}\right) = -\lambda t$$

$$(1) \qquad t = -\frac{1}{\lambda}\ln\left(\frac{R}{R_0}\right)$$

To find λ, we use Equation 44.9:

$$(2) \qquad \lambda = \frac{0.693}{T_{1/2}} = \frac{0.693}{8.04\ \text{days}}$$

Substituting Equation (2) into Equation (1) gives

$$t = -\left(\frac{8.04\ \text{days}}{0.693}\right)\ln\left(\frac{4.2\ \text{mCi}}{5.0\ \text{mCi}}\right) = \boxed{2.0\ \text{days}}$$

44.6 THE DECAY PROCESSES

As we stated in the preceding section, a radioactive nucleus spontaneously decays by one of three processes: alpha decay, beta decay, or gamma decay. Let us discuss these three processes in more detail.

Alpha Decay

A nucleus emitting an alpha particle ($^4_2$He) loses two protons and two neutrons. Therefore, the atomic number Z decreases by 2, the mass number A decreases by 4, and the neutron number N decreases by 2. The decay can be written

$$^A_Z\text{X} \longrightarrow\ ^{A-4}_{Z-2}\text{Y} + ^4_2\text{He} \qquad \textbf{(44.10)}$$

Alpha decay

where X is called the **parent nucleus** and Y the **daughter nucleus.** As a general rule in any decay equation such as this, (1) the sum of the mass numbers A must be the same on both sides of the equation and (2) the sum of the atomic numbers Z must be the same on both sides of the equation. As examples, ^{238}U and ^{226}Ra are both alpha emitters and decay according to the schemes

$$^{238}_{92}\text{U} \longrightarrow\ ^{234}_{90}\text{Th} + ^4_2\text{He} \qquad \textbf{(44.11)}$$

$$^{226}_{88}\text{Ra} \longrightarrow\ ^{222}_{86}\text{Rn} + ^4_2\text{He} \qquad \textbf{(44.12)}$$

The half-life for ^{238}U decay is 4.47×10^9 years, and that for ^{226}Ra decay is 1.60×10^3 years. The decay of ^{226}Ra is shown in Figure 44.14.

When one element changes into another, as happens in alpha decay, the process is called **spontaneous decay.** In any spontaneous decay, relativistic energy and momentum must be conserved. If we call M_X the mass of the parent nucleus, M_Y the mass of the daughter nucleus, and M_α the mass of the alpha particle, we can define the **disintegration energy** Q as

$$Q = (M_X - M_Y - M_\alpha)c^2 \qquad \textbf{(44.13)}$$

The disintegration energy Q

The energy Q is in joules when the masses are in kilograms and c is the speed of light, 3.00×10^8 m/s. However, when the masses are expressed in the more conve-

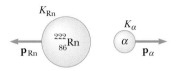

Figure 44.14 The alpha decay of radium. The radium nucleus is initially at rest. After the decay, the radon nucleus has kinetic energy K_{Rn} and momentum $\mathbf{p}_{Rn}$, and the alpha particle has kinetic energy K_α and momentum $\mathbf{p}_\alpha$.

nient unit u, Q can be calculated in MeV using the expression

$$Q = (M_X - M_Y - M_\alpha) \times 931.494 \text{ MeV/u} \qquad \textbf{(44.14)}$$

The disintegration energy Q appears in the form of kinetic energy in the daughter nucleus and the alpha particle and is sometimes referred to as the Q value of the nuclear reaction. In the case of the ^{226}Ra decay described in Figure 44.14, if the parent nucleus decays at rest, the kinetic energy of the products is 4.87 MeV. Most of this kinetic energy is associated with the alpha particle because this particle is much less massive than the daughter nucleus ^{222}Rn. That is, because momentum must be conserved, the lighter alpha particle recoils with a much higher speed than the daughter nucleus. Generally, less massive particles carry off most of the energy in nuclear decays.

Finally, it is interesting to note that if one assumed that ^{238}U (or any other alpha emitter) decayed by emitting either a proton or a neutron, the mass of the decay products would exceed that of the parent nucleus, corresponding to a negative Q value. A negative Q value indicates that such a proposed decay does not occur spontaneously.

EXAMPLE 44.7   The Energy Liberated When Radium Decays

The ^{226}Ra nucleus undergoes alpha decay according to Equation 44.12. Calculate the Q value for this process. Take the masses to be 226.025 402 u for ^{226}Ra, 222.017 571 u for ^{222}Rn, and 4.002 602 u for $^4_2$He, as found in Table A.3.

Solution We may add 88 electrons to both sides of the reaction 44.12. The differences in electron binding energies are negligible when compared with the Q value for the nuclear decay process. Then, we may use the masses of neutral atoms in Equation 44.14 to see that

$$Q = (M_X - M_Y - M_\alpha) \times 931.494 \text{ MeV/u}$$
$$= (226.025\ 402 \text{ u} - 222.017\ 571 \text{ u} - 4.002\ 602 \text{ u})$$
$$\times 931.494 \text{ MeV/u}$$
$$= (0.005\ 229 \text{ u}) \times (931.494 \text{ MeV/u}) = \boxed{4.87 \text{ MeV}}$$

It is left as a problem (Problem 57) to show that the kinetic energy of the alpha particle is about 4.8 MeV, whereas that of the recoiling daughter nucleus is only about 0.1 MeV.

To understand the mechanism of alpha decay, let us imagine a system consisting of (1) the alpha particle, already formed as an entity within the nucleus, and (2) the daughter nucleus that will result when the alpha particle is emitted. Figure 44.15 shows a plot of potential energy versus separation distance r between the alpha particle and the daughter nucleus, where the distance marked R is the range of the nuclear force. The curve represents the combined effects of (1) the Coulomb repulsive energy, which gives the positive peak for $r > R$, and (2) the nuclear attractive force, which causes the curve to be negative for $r < R$. As we saw in Example 44.7, the disintegration energy Q is about 5 MeV, which is the approximate kinetic energy of the alpha particle, represented by the lower dashed line in Figure 44.15.

According to classical physics, the alpha particle is trapped in a potential well. How, then, does it ever escape from the nucleus? The answer to this question was first provided by George Gamow (1904–1968) in 1928 and independently by R. W. Gurney and E. U. Condon in 1929, using quantum mechanics. Briefly, the view of quantum mechanics is that there is always some probability that the particle can tunnel through the barrier (see Section 41.7). Recall that the probability of locating the particle depends on its wave function ψ and that the tunneling probability is measured by $|\psi|^2$. Figure 44.16 is a sketch of the wave function for a particle of

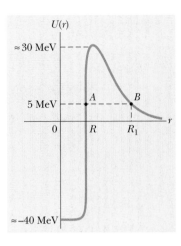

Figure 44.15 Potential energy versus separation distance for a system consisting of an alpha particle and a daughter nucleus. Classically, the energy of the alpha particle is not sufficiently large to overcome the energy barrier, and so the particle should not be able to escape from the nucleus. In reality, the alpha particle does escape by tunneling through the barrier.

energy E meeting a rectangular barrier of finite height, a shape that approximates the nuclear barrier. Note that the wave function exists both inside and outside the barrier. Although the amplitude of the wave function is greatly reduced on the far side of the barrier, its finite value in this region indicates a small but finite probability that the particle can penetrate the barrier. Outside the range of the nuclear force, the function ψ correctly describes the probability that the nucleus will decay. Although the decay probability is constant in time, *the precise moment of decay cannot be predicted.* In general, quantum mechanics implies that the future is indeterminate. (This is in contrast to classical mechanics, where the trajectory of an object can in principle be calculated to an arbitrarily high precision from precise knowledge of its initial coordinates and velocity and of the forces acting on it.) Thus, the fundamental laws of nature are probabilistic and it appears that Einstein was wrong in his famous statement, "God does not roll dice."

Schrödinger's Cat and the Probability of Decay. A radiation detector (see Section 45.6) can be used to show that a radioactive nucleus decays by radiating a particle at a particular moment and in a particular direction. To point out the contrast between this experimental result and its wave function, Erwin Schrödinger imagined a box containing a cat, a radioactive sample, a radiation counter, and a vial of poison. When a nucleus in the sample decays, the counter triggers the administration of lethal poison to the cat. Quantum mechanics correctly predicts the probability of finding the cat dead when the box is opened. However, many questions arise regarding this intriguing thought experiment: Before the box is opened, does the cat have a wave function describing the cat as fractionally dead, with some chance of being alive? Does the act of measurement change the system from a probabilistic state to a definite state? When a particle emitted by a radioactive nucleus is detected at one particular location, does the wave function describing the particle drop to zero instantaneously everywhere else in the Universe? Is there a fundamental difference between a quantum system and a macroscopic system? The answers to such questions are basically unknown.

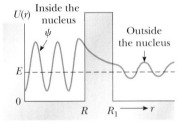

Figure 44.16 The nuclear potential energy is modeled as a rectangular barrier—the blue region extending from R to R_1. The energy of the alpha particle is E, which is less than the height of the barrier. According to quantum mechanics, the alpha particle has some chance of tunneling through the barrier, as indicated by the finite size of the wave function for $r > R_1$.

Quick Quiz 44.5

In alpha decay, the half-life of the decay goes down as the energy of the decay goes up. Why is this?

Artist's rendering of the *Pioneer 10* spacecraft at the end of its useful life, after a 25-year mission. The view is of the Sun and part of the Milky Way galaxy. On the end of the two larger booms are thermoelectric generators that create electricity from the energy given off by ^{238}Pu as the alpha particles it emits collide with surrounding materials. At launch, the generators produced a total power of 160 W. This radioactive power source was necessary because the great distance from the Sun precluded the use of solar panels. *(Courtesy of NASA Ames Home Page)*

The Smoke Detector.   A life-saving application of alpha decay is in the household smoke detector, shown in Figure 44.17. Most of the common ones use a radioactive material. The detector consists of an ionization chamber, a sensitive current detector, and an alarm. A weak radioactive source (usually $^{241}_{95}$Am) ionizes the air in the chamber of the detector, creating charged particles. A voltage is maintained between the plates inside the chamber, setting up a small but detectable current in the external circuit. As long as the current is maintained, the alarm is deactivated. However, if smoke drifts into the chamber, the ions become attached to the smoke particles. These heavier particles do not drift as readily as do the lighter ions, which causes a decrease in the detector current. The external circuit senses this decrease in current and sets off the alarm.

Beta Decay

When a radioactive nucleus undergoes beta decay, the daughter nucleus contains the same number of nucleons as the parent nucleus but the atomic number is

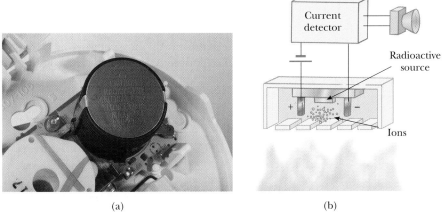

Figure 44.17 (a) A smoke detector uses alpha decay to determine whether smoke is in the air. (b) Smoke entering the chamber reduces the detected current, causing the alarm to sound. *(a, George Semple)*

changed by 1, which means that the number of protons changes:

$$_Z^A X \longrightarrow _{Z+1}^A Y + e^-$$ **(44.15)**

$$_Z^A X \longrightarrow _{Z-1}^A Y + e^+$$ **(44.16)**

Beta decay

where, as we discussed in Section 44.5, the symbol e⁻ is used to designate an electron and e⁺ designates a positron, with *beta particle* being the general term referring to either. As with alpha decay, the nucleon number and total charge are both conserved in beta decays. From the fact that *A* does not change but *Z* does, we conclude that in beta decay, either a neutron changes to a proton (Eq. 44.15) or a proton changes to a neutron (Eq. 44.16). It is also important to note that the electron or positron emitted in these decays is not present beforehand in the nucleus; it is created at the moment of decay from the rest energy of the decaying nucleus.

Two typical beta decay processes are

$$_6^{14} C \longrightarrow _7^{14} N + e^-$$ **(44.17)**

$$_7^{12} N \longrightarrow _6^{12} C + e^+$$ **(44.18)**

As we shall see later, *beta decay is not described completely by these expressions.* We shall give reasons for this shortly.

Let us consider the energy of the system undergoing beta decay before and after the decay. As with alpha decay, energy must be conserved. Experimentally, it is found that beta particles from a single type of nucleus are emitted over a continuous range of energies (Fig. 44.18). The kinetic energy of the system after the decay is equal to the decrease in mass of the system, that is, the *Q* value. However, because all decaying nuclei in the sample have the same initial mass, *the Q value must be the same for each decay.* In view of this, why do the emitted particles have the range of kinetic energies shown in Figure 44.18? The law of conservation of energy seems to be violated! And it gets worse: Further analysis of the decay processes described by Equations 44.15 and 44.16 shows that the laws of conservation of both angular momentum (spin) and linear momentum are also violated!

After a great deal of experimental and theoretical study, Pauli in 1930 proposed that a third particle must be present to carry away the "missing" energy and momentum. Fermi later named this particle the **neutrino** (little neutral one) because it had to be electrically neutral and have little or no mass. Although it

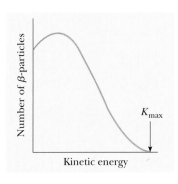

Figure 44.18 A typical beta-decay curve. The maximum kinetic energy observed for the beta particles corresponds to the *Q* value for the reaction.

eluded detection for many years, the neutrino (symbol ν, Greek letter nu) was finally detected experimentally in 1956. It has the following properties:

Properties of the neutrino

- It has zero electric charge.
- Its mass is either zero (in which case it travels at the speed of light) or very small; there is much recent persuasive experimental evidence that suggests that the neutrino mass is not zero.
- It has a spin of $\frac{1}{2}$, which allows the law of conservation of angular momentum to be satisfied in beta decay.
- It interacts very weakly with matter and is therefore very difficult to detect.

We can now write the beta-decay processes for carbon-14 and nitrogen-12 (Eqs. 44.17 and 44.18) in their correct form:

$$^{14}_{6}\text{C} \longrightarrow {}^{14}_{7}\text{N} + \text{e}^- + \bar{\nu} \qquad \textbf{(44.19)}$$

$$^{12}_{7}\text{N} \longrightarrow {}^{12}_{6}\text{C} + \text{e}^+ + \nu \qquad \textbf{(44.20)}$$

where the symbol $\bar{\nu}$ represents the **antineutrino,** the antiparticle to the neutrino. We shall discuss antiparticles further in Chapter 46. For now, it suffices to say that **a neutrino is emitted in positron decay and an antineutrino is emitted in electron decay.** As with alpha decay, the decays listed above are analyzed by applying conservation laws, but relativistic expressions must be used for beta particles because their kinetic energy is large (typically 1 MeV) compared with their rest energy of 0.511 MeV.

In Equation 44.19, the number of protons has increased by one and the number of neutrons has decreased by one. We can write the fundamental process of e^- decay in terms of a neutron changing into a proton as follows:

$$\text{n} \longrightarrow \text{p} + \text{e}^- + \bar{\nu} \qquad \textbf{(44.21)}$$

The electron and the antineutrino are ejected from the nucleus, with the net result that there is one more proton and one fewer neutron, consistent with the changes in Z and $A - Z$. A similar process occurs in e^+ decay, with a proton changing into a neutron, a positron, and a neutrino.

A process that competes with e^+ decay is **electron capture.** This occurs when a parent nucleus captures one of its own orbital electrons and emits a neutrino. The final product after decay is a nucleus whose charge is $Z - 1$:

Electron capture

$$^{A}_{Z}\text{X} + {}^{0}_{-1}\text{e} \longrightarrow {}^{A}_{Z-1}\text{Y} + \nu \qquad \textbf{(44.22)}$$

In most cases, it is a K-shell electron that is captured, and for this reason the process is referred to as **K capture.** One example is the capture of an electron by $^{7}_{4}\text{Be}$:

$$^{7}_{4}\text{Be} + {}^{0}_{-1}\text{e} \longrightarrow {}^{7}_{3}\text{Li} + \nu$$

Because the neutrino is very difficult to detect, electron capture is usually observed by the x-rays given off as higher-shell electrons cascade downward to fill the vacancy created in the K shell.

Finally, we specify Q values for the beta-decay processes. The Q values for e^- decay and electron capture are given by $Q = (M_X - M_Y)c^2$, where M_X and M_Y are the masses of neutral atoms. The Q values for e^+ decay are given by $Q = (M_X - M_Y - 2m_e)c^2$. These relationships are useful in determining whether a process is energetically possible.

Quick Quiz 44.6

In beta decay, the kinetic energy of the emitted electron or positron lies somewhere in a relatively large range of possibilities. In alpha decay, the kinetic energy of the emitted alpha particle can have only discrete values. Why is there this difference?

Carbon Dating

The beta decay of ^{14}C (Eq. 44.19) is commonly used to date organic samples. Cosmic rays in the upper atmosphere cause nuclear reactions that create ^{14}C. The ratio of ^{14}C to ^{12}C in the carbon dioxide molecules of our atmosphere has a constant value of approximately 1.3×10^{-12}. The carbon atoms in all living organisms have this same ^{14}C/^{12}C ratio because the organisms continuously exchange carbon dioxide with their surroundings. When an organism dies, however, it no longer absorbs ^{14}C from the atmosphere, and so the ^{14}C/^{12}C ratio decreases as the ^{14}C decays with a half-life of 5 730 years. It is therefore possible to measure the age of a material by measuring its ^{14}C activity. Using this technique, scientists have been able to identify samples of wood, charcoal, bone, and shell as having lived from 1 000 to 25 000 years ago. This knowledge has helped us reconstruct the history of living organisms—including humans—during this time span.

A particularly interesting example is the dating of the Dead Sea Scrolls. This group of manuscripts was discovered by a shepherd in 1947. Translation showed them to be religious documents, including most of the books of the Old Testament. Because of their historical and religious significance, scholars wanted to know their age. Carbon dating applied to the material in which they were wrapped established their age at approximately 1 950 years.

CONCEPTUAL EXAMPLE 44.8   **The Age of Ice Man**

In 1991, a German tourist discovered the well-preserved remains of a man, now called the Ice Man, trapped in a glacier in the Italian Alps (Fig. 44.19). Radioactive dating with ^{14}C revealed that this person was alive about 5 300 years ago. Why did scientists date the sample using ^{14}C rather than ^{11}C, which is a beta emitter having a half-life of 20.4 min?

Solution   Because ^{14}C has a half-life of 5 730 years, the fraction of ^{14}C nuclei remaining after one half-life is high enough to allow accurate measurements of changes in the sample's activity. Because ^{11}C has a very short half-life, it is not useful—its activity decreases to a vanishingly small value over the age of the sample, making it impossible to detect.

As a general rule, the isotope chosen to date a sample should have a half-life that is of the same order of magnitude as the age of the sample. If the half-life is much less than the age of the sample, there won't be enough activity left to measure because almost all of the original radioactive nuclei will have decayed away. If the half-life is much greater than the age of the sample, the amount of decay that has taken place since the sample died will be too small to measure. For example, if you have a specimen estimated to have died 50 years ago, neither ^{14}C (5 730 years) nor ^{11}C (20 min) is suitable. If you know your sample contains hydrogen, however, you can measure the activity of ^{3}H (tritium), a beta emitter that has a half-life of 12.3 years.

Figure 44.19   The Ice Man, discovered in 1991 when an Italian glacier melted enough to expose his remains. His possessions, particularly his tools, have shed light on the way people lived in the Bronze Age. Carbon-14 dating was used to determine how long ago this person lived. *(Paul Hanny/Gamma Liaison)*

EXAMPLE 44.9   Radioactive Dating

A 25.0-g piece of charcoal is found in some ruins of an ancient city. The sample shows a ^{14}C activity R of 250 decays/min. How long has the tree this charcoal came from been dead?

Solution   First, let us calculate the decay constant λ for ^{14}C, which has a half-life of 5 730 years.

$$\lambda = \frac{0.693}{T_{1/2}} = \frac{0.693}{(5\ 730\ \text{yr})(3.15 \times 10^7\ \text{s/yr})}$$

$$= 3.84 \times 10^{-12}\ \text{s}^{-1}$$

The number of ^{14}C nuclei can be calculated in two steps. First, the number of ^{12}C nuclei in 25.0 g of carbon is

$$N(^{12}\text{C}) = \frac{6.02 \times 10^{23}\ \text{nuclei/mol}}{12.0\ \text{g/mol}}\ (25.0\ \text{g})$$

$$= 1.25 \times 10^{24}\ \text{nuclei}$$

Knowing that the ratio of ^{14}C to ^{12}C in the live sample was 1.3×10^{-12}, we see that the number of ^{14}C nuclei in 25.0 g *before* decay was

$$N_0(^{14}\text{C}) = (1.3 \times 10^{-12})(1.25 \times 10^{24}) = 1.6 \times 10^{12}\ \text{nuclei}$$

Hence, the initial activity of the sample was

$$R_0 = N_0\lambda = (1.6 \times 10^{12}\ \text{nuclei})(3.84 \times 10^{-12}\ \text{s}^{-1})$$

$$= 6.14\ \text{decays/s} = 370\ \text{decays/min}$$

We now use Equation 44.8, which relates the activity R at any time t to the initial activity R_0:

$$R = R_0 e^{-\lambda t}$$

$$e^{-\lambda t} = \frac{R}{R_0}$$

Using $R = 250$ decays/min and $R_0 = 370$ decays/min, we calculate t by taking the natural logarithm of both sides of this expression:

$$-\lambda t = \ln\left(\frac{R}{R_0}\right) = \ln\left(\frac{250}{370}\right) = -0.39$$

$$t = \frac{0.39}{\lambda} = \frac{0.39}{3.84 \times 10^{-12}\ \text{s}^{-1}}$$

$$= 1.0 \times 10^{11}\ \text{s} = \boxed{3\ 200\ \text{yr}}$$

Gamma Decay

Very often, a nucleus that undergoes radioactive decay is left in an excited energy state. The nucleus can then undergo a second decay to a lower energy state, perhaps to the ground state, by emitting a high-energy photon:

$$^{A}_{Z}\text{X}^* \longrightarrow {}^{A}_{Z}\text{X} + \gamma \tag{44.23}$$

where X* indicates a nucleus in an excited state. The typical half-life of an excited nuclear state is 10^{-10} s. Photons emitted in such a de-excitation process are called gamma rays. Such photons have very high energy (1 MeV to 1 GeV) relative to the energy of visible light (about 1 eV). Recall from Sections 40.5 and 42.7 that the energy of a photon emitted or absorbed by an atom equals the difference in energy between the two electronic states involved in the transition. Similarly, a gamma ray photon has an energy hf that equals the energy difference ΔE between two nuclear energy levels. When a nucleus decays by emitting a gamma ray, the only change in the nucleus is that it ends up in a lower energy state.

A nucleus may reach an excited state as the result of a violent collision with another particle. However, it is more common for a nucleus to be in an excited state after it has undergone alpha or beta decay. The following sequence of events represents a typical situation in which gamma decay occurs:

$$^{12}_{5}\text{B} \longrightarrow {}^{12}_{6}\text{C}^* + e^- + \bar{\nu} \tag{44.24}$$

$$^{12}_{6}\text{C}^* \longrightarrow {}^{12}_{6}\text{C} + \gamma \tag{44.25}$$

Figure 44.20 shows the decay scheme for ^{12}B, which undergoes beta decay to either of two levels of ^{12}C. It can either (1) decay directly to the ground state of

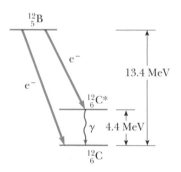

Gamma decay

Figure 44.20 The ^{12}B nucleus undergoes beta decay to either of two levels of ^{12}C: directly to the ground state or to an excited state. Decay to the excited level ^{12}C* is followed by gamma decay to the ground state.

TABLE 44.3	Various Decay Pathways
Alpha decay	$^{A}_{Z}X \longrightarrow {}^{A-4}_{Z-2}Y + {}^{4}_{2}He$
Beta decay (e^{-})	$^{A}_{Z}X \longrightarrow {}^{A}_{Z+1}Y + e^{-} + \overline{\nu}$
Beta decay (e^{+})	$^{A}_{Z}X \longrightarrow {}^{A}_{Z-1}Y + e^{+} + \nu$
Electron capture	$^{A}_{Z}X + e^{-} \longrightarrow {}^{A}_{Z-1}Y + \nu$
Gamma decay	$^{A}_{Z}X^{*} \longrightarrow {}^{A}_{Z}X + \gamma$

^{12}C by emitting a 13.4-MeV electron or (2) undergo beta decay to an excited state of ^{12}C* followed by gamma decay to the ground state. The latter process results in the emission of a 9.0-MeV electron and a 4.4-MeV photon.

The various pathways by which a radioactive nucleus can undergo decay are summarized in Table 44.3.

44.7 NATURAL RADIOACTIVITY

Radioactive nuclei are generally classified into two groups: (1) unstable nuclei found in nature, which give rise to **natural radioactivity,** and (2) unstable nuclei produced in the laboratory through nuclear reactions, which exhibit **artificial radioactivity.**

As Table 44.4 shows, there are three series of naturally occurring radioactive nuclei. Each series starts with a specific long-lived radioactive isotope whose half-life exceeds that of any of its descendants. The three natural series begin with the isotopes ^{238}U, ^{235}U, and ^{232}Th, and the corresponding stable end products are three isotopes of lead: ^{206}Pb, ^{207}Pb, and ^{208}Pb. The fourth series in Table 44.4 begins with ^{237}Np and has as its stable end product ^{209}Bi. The element ^{237}Np is a *transuranic* element (one having an atomic number greater than that of uranium) not found in nature. This element has a half-life of "only" 2.14×10^{6} years.

Figure 44.21 shows the successive decays for the ^{232}Th series. Note that ^{232}Th first undergoes alpha decay to ^{228}Ra. Next, ^{228}Ra undergoes two successive beta decays to ^{228}Th. The series continues and finally branches when it reaches ^{212}Bi. At this point, there are two decay possibilities. The end of the decay series is the stable isotope ^{208}Pb. The sequence shown in Figure 44.21 is characterized by a mass-number decrease of either 4 (for alpha decays) or 0 (for beta or gamma decays). The two uranium series are more complex than the ^{232}Th series. Also, there are several naturally occurring radioactive isotopes, such as ^{14}C and ^{40}K, that are not part of any decay series.

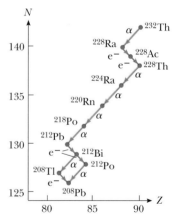

Figure 44.21 Successive decays for the ^{232}Th series.

	TABLE 44.4	The Four Radioactive Series		
Series		**Starting Isotope**	**Half-Life (years)**	**Stable End Product**
Uranium	Natural	$^{238}_{92}$U	4.47×10^{9}	$^{206}_{82}$Pb
Actinium		$^{235}_{92}$U	7.04×10^{8}	$^{207}_{82}$Pb
Thorium		$^{232}_{90}$Th	1.41×10^{10}	$^{208}_{82}$Pb
Neptunium		$^{237}_{93}$Np	2.14×10^{6}	$^{209}_{83}$Bi

Because of these radioactive series, our environment is constantly replenished with radioactive elements that would otherwise have disappeared long ago. For example, because the Solar System is approximately 5×10^9 years old, the supply of ^{226}Ra (whose half-life is only 1 600 years) would have been depleted by radioactive decay long ago if it were not for the radioactive series starting with ^{238}U.

44.8 NUCLEAR REACTIONS

It is possible to change the structure of nuclei by bombarding them with energetic particles. Such collisions, which change the identity of the target nuclei, are called **nuclear reactions.** Rutherford was the first to observe them, in 1919, using naturally occurring radioactive sources for the bombarding particles. Since then, thousands of nuclear reactions have been observed following the development of charged-particle accelerators in the 1930s. With today's advanced technology in particle accelerators and particle detectors, it is possible to achieve particle energies of at least 1 000 GeV = 1 TeV. These high-energy particles are used to create new particles whose properties are helping to solve the mysteries of the nucleus.

Consider a reaction in which a target nucleus X is bombarded by a particle a, resulting in a daughter nucleus Y and a particle b:

Nuclear reaction

$$a + X \longrightarrow Y + b \tag{44.26}$$

Sometimes this reaction is written in the more compact form

$$X(a, b)Y$$

In Section 44.6, the Q value, or disintegration energy, of a radioactive decay was defined as the energy released as a result of the decay process. Likewise, we define the **reaction energy** Q associated with a nuclear reaction as *the total energy released as the result of the reaction:*

Reaction energy Q

$$Q = (M_a + M_X - M_Y - M_b)c^2 \tag{44.27}$$

As an example, consider the reaction ^{7}Li $(p, \alpha)^4$He. The notation p indicates a proton, which is a hydrogen nucleus. Thus, we can write this reaction in the expanded form

$$^1_1\text{H} + ^7_3\text{Li} \longrightarrow ^4_2\text{He} + ^4_2\text{He}$$

Exothermic reaction
Endothermic reaction

Threshold energy

The Q value for this reaction is 17.3 MeV. A reaction such as this, for which Q is positive, is called **exothermic.** A reaction for which Q is negative is called **endothermic.** An endothermic reaction does not occur unless the bombarding particle has a kinetic energy greater than Q. The minimum energy necessary for such a reaction to occur is called the **threshold energy.**

Nuclear reactions must obey the law of conservation of linear momentum. Generally the only force acting on the interacting particles is their mutual force of interaction; that is, there are no external accelerating electric fields present near the colliding particles.

If particles a and b in a nuclear reaction are identical, so that X and Y are also necessarily identical, the reaction is called a **scattering event.** If kinetic energy is conserved as a result of the reaction (that is, if $Q = 0$), it is classified as *elastic scattering.* If kinetic energy is not conserved, $Q \neq 0$ and the reaction is described as *in-*

TABLE 44.5	Q Values for Nuclear Reactions Involving Light Nuclei
Reaction[a]	**Measured Q-Value (MeV)**
^{2}H(n, γ)^{3}H	6.257 ± 0.004
^{2}H(d, p)^{3}H	4.032 ± 0.004
^{6}Li(p, α)^{3}H	4.016 ± 0.005
^{6}Li(d, p)^{7}Li	5.020 ± 0.006
^{7}Li(p, n)^{7}Be	-1.645 ± 0.001
^{7}Li(p, α)^{4}He	17.337 ± 0.007
^{9}Be(n, γ)^{10}Be	6.810 ± 0.006
^{9}Be(γ, n)^{8}Be	-1.666 ± 0.002
^{9}Be(d, p)^{10}Be	4.585 ± 0.005
^{9}Be(p, α)^{6}Li	2.132 ± 0.006
^{10}B(n, α)^{7}Li	2.793 ± 0.003
^{10}B(p, α)^{7}Be	1.148 ± 0.003
^{12}C(n, γ)^{13}C	4.948 ± 0.004
^{13}C(p, n)^{13}N	-3.003 ± 0.002
^{14}N(n, p)^{14}C	0.627 ± 0.001
^{14}N(n, γ)^{15}N	10.833 ± 0.007
^{18}O(p, n)^{18}F	-2.453 ± 0.002
^{19}F(p, α)^{16}O	8.124 ± 0.007

From C. W. Li, W. Whaling, W. A. Fowler, and C. C. Lauritsen, *Phys. Rev.* 83:512, 1951.

[a]The symbols n, p, d, α, and γ denote the neutron, proton, deuteron, alpha particle, and photon, respectively.

elastic scattering. This terminology is identical to that used in describing collisions between macroscopic objects (Section 9.4).

Measured Q values for a number of nuclear reactions involving light nuclei are given in Table 44.5.

In addition to energy and momentum, the total charge and total number of nucleons must be conserved in any nuclear reaction. For example, consider the reaction ^{19}F(p, α)^{16}O, which has a Q value of 8.124 MeV. We can show this reaction more completely as

$$^1_1\text{H} + {}^{19}_9\text{F} \longrightarrow {}^{16}_8\text{O} + {}^4_2\text{He}$$

The total number of nucleons before the reaction $(1 + 19 = 20)$ is equal to the total number after the reaction $(16 + 4 = 20)$. Furthermore, the total charge $(Z = 10)$ is the same before and after the reaction.

SUMMARY

A nucleus is represented by the symbol ^A_ZX, where A is the **mass number** (the total number of nucleons) and Z is the **atomic number** (the total number of protons). The total number of neutrons in a nucleus is the **neutron number** N, where $A = N + Z$. Nuclei having the same Z value but different A and N values are **isotopes** of one another.

Assuming that nuclei are spherical, their radius is given by

$$r = r_0 A^{1/3} \tag{44.1}$$

where $r_0 = 1.2$ fm.

Nuclei are stable because of the **nuclear force** between nucleons. This short-range force dominates the Coulomb repulsive force at distances of less than about 2 fm and is independent of charge. Light nuclei are most stable when the number of protons they contain equals the number of neutrons. Heavy nuclei are most stable when the number of neutrons they contain exceeds the number of protons. The most stable nuclei have Z and N values that are both even.

Nuclei have an intrinsic spin angular momentum of magnitude $\sqrt{I(I + 1)}\,\hbar$, where I is the **nuclear spin quantum number.** The magnetic moment of a nucleus is measured in terms of the **nuclear magneton** μ_n, where

$$\mu_n \equiv \frac{e\hbar}{2m_p} = 5.05 \times 10^{-27}\,\text{J/T} \tag{44.3}$$

When a nuclear spin magnetic moment is placed in an external magnetic field, it precesses about the field with a frequency (the **Larmor precessional frequency**) that is proportional to the magnitude of the field.

The difference between the sum of the masses of a group of separate nucleons and the mass of the compound nucleus containing these nucleons, when multiplied by c^2, gives the **binding energy** E_b of the nucleus. We can calculate the binding energy of the nucleus of an atom of mass M_A using the expression

$$E_b(\text{MeV}) = (Z m_p + N m_n - M_A) \times 931.494\,\text{MeV/u} \tag{44.4}$$

where m_p is the mass of the proton and m_n is the mass of the neutron.

The **liquid-drop model** of nuclear structure treats the nucleons as molecules in a drop of liquid. The three main contributions influencing binding energy are the volume effect, the surface effect, and the Coulomb repulsion effect. Summing such contributions results in the **semiempirical binding energy formula:**

$$E_b = C_1 A - C_2 A^{2/3} - C_3 \frac{Z(Z - 1)}{A^{1/3}} - C_4 \frac{(N - Z)^2}{A} \tag{44.5}$$

The **independent-particle model** assumes that each nucleon exists in a shell and can only have discrete energy values. The stability of certain nuclei can be explained with this model.

A radioactive substance decays by **alpha decay, beta decay,** or **gamma decay.** An alpha particle is the ^{4}He nucleus; a beta particle is either an electron (e^-) or a positron (e^+); a gamma particle is a high-energy photon.

If a radioactive material contains N_0 radioactive nuclei at $t = 0$, the number N of nuclei remaining after a time t has elapsed is

$$N = N_0 e^{-\lambda t} \tag{44.7}$$

where λ is the **decay constant,** a number equal to the probability per second that a nucleus will decay. The **decay rate,** or **activity,** of a radioactive substance is

$$R = \left| \frac{dN}{dt} \right| = R_0 e^{-\lambda t} \tag{44.8}$$

where $R_0 = N_0 \lambda$ is the activity at $t = 0$. The **half-life** $T_{1/2}$ is defined as the time it takes half of a given number of radioactive nuclei to decay, where

$$T_{1/2} = \frac{0.693}{\lambda} \tag{44.9}$$

In alpha decay, a helium nucleus is ejected from the parent nucleus with a definite kinetic energy. A nucleus undergoing beta decay emits either an electron (e^-) and an antineutrino ($\bar{\nu}$) or a positron (e^+) and a neutrino (ν). The electron or positron is ejected with a range of energies. In **electron capture,** the nucleus of an atom absorbs one of its own electrons and emits a neutrino. In gamma decay, a nucleus in an excited state decays to its ground state and emits a gamma ray.

Nuclear reactions can occur when a target nucleus X is bombarded by a particle a, resulting in a daughter nucleus Y and a particle b:

$$a + X \longrightarrow Y + b \qquad \textbf{(44.26)}$$

The energy released in such a reaction, called the **reaction energy** Q, is

$$Q = (M_a + M_X - M_Y - M_b)c^2 \qquad \textbf{(44.27)}$$

QUESTIONS

1. Why are heavy nuclei unstable?

2. The magnetic moment of a proton precesses with a frequency ω_p in the presence of a magnetic field. If the magnetic field magnitude is doubled, what happens to the precessional frequency?

3. Explain why nuclei that are well off the line of stability in Figure 44.3 tend to be unstable.

4. Why do nearly all the naturally occurring isotopes lie above the $N = Z$ line in Figure 44.3?

5. Consider two heavy nuclei X and Y having similar mass numbers. If X has the higher binding energy, which nucleus tends to be more unstable?

6. Discuss the differences between the liquid-drop model and the independent-particle model of the nucleus.

7. How many values of I_z are possible for $I = 5/2$? for $I = 3$?

8. In nuclear magnetic resonance, how does increasing the value of the constant magnetic field change the frequency of the radio-frequency field that excites a particular transition?

9. Would the liquid-drop or independent-particle model be more appropriate to predict the behavior of a nucleus in a fission reaction? Which would be more successful in predicting the magnetic moment of a given nucleus? Which could better explain the γ-ray spectrum of an excited nucleus?

10. If a nucleus has a half-life of 1 year, does this mean it will be completely decayed after 2 years? Explain.

11. What fraction of a radioactive sample has decayed after two half-lives have elapsed?

12. Two samples of the same radioactive nuclide are prepared. Sample A has twice the initial activity of sample B. How does the half-life of A compare with the half-life of B? After each has passed through five half-lives, what is the ratio of their activities?

13. Explain why the half-lives for radioactive nuclei are essentially independent of temperature.

14. The radioactive nucleus $^{226}_{88}$Ra has a half-life of approximately 1.6×10^3 years. Being that the Solar System is about 5 billion years old, why do we still find this nucleus in nature?

15. Why is the electron involved in the reaction

$$^{14}_{6}C \longrightarrow {}^{14}_{7}N + e^- + \bar{\nu}$$

written as e^-, while the electron involved in the reaction

$$^{7}_{4}Be + {}^{0}_{-1}e \longrightarrow {}^{7}_{3}Li + \nu$$

is written as $_{-1}^{0}e$?

16. A free neutron undergoes beta decay with a half-life of about 15 min. Can a free proton undergo a similar decay?

17. Explain how you can carbon date the age of a sample.

18. What is the difference between a neutrino and a photon?

19. Does the Q in Equation 44.27 represent the quantity (final mass − initial mass)c^2, or does it represent the quantity (initial mass − final mass)c^2?

20. Use Equations 44.19 to 44.21 to explain why the neutrino must have a spin of $\frac{1}{2}$.

21. If a nucleus such as ^{226}Ra initially at rest undergoes alpha decay, which has more kinetic energy after the decay, the alpha particle or the daughter nucleus?

22. Can a nucleus emit alpha particles that have different energies? Explain.

23. Explain why many heavy nuclei undergo alpha decay but do not spontaneously emit neutrons or protons.

24. If an alpha particle and an electron have the same kinetic energy, which undergoes the greater deflection when passed through a magnetic field?

25. If film is kept in a wooden box, alpha particles from a radioactive source outside the box cannot expose the film but beta particles can. Explain.

26. Pick any beta decay process and show that the neutrino must have zero charge.

27. Suppose it could be shown that the cosmic ray intensity at the Earth's surface was much greater 10 000 years ago. How would this difference affect what we accept as valid carbon-dated values of the age of ancient samples of once-living matter?

28. Why is carbon dating unable to provide accurate estimates of very old material?

29. Element X has several isotopes. What do these isotopes have in common? How do they differ?

30. Explain the main differences between alpha, beta, and gamma rays.

31. How many protons are there in the nucleus $^{222}_{86}$Rn? How many neutrons? How many orbiting electrons are there in the neutral atom?

PROBLEMS

1, 2, 3 = straightforward, intermediate, challenging   ☐ = full solution available in the *Student Solutions Manual and Study Guide*
WEB = solution posted at **http://www.saunderscollege.com/physics/**   ⌨ = Computer useful in solving problem   ⌨ = Interactive Physics
☐ = paired numerical/symbolic problems

Note: Table 44.1 and 44.6 will be useful for many of these problems. A more complete list of atomic masses is given in Table A.3 in Appendix A.

Section 44.1   Some Properties of Nuclei

1. What is the order of magnitude of the number of protons in your body? Of the number of neutrons? Of the number of electrons?

2. Review Problem.   Singly ionized carbon is accelerated through 1 000 V and passed into a mass spectrometer to determine the isotopes present (see Chapter 29). The magnitude of the magnetic field in the spectrometer is 0.200 T. (a) Determine the orbit radii for the ^{12}C and the ^{13}C isotopes as they pass through the field. (b) Show that the ratio of radii may be written in the form

$$\frac{r_1}{r_2} = \sqrt{\frac{m_1}{m_2}}$$

and verify that your radii in part (a) agree with this.

3. An α particle ($Z = 2$, mass 6.64×10^{-27} kg) approaches to within 1.00×10^{-14} m of a carbon nucleus ($Z = 6$). What are (a) the maximum Coulomb force on the α particle, (b) the acceleration of the α particle at this point, and (c) the potential energy of the α particle at this point?

4. In a Rutherford scattering experiment, alpha particles having kinetic energy of 7.70 MeV are fired toward a gold nucleus. (a) Use energy conservation to determine the distance of closest approach between the alpha particle and gold nucleus. Assume the nucleus remains at rest. (b) Calculate the de Broglie wavelength for the 7.70-MeV alpha particle and compare it with the distance obtained in part (a). (c) Based on this comparison, why is it proper to treat the alpha particle as a particle and not as a wave in the Rutherford scattering experiment?

5. (a) Use energy methods to calculate the distance of closest approach for a head-on collision between an alpha particle having an initial energy of 0.500 MeV and a gold nucleus (^{197}Au) at rest. (Assume the gold nucleus remains at rest during the collision.) (b) What minimum initial speed must the alpha particle have in order to get as close as 300 fm?

6. How much energy (in MeV units) must an α particle have to reach the surface of a gold nucleus ($Z = 79$, $A = 197$)?

7. Find the radius of (a) a nucleus of $^4_2$He and (b) a nucleus of $^{238}_{92}$U.

8. Find the nucleus that has a radius approximately equal to one-half the radius of uranium $^{238}_{92}$U.

9. A star ending its life with a mass of two times the mass of the Sun is expected to collapse, combining its protons and electrons to form a neutron star. Such a star could be thought of as a gigantic atomic nucleus. If a star of mass $2 \times 1.99 \times 10^{30}$ kg collapsed into neutrons ($m_n = 1.67 \times 10^{-27}$ kg), what would its radius be? (Assume that $r = r_0 A^{1/3}$.)

10. Review Problem.   What would be the gravitational force between two golf balls (each with a 4.30-cm diameter), 1.00 meter apart, if they were made of nuclear matter?

11. From Table A.3, identify the stable nuclei that correspond to the magic numbers given by Equation 44.2.

12. For the stable nuclei in Table A.3, identify the number of stable nuclei that are even Z, even N; even Z, odd N; odd Z, even N; and odd Z, odd N.

13. Construct a diagram like that of Figure 44.4 for the case when I equals (a) 5/2 and (b) 4.

Section 44.2   Nuclear Magnetic Resonance and Magnetic Resonance Imaging

14. The Larmor precessional frequency is

$$f = \frac{\Delta E}{h} = \frac{2\mu B}{h}$$

Calculate the radio-wave frequency at which resonance absorption will occur for (a) free neutrons in a magnetic field of 1.00 T, (b) free protons in a magnetic field of 1.00 T, and (c) free protons in the Earth's mag-

TABLE 44.6   Some Atomic Masses

Element	Atomic Mass (u)	Element	Atomic Mass (u)
$^4_2$He	4.002 602	$^{27}_{13}$Al	26.981 538
$^7_3$Li	7.016 003	$^{30}_{15}$P	29.978 307
$^9_4$Be	9.012 174	$^{40}_{20}$Ca	39.962 591
$^{10}_5$B	10.012 936	$^{42}_{20}$Ca	41.958 618
$^{12}_6$C	12.000 000	$^{43}_{20}$Ca	42.958 767
$^{13}_6$C	13.003 355	$^{56}_{26}$Fe	55.934 940
$^{14}_7$N	14.003 074	$^{64}_{30}$Zn	63.929 144
$^{15}_7$N	15.000 108	$^{64}_{29}$Cu	63.929 599
$^{15}_8$O	15.003 065	$^{93}_{41}$Nb	92.906 376
$^{17}_8$O	16.999 132	$^{197}_{79}$Au	196.966 543
$^{18}_8$O	17.999 160	$^{202}_{80}$Hg	201.970 617
$^{18}_9$F	18.000 937	$^{216}_{84}$Po	216.001 889
$^{20}_{10}$Ne	19.992 435	$^{220}_{86}$Rn	220.011 369
$^{23}_{11}$Na	22.989 770	$^{234}_{90}$Th	234.043 593
$^{23}_{12}$Mg	22.994 124	$^{238}_{92}$U	238.050 784

netic field at a location where the magnitude of the field is 50.0 μT.

Section 44.3   Binding Energy and Nuclear Forces

15. Calculate the binding energy per nucleon for (a) ^{2}H, (b) ^{4}He, (c) ^{56}Fe, and (d) ^{238}U.

16. The peak of the stability curve occurs at ^{56}Fe. This is why iron is prominent in the spectrum of the Sun and stars. Show that ^{56}Fe has a higher binding energy per nucleon than its neighbors ^{55}Mn and ^{59}Co. Compare your results with Figure 44.8.

WEB **17.** Nuclei having the same mass numbers are called *isobars*. The isotope $^{139}_{57}$La is stable. A radioactive isobar $^{139}_{59}$Pr is located below the line of stable nuclei in Figure 44.3 and decays by e^+ emission. Another radioactive isobar of ^{139}La, $^{139}_{55}$Cs, decays by e^- emission and is located above the line of stable nuclei in Figure 44.3. (a) Which of these three isobars has the highest neutron-to-proton ratio? (b) Which has the greatest binding energy per nucleon? (c) Which do you expect to be heavier, ^{139}Pr or ^{139}Cs?

18. Two nuclei having the same mass number are known as *isobars*. Calculate the difference in binding energy per nucleon for the isobars $^{23}_{11}$Na and $^{23}_{12}$Mg. How do you account for the difference?

WEB **19.** A pair of nuclei for which $Z_1 = N_2$ and $Z_2 = N_1$ are called mirror isobars (the atomic and neutron numbers are interchanged). Binding energy measurements on these nuclei can be used to obtain evidence of the charge independence of nuclear forces (that is, proton–proton, proton–neutron, and neutron–neutron nuclear forces are equal). Calculate the difference in binding energy for the two mirror isobars $^{15}_8$O and $^{15}_7$N.

20. The energy required to construct a uniformly charged sphere of total charge Q and radius R is $U = 3k_eQ^2/5R$, where k_e is the Coulomb constant (see Problem 67). Assume that a ^{40}Ca nucleus contains 20 protons uniformly distributed in a spherical volume. (a) How much energy is required to counter the electrostatic repulsion given by the above equation? (*Hint:* First calculate the radius of a ^{40}Ca nucleus.) (b) Calculate the binding energy of ^{40}Ca. (c) Explain what you can conclude from comparing the result of part (b) and that of part (a).

21. Calculate the minimum energy required to remove a neutron from the $^{43}_{20}$Ca nucleus.

Section 44.4   Nuclear Models

22. (a) In the liquid-drop model of nuclear structure, why does the surface-effect term $-C_2A^{2/3}$ have a negative sign? (b) The binding energy of the nucleus increases as the volume-to-surface ratio increases. Calculate this ratio for both spherical and cubical shapes and explain which is more plausible for nuclei.

23. Using the graph in Figure 44.8, estimate how much energy is released when a nucleus of mass number 200 is split into two nuclei each of mass number 100.

24. (a) Use Equation 44.5 to compute the binding energy for $^{56}_{26}$Fe. (b) What percentage is contributed to the binding energy by each of the four terms?

Section 44.5   Radioactivity

25. A radioactive sample contains 1.00×10^{15} atoms and has an activity of 6.00×10^{11} Bq. What is its half-life?

26. The half-life of ^{131}I is 8.04 days. On a certain day, the activity of an iodine-131 sample is 6.40 mCi. What is its activity 40.2 days later?

WEB **27.** A freshly prepared sample of a certain radioactive isotope has an activity of 10.0 mCi. After 4.00 h, its activity is 8.00 mCi. (a) Find the decay constant and half-life. (b) How many atoms of the isotope were contained in the freshly prepared sample? (c) What is the sample's activity 30.0 h after it is prepared?

28. How much time elapses before 90.0% of the radioactivity of a sample of $^{72}_{33}$As disappears, as measured by its activity? The half-life of $^{72}_{33}$As is 26 h.

29. The radioactive isotope ^{198}Au has a half-life of 64.8 h. A sample containing this isotope has an initial activity ($t = 0$) of 40.0 μCi. Calculate the number of nuclei that decay in the time interval between $t_1 = 10.0$ h and $t_2 = 12.0$ h.

30. A radioactive nucleus has half-life $T_{1/2}$. A sample containing these nuclei has initial activity R_0. Calculate the number of nuclei that decay in the time interval between the times t_1 and t_2.

31. Determine the activity of 1.00 g of ^{60}Co. The half-life of ^{60}Co is 5.27 yr.

Section 44.6   The Decay Processes

32. Identify the missing nuclide or particle (X):
(a) X → $^{65}_{28}$Ni + γ
(b) $^{215}_{84}$Po → X + α
(c) X → $^{55}_{26}$Fe + e^+ + ν
(d) $^{109}_{48}$Cd + X → $^{109}_{47}$Ag + ν
(e) $^{14}_{11}$Na + $^4_2$He → X + $^{17}_8$O

33. Find the energy released in the alpha decay

$$^{238}_{92}\text{U} \longrightarrow {}^{234}_{90}\text{Th} + {}^4_2\text{He}$$

You will find the following mass values useful:

$$M(^{238}_{92}\text{U}) = 238.050\ 784\ \text{u}$$

$$M(^{234}_{90}\text{Th}) = 234.043\ 593\ \text{u}$$

$$M(^4_2\text{He}) = 4.002\ 602\ \text{u}$$

34. A living specimen in equilibrium with the atmosphere contains one atom of ^{14}C (half-life = 5 730 yr) for every 7.7×10^{11} stable carbon atoms. An archeological sample of wood (cellulose, $C_{12}H_{22}O_{11}$) contains 21.0 mg of carbon. When the sample is placed inside a shielded beta counter with 88.0% counting efficiency, 837 counts are accumulated in one week. Assuming that the cosmic-ray flux and the Earth's atmosphere have not changed appreciably since the sample was formed, find the age of the sample.

35. A ^{3}H nucleus beta decays into ^{3}He by creating an electron and an antineutrino according to the reaction

$$^3_1\text{H} \longrightarrow {}^3_2\text{He} + e^- + \bar{\nu}$$

Use Table A.3 to determine the total energy released in this reaction.

36. Determine which decays can occur spontaneously:
(a) $^{40}_{20}$Ca → e^+ + $^{40}_{19}$K

(b) $^{98}_{44}$Ru → $^4_2$He + $^{94}_{42}$Mo
(c) $^{144}_{60}$Nd → $^4_2$He + $^{140}_{58}$Ce

37. The nucleus $^{15}_8$O decays by electron capture. Write (a) the basic nuclear process and (b) the decay process referring to neutral atoms. (c) Determine the energy of the neutrino. Disregard the daughter's recoil.

Section 44.7   Natural Radioactivity

38. A rock sample contains traces of ^{238}U, ^{235}U, ^{232}Th, ^{208}Pb, ^{207}Pb, and ^{206}Pb. Careful analysis shows that the ratio of the amount of ^{238}U to ^{206}Pb is 1.164. (a) Assume that the rock originally contained no lead, and determine the age of the rock. (b) What should be the ratios of ^{235}U to ^{207}Pb and of ^{232}Th to ^{208}Pb so that they would yield the same age for the rock? Neglect the minute amounts of the intermediate decay products in the decay chains. Note that this form of multiple dating gives reliable geological dates.

39. Enter the correct isotope symbol in each open square in Figure P44.39, which shows the sequences of decays starting with uranium-235 and ending with the stable isotope lead-207.

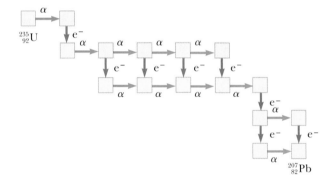

Figure P44.39

40. *Indoor air pollution.* Uranium is naturally present in rock and soil. At one step in its series of radioactive decays, ^{238}U produces the chemically inert gas radon-222, with a half-life of 3.82 days. The radon seeps out of the ground to mix into the atmosphere, typically making open air radioactive with activity 0.3 pCi/L. In homes ^{222}Rn can be a serious pollutant, accumulating to reach much higher activities in enclosed spaces. If the radon radioactivity exceeds 4 pCi/L, the Environmental Protection Agency suggests taking action to reduce it, by reducing infiltration of air from the ground. (a) Convert the activity 4 pCi/L to units of becquerel per cubic meter. (b) How many ^{222}Rn atoms are in one cubic meter of air displaying this activity? (c) What fraction of the mass of the air does the radon constitute?

41. The most common isotope of radon is ^{222}Rn, which has half-life 3.82 days. (a) What fraction of the nuclei that

were on Earth one week ago are now undecayed?
(b) What fraction of those that existed one year ago?
(c) In view of these results, explain why radon remains a problem, contributing significantly to our background radiation exposure.

Section 44.8   Nuclear Reactions

42. The reaction $^{27}_{13}\text{Al}(\alpha, n)^{30}_{15}\text{P}$, achieved in 1934, was the first known in which the product nucleus is radioactive. Calculate the Q value of this reaction.

WEB 43. Natural gold has only one isotope, $^{197}_{79}\text{Au}$. If natural gold is irradiated by a flux of slow neutrons, e^- particles are emitted. (a) Write the reaction equation. (b) Calculate the maximum energy of the emitted beta particles. The mass of $^{198}_{80}\text{Hg}$ is 197.966 743 u.

44. Identify the unknown particles X and X′ in the following nuclear reactions:
(a) $\text{X} + {}^4_2\text{He} \rightarrow {}^{24}_{12}\text{Mg} + {}^1_0\text{n}$
(b) $^{235}_{92}\text{U} + {}^1_0\text{n} \rightarrow {}^{90}_{38}\text{Sr} + \text{X} + 2{}^1_0\text{n}$
(c) $2{}^1_1\text{H} \rightarrow {}^2_1\text{H} + \text{X} + \text{X}'$

45. A beam of 6.61-MeV protons is incident on a target of $^{27}_{13}\text{Al}$. Those that collide produce the reaction

$$\text{p} + {}^{27}_{13}\text{Al} \longrightarrow {}^{27}_{14}\text{Si} + \text{n}$$

($^{27}_{14}\text{Si}$ has mass 26.986 721 u.) Neglecting any recoil of the product nucleus, determine the kinetic energy of the emerging neutrons.

46. (a) Suppose $^{10}_5\text{B}$ is struck by an alpha particle, releasing a proton and a product nucleus in the reaction. What is the product nucleus? (b) An alpha particle and a product nucleus are produced when $^{13}_6\text{C}$ is struck by a proton. What is the product nucleus?

47. Using the Q values of appropriate reactions from Table 44.5, calculate the masses of ^8Be and ^{10}Be in atomic mass units to four decimal places.

48. Determine the Q value associated with the spontaneous fission of ^{236}U into the fragments ^{90}Rb and ^{143}Cs, which have mass 89.914 811 u and 142.927 220 u, respectively. The masses of the other particles involved in the reaction are given in Appendix A.3.

ADDITIONAL PROBLEMS

49. Consider a radioactive sample. Determine the ratio of the number of atoms decaying during the first half of its half-life to the number of atoms decaying during the second half of its half-life.

50. One method of producing neutrons for experimental use is to bombard ^7_3Li with protons. The neutrons are emitted according to the reaction

$$^1_1\text{H} + {}^7_3\text{Li} \longrightarrow {}^7_4\text{Be} + {}^1_0\text{n}$$

What is the minimum kinetic energy the incident proton must have if this reaction is to occur? You may use the result of Problem 70.

51. A by-product of some fission reactors is the isotope $^{239}_{94}\text{Pu}$, an alpha emitter having a half-life of 24 120 years:

$$^{239}_{94}\text{Pu} \longrightarrow {}^{235}_{92}\text{U} + \alpha$$

Consider a sample of 1.00 kg of pure $^{239}_{94}\text{Pu}$ at $t = 0$. Calculate (a) the number of $^{239}_{94}\text{Pu}$ nuclei present at $t = 0$ and (b) the initial activity in the sample. (c) How long does the sample have to be stored if a "safe" activity level is 0.100 Bq?

52. (a) The atomic mass of ^{57}Co is 56.936 294 u. Can ^{57}Co decay by e^+ emission? Explain. (b) Can ^{14}C decay by e^- emission? Explain. (c) If either answer is yes, what is the range of kinetic energies available for the beta particle?

53. (a) Find the radius of the $^{12}_6\text{C}$ nucleus. (b) Find the force of repulsion between a proton at the surface of a $^{12}_6\text{C}$ nucleus and the remaining five protons. (c) How much work (in MeV) has to be done to overcome this electrostatic repulsion to put the last proton into the nucleus? (d) Repeat parts (a), (b), and (c) for $^{238}_{92}\text{U}$.

54. The activity of a radioactive sample was measured over 12 h, with the following net count rates:

Time (h)	Counting Rate (counts/min)
1.00	3 100
2.00	2 450
4.00	1 480
6.00	910
8.00	545
10.0	330
12.0	200

(a) Plot the logarithm of counting rate as a function of time. (b) Determine the disintegration constant and half-life of the radioactive nuclei in the sample. (c) What counting rate would you expect for the sample at $t = 0$? (d) Assuming the efficiency of the counting instrument to be 10.0%, calculate the number of radioactive atoms in the sample at $t = 0$.

55. (a) Why is the beta decay $\text{p} \rightarrow \text{n} + e^+ + \nu$ forbidden for a free proton? (b) Why is the same reaction possible if the proton is bound in a nucleus? For example, the following reaction occurs:

$$^{13}_7\text{N} \longrightarrow {}^{13}_6\text{C} + e^+ + \nu$$

(c) How much energy is released in the reaction given in part (b)? [$m(e^+) = 0.000\ 549$ u, $M(^{13}\text{C}) = 13.003\ 355$ u, $M(^{13}\text{N}) = 13.005\ 738$ u]

56. In a piece of rock from the Moon, the ^{87}Rb content is assayed to be 1.82×10^{10} atoms per gram of material, and the ^{87}Sr content is found to be 1.07×10^9 atoms per gram. (a) Calculate the age of the rock. (b) Could the material in the rock actually be much older? What assumption is implicit in using the radioactive dating method? (The relevant decay is $^{87}\text{Rb} \rightarrow {}^{87}\text{Sr} + e^-$. The half-life of the decay is 4.75×10^{10} yr.)

WEB 57. The decay of an unstable nucleus by alpha emission is represented by Equation 44.10. The disintegration energy Q given by Equation 44.13 must be shared by the alpha particle and the daughter nucleus in order to conserve both energy and momentum in the decay process. (a) Show that Q and K_α, the kinetic energy of the alpha particle, are related by the expression

$$Q = K_\alpha \left(1 + \frac{M_\alpha}{M} \right)$$

where M is the mass of the daughter nucleus. (b) Use the result of part (a) to find the energy of the alpha particle emitted in the decay of ^{226}Ra. (See Example 44.7 for the calculation of Q.)

58. The ^{145}Pm nucleus decays by alpha emission. (a) Determine the daughter nucleus. (b) Using the values given in Table A.3, determine the energy released in this decay. (c) What fraction of this energy is carried away by the alpha particle when the recoil of the daughter is taken into account?

59. When, after a reaction or disturbance of any kind, a nucleus is left in an excited state, it can return to its normal (ground) state by emission of a gamma-ray photon (or several photons). This process is illustrated by Equation 44.23. The emitting nucleus must recoil to conserve both energy and momentum. (a) Show that the recoil energy of the nucleus is

$$E_r = \frac{(\Delta E)^2}{2Mc^2}$$

where ΔE is the difference in energy between the excited and ground states of a nucleus of mass M. (b) Calculate the recoil energy of the ^{57}Fe nucleus when it decays by gamma emission from the 14.4-keV excited state. For this calculation, take the mass to be 57 u. (*Hint:* When writing the equation for conservation of energy, use $(Mv)^2/2M$ for the kinetic energy of the recoiling nucleus. Also, assume that $hf \ll Mc^2$ and use the binomial expansion.)

60. After the sudden release of radioactivity from the Chernobyl nuclear reactor accident in 1986, the radioactivity of milk in Poland rose to 2 000 Bq/L due to iodine-131, with half-life 8.04 days. Radioactive iodine is particularly hazardous, because the thyroid gland concentrates iodine. The Chernobyl accident caused a measurable increase in thyroid cancers among children in Belarus. (a) For comparison, find the activity of milk due to potassium. Assume that 1 L of milk contains 2.00 g of potassium, of which 0.011 7% is the isotope ^{40}K that has a half-life 1.28×10^9 yr. (b) After what time would the activity due to iodine fall below that due to potassium?

61. Europeans named a certain direction in the sky as between the horns of Taurus the Bull. On the day they named as July 4, 1054 A.D., a brilliant light appeared there. Europeans left no surviving record of the supernova, which could be seen in daylight for some days. It faded but remained visible for years, dimming for a time with the 77.1-day half-life of the radioactive cobalt-56 that had been created in the explosion. (a) The remains of the star now form the Crab Nebula. In it, the cobalt-56 has now decreased to what fraction of its original activity? (b) Suppose that an American, of the people called Anasazi, made a charcoal drawing of the supernova. The carbon-14 in the charcoal has now decayed to what fraction of its original activity?

62. A theory of nuclear astrophysics proposes that all the heavy elements, such as uranium, are formed in supernova explosions ending the lives of massive stars. If we assume that at the time of the explosion there were equal amounts of ^{235}U and ^{238}U, how long ago did the star(s) explode that released the elements that formed our Earth? The present ^{235}U/^{238}U ratio is 0.007 25. The half-lives of ^{235}U and ^{238}U are 0.704×10^9 years and 4.47×10^9 years.

63. After determining that the Sun has existed for hundreds of millions of years, but before the discovery of nuclear physics, scientists could not explain why the Sun has continued to burn for such a long time. For example, if it were a coal fire, it would have burned up in about 3 000 years. Assume that the Sun, whose mass is 1.99×10^{30} kg, originally consisted entirely of hydrogen and that its total power output is 3.77×10^{26} W. (a) If the energy-generating mechanism of the Sun is the transforming of hydrogen into helium via the net reaction

$$4 {}^{1}_{1}\text{H} + 2 {}^{0}_{-1}\text{e} \longrightarrow {}^{4}_{2}\text{He} + 2\nu + \gamma$$

calculate the energy (in joules) given off by this reaction. (b) Determine how many hydrogen atoms constitute the Sun. Take the mass of one hydrogen atom to be 1.67×10^{-27} kg. (c) Assuming that the total power output remains constant, after what time will all the hydrogen be converted into helium, making the Sun die? The actual projected lifetime of the Sun is about 10 billion years, because only the hydrogen in a relatively small core is available as a fuel. Only in the core are temperatures and densities high enough for the fusion reaction to be self-sustaining.

64. (a) One method of producing neutrons for experimental use is bombardment of light nuclei with alpha particles. In one particular arrangement, alpha particles emitted by polonium are incident on beryllium nuclei:

$$ {}^{4}_{2}\text{He} + {}^{9}_{4}\text{Be} \longrightarrow {}^{12}_{6}\text{C} + {}^{1}_{0}\text{n}$$

What is the Q value? (b) Neutrons are also often produced by small-particle accelerators. In one design, deuterons accelerated in a Van de Graaff generator bombard other deuterium nuclei:

$$ {}^{2}_{1}\text{H} + {}^{2}_{1}\text{H} \longrightarrow {}^{3}_{2}\text{He} + {}^{1}_{0}\text{n}$$

Is this reaction exothermic or endothermic? Calculate its Q value.

65. **Review Problem.** Consider the Bohr model of the hydrogen atom, with the electron in the ground state. The magnetic field at the nucleus produced by the orbiting electron has a value of 12.5 T (see Chapter 30, Problem 1). The proton can have its magnetic moment aligned in either of two directions perpendicular to the plane of the electron's orbit. Because of the interaction of the proton's magnetic moment with the electron's magnetic field, there will be a difference in energy between the states with the two different orientations of the proton's magnetic moment. Find that energy difference in eV.

66. Many radioisotopes have important industrial, medical, and research applications. One of these is ^{60}Co, which has a half-life of 5.27 years and decays by the emission of a beta particle (energy 0.31 MeV) and two gamma photons (energies 1.17 MeV and 1.33 MeV). A scientist wishes to prepare a ^{60}Co sealed source that will have an activity of 10.0 Ci after 30.0 months of use. (a) What is the initial mass of ^{60}Co required? (b) At what rate will the source emit energy after 30.0 months?

67. **Review Problem.** Consider a model of the nucleus in which the positive charge (Ze) is uniformly distributed throughout a sphere of radius R. By integrating the energy density, $\frac{1}{2}\epsilon_0 E^2$, over all space, show that the electrostatic energy may be written

$$U = \frac{3Z^2 e^2}{20\pi\epsilon_0 R}$$

68. The ground state of $^{93}_{43}$Tc (molar mass, 92.910 2 g/mol) decays by electron capture and e^+ emission to energy levels of the daughter (molar mass in ground state, 92.906 8 g/mol) at 2.44 MeV, 2.03 MeV, 1.48 MeV, and 1.35 MeV. (a) For which of these levels are electron capture and e^+ decay allowed? (b) Identify the daughter and sketch the decay scheme, assuming all excited states de-excite by direct γ decay to the ground state.

69. Free neutrons have a characteristic half-life of 10.4 min. What fraction of a group of free neutrons with kinetic energy 0.040 0 eV will decay before traveling a distance of 10.0 km?

70. When the nuclear reaction represented by Equation 44.26 is endothermic, the disintegration energy Q is negative. For the reaction to proceed, the incoming particle must have a minimum energy called the threshold energy, E_{th}. Some fraction of the energy of the incident particle is transferred to the compound nucleus to conserve momentum. Therefore, E_{th} must be greater than Q. (a) Show that

$$E_{th} = -Q\left(1 + \frac{M_a}{M_X}\right)$$

(b) Calculate the threshold energy of the incident alpha particle in the reaction

$$^4_2\text{He} + {}^{14}_7\text{N} \longrightarrow {}^{17}_8\text{O} + {}^1_1\text{H}$$

71. *Student determination of the half-life of ^{137}Ba.* The radioactive barium isotope ^{137}Ba has a relatively short half-life and can be easily extracted from a solution containing radioactive cesium (^{137}Cs). This barium isotope is commonly used in an undergraduate laboratory exercise for demonstrating the radioactive decay law. The data presented in Figure P45.71 were taken by undergraduate students using modest experimental equipment. Determine the half-life for the decay of ^{137}Ba using their data.

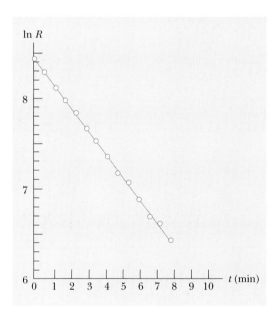

Figure P44.71

---

force. Therefore, a given nucleon interacts only with its nearest neighbors rather than with all other nucleons in the nucleus. No matter how many nucleons are present, pulling one out involves separating it only from its nearest neighbors. The energy to do this is therefore approximately independent of how many nucleons are present.

The electric force binding the electrons in an atom to the nucleus is a long-range force, and so each electron interacts with all the protons in the nucleus. More protons in the nucleus thus means a stronger electron-nucleus attraction. As a result, the energy needed to remove an electron from the atom varies with atomic number.

44.4 No. The negative of the slope of a potential energy–versus–position graph is force. For nucleon separation distances of less than 1 fm, both slopes in Figure 44.9 are negative and so the force is positive, meaning the particles repel each other. From 1 fm to 3 fm the slope is positive, indicating an attractive force (of very short range). If a tangent to the graph were vertical at some point, the force would have to be infinite—a physical impossibility.

44.5 Figure 44.15 shows that the higher the energy of the alpha particle, the thinner the potential barrier. The thinner barrier translates to a higher probability of escape. The higher probability of escape translates to a faster rate of decay, which appears as a shorter half-life.

44.6 In alpha decay, there are only two products—the alpha particle and the daughter nucleus. There are also two conservation principles involved—energy and linear momentum. As a result, the alpha particle must be ejected with a discrete energy to satisfy both conservation principles. There are a small number of discrete energies of the alpha particle, as the daughter nucleus can be left in various excited states, but the allowed energies of the alpha particle are not continuous.

In beta decay, we have the same two conservation principles but three products—the beta particle, the daughter nucleus, and the neutrino. There are many ways that the energy can be divided among the three particles to satisfy the two conservation principles, and as a result the beta particle is emitted over a continuous range of energies.

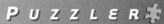

PUZZLER

Nearly everything on the Earth gets its energy from the Sun. Plants use sunlight to make carbohydrates, which are a source of energy for animals. The wind blows because of solar heating of the atmosphere. Automobiles are powered by fossil fuels, which are essentially stored solar energy. All these things get their energy from the Sun, but what is the source of the Sun's energy? *(European Space Agency/Science Photo Library/Photo Researchers, Inc.)*

c h a p t e r

45

Nuclear Fission and Fusion

Chapter Outline

*I*n this chapter, we are concerned primarily with the two means by which energy can be derived from nuclear reactions: fission, in which a large nucleus splits (or fissions) into two smaller nuclei, and fusion, in which two small nuclei fuse to form a larger one. In either case, there is a release of energy that can be used either destructively (as in bombs) or constructively (as in the production of electric power). We also examine the ways in which radiation interacts with matter and look at several devices used to detect radiation. The chapter concludes with a discussion of some industrial and biological applications of radiation.

45.1 ► INTERACTIONS INVOLVING NEUTRONS

To understand nuclear fission and the physics of nuclear reactors, we must first understand how neutrons interact with nuclei. Because of their charge neutrality, neutrons are not subject to Coulomb forces and as a result do not interact electrically with electrons. Therefore, because any piece of matter consists of electrons orbiting tiny atomic nuclei, matter appears quite "open" to free neutrons.

In general, the rate of neutron-induced reactions increases as the neutron kinetic energy decreases. Free neutrons undergo beta decay with a mean lifetime of about 10 min. Once free neutrons enter matter, however, many of them are absorbed by atomic nuclei and are stabilized from decay by the nuclear force of other nucleons.

A **fast neutron** (energy greater than about 1 MeV) traveling through matter undergoes many scattering events with the nuclei. In each event, the neutron gives up some of its kinetic energy to a nucleus. Once the neutron energy is sufficiently low, there is a high probability that the neutron will be captured by a nucleus, an event that is accompanied by the emission of a gamma ray. This **neutron capture** can be written

Neutron capture

$$\,^1_0n + \,^A_ZX \longrightarrow \,^{A+1}_ZX^* \longrightarrow \,^{A+1}_ZX + \gamma \qquad\qquad \textbf{(45.1)}$$

Once the neutron is captured, the nucleus $^{A+1}_ZX^*$ is in an excited state for a very short time before it undergoes gamma decay. Also, the product nucleus $^{A+1}_ZX$ is usually radioactive and decays by beta emission.

The neutron-capture rate as neutrons pass through any sample depends on which atoms are contained in the sample and on the energy of the incident neutrons. In addition, the capture rate also depends on the type of atoms in the sample. For some materials and for fast neutrons, elastic collisions dominate. Materials for which this occurs are called **moderators** because they slow down (or moderate) the originally energetic neutrons very effectively. The interaction of neutrons with matter increases with decreasing neutron energy because a slow neutron spends more time in the vicinity of target nuclei. A good moderator should be composed of nuclei that have a low tendency to capture fast neutrons. Moderator nuclei should be of low mass so that more kinetic energy is transferred to them in elastic collisions. For this reason, materials that are abundant in hydrogen, such as paraffin and water, are good moderators for neutrons.

Moderator

Quick Quiz 45.1 ►

What would be the ideal target particle in a neutron moderator if we were trying to stop the incoming neutrons completely?

Sooner or later, most neutrons bombarding a moderator become **thermal neutrons,** which means they are in thermal equilibrium with the moderator material. Their average kinetic energy at room temperature is, from Equation 21.4,

$$K_{av} = \tfrac{3}{2}k_{B}T \approx \tfrac{3}{2}k_{B}(300 \text{ K}) \approx 0.04 \text{ eV}$$

which corresponds to a neutron root-mean-square speed of about 2 800 m/s. Thermal neutrons have a distribution of speeds, just as the molecules in a container of gas do (see Chapter 21). A high-energy neutron, one whose energy is several MeV, *thermalizes* (that is, reaches K_{av}) in less than 1 ms when incident on a moderator. Thermal neutrons have a very high probability of being captured by the moderator nuclei.

45.2 NUCLEAR FISSION

As we saw in Section 44.3, nuclear **fission** occurs when a heavy nucleus, such as ^{235}U, splits into two smaller nuclei. In such a reaction, **the combined mass of the daughter nuclei is less than the mass of the parent nucleus,** and the difference in mass is called the **mass defect.** Fission is initiated when a heavy nucleus captures a thermal neutron. Multiplying the mass defect by c^{2} gives the numerical value of the released energy. Energy is released because the binding energy per nucleon of the daughter nuclei is about 1 MeV greater than that of the parent nucleus (see Fig. 44.8).

Nuclear fission was first observed in 1938 by Otto Hahn (1879–1968) and Fritz Strassman (b. 1902) following some basic studies by Fermi. After bombarding uranium ($Z = 92$) with neutrons, Hahn and Strassman discovered among the reaction products two medium-mass elements, barium and lanthanum. Shortly thereafter, Lise Meitner (1878–1968) and her nephew Otto Frisch (1904–1979) explained what had happened. The uranium nucleus had split into two nearly equal fragments after absorbing a neutron. Such an occurrence was of considerable interest to physicists attempting to understand the nucleus, but it was to have even more far-reaching consequences. Measurements showed that about 200 MeV of energy was released in each fission event, and this fact was to affect the course of history.

The fission of ^{235}U by thermal neutrons can be represented by the equation

$$_{0}^{1}\text{n} + _{92}^{235}\text{U} \longrightarrow _{92}^{236}\text{U}^{*} \longrightarrow \text{X} + \text{Y} + \text{neutrons} \qquad \textbf{(45.2)}$$

where ^{236}U* is an intermediate excited state that lasts only for about 10^{-12} s before splitting into nuclei X and Y, which are called **fission fragments.** In any fission equation, there are many combinations of X and Y that satisfy the requirements of conservation of energy and charge. With uranium, for example, there are about 90 daughter nuclei that can be formed.

Fission also results in the production of several neutrons, typically two or three. On the average, about 2.5 neutrons are released per event. A typical fission reaction for uranium is

$$_{0}^{1}\text{n} + _{92}^{235}\text{U} \longrightarrow _{56}^{141}\text{Ba} + _{36}^{92}\text{Kr} + 3(_{0}^{1}\text{n}) \qquad \textbf{(45.3)}$$

The breakup of the uranium nucleus can be compared to what happens to a drop of water when excess energy is added to it. (Recall the liquid-drop model of the nucleus described in Section 44.4.) Initially, all the atoms in the drop have some energy, but this is not enough to break up the drop. However, if enough energy is added to set the drop into vibration, the drop elongates and compresses

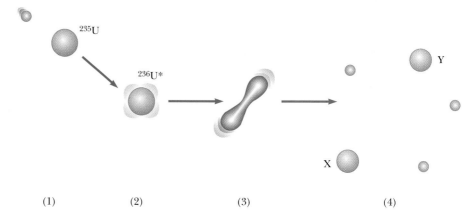

Figure 45.1   Nuclear fission event as described by the liquid-drop model of the nucleus.

until the amplitude of vibration becomes large enough to cause it to break. In the uranium nucleus, a similar process occurs (Fig. 45.1):

Stage 1.   The ^{235}U nucleus captures a thermal neutron.
Stage 2.   This capture results in the formation of ^{236}U*, and the excess energy of this nucleus causes it to oscillate violently.
Stage 3.   The ^{236}U* nucleus becomes highly distorted, and the force of repulsion between protons in the two halves of the dumbbell shape tends to increase the distortion.
Stage 4.   The nucleus splits into two fragments, emitting several neutrons in the process.

Quick Quiz 45.2

Which of the following is true for stage 3 of a ^{235}U fission event relative to stages 1 and 2? (Refer to Fig. 45.1.) (a) Both the nuclear force and the electrostatic force are smaller. (b) Both forces are greater. (c) The nuclear force is greater and the electrostatic force is smaller. (d) The nuclear force is smaller and the electrostatic force is greater.

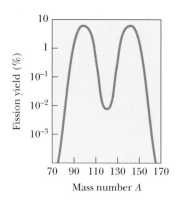

Figure 45.2   Distribution of fission products versus mass number for the fission of ^{235}U bombarded with thermal neutrons. Note that the ordinate scale is logarithmic.

Figure 45.2 is a graph of the distribution of fission products versus mass number A. The most probable products have mass numbers $A \approx 140$ and $A \approx 95$, both of which fall to the left of the stability line shown in Figure 44.3, meaning they contain more neutrons than protons. **These fragments, because they are unstable owing to their excess of neutrons, almost instantaneously release two or three neutrons.** Fragments having values of A other than 140 and 95, but still rich in neutrons, decay to more stable nuclei through a succession of e^- decays, emitting gamma rays in the process.

Let us estimate the disintegration energy Q released in a typical fission process. From Figure 44.8 we see that the binding energy per nucleon is about 7.2 MeV for heavy nuclei ($A \approx 240$) and about 8.2 MeV for nuclei of intermediate mass. This means that the nucleons in fission fragments are more tightly bound and therefore have less mass than the nucleons in a parent nucleus. This decrease in nucleon mass appears as released energy when fission occurs. The amount of energy released is $(8.2 - 7.2)$ MeV per nucleon. Assuming a total of 240 nucleons, we find that the energy released per fission event is

$$Q = (240 \text{ nucleons})\left(8.2 \frac{\text{MeV}}{\text{nucleon}} - 7.2 \frac{\text{MeV}}{\text{nucleon}}\right) = 240 \text{ MeV}$$

This is a very large amount of energy relative to the amount released in chemical processes. For example, the energy released in the combustion of one molecule of octane used in gasoline engines is about one-millionth the energy released in a single fission event!

Quick Quiz 45.3

If a heavy nucleus were to fission into just two daughter nuclei, they would be unstable. Why?

Quick Quiz 45.4

Which of the following are possible fission reactions?
(a) $_{0}^{1}n + _{92}^{235}U \rightarrow _{54}^{140}Xe + _{38}^{94}Sr + 2(_{0}^{1}n)$
(b) $_{0}^{1}n + _{92}^{235}U \rightarrow _{50}^{132}Sn + _{42}^{101}Mo + 3(_{0}^{1}n)$
(c) $_{0}^{1}n + _{94}^{239}Pu \rightarrow _{53}^{127}I + _{41}^{93}Nb + 3(_{0}^{1}n)$

EXAMPLE 45.1 The Energy Released in the Fission of ^{235}U

Calculate the energy released when 1.00 kg of ^{235}U fissions, taking the disintegration energy per event to be $Q = 208$ MeV.

Solution We need to know the number of nuclei in 1.00 kg of uranium. Because $A = 235$, we know that the molar mass of this isotope is 235 g/mol. Therefore, the number of nuclei in our sample is

$$N = \left(\frac{6.02 \times 10^{23} \text{ nuclei/mol}}{235 \text{ g/mol}}\right)(1.00 \times 10^{3} \text{ g})$$

$$= 2.56 \times 10^{24} \text{ nuclei}$$

Hence, the total disintegration energy is

$$E = NQ = (2.56 \times 10^{24} \text{ nuclei})\left(208 \frac{\text{MeV}}{\text{nucleus}}\right)$$

$$= 5.32 \times 10^{26} \text{ MeV}$$

Because 1 MeV is equivalent to 4.45×10^{-20} kWh, we find that $E = 2.37 \times 10^{7}$ kWh. This is enough energy to keep a 100-W lightbulb burning for 30 000 years! If the energy in 1 kg of ^{235}U were suddenly released, it would be equivalent to detonating about 20 000 tons of TNT.

45.3 NUCLEAR REACTORS

In the preceding section, we learned that, when ^{235}U fissions, an average of 2.5 neutrons are emitted per event. These neutrons can in turn trigger other nuclei to fission, with the possibility of a chain reaction (Fig. 45.3). Calculations show that if the chain reaction is not controlled (that is, if it does not proceed slowly), it can result in a violent explosion, with the release of an enormous amount of energy. This is the principle behind the first type of nuclear bomb exploded in 1945, an uncontrolled fission reaction. When the reaction is controlled, however, the energy released can be put to less destructive use. In the United States, for example, nearly 20% of the electricity generated each year comes from nuclear power plants, and nuclear power is used extensively in many countries, including France, Japan, and Germany.

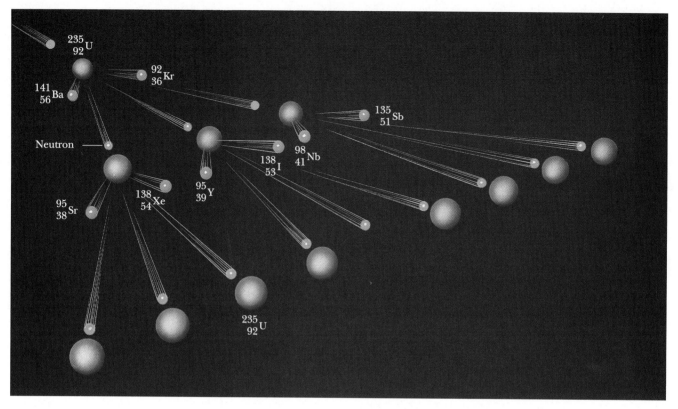

Figure 45.3   A nuclear chain reaction initiated by the capture of a neutron.

Chain reaction

A nuclear reactor is a system designed to maintain what is called a **self-sustained chain reaction.** This important process was first achieved in 1942 by Enrico Fermi and his team at the University of Chicago, with naturally occurring uranium as the fuel.[1] Most reactors in operation today also use uranium as fuel. However, naturally occurring uranium contains only about 0.7% of the ^{235}U isotope, with the remaining 99.3% being ^{238}U. This fact is important to the operation of a reactor because ^{238}U almost never fissions. Instead, it tends to absorb neutrons, producing neptunium and plutonium. For this reason, reactor fuels must be artificially *enriched* to contain at least a few percent ^{235}U.

To achieve a self-sustained chain reaction, an average of one neutron emitted in each ^{235}U fission must be captured by another ^{235}U nucleus and cause that nucleus to undergo fission. A useful parameter for describing the level of reactor operation is the **reproduction constant** K, defined as **the average number of neutrons from each fission event that cause another fission event.** As we have seen, K has an average value of 2.5 in the fission of uranium. However, in practice K is less than this because of several factors discussed in the following paragraphs.

A self-sustained chain reaction is achieved when $K = 1$. Under this condition, the reactor is said to be **critical.** When $K < 1$, the reactor is subcritical and the re-

Reproduction constant

---

[1] Although Fermi's reactor was the first manufactured nuclear reactor, there is evidence that a natural fission reaction may have sustained itself for perhaps hundreds of thousands of years in a deposit of uranium in Gabon, West Africa. See G. Cowan, "A Natural Fission Reactor," *Sci. Am.* 235(5):36, 1976.

action dies out. When $K > 1$, the reactor is supercritical and a runaway reaction occurs. In a nuclear reactor used to furnish power to a utility company, it is necessary to maintain a value of K slightly greater than unity.

In any reactor, a fraction of the neutrons produced in fission leak out of the core before inducing other fission events. If the fraction leaking out is too large, the reactor will not operate. The percentage lost is large if the reactor is very small because leakage is a function of the ratio of surface area to volume. Therefore, a critical feature of the reactor design is an optimal surface area-to-volume ratio.

The neutrons released in fission events are very energetic, having kinetic energies of about 2 MeV. Because the probability of neutron capture increases with decreasing energy, it is necessary to slow these neutrons to thermal energies if they are to be captured and cause other ^{235}U nuclei to fission (see Example 9.8). The energetic neutrons are slowed down by a moderator substance surrounding the fuel.

In the first nuclear reactor ever constructed (Fig. 45.4), Fermi placed bricks of graphite (carbon) between the fuel elements. Carbon nuclei are about 12 times more massive than neutrons, but after several collisions with carbon nuclei, a neutron is slowed sufficiently to increase its likelihood of fission with ^{235}U. In this design, carbon is the moderator; most modern reactors use water as the moderator.

In the process of slowing down, neutrons may be captured by nuclei that do not fission. The most common event of this type is neutron capture by ^{238}U, which constitutes more than 90% of the uranium in the fuel elements. The probability of neutron capture by ^{238}U is very high when the neutrons have high kinetic energies and very low when they have low kinetic energies. Thus, the slowing down of the neutrons by the moderator serves the secondary purpose of making them available for reaction with ^{235}U and decreasing their chances of being captured by ^{238}U.

Enrico Fermi (1901–1954)
Fermi, an Italian physicist, was awarded the Nobel Prize in 1938 for producing transuranic elements by neutron irradiation and for his discovery of nuclear reactions brought about by thermal neutrons. He made many other outstanding contributions to physics, including his theory of beta decay, the free-electron theory of metals, and the development of the world's first fission reactor in 1942. Fermi was truly a gifted theoretical and experimental physicist. He was also well known for his ability to present physics in a clear and exciting manner. "Whatever Nature has in store for mankind, unpleasant as it may be, men must accept, for ignorance is never better than knowledge." *(National Accelerator Laboratory)*

Figure 45.4 Artist's rendition of the world's first nuclear reactor. Because of wartime secrecy, there are few photographs of the completed reactor, which was composed of layers of moderating graphite interspersed with uranium. A self-sustained chain reaction was first achieved on December 2, 1942. Word of the success was telephoned immediately to Washington with this message: "The Italian navigator has landed in the New World and found the natives very friendly." The historic event took place in an improvised laboratory in the racquet court under the stands of the University of Chicago's Stagg Field, and the Italian navigator was Enrico Fermi. *(Courtesy of Chicago Historical Society)*

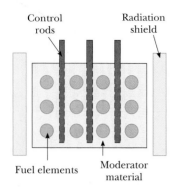

Figure 45.5 Cross-section of a reactor core showing the control rods, fuel elements containing enriched fuel, and moderating material, all surrounded by a radiation shield.

Control of Power Level

It is possible for a reactor to reach the critical stage ($K = 1$) after all the neutron losses described previously are minimized. However, a method of control is needed to maintain a K value near unity. If K rises above this value, the internal energy produced in the reaction could melt the reactor.

The basic design of a nuclear reactor core is shown in Figure 45.5. The fuel elements consist of uranium that has been enriched in the ^{235}U isotope. To control the power level, control rods are inserted into the reactor core. These rods are made of materials, such as cadmium, that are very efficient in absorbing neutrons. By adjusting the number and position of the control rods in the reactor core, the K value can be varied and any power level within the design range of the reactor can be achieved.

Several types of reactor systems convert the kinetic energy of fission fragments to electrical energy. The most common reactor in use in the United States is the pressurized-water reactor (Fig. 45.6), and we shall examine this type because its main parts are common to all reactor designs. Fission events in the reactor core raise the temperature of the water contained in the primary (closed) loop, which is maintained at high pressure to keep the water from boiling. (This water also serves as the moderator.) The hot water is pumped through a heat exchanger, where the internal energy of the water is transferred to the water contained in the secondary loop. The hot water in the secondary loop is converted to steam, which drives a turbine–generator system to create electric power. The water in the secondary loop is isolated from the water in the primary loop to avoid contamination of the secondary water and the steam by radioactive nuclei from the reactor core.

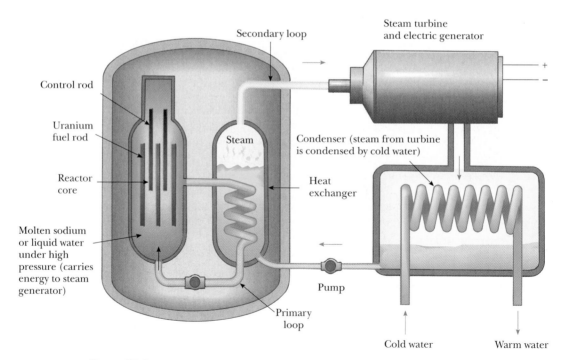

Figure 45.6 Main components of a pressurized-water reactor.

Safety and Waste Disposal

The 1979 near-disaster at a nuclear plant at Three Mile Island in Pennsylvania and the 1986 accident at the Chernobyl reactor in Ukraine rightfully focused attention on reactor safety. The Three Mile Island accident was the result of inadequate control-room instrumentation and poor emergency-response training. There were no injuries or detectable health impacts from the event, even though more than one third of the fuel melted.

This unfortunately was not the case at Chernobyl, where the activity of the materials released immediately after the accident totaled approximately 12×10^{18} Bq and resulted in the evacuation of 116 000 people. At least 237 people suffered from acute radiation sickness and about 800 children later contracted thyroid cancer from the ingestion of radioactive iodine in milk from cows that ate contaminated grass. One conclusion of an international conference studying the Ukraine accident was that "the main causes of the Chernobyl accident were the coincidence of severe deficiencies in the reactor physical design and in the design of the shut-down system and a violation of procedures." Most of these deficiencies have been addressed at plants of similar design in Russia and neighboring countries of the former Soviet Union.

There are no plants of the Chernobyl type in the United States. Many U. S. plants are of the pressurized-water design, as noted earlier.

Commercial reactors achieve safety through careful design and rigid operating protocol, and it is only when these variables are compromised that reactors pose a danger. Radiation exposure and the potential health risks associated with such exposure are controlled by three layers of containment. The fuel and radioactive fission products are contained inside the reactor vessel. Should this vessel rupture, the reactor building acts as a second containment structure to prevent radioactive material from contaminating the environment. Finally, the reactor facilities must be in a remote location to protect the general public from exposure should radiation escape the reactor building.

A continuing concern about nuclear fission reactors is the safe disposal of radioactive material when the reactor core is replaced. This waste material contains long-lived, highly radioactive isotopes and must be stored over long periods of time in such a way that there is no chance of environmental contamination. At present, sealing radioactive wastes in waterproof containers and burying them in deep salt mines seems to be the most promising solution.

Transport of reactor fuel and reactor wastes poses additional safety risks. Accidents during transport of nuclear fuel could expose the public to harmful levels of radiation. The Department of Energy requires stringent crash tests of all containers used to transport nuclear materials. Container manufacturers must demonstrate that their containers will not rupture even in high-speed collisions.

The safety issues associated with nuclear power reactors are complex and often emotional. All sources of energy have associated risks. In each case, the risks must be weighed against the benefits and the availability of the energy source.

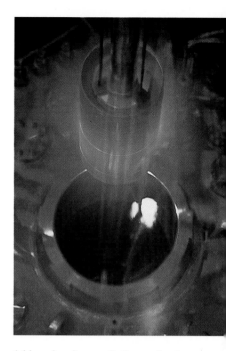

A blue glow from radiation emitted by a fuel element of a reactor at Oak Ridge National Laboratory. The glow results when energetic particles move through the water faster than the speed of light in that medium. *(U.S. Department of Energy/Photo Researchers, Inc.)*

45.4 NUCLEAR FUSION

In Chapter 44 we found that the binding energy for light nuclei ($A < 20$) is much smaller than the binding energy for heavier nuclei. This suggests a process that is the reverse of fission. As we saw in Section 39.8, when two light nuclei combine to form a heavier nucleus, the process is called nuclear **fusion.** Because the mass of the final nucleus is less than the combined masses of the original nuclei, there is a loss of mass accompanied by a release of energy.

Two examples of such energy-liberating fusion reactions are

$$^1_1\text{H} + {}^1_1\text{H} \longrightarrow {}^2_1\text{H} + e^+ + \nu$$

and

$$^1_1\text{H} + {}^2_1\text{H} \longrightarrow {}^3_2\text{He} + \gamma$$

This second reaction is followed by either hydrogen–helium fusion or helium–helium fusion:

$$^1_1\text{H} + {}^3_2\text{He} \longrightarrow {}^4_2\text{He} + e^+ + \nu$$

$$^3_2\text{He} + {}^3_2\text{He} \longrightarrow {}^4_2\text{He} + {}^1_1\text{H} + {}^1_1\text{H}$$

These are the basic reactions in the **proton–proton cycle,** believed to be one of the basic cycles by which energy is generated in the Sun and other stars that contain an abundance of hydrogen. Most of the energy production takes place in the Sun's interior, where the temperature is approximately 1.5×10^7 K. As we shall see later, such high temperatures are required to drive these reactions, and they are therefore called **thermonuclear fusion reactions.** The hydrogen (fusion) bomb, first exploded in 1952, is an example of an uncontrolled thermonuclear fusion reaction. It uses a fission bomb as the "trigger" to create the high temperatures needed for fusion.

All of the reactions in the proton–proton cycle are exothermic. An overview of the cycle is that four protons combine to form an alpha particle and two positrons, with the release of 25 MeV of energy.

Thermonuclear reaction

Fusion Reactions

The enormous amount of energy released in fusion reactions suggests the possibility of harnessing this energy for useful purposes. A great deal of effort is currently under way to develop a sustained and controllable thermonuclear reactor—a fusion power reactor. Controlled fusion is often called the ultimate energy source because of the availability of its fuel source: water. For example, if deuterium were used as the fuel, 0.12 g of it could be extracted from 1 gal of water at a cost of about four cents. Such rates would make the fuel costs of even an inefficient reactor almost insignificant. An additional advantage of fusion reactors is that comparatively few radioactive by-products are formed. For the proton-proton cycle, for instance, the end product is safe, nonradioactive helium. Unfortunately, a thermonuclear reactor that can deliver a net power output spread out over a reasonable time interval is not yet a reality, and many difficulties must be resolved before a successful device is constructed.

The Sun's energy is based, in part, upon a set of reactions in which hydrogen is converted to helium. Unfortunately, the proton–proton interaction is not suitable for use in a fusion reactor because the event requires very high pressures and densities. The process works in the Sun only because of the extremely high density of protons in the Sun's interior.

The reactions that appear most promising for a fusion power reactor involve deuterium (^2_1H) and tritium (^3_1H):

$$^2_1\text{H} + {}^2_1\text{H} \longrightarrow {}^3_2\text{He} + {}^1_0\text{n} \qquad Q = 3.27 \text{ MeV}$$

$$^2_1\text{H} + {}^2_1\text{H} \longrightarrow {}^3_1\text{H} + {}^1_1\text{H} \qquad Q = 4.03 \text{ MeV} \qquad \textbf{(45.4)}$$

$$^2_1\text{H} + {}^3_1\text{H} \longrightarrow {}^4_2\text{He} + {}^1_0\text{n} \qquad Q = 17.59 \text{ MeV}$$

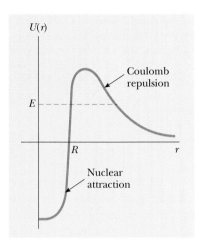

Figure 45.7   Potential energy as a function of separation distance between two deuterons. The Coulomb repulsive force is dominant at long range, and the nuclear force is dominant at short range, where R is of the order of 1 fm. If we neglect tunneling, to undergo fusion the two deuterons require an energy E greater than the height of the barrier.

As noted earlier, deuterium is available in almost unlimited quantities from our lakes and oceans and is very inexpensive to extract. Tritium, however, is radioactive ($T_{1/2} = 12.3$ years) and undergoes beta decay to ^{3}He. For this reason, tritium does not occur naturally to any great extent and must be artificially produced.

One of the major problems in obtaining energy from nuclear fusion is that the Coulomb repulsive force between two charged nuclei must be overcome before they can fuse. Potential energy as a function of the separation distance between two deuterons (deuterium nuclei, each having charge $+ e$) is shown in Figure 45.7. The potential energy is positive in the region $r > R$, where the Coulomb repulsive force dominates ($R \approx 1$ fm), and negative in the region $r < R$, where the nuclear force dominates. The fundamental problem then is to give the two nuclei enough kinetic energy to overcome this repulsive force. This can be accomplished by raising the fuel to extremely high temperatures (to about 10^8 K, far greater than the interior temperature of the Sun). At these high temperatures, the atoms are ionized and the system consists of a collection of electrons and nuclei, commonly referred to as a *plasma*.

> High temperatures are required to overcome the large Coulomb barrier

EXAMPLE 45.2   The Fusion of Two Deuterons

The separation distance between two deuterons must be about 1.0×10^{-14} m in order for the nuclear force to overcome the repulsive Coulomb force. (a) Calculate the height of the potential barrier due to the repulsive force.

Solution   The potential energy associated with two charges separated by a distance r is, from Equation 25.13,

$$U = k_e \frac{q_1 q_2}{r}$$

where k_e is the Coulomb constant. For the case of two deuterons, $q_1 = q_2 = + e$, so that

$$U = k_e \frac{e^2}{r} = \left(8.99 \times 10^9 \ \frac{\text{N} \cdot \text{m}^2}{\text{C}^2} \right) \frac{(1.60 \times 10^{-19} \ \text{C})^2}{1.0 \times 10^{-14} \ \text{m}}$$

$$= 2.3 \times 10^{-14} \text{J} = \boxed{0.14 \text{ MeV}}$$

(b) Estimate the temperature required for a deuteron to overcome the potential barrier, assuming an energy of $\frac{3}{2} k_B T$ per deuteron (where k_B is Boltzmann's constant).

Solution   Because the total Coulomb energy of the pair is 0.14 MeV, the Coulomb energy per deuteron is 0.07 MeV = 1.1×10^{-14} J. Setting this energy equal to the average energy per deuteron gives

$$\tfrac{3}{2} k_B T = 1.1 \times 10^{-14} \text{J}$$

Solving for T gives

$$T = \frac{2(1.1 \times 10^{-14} \text{J})}{3(1.38 \times 10^{-23} \text{J/K})} = \boxed{5.3 \times 10^8 \text{ K}}$$

This calculated temperature is too high because the particles in the plasma have a Maxwellian speed distribution, and therefore some fusion reactions are caused by particles in the high-energy tail of this distribution. Furthermore, even those particles that do not have enough energy to overcome the barrier have some probability of tunneling through. When these effects are taken into account, a temperature of "only" 4×10^8 K appears adequate to fuse the two deuterons.

(c) Find the energy released in the deuterium–deuterium reaction

$$^2_1\text{H} + {}^2_1\text{H} \longrightarrow {}^3_1\text{H} + {}^1_1\text{H}$$

Solution The mass of a single deuterium atom is equal to 2.014 102 u. Thus, the total mass before the reaction is 4.028 204 u. After the reaction, the sum of the masses is 3.016 049 u + 1.007 825 u = 4.023 874 u. The excess mass is 0.004 33 u, equivalent to 4.03 MeV, as was noted in Equation 45.4.

Critical ignition temperature

The temperature at which the power generation rate in any fusion reaction exceeds the loss rate (due to mechanisms such as radiation losses) is called the **critical ignition temperature** T_{ignit}. This temperature for the deuterium–deuterium (D–D) reaction is 4×10^8 K. From the relationship $E \approx k_B T$, this temperature is equivalent to approximately 35 keV. It turns out that the critical ignition temperature for the deuterium–tritium (D–T) reaction is about 4.5×10^7 K, or only 4 keV. A plot of the power $\mathscr{P}_{\text{gen}}$ generated by fusion versus temperature for the two reactions is shown in Figure 45.8. The green straight line represents the power $\mathscr{P}_{\text{lost}}$ lost via the radiation mechanism known as **bremsstrahlung.** In this principal mechanism of energy loss, radiation (primarily x-rays) is emitted as the result of electron–ion collisions within the plasma. The intersections of the $\mathscr{P}_{\text{lost}}$ line with the $\mathscr{P}_{\text{gen}}$ curves give the critical ignition temperatures.

Confinement time

In addition to the high temperature requirements, there are two other critical parameters that determine whether or not a thermonuclear reactor is successful: the **ion density** n and **confinement time** τ, which is the length of time the ions are maintained at $T > T_{\text{ignit}}$. The British physicist J. D. Lawson has shown that the ion density and confinement time must both be large enough to ensure that more fusion energy is released than the amount required to raise the temperature of the plasma. In particular, **Lawson's criterion** states that a net energy output is possible under the following conditions:

Lawson's criterion

$$
\begin{aligned}
n\tau &\geq 10^{14} \text{ s/cm}^3 \quad \text{(D–T)} \\
n\tau &\geq 10^{16} \text{ s/cm}^3 \quad \text{(D–D)}
\end{aligned}
\tag{45.5}
$$

A graph of $n\tau$ versus temperature for the D–T and D–D reactions is given in Figure 45.9. The product $n\tau$ is referred to as the **Lawson number** of a reaction.

Lawson's criterion was arrived at by comparing the energy required to heat a given plasma with the energy generated by the fusion process.[2] The energy E_{in} re-

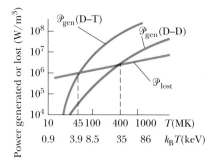

Figure 45.8 Power generated versus temperature for deuterium–deuterium (D–D) and deuterium–tritium (D–T) fusion. The green line represents power lost as a function of temperature. When the generation rate exceeds the loss rate, ignition takes place.

---

[2] Lawson's criterion neglects the energy needed to set up the strong magnetic field used to confine the hot plasma. This energy is expected to be about 20 times greater than the energy required to raise the temperature of the plasma. For this reason, it is necessary either to have a magnetic energy recovery system or to use superconducting magnets.

quired to raise the temperature of the plasma is proportional to the ion density n, which we can express as $E_{in} = C_1 n$, where C_1 is some constant. The energy generated by the fusion process is proportional to $n^2 \tau$, or $E_{gen} = C_2 n^2 \tau$. This may be understood by realizing that the fusion energy released is proportional to both the rate at which interacting ions collide ($\propto n^2$) and the confinement time τ. Net energy is produced when $E_{gen} > E_{in}$. When the constants C_1 and C_2 are calculated for different reactions, the condition that $E_{gen} \geq E_{in}$ leads to Lawson's criterion.

In summary, the three basic requirements of a successful thermonuclear power reactor are

- The plasma temperature must be very high—about 4.5×10^7 K for the D–T reaction and 4×10^8 K for the D–D reaction.
- The ion density must be high. It is necessary to have a high density of interacting nuclei to increase the collision rate between particles.
- The confinement time of the plasma must be long. To meet Lawson's criterion, the product $n\tau$ must be large. For a given value of n, the probability of fusion between two particles increases as τ increases.

Requirements for a fusion power reactor

Current efforts are aimed at meeting Lawson's criterion at temperatures exceeding T_{ignit}. Although the minimum required plasma densities have been achieved, the problem of confinement time is more difficult. How can a plasma be confined at 10^8 K for 1 s? The two basic techniques under investigation are magnetic confinement and inertial confinement.

Magnetic Confinement

Many fusion-related plasma experiments use **magnetic confinement** to contain the plasma. A toroidal device called a **tokamak,** first developed in Russia, is shown in Figure 45.10a. A combination of two magnetic fields is used to confine and stabilize the plasma: (1) a strong toroidal field produced by the current in the toroidal windings surrounding a donut-shaped vacuum chamber and (2) a weaker "poloidal" field produced by the toroidal current. In addition to confining the plasma, the toroidal current is used to raise its temperature. The resultant helical magnetic field lines spiral around the plasma and keep it from touching the walls of the vacuum chamber. (If the plasma touches the walls, its temperature is reduced and heavy impurities sputtered from the walls "poison" it and lead to large power losses.)

One of the major breakthroughs in magnetic confinement in the 1980s was in the area of auxiliary energy input to reach ignition temperatures. Experiments have shown that injecting a beam of energetic neutral particles into the plasma is a very efficient method of raising it to ignition temperatures (5 to 10 keV). Radio-frequency energy input will probably be needed for reactor-size plasmas.

When it was in operation, the Tokamak Fusion Test Reactor (TFTR, Fig. 45.10b) at Princeton reported central ion temperatures of 510 million degrees Celsius, more than 30 times hotter than the center of the Sun. The $n\tau$ values in the TFTR for the D–T reaction were well above 10^{13} s/cm³ and close to the value required by Lawson's criterion. In 1991, reaction rates of 6×10^{17} D–T fusions per second were reached in the JET tokamak at Abington, England.

One of the new generations of fusion experiments is the National Spherical Torus Experiment (NSTX) shown in Figure 45.10c. Rather than the donut-shaped plasma of a tokamak, the NSTX produces a spherical plasma that has a hole through its center. The major advantage of the spherical configuration is its ability to confine the plasma at a higher pressure in a given magnetic field. This approach could lead to development of smaller, more economical fusion reactors.

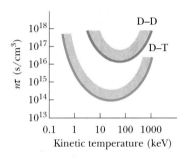

Figure 45.9 The Lawson number $n\tau$ versus temperature for the D–T and D–D fusion reactions. The regions above the colored curves represent favorable conditions for fusion.

web

For more information, visit the Princeton Plasma Physics Laboratory at **www.pppl.gov**

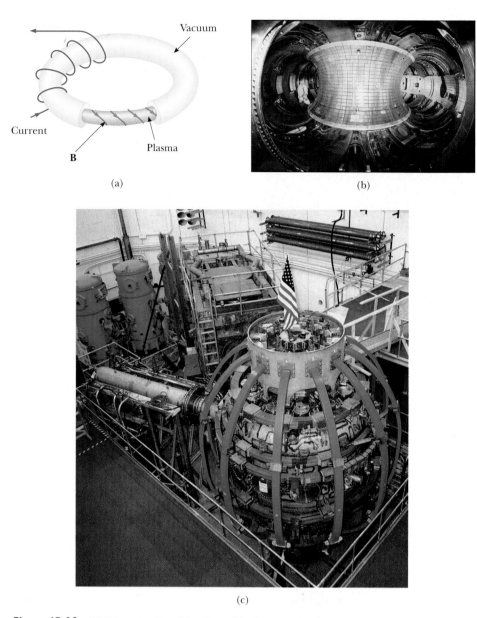

Figure 45.10   (a) Diagram of a tokamak used in the magnetic confinement scheme. (b) Interior view of the recently closed Tokamak Fusion Test Reactor (TFTR) vacuum vessel at the Princeton Plasma Physics Laboratory. *(Courtesy of Princeton Plasma Physics Laboratory)*   (c) The National Spherical Torus Experiment (NSTX) that began operation in March 1999. *(Courtesy of Princeton University)*

An international collaborative effort involving four major fusion programs is currently under way to build a fusion reactor called ITER (International Thermonuclear Experimental Reactor). This facility will address the remaining technological and scientific issues concerning the feasibility of fusion power. The design is completed (Fig. 45.11), and site and construction negotiations are under way. If the planned device works as expected, the Lawson number for ITER will be about six times greater than the current record holder, the JT-60U tokamak in Japan. ITER will produce 1.5 GW of power, and the energy content of the alpha particles

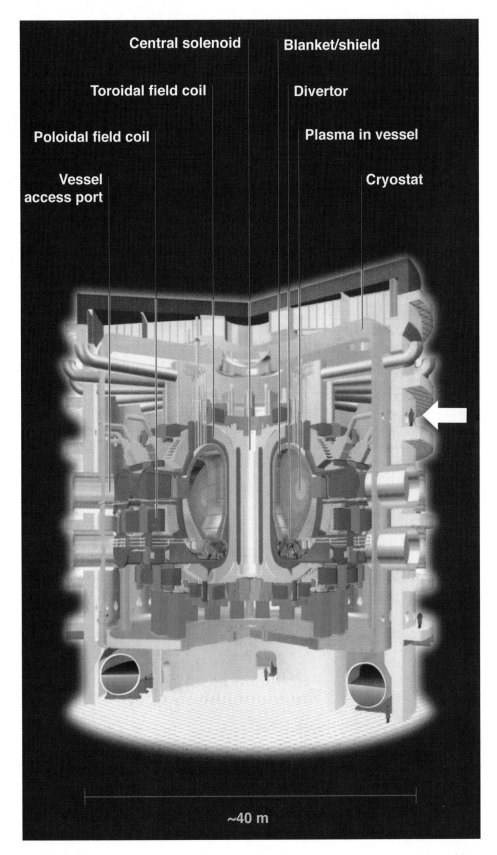

Central solenoid

Toroidal field coil

Poloidal field coil

Vessel access port

Blanket/shield

Divertor

Plasma in vessel

Cryostat

~40 m

Figure 45.11 Cutaway diagram of the ITER (International Thermonuclear Experimental Reactor). Note the size of the reactor relative to that of a person *(arrow)*. *(Courtesy of ITER)*

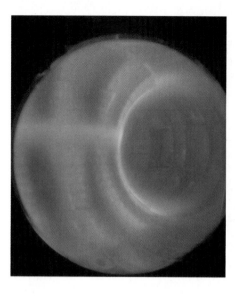

During the operation of the Tokamak Fusion Test Reactor, the plasma discharge was monitored using an optical system that showed the interior of the vacuum vessel. This view of a high-temperature deuterium plasma shows a bright radiation belt in the foreground. *(Courtesy of Princeton Plasma Physics Laboratory)*

inside the reactor will be so intense that they will sustain the fusion reaction, allowing the auxiliary energy sources to be turned off once the reaction is initiated. Such a state of sustained burn is referred to as *ignition*.

EXAMPLE 45.3 ▶ Inside a Fusion Reactor

In 1998 the JT-60U tokamak in Japan was operated with a D–T plasma density of 4.8×10^{13} cm^{-3} at a temperature (in energy units) of 16.1 keV. It was able to confine this plasma inside a magnetic field for 1.1 s. (a) Does this meet Lawson's criterion?

Solution   Equation 45.5 says that for a D–T plasma, the Lawson number $n\tau$ must be greater than 10^{14} s/cm^3. For the JT-60U,

$$n\tau = (4.8 \times 10^{13} \text{ cm}^{-3})(1.1 \text{ s}) = 5.3 \times 10^{13} \text{ s/cm}^3$$

which is close to meeting Lawson's criterion. In fact, scientists recorded a power gain of 1.25, indicating the reactor was operating slightly past the break-even point and was producing more energy than it required to maintain the plasma.

(b) How does the plasma density compare with the density of atoms in an ideal gas when the gas is at room temperature and pressure?

Solution   The density of atoms in a sample of ideal gas is given by N_A/V_m, where N_A is Avogadro's number and V_m is the molar volume of an ideal gas under standard conditions, 2.24×10^{-2} m^3/mol. Thus, the density of the gas is

$$\frac{N_A}{V_m} = \frac{6.02 \times 10^{23} \text{ atoms/mol}}{2.24 \times 10^{-2} \text{ m}^3/\text{mol}} = 2.7 \times 10^{25} \text{ atoms/m}^3$$

This is more than 500 000 times greater than the plasma density in the reactor.

Inertial Confinement

The second technique for confining a plasma is called **inertial confinement** and makes use of a D–T target that has a very high particle density. In this scheme, the confinement time is very short (typically 10^{-11} to 10^{-9} s), so that, because of their own inertia, the particles do not have a chance to move appreciably from their initial positions. Thus, Lawson's criterion can be satisfied by combining a high particle density with a short confinement time.

Laser fusion is the most common form of inertial confinement. A small D–T pellet, about 1 mm in diameter, is struck simultaneously by several focused, high-

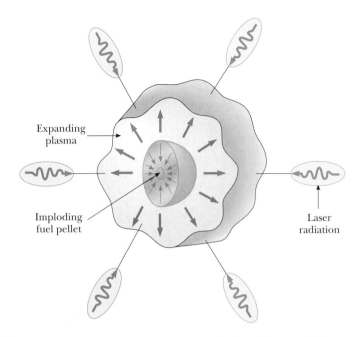

Figure 45.12   In inertial confinement, a D–T fuel pellet fuses when struck by several high-intensity laser beams simultaneously.

intensity laser beams, resulting in a large pulse of input energy that causes the surface of the fuel pellet to evaporate (Fig. 45.12). The escaping particles exert a third-law reaction force on the core of the pellet, resulting in a strong, inwardly moving compressive shock wave. This shock wave increases the pressure and density of the core and produces a corresponding increase in temperature. When the temperature of the core reaches ignition temperature, fusion reactions occur.

Two of the leading laser fusion laboratories in the United States are the Omega facility at the University of Rochester in New York and the Nova facility at Lawrence Livermore National Laboratory in California. The Omega facility focuses 24 laser beams on the target, and the Nova facility employs 10 beams. Figure 45.13a shows the target chamber at Nova, and Figure 45.13b shows the tiny, spherical D–T pellets used. Nova is capable of injecting a power of 2×10^{14} W into a 0.5-mm D–T pellet and has achieved values of $n\tau \approx 5 \times 10^{14}$ s/cm³ and ion temperatures of 5.0 keV. These values are close to those required for D–T ignition. This steady progress has led the U. S. Department of Energy and other groups to plan a national facility that will involve a laser fusion device with an input energy in the 5–10 MJ range.

Fusion Reactor Design

In the D–T fusion reaction shown in Figure 45.14,

$$^{2}_{1}\text{H} + {}^{3}_{1}\text{H} \longrightarrow {}^{4}_{2}\text{He} + {}^{1}_{0}\text{n} \qquad Q = 17.59 \text{ MeV}$$

the alpha particle carries 20% of the energy and the neutron carries 80%, or about 14 MeV. The alpha particles, because they are charged, are primarily absorbed by the plasma; this causes the plasma's temperature to increase. In contrast, the 14-MeV neutrons, being electrically neutral, pass through the plasma and

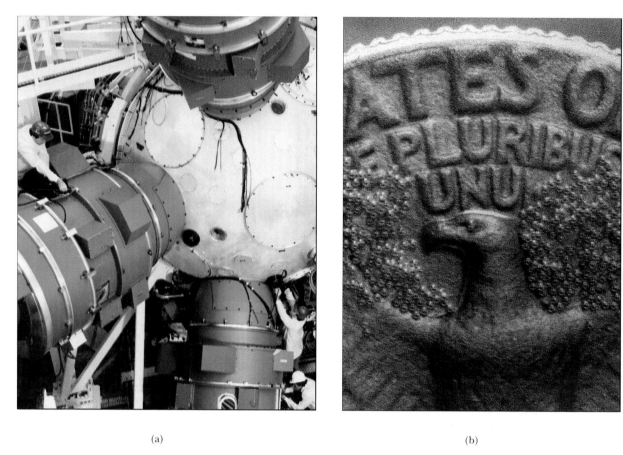

(a)   (b)

Figure 45.13   (a) The target chamber of the Nova Laser Facility at Lawrence Livermore Laboratory. *(Courtesy of University of California Lawrence Livermore National Laboratory and the U. S. Department of Energy)*   (b) Spherical plastic target shells used to contain the D–T fuel, shown clustered on a quarter. The shells have very smooth surfaces and are about 100 nm thick. *(Courtesy of Los Alamos National Laboratory)*

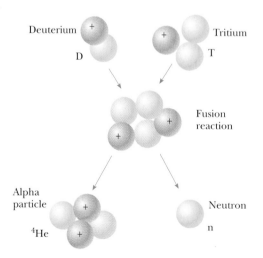

Figure 45.14   Deuterium–tritium fusion. Eighty percent of the energy released is in the 14-MeV neutron.

must be absorbed into a surrounding blanket material, where their large kinetic energy is extracted and used to generate electric power.

One scheme is to use molten lithium metal as the neutron-absorbing material and to circulate the lithium in a closed heat-exchange loop to produce steam and drive turbines as in a conventional power plant. Figure 45.15 shows a diagram of such a reactor. It is estimated that a blanket of lithium about 1 m thick will capture nearly 100% of the neutrons from the fusion of a small D–T pellet.

The capture of neutrons by lithium is described by the reaction

$$_0^1n + {}_3^6Li \longrightarrow {}_1^3H + {}_2^4He$$

where the kinetic energies of the charged tritium $_1^3H$ and alpha particle are converted to internal energy in the molten lithium. An extra advantage of using lithium as the energy-transfer medium is that the tritium produced can be separated from the lithium and returned as fuel to the reactor.

Advantages and Problems of Fusion

If fusion power can ever be harnessed, it will offer several advantages over fission-generated power: (1) low cost and abundance of fuel (deuterium), (2) impossibility of runaway accidents, and (3) lesser radiation hazard. Some of the anticipated problems and disadvantages include (1) scarcity of lithium, (2) limited supply of helium, which is needed for cooling the superconducting magnets used to produce strong confining fields, and (3) structural damage and induced radioactivity caused by neutron bombardment. If such problems and the engineering design factors can be resolved, nuclear fusion will become a feasible source of energy by the middle of the 21st century.

> Advantages of fusion

> Problem areas and disadvantages of fusion

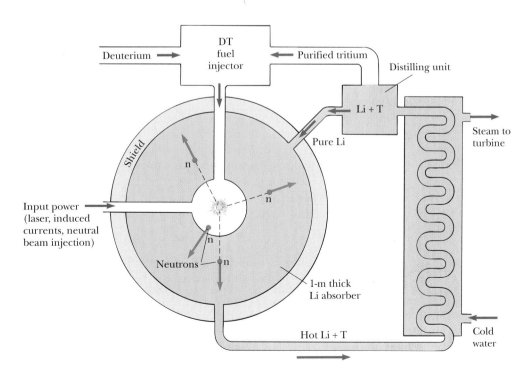

Figure 45.15   Diagram of a fusion reactor.

Optional Section

45.5 ▷ RADIATION DAMAGE IN MATTER

In Chapter 34 we learned that electromagnetic radiation is all around us in the form of radio waves, microwaves, light waves, and so on. In this section, we turn to forms of radiation that can cause severe damage as they pass through matter. These include radiation resulting from radioactive processes and radiation in the form of energetic particles such as neutrons and protons. It is these forms that we refer to here and in the following two sections when we use the word *radiation*.

The degree and type of damage depend upon several factors, including the type and energy of the radiation and the properties of the matter. The metals used in nuclear reactor structures can be severely weakened by high fluxes of energetic neutrons because these high fluxes often lead to metal fatigue. The damage in such situations is in the form of atomic displacements, often resulting in major alterations in the properties of the material. Materials can also be damaged by electromagnetic radiation, such as gamma rays or x-rays, that doesn't displace the atoms in a material but rather strips them of electrons so that they become ions. For example, defects called color centers can be produced in inorganic crystals by irradiating the crystals with x-rays. One extensively studied color center has been identified as an electron trapped in a Cl^- ion vacancy.

Radiation damage in organisms is primarily due to ionization effects in cells. The normal operation of a cell may be disrupted when highly reactive ions are formed as the result of ionizing radiation. For example, hydrogen and the hydroxyl radical OH^- produced from water molecules can induce chemical reactions that may break bonds in proteins and other vital molecules. Furthermore, the ionizing radiation may affect vital molecules directly by removing electrons from their structure. Large doses of radiation are especially dangerous because damage to a great number of molecules in a cell may cause the cell to die. Although the death of a single cell is usually not a problem, the death of many cells may result in irreversible damage to the organism. Cells that divide rapidly, such as those of the digestive tract, reproductive organs, and hair follicles, are especially susceptible. Also, cells that do survive the radiation may become defective. These defective cells can produce more defective cells and lead to cancer.

In biological systems, it is common to separate radiation damage into two categories: somatic damage and genetic damage. *Somatic damage* is that associated with any body cell except the reproductive cells. Somatic damage can lead to cancer or seriously alter the characteristics of specific organisms. *Genetic damage* affects only reproductive cells. Damage to the genes in reproductive cells can lead to defective offspring. Clearly, we must be concerned about the effect of diagnostic treatments such as x-rays and other forms of radiation exposure.

There are several units used to quantify the amount, or dose, of any radiation that interacts with a substance.

The roentgen

> The **roentgen** (R) is that amount of ionizing radiation that produces an electric charge of 3.33×10^{-10} C in 1 cm^3 of air under standard conditions.

Equivalently, the roentgen is that amount of radiation that increases the energy of 1 kg of air by 8.76×10^{-3} J.

For most applications, the roentgen has been replaced by the rad (an acronym for *radiation absorbed dose*):

One **rad** is that amount of radiation that increases the energy of 1 kg of absorbing material by 1×10^{-2} J.

Although the rad is a perfectly good physical unit, it is not the best unit for measuring the degree of biological damage produced by radiation because damage depends not only on the dose but also on the type of the radiation. For example, a given dose of alpha particles causes about ten times more biological damage than an equal dose of x-rays. The **RBE** (relative biological effectiveness) factor for a given type of radiation is **the number of rads of x-radiation or gamma radiation that produces the same biological damage as 1 rad of the radiation being used.** The RBE factors for different types of radiation are given in Table 45.1. The values are only approximate because they vary with particle energy and with the form of the damage. The RBE factor should be considered only a first-approximation guide to the actual effects of radiation.

Finally, the **rem** (radiation equivalent in man) is the product of the dose in rad and the RBE factor:

$$\text{Dose in rem} \equiv \text{dose in rad} \times \text{RBE} \qquad \textbf{(45.6)}$$

According to this definition, 1 rem of any two radiations produces the same amount of biological damage. From Table 45.1, we see that a dose of 1 rad of fast neutrons represents an effective dose of 10 rem, but 1 rad of gamma radiation is equivalent to a dose of only 1 rem.

Low-level radiation from natural sources, such as cosmic rays and radioactive rocks and soil, delivers to each of us a dose of about 0.13 rem/yr; this radiation is called *background radiation*. It is important to note that background radiation varies with geography, with the main factors being altitude (exposure to cosmic rays) and geology (radon gas released by some rock formations, deposits of naturally radioactive minerals).

The upper limit of radiation dose rate recommended by the U.S. government (apart from background radiation) is about 0.5 rem/yr. Many occupations involve much higher radiation exposures, and so an upper limit of 5 rem/yr has been set for combined whole-body exposure. Higher upper limits are permissible for certain parts of the body, such as the hands and the forearms. A dose of 400 to 500 rem results in a mortality rate of about 50% (which means that half the people exposed to this radiation level die). The most dangerous form of exposure is either

TABLE 45.1   RBE[a] Factors for Several Types of Radiation

Radiation	RBE Factor
X-rays and gamma rays	1.0
Beta particles	1.0–1.7
Alpha particles	10–20
Thermal neutrons	4–5
Fast neutrons and protons	10
Heavy ions	20

[a]RBE = relative biological effectiveness.

ingestion or inhalation of radioactive isotopes, especially isotopes of those elements the body retains and concentrates, such as ^{90}Sr. In some cases, a dose of 1 000 rem can result from ingesting 1 mCi of radioactive material.

Optional Section

45.6 ▶ RADIATION DETECTORS

Various devices have been developed for detecting radiation. These devices are used for a variety of purposes, including medical diagnoses, radioactive dating measurements, measuring background radiation, and measuring the mass, energy, and momentum of particles created in high-energy nuclear reactions.

Ion chamber

In an **ion chamber** (Fig. 45.16), electron–ion pairs are generated as radiation passes through a gas and produce an electrical signal. Two plates in the chamber are connected to a voltage supply and thereby maintained at different electric potentials. The positive plate attracts the electrons, and the negative plate attracts positive ions, causing a current pulse that is proportional to the number of electron–ion pairs produced when a radioactive particle enters the chamber. When an ion chamber is used both to detect the presence of a radioactive particle and to measure its energy, it is called a **proportional counter.**

Geiger counter

The **Geiger counter** (Fig. 45.17) is perhaps the most common form of ion chamber used to detect radioactivity. It can be considered the prototype of all counters that use the ionization of a medium as the basic detection process. It consists of a thin wire electrode aligned along the central axis of a cylindrical metallic tube filled with a gas at low pressure. The wire is maintained at a high positive electric potential (about 10^3 V) relative to the tube. When a high-energy particle resulting, for example, from a radioactive decay enters the tube through a thin

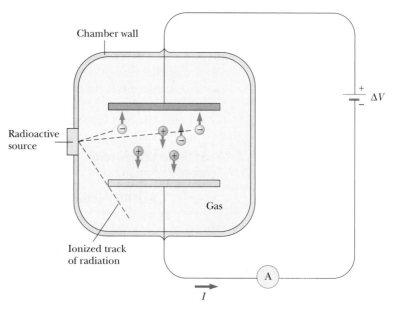

Figure 45.16 Simplified diagram of an ion chamber. The radioactive source creates electrons and positive ions that are collected by the charged plates. The current set up in the external circuit is proportional to a radioactive particle's kinetic energy if the particle stops in the chamber.

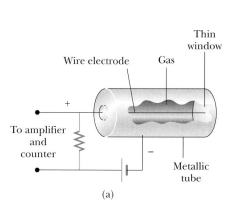

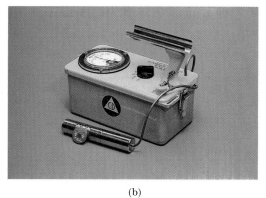

(a)   (b)

Figure 45.17   (a) Diagram of a Geiger counter. The voltage between the wire electrode and the metallic tube is usually about 1 000 V. (b) A Geiger counter. *(David Rogers)*

window at one end, some of the gas atoms are ionized. The electrons removed from these atoms are attracted toward the wire electrode, and in the process they ionize other atoms in their path. This sequential ionization results in an *avalanche* of electrons that produces a current pulse. After the pulse has been amplified, it can either be used to trigger an electronic counter or be delivered to a loudspeaker that clicks each time a particle is detected. Although a Geiger counter easily detects the presence of a radioactive particle, the energy lost by the particle in the counter is *not* proportional to the current pulse produced. Thus, a Geiger counter cannot be used to measure the energy of a radioactive particle.

A **semiconductor diode detector** is essentially a reverse-bias p–n junction. Recall from Section 43.7 that a p–n junction passes current readily when forward-biased and prohibits a current when reverse-biased. As an energetic particle passes through the junction, electrons are excited into the conduction band and holes are formed in the valence band. The internal electric field sweeps the electrons toward the positive (n) side of the junction and the holes toward the negative (p) side. This movement of electrons and holes creates a pulse of current that is measured with an electronic counter. In a typical device, the duration of the pulse is 10^{-8} s.

Semiconductor diode detector

A **scintillation counter** usually uses a solid or liquid material whose atoms are easily excited by radiation. The excited atoms then emit visible-light photons when they return to their ground state. Common materials used as scintillators are transparent crystals of sodium iodide and certain plastics. If the scintillator material is attached to one end of a device called a **photomultiplier** (PM) tube as shown in Figure 45.18, the photons emitted by the scintillator can be converted to an electrical signal. The PM tube consists of numerous electrodes, called *dynodes,* whose electric potentials increase in succession along the length of the tube. Between the top of the tube and the scintillator material is a plate called a photocathode. When photons leaving the scintillator hit this plate, electrons are emitted (because of the photoelectric effect). As one of these emitted electrons strikes the first dynode, the electron has sufficient kinetic energy to eject several other electrons from the dynode surface. When these electrons are accelerated to the second dynode, many more electrons are ejected, and thus a multiplication process occurs. The end result is 1 million or more electrons striking the last dynode. Hence, one particle striking the scintillator produces a sizable electrical pulse at the PM output, and this pulse is sent to an electronic counter.

Scintillation counter

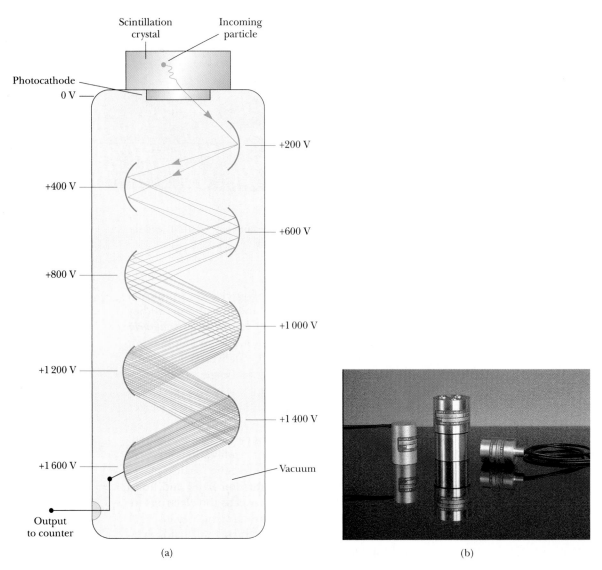

Figure 45.18   (a) Diagram of a scintillation counter connected to a photomultiplier tube. (b) The sodium iodide in these scintillation crystals flashes when an energetic particle passes through, something like the way the atmosphere flashes when a meteor passes through it.

Both the scintillator and the semiconductor diode detector are much more sensitive than a Geiger counter, mainly because of the higher density of the detecting medium. Both can also be used to measure particle energy if the particle stops in the detector.

Track detectors are various devices used to view the tracks of charged particles directly. High-energy particles produced in particle accelerators may have energies ranging from 10^9 to 10^{12} eV. Thus, they cannot be stopped and cannot have their energy measured with the small detectors already mentioned. Instead, the energy and momentum of these energetic particles are found from the curvature of their path in a magnetic field of known magnitude and direction.

A **photographic emulsion** is the simplest example of a track detector. A charged particle ionizes the atoms in an emulsion layer. The path of the particle

Track detector

corresponds to a family of points at which chemical changes have occurred in the emulsion. When the emulsion is developed, the particle's track becomes visible.

A **cloud chamber** contains a gas that has been supercooled to just below its usual condensation point. An energetic radioactive particle passing through ionizes the gas along the particle's path. The ions serve as centers for condensation of the supercooled gas. The track can be seen with the naked eye and can be photographed. A magnetic field can be applied to determine the charges of the radioactive particles, as well as their momentum and energy.

Cloud chamber

A device called a **bubble chamber,** invented in 1952 by D. Glaser, uses a liquid (usually liquid hydrogen) maintained near its boiling point. Ions produced by incoming charged particles leave bubble tracks, which can be photographed (Fig. 45.19). Because the density of the detecting medium in a bubble chamber is much higher than the density of the gas in a cloud chamber, the bubble chamber has a much higher sensitivity.

Bubble chamber

A **spark chamber** is a counting device that consists of an array of conducting parallel plates and is capable of recording a three-dimensional track record. Even-numbered plates are grounded, and odd-numbered plates are maintained at a high electric potential (about 10 kV). The spaces between the plates contain an inert gas at atmospheric pressure. When a charged particle passes through the chamber, gas atoms are ionized, resulting in a current surge and visible sparks along the particle path. These sparks may be photographed or electronically detected and sent to a computer for path reconstruction and determination of particle mass, momentum, and energy.

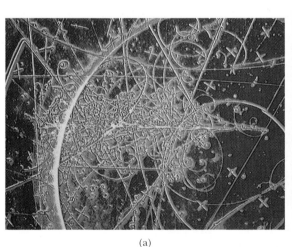
(a)

(b)

Figure 45.19 (a) Artificially colored bubble-chamber photograph showing tracks of particles that have passed through the chamber. *(Photo Researchers, Inc./Science Photo Library)* (b) This research scientist is studying a photograph of particle tracks made in a bubble chamber at Fermilab. The curved tracks are produced by charged particles moving through the chamber in the presence of an applied magnetic field. Negatively charged particles deflect in one direction, while positively charged particles deflect in the opposite direction. *(Dan McCoy/Rainbow)*

Neutron detectors are more difficult to construct than charged-particle detectors because neutrons do not interact electrically with atoms as they pass through matter. Fast neutrons, however, can be detected by filling an ion chamber with hydrogen gas and detecting the resulting ionization by high-speed recoiling protons produced in neutron–proton collisions. Thermal neutrons having energies less than about 1 MeV do not transfer sufficient energy to the protons to be detected in this way; however, they can be detected by using an ion chamber filled with BF_3 gas. In this case, the boron nuclei disintegrate during neutron capture, emitting alpha particles that, because of their charge, are easily detected in the ion chamber.

Optional Section

45.7 USES OF RADIATION

Nuclear physics applications are extremely widespread in manufacturing, medicine, and biology. Even a brief discussion of all the possibilities would fill an entire book, and to keep such a book up to date would require a number of revisions each year. In this section, we present a few of these applications and the underlying theories supporting them.

Tracing

Radioactive tracers are used to track chemicals participating in various reactions. One of the most valuable uses of radioactive tracers is in medicine. For example, iodine, a nutrient needed by the human body, is obtained largely through the intake of iodized salt and seafood. To evaluate the performance of the thyroid, the patient drinks a very small amount of radioactive sodium iodide containing ^{131}I, an artificially produced isotope of iodine (the natural, nonradioactive isotope is ^{127}I). Two hours later, the amount of iodine in the thyroid gland is determined by measuring the radiation intensity at the neck area. How much or how little ^{131}I is still in the thyroid is a measure of how well that gland is functioning.

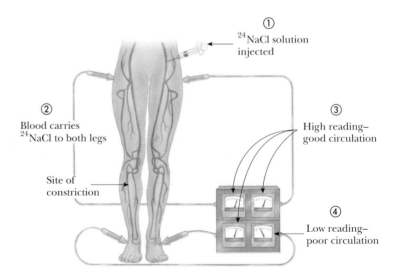

Figure 45.20 A tracer technique for determining the condition of the human circulatory system.

A second medical application is indicated in Figure 45.20. A solution containing radioactive sodium is injected into a vein in the leg, and the time at which the radioisotope arrives at another part of the body is detected with a radiation counter. The elapsed time is a good indication of the presence or absence of constrictions in the circulatory system.

Tracers are also useful in agricultural research. Suppose the best method of fertilizing a plant is to be determined. A certain element in a fertilizer, such as nitrogen, can be *tagged* (identified) with one of its radioactive isotopes. The fertilizer is then sprayed on one group of plants, sprinkled on the ground for a second group, and raked into the soil for a third. A Geiger counter is then used to track the nitrogen through the three groups.

Tracing techniques are as wide ranging as human ingenuity can devise. Present applications range from checking how teeth absorb fluoride to monitoring how cleansers contaminate food-processing equipment to studying deterioration inside an automobile engine. In the latter case, a radioactive material is used in the manufacture of the piston rings, and the oil is checked for radioactivity to determine the amount of wear on the rings.

Materials Analysis

For centuries, a standard method of identifying the elements in a sample of material has been chemical analysis, which involves determining how the material reacts with various chemicals. A second method is spectral analysis, which uses the fact that, when excited, each element emits its own characteristic set of electromagnetic wavelengths. These methods are now supplemented by a third technique, **neutron activation analysis.** Both chemical and spectral methods have the disadvantage that a fairly large sample of the material must be destroyed for the analysis. In addition, extremely small quantities of an element may go undetected by either method. Neutron activation analysis has an advantage over the other two methods in both respects.

When a material is irradiated with neutrons, nuclei in the material absorb the neutrons and are changed to different isotopes, most of which are radioactive. For example, ^{65}Cu absorbs a neutron to become ^{66}Cu, which undergoes beta decay:

$$^{1}_{0}n + ^{65}_{29}Cu \longrightarrow ^{66}_{29}Cu \longrightarrow ^{66}_{30}Zn + e^{-} + \overline{\nu}$$

The presence of the copper can be deduced because it is known that ^{66}Cu has a half-life of 5.1 min and decays with the emission of beta particles having maximum energies of 2.63 and 1.59 MeV. Also emitted in the decay of ^{66}Cu is a 1.04-MeV gamma ray. By examining the radiation emitted by a substance after it has been exposed to neutron irradiation, one can detect extremely small amounts of an element in that substance.

Neutron activation analysis is used routinely in a number of industries, for example in commercial aviation for the checking of airline luggage for hidden explosives (Fig. 45.21). The following nonroutine use is of interest. Napoleon died on the island of St. Helena in 1821, supposedly of natural causes. Over the years, suspicion has existed that his death was not all that natural. After his death, his head was shaved and locks of his hair were sold as souvenirs. In 1961, the amount of arsenic in a sample of this hair was measured by neutron activation analysis, and an unusually large quantity of arsenic was found. (Activation analysis is so sensitive that very small pieces of a single hair could be analyzed.) Results showed that the arsenic was fed to him irregularly. In fact, the arsenic concentration pattern corre-

Figure 45.21   This bomb detector irradiates luggage with neutrons. If there are hidden explosives inside, chemicals within the explosive materials become radioactive and can be easily detected. The half-life of the resulting radiation is so short that there is no danger to the security personnel removing the luggage for further inspection. *(Shahn Kermani/Gamma Liaison)*

sponded to the fluctuations in the severity of Napoleon's illness as determined from historical records.

Art historians use neutron activation analysis to detect forgeries. The pigments used in paints have changed throughout history, and old and new pigments react differently to neutron activation. The method can even reveal hidden works of art behind existing paintings.

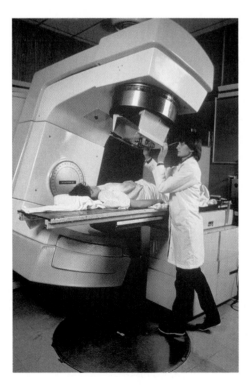

Figure 45.22   This large machine is being set to deliver a dose of radiation from ^{60}Co in an effort to destroy a cancerous tumor. Cancer cells are especially susceptible to this type of therapy because they tend to divide more often than cells of healthy tissue nearby. *(Science/Visuals Unlimited)*

Figure 45.23   The irradiated strawberries on the right have *not* become radioactive. The radiation has killed or incapacitated the mold spores that have spoiled the strawberries on the left. *(Council for Agricultural Science & Technology)*

Another analysis technique that takes advantage of radioactivity is PIXE, **photon-induced x-ray emission.** This process uses x-ray photons to excite the innermost electrons of an atom in the material being analyzed. When the electrons fall back down from their excited states, they give off characteristic spectra that can be evaluated to reveal the types of elements present.

Radiation Therapy

Because radiation causes the most damage to rapidly dividing cells (as discussed in Section 45.5), it is useful in cancer treatment because tumor cells divide extremely rapidly. Several mechanisms can be used to deliver radiation to a tumor. In some cases, a narrow beam of x-rays or radiation from a source such as ^{60}Co is used, as shown in Figure 45.22. In other situations, thin radioactive needles called *seeds* are implanted in the cancerous tissue. The radioactive isotope ^{151}I is used to treat cancer of the thyroid.

Food Preservation

Radiation is finding increasing use as a means of preserving food because exposure to high levels of radiation can destroy or incapacitate bacteria and mold spores (Fig. 45.23). Food preserved this way can be placed in a sealed container (to keep out new spoiling agents) and stored for long periods of time.

SUMMARY

The probability that neutrons are captured as they move through matter generally increases with decreasing neutron energy. A **thermal neutron** is a slow-moving neutron that has a high probability of being captured by a nucleus in a **neutron capture event:**

$$\_{0}^{1}n + \_{Z}^{A}X \longrightarrow \_{Z}^{A+1}X + \gamma \tag{45.1}$$

Nuclear fission occurs when a very heavy nucleus, such as ^{235}U, splits into two smaller **fission fragments.** Thermal neutrons can create fission in ^{235}U:

$$^{1}_{0}\text{n} + ^{235}_{92}\text{U} \longrightarrow ^{236}_{92}\text{U}^* \longrightarrow \text{X} + \text{Y} + \text{neutrons} \tag{45.2}$$

where X and Y are the fission fragments and ^{236}U* is an intermediate excited state. On the average, 2.5 neutrons are released per fission event. The fragments then undergo a series of beta and gamma decays to various stable isotopes. The energy released per fission event is about 200 MeV.

The **reproduction constant** K is the average number of neutrons released from each fission event that cause another event. In a fission reactor, it is necessary to maintain $K \approx 1$. The value of K is affected by such factors as reactor geometry, mean neutron energy, and probability of neutron capture. Neutron energies are regulated with a moderator material that slows down energetic neutrons and therefore increases the probability of neutron capture by other ^{235}U nuclei. The power level of the reactor is adjusted with control rods made of a material that is very efficient in absorbing neutrons.

In **nuclear fusion,** two light nuclei fuse to form a heavier nucleus and release energy. The major obstacle in obtaining useful energy from fusion is the large Coulomb repulsive force between the charged nuclei at small separation distances. Sufficient energy must be supplied to the particles to overcome this Coulomb barrier. The temperature required to produce fusion is of the order of 10^8 K, and at this temperature all matter occurs as a plasma.

In a fusion reactor, the plasma temperature must reach the **critical ignition temperature,** the temperature at which the power generated by the fusion reactions exceeds the power lost in the system. The most promising fusion reaction is the D–T reaction, which has a critical ignition temperature of approximately 4.5×10^7 K. Two critical parameters in fusion reactor design are **ion density** n and **confinement time** τ, the length of time the interacting particles must be maintained at $T > T_{\text{ignit}}$. **Lawson's criterion** states that for the D–T reaction, $n\tau \geq 10^{14}$ s/cm^3.

QUESTIONS

1. Explain the function of a moderator in a fission reactor.
2. Why is water a better shield against neutrons than lead or steel?
3. Discuss the advantages and disadvantages of fission reactors from the point of view of safety, pollution, and resources. Make a comparison with power generated from the burning of fossil fuels.
4. Why would a fusion reactor produce less radioactive waste than a fission reactor?
5. Lawson's criterion states that the product of ion density and confinement time must exceed a certain number before a break-even fusion reaction can occur. Why should these two parameters determine the outcome?
6. Why is the temperature required for the D–T fusion less than that needed for the D–D fusion? Estimate the relative importance of Coulomb repulsion and nuclear attraction in each case.

7. What factors make a fusion reaction difficult to achieve?
8. Discuss the similarities and differences between fusion and fission.
9. Discuss the advantages and disadvantages of fusion power from the point of safety, pollution, and resources.
10. Discuss three major problems associated with the development of a controlled fusion reactor.
11. Describe two techniques being pursued in an effort to obtain power from nuclear fusion.
12. If two radioactive samples have the same activity measured in curies, will they necessarily create the same damage to a medium? Explain.
13. Why should a radiologist be extremely cautious about x-ray doses when treating pregnant women?
14. The design of a PM tube might suggest that any number of dynodes may be used to amplify a weak signal. What factors do you suppose would limit the amplification in this device?

15.   *And swift, and swift past comprehension*
        Turn round Earth's beauty and her might.
        The heavens blaze in alternation
        With deep and chill and rainy night.
        In mighty currents foams the ocean
        Up from the rocks' abyssal base.
        With rock and sea torn into motion
        In ever-swift celestial race.
        And tempests bluster in a contest
        From sea to land, from land to sea.
        In rage they forge a chain around us
        Of deepest meaning, energy.

There flames a lightning disaster
Before the thunder, in its way.
But all Your servants honor, Master,
The gentle order of Your day

Johann Wolfgang von Goethe wrote this song of the archangels in *Faust* a half-century before the law of conservation of energy was recognized. Students often find it convenient to think of a list of several "forms of energy," from kinetic to nuclear. Argue for or against the view that these lines of poetry make an obvious or an oblique reference to every form of energy.

PROBLEMS

1, 2, 3 = straightforward, intermediate, challenging   ☐ = full solution available in the *Student Solutions Manual and Study Guide*
WEB = solution posted at **http://www.saunderscollege.com/physics/**   🖥 = Computer useful in solving problem   🔵 = Interactive Physics
☐ = paired numerical/symbolic problems

Section 45.2   Nuclear Fission

1. Find the energy released in the fission reaction

$$^{1}_{0}n + ^{235}_{92}U \longrightarrow ^{98}_{40}Zr + ^{135}_{52}Te + 3(^{1}_{0}n)$$

The atomic masses of the fission products are: $^{98}_{40}Zr$, 97.912 0 u; $^{135}_{52}Te$, 134.908 7 u.

2. Strontium-90 is a particularly dangerous fission product of ^{235}U because it is radioactive and it substitutes for calcium in bones. What other direct fission products would accompany it in the neutron-induced fission of ^{235}U? (*Note:* This reaction may release two, three, or four free neutrons.)

WEB 3. List the nuclear reactions required to produce ^{233}U from ^{232}Th under fast neutron bombardment.

4. List the nuclear reactions required to produce ^{239}Pu from ^{238}U under fast neutron bombardment.

5. (a) Find the energy released in the following fission reaction:

$$^{1}_{0}n + ^{235}_{92}U \longrightarrow ^{141}_{56}Ba + ^{92}_{36}Kr + 3(^{1}_{0}n)$$

The required masses are

$$M(^{1}_{0}n) = 1.008\ 665\ u$$

$$M(^{235}_{92}U) = 235.043\ 924\ u$$

$$M(^{141}_{56}Ba) = 140.913\ 9\ u$$

$$M(^{92}_{36}Kr) = 91.897\ 3\ u$$

(b) What fraction of the initial mass of the system is given off?

6. A typical nuclear fission power plant produces about 1.00 GW of electric power. Assume that the plant has an overall efficiency of 40.0% and that each fission releases 200 MeV of energy. Calculate the mass of ^{235}U consumed each day.

7. **Review Problem.** Suppose enriched uranium containing 3.40% of the fissionable isotope $^{235}_{92}U$ is used as fuel for a ship. The water exerts an average frictional drag of 1.00×10^{5} N on the ship. How far can the ship travel per kilogram of fuel? Assume that the energy released per fission event is 208 MeV and that the ship's engine has an efficiency of 20.0%.

Section 45.3   Nuclear Reactors

8. To minimize neutron leakage from a reactor, the surface area-to-volume ratio should be a minimum. For a given volume *V*, calculate this ratio for (a) a sphere, (b) a cube, and (c) a parallelepiped of dimensions $a \times a \times 2a$. (d) Which of these shapes would have minimum leakage? Which would have maximum leakage?

WEB 9. It has been estimated that there is on the order of 10^{9} tons of natural uranium available at concentrations exceeding 100 parts per million, of which 0.7% is ^{235}U. If all the world's energy use (7×10^{12} J/s) were to be supplied by ^{235}U fission, how long would this supply last? (This estimate of uranium supply was taken from K. S. Deffeyes and I. D. MacGregor, *Sci. Am.* 242:66, 1980.)

10. If the reproduction constant (neutron multiplication factor) is 1.000 25 for a chain reaction in a fission reactor and the average time between successive fissions is 1.20 ms, by what factor will the reaction rate increase in 1 min?

11. A large nuclear power reactor produces about 3 000 MW of power in its core. Three months after a reactor is shut down, the core power from radioactive by-products is 10.0 MW. Assuming that each emission delivers 1.00 MeV of energy to the power, find the activity in becquerels three months after the reactor is shut down.

Section 45.4   Nuclear Fusion

12. (a) If a fusion generator were built to create 3.00 GW of power, determine the rate of fuel burning in grams per hour if the D–T reaction is used. (b) Do the same for the D–D reaction assuming that the reaction products are split evenly between (n, ^{3}He) and (p, ^{3}H).

13. Two nuclei having atomic numbers Z_1 and Z_2 approach each other with a total energy E. (a) Suppose they will spontaneously fuse if they approach within a distance of 1.00×10^{-14} m. Find the minimum value of E required to produce fusion, in terms of Z_1 and Z_2. (b) Evaluate the minimum energy for fusion for the D–D and D–T reactions (the first and third reactions in Eq. 45.4).

14. **Review Problem.**   Consider the deuterium-tritium fusion reaction with the tritium nucleus at rest:

$$^2_1\text{H} + {}^3_1\text{H} \longrightarrow {}^4_2\text{He} + {}^1_0\text{n}$$

(a) Suppose that the reactant nuclei will spontaneously fuse if their surfaces touch. From Equation 44.1, determine the required distance of closest approach between their centers. (b) What is the Coulomb potential energy (in eV) at this distance? (c) Suppose the deuteron is fired straight at an originally stationary tritium nucleus with just enough energy to reach the required distance of closest approach. What is the common speed of the deuterium and tritium nuclei as they touch, in terms of the initial deuteron speed v_i? (*Hint:* At this point, the two nuclei have a common speed equal to the center-of-mass speed.) (d) Use energy methods to find the minimum initial deuteron energy required to achieve fusion. (e) Why does the fusion reaction actually occur at much lower deuteron energies than that calculated in part (d)?

web 15. To understand why plasma containment is necessary, consider the rate at which an unconfined plasma would be lost. (a) Estimate the rms speed of deuterons in a plasma at 4.00×10^8 K. (b) Estimate the order of magnitude of the time such a plasma would remain in a 10-cm cube if no steps were taken to contain it.

16. Of all the hydrogen in the oceans, 0.030 0% of the mass is deuterium. The oceans have a volume of 317 million mi^3. (a) If all the deuterium in the oceans were fused to $^4_2$He, how many joules of energy would be released? (b) World energy consumption is about 7.00×10^{12} W. If consumption were 100 times greater, how many years would the energy calculated in (a) last?

17. It has been pointed out that fusion reactors are safe from explosion because there is never enough energy in the plasma to do much damage. (a) In 1992, the TFTR reactor achieved an ion temperature of 4.0×10^8 K, an ion density of 2.0×10^{13} cm^{-3}, and a confinement time of 1.4 s. Calculate the amount of energy stored in the plasma of the TFTR reactor. (b) How many kilograms of water could be boiled away by this much energy? (The plasma volume of the TFTR reactor is about 50 m^3.)

18. **Review Problem.**   To confine a stable plasma, the magnetic energy density in the magnetic field (Eq. 32.14) must exceed the pressure $2nk_\text{B}T$ of the plasma by a factor of at least 10. In the following, assume a confinement time $\tau = 1.00$ s. (a) Using Lawson's criterion, determine the ion density required for the D–T reaction. (b) From the ignition temperature criterion, determine the required plasma pressure. (c) Determine the magnitude of the magnetic field required to contain the plasma.

19. Find the number of ^{6}Li and the number of ^{7}Li nuclei present in 2.00 kg of lithium. (The natural abundance of ^{6}Li is 7.5%; the remainder is ^{7}Li.)

20. One old prediction for the future was to have a fusion reactor supply energy to dissociate the molecules in garbage into separate atoms and then to ionize the atoms. This material could be put through a giant mass spectrometer, so that trash would be a new source of isotopically pure elements—the mine of the future. Assuming an average atomic mass of 56 and an average charge of 26 (a high estimate, considering all the organic materials), at a beam current of 1.00 MA, how long would it take to process 1.00 metric ton of trash?

(Optional)
Section 45.5   Radiation Damage in Matter

21. **Review Problem.**   A building has become accidentally contaminated with radioactivity. The longest-lived material in the building is strontium-90. ($^{90}_{38}$Sr has an atomic mass 89.907 7 u, and its half-life is 29.1 yr.) If the building initially contained 5.00 kg of this substance uniformly distributed throughout the building (a very unlikely situation) and the safe level is less than 10.0 counts/min, how long will the building be unsafe?

22. **Review Problem.**   A particular radioactive source produces 100 mrad of 2-MeV gamma rays per hour at a distance of 1.00 m. (a) How long could a person stand at this distance before accumulating an intolerable dose of 1 rem? (b) Assuming the radioactive source is a point source, at what distance would a person receive a dose of 10.0 mrad/h?

23. Assume that an x-ray technician takes an average of eight x-rays per day and receives a dose of 5 rem/yr as a result. (a) Estimate the dose in rem per photograph taken. (b) How does the technician's exposure compare with low-level background radiation?

24. When gamma rays are incident on matter the intensity of the gamma rays passing through the material varies with depth x as $I(x) = I_0 e^{-\mu x}$, where μ is the absorption coefficient and I_0 is the intensity of the radiation at the surface of the material. For 0.400-MeV gamma rays in lead, the absorption coefficient is 1.59 cm^{-1}. (a) Determine the "half-thickness" for lead—that is, the thickness of lead that would absorb half the incident gamma rays. (b) What thickness will reduce the radiation by a factor of 10^4?

WEB 25. A "clever" technician decides to warm some water for his coffee with an x-ray machine. If the machine produces 10.0 rad/s, how long will it take to raise the temperature of a cup of water by 50.0°C?

26. **Review Problem.** The danger to the body from a high dose of gamma rays is not due to the amount of energy absorbed but occurs because of the ionizing nature of the radiation. To illustrate this, calculate the rise in body temperature that would result if a "lethal" dose of 1 000 rad were absorbed strictly as internal energy. Take the specific heat of living tissue as 4186 J/kg · °C.

27. Technetium-99 is used in certain medical diagnostic procedures. If 1.00×10^{-8} g of ^{99}Tc is injected into a 60.0-kg patient and half of the 0.140-MeV gamma rays are absorbed in the body, determine the total radiation dose received by the patient.

28. Strontium-90 from the testing of atomic bombs can still be found in the atmosphere. Each decay of ^{90}Sr releases 1.1 MeV of energy into the bones of a person who has had strontium replace the calcium. If a 70.0-kg person receives 1.00 ng of ^{90}Sr from contaminated milk, calculate the absorbed dose rate (in J/kg) in one year. Assume the half-life of ^{90}Sr to be 29.1 yr.

(Optional)
Section 45.6 **Radiation Detectors**

29. In a Geiger tube, the voltage between the electrodes is typically 1.00 kV and the current pulse discharges a 5.00-pF capacitor. (a) What is the energy amplification of this device for a 0.500-MeV electron? (b) How many electrons are avalanched by the initial electron?

30. In a Geiger tube, the voltage between the electrodes is ΔV and the current pulse discharges a capacitor having capacitance C. (a) What is the energy amplification of this device for an electron of energy E? (b) How many electrons are avalanched by the initial electron?

31. In a certain photomultiplier tube there are seven dynodes, having potentials of 100, 200, 300, . . . , 700 V. The average energy required to free an electron from a dynode surface is 10.0 eV. Assume that just one electron is incident and that the tube functions with 100% efficiency. (a) How many electrons are freed at the first dynode? (b) How many electrons are collected by the last dynode? (c) What is the energy available to the counter?

32. (a) Your grandmother recounts to you how, as young children, your father, aunts, and uncles made the screen door slam continually as they ran between the house and the back yard. The time interval between slams varied randomly, but the average slamming rate stayed constant at 38.0/h from dawn to dusk every summer day. If the slamming rate suddenly dropped to zero, the children would have found a nest of baby field mice or gotten into some other mischief requiring adult intervention. How long after the last screen-door slam would a prudent and attentive parent wait before leaving her or his work to see about the children? Explain your reasoning. (b) A student wishes to measure the half-life of a radioactive substance, using a small sample. The clicks of her Geiger counter are randomly spaced in time. The counter registers 372 counts during one 5.00-min interval, and 337 counts during the next 5.00 min. The average background rate is 15 counts/min. Find the most probable value for the half-life. (c) Estimate the uncertainty in the half-life determination. Explain your reasoning.

(Optional)
Section 45.7 **Uses of Radiation**

33. During the manufacture of a steel engine component, radioactive iron (^{59}Fe) is included in the total mass of 0.200 kg. The component is placed in a test engine when the activity due to this isotope is 20.0 μCi. After a 1 000-h test period, oil is removed from the engine and found to contain enough ^{59}Fe to produce 800 disintegrations/min/L of oil. The total volume of oil in the engine is 6.50 L. Calculate the total mass worn from the engine component per hour of operation. (The half-life for ^{59}Fe is 45.1 days.)

34. At some time in your past or future, you may find yourself in a hospital to have a PET scan. The acronym stands for *positron-emission tomography*. In the procedure, a radioactive element that undergoes e^+ decay is introduced into your body. The equipment detects the gamma rays that result from pair annihilation when the emitted positron encounters an electron in your body's tissue. Suppose that you receive an injection of glucose containing on the order of 10^{10} atoms of ^{14}O. Assume that the oxygen is uniformly distributed through 2 L of blood after 5 min. What will be the order of magnitude of the activity of the oxygen atoms in 1 cm^3 of the blood?

35. You want to find out how many ^{65}Cu atoms are in a small sample of material. You bombard it with neutrons to ensure that on the order of 1% of these copper nuclei absorb a neutron. After activation you turn off the neutron flux and then use a highly efficient detector to monitor the gamma radiation that comes out of the sample. Assume that one half of the ^{66}Cu nuclei emit a 1.04-MeV gamma ray in their decay. (The other half of the activated nuclei decay directly to the ground state of ^{66}Ni.) If after 10 min (two half-lives) you have detected 10^4 MeV of photon energy at 1.04 MeV, (a) about how many ^{65}Cu atoms are in the sample? (b) Assume the sample contains natural copper. Refer to the isotopic abundances listed in Table A.3 and estimate the total mass of copper in the sample.

36. When a material of interest is irradiated by neutrons, radioactive atoms are produced continually and some decay according to their characteristic half-lives. (a) If one species of a radioactive nucleus is produced at a constant rate R and its decay is governed by the conven-

tional radioactive decay law, show that the number of radioactive atoms accumulated after an irradiation time t is

$$N = \frac{R}{\lambda} (1 - e^{-\lambda t})$$

(b) What is the maximum number of radioactive atoms that can be produced?

ADDITIONAL PROBLEMS

37. Carbon detonations are powerful nuclear reactions that temporarily tear apart the cores inside massive stars late in their lives. These blasts are produced by carbon fusion, which requires a temperature of about 6×10^8 K to overcome the strong Coulomb repulsion between carbon nuclei. (a) Estimate the repulsive energy barrier to fusion, using the required ignition temperature for carbon fusion. (In other words, what is the average kinetic energy for a carbon nucleus at 6×10^8 K?) (b) Calculate the energy (in MeV) released in each of these "carbon-burning" reactions:

$$^{12}\text{C} + ^{12}\text{C} \longrightarrow ^{20}\text{Ne} + ^{4}\text{He}$$

$$^{12}\text{C} + ^{12}\text{C} \longrightarrow ^{24}\text{Mg} + \gamma$$

(c) Calculate the energy (in kWh) given off when 2.00 kg of carbon completely fuses according to the first reaction.

38. The atomic bomb dropped on Hiroshima on August 6, 1945 released 5×10^{13} J of energy (equivalent to that from 12 000 tons of TNT). Estimate (a) the number of $^{235}_{92}\text{U}$ nuclei fissioned and (b) the mass of this $^{235}_{92}\text{U}$.

39. Compare the fractional mass loss in a typical ^{235}U fission reaction with the fractional mass loss in D–T fusion.

40. **Review Problem.**   Consider a nucleus at rest, which then spontaneously splits into two fragments of masses m_1 and m_2. Show that the fraction of the total kinetic energy that is carried by m_1 is

$$\frac{K_1}{K_{\text{tot}}} = \frac{m_2}{m_1 + m_2}$$

and the fraction carried by m_2 is

$$\frac{K_2}{K_{\text{tot}}} = \frac{m_1}{m_1 + m_2}$$

assuming relativistic corrections can be ignored. (*Note:* If the parent nucleus was moving before the decay, then m_1 and m_2 still divide the kinetic energy as shown, as long as all velocities are measured in the center-of-mass frame of reference, in which the total momentum of the system is zero.)

41. The half-life of tritium is 12.3 yr. If the TFTR fusion reactor contained 50.0 m^3 of tritium at a density equal to 2.00×10^{14} ions/cm^3, how many curies of tritium were

in the plasma? Compare this value with a fission inventory (the estimated supply of fissionable material) of 4×10^{10} Ci.

42. **Review Problem.**   A very slow neutron (with speed approximately equal to zero) can initiate the reaction

$$\text{n} + ^{10}_{5}\text{B} \longrightarrow ^{7}_{3}\text{Li} + ^{4}_{2}\text{He}$$

If the alpha particle moves away with a speed of 9.30×10^6 m/s, calculate the kinetic energy of the lithium nucleus. Use nonrelativistic formulas.

43. **Review Problem.**   A nuclear power plant operates by using the energy released in nuclear fission to convert 20°C water into 400°C steam. How much water could theoretically be converted to steam by the complete fissioning of 1.00 g of ^{235}U at 200 MeV/fission?

44. **Review Problem.**   A nuclear power plant operates by using the energy released in nuclear fission to convert liquid water at T_c into steam at T_h. How much water could theoretically be converted to steam by the complete fissioning of a mass m of ^{235}U at 200 MeV/fission?

45. About 1 of every 3 300 water molecules contains one deuterium atom. (a) If all the deuterium nuclei in 1 L of water are fused in pairs according to the D–D reaction $^2\text{H} + ^2\text{H} \rightarrow ^3\text{He} + \text{n} + 3.27$ MeV, how much energy in joules is liberated? (b) Burning gasoline produces about 3.40×10^7 J/L. Compare the energy obtainable from the fusion of the deuterium in a liter of water with the energy liberated from the burning of a liter of gasoline.

46. The alpha-emitter polonium-210 ($^{210}_{84}\text{Po}$) is used in a nuclear energy source on a spacecraft. Determine the initial power output of the energy source if it contains 0.155 kg of ^{210}Po. Assume that the efficiency for conversion of radioactive decay energy to electrical energy is 1.00%.

47. A certain nuclear plant generates 3.065 GW of nuclear power to create 1.000 GW of electric power. Of the waste energy, 3.0% is ejected to the atmosphere and the remainder is passed into a river. A state law requires that the river water be warmed no more than 3.50°C when it is returned to the river. (a) Determine the amount of cooling water necessary (in kg/h and m^3/h) to cool the plant. (b) If fission generates 7.80×10^{10} J/g of ^{235}U, determine the rate of fuel burning (in kg/h) of ^{235}U.

48. **Review Problem.**   The first nuclear bomb was a fissioning mass of plutonium-239, exploded in the Trinity test, before dawn on July 16, 1945 at Alamogordo, New Mexico. Enrico Fermi was 14 km away, lying on the ground facing away from the bomb. After the whole sky had flashed with unbelievable brightness, Fermi stood up and began dropping bits of paper to the ground. They first fell at his feet in the calm and silent air. As the shock wave passed, about 40 s after the explosion, the paper then in flight jumped about 5 cm away from

ground zero. (a) Assume the shock wave in air propagated equally in all directions without absorption. Find the change in volume of a sphere of radius 14 km as it expands by 5 cm. (b) Find the work $P \Delta V$ done by the air in this sphere on the next layer of air farther from the center. (c) Assume the shock wave carried on the order of one tenth of the energy of the explosion. Make an order-of-magnitude estimate of the bomb yield. (d) One ton of exploding trinitrotoluene (TNT) releases energy of 4.2 GJ. What was the order of magnitude of the energy of the Trinity test in equivalent tons of TNT? The dawn revealed the mushroom cloud. Fermi's immediate knowledge of the bomb yield agreed with that determined days later by analysis of elaborate measurements.

49. Natural uranium must be processed to produce uranium enriched in ^{235}U for bombs and power plants. The processing yields a large quantity of nearly pure ^{238}U as a by-product. Because of its high mass density, it is used in armor-piercing artillery shells. (a) Find the edge dimension of a 70.0-kg cube of ^{238}U. (Refer to Table 1.5.) (b) The isotope ^{238}U has a long half-life of 4.47×10^9 yr. As soon as one nucleus decays, it begins a relatively rapid series of 14 steps, which together constitute the net reaction

$$^{238}_{92}\text{U} \longrightarrow 8(^{4}_{2}\text{He}) + 6(^{0}_{-1}\text{e}) + ^{206}_{82}\text{Pb} + 6\bar{\nu} + Q_{net}$$

Find the net decay energy. (Refer to Table A.3.) (c) Argue that a radioactive sample of decay rate R and releasing energy Q per decay has power output $\mathcal{P} = QR$. (d) Consider an artillery shell with a jacket of 70.0 kg of ^{238}U. Find its power output due to the radioactivity of the uranium and its daughters. Assume the shell is old enough that the daughters have reached steady-state amounts. Express the power in joules per year. (e) Assume that a 17-year-old soldier of mass 70.0 kg works in an arsenal where many such artillery shells are stored. If his radiation exposure is limited to 5.00 rem per year, find the rate at which he can absorb the energy of radiation, in joules per year. Assume an average RBE factor of 1.10.

50. A 2.0-MeV neutron is emitted in a fission reactor. If it loses one half its kinetic energy in each collision with a moderator atom, how many collisions must it undergo in order to become a thermal neutron (with energy 0.039 eV)?

WEB 51. Assuming that a deuteron and a triton are at rest when they fuse according to $^2\text{H} + ^3\text{H} \rightarrow ^4\text{He} + n + 17.6$ MeV, determine the kinetic energy acquired by the neutron.

52. A sealed capsule containing the radiopharmaceutical phosphorus-32 ($^{32}_{15}$P), an e^- emitter, is implanted into a patient's tumor. The average kinetic energy of the beta particles is 700 keV. If the initial activity is 5.22 MBq, determine the absorbed dose during a 10.0-day period. Assume the beta particles are completely absorbed in 100 g of tissue. (*Hint:* Find the number of beta particles emitted.)

53. (a) Calculate the energy (in kilowatt-hours) released if 1.00 kg of ^{239}Pu undergoes complete fission and the energy released per fission event is 200 MeV. (b) Calculate the energy (in electron volts) released in the D–T fusion:

$$^2_1\text{H} + ^3_1\text{H} \longrightarrow ^4_2\text{He} + ^1_0\text{n}$$

(c) Calculate the energy (in kilowatt-hours) released if 1.00 kg of deuterium undergoes fusion according to this reaction. (d) Calculate the energy (in kilowatt-hours) released by the combustion of 1.00 kg of coal if each $C + O_2 \rightarrow CO_2$ reaction yields 4.20 eV. (e) List advantages and disadvantages of each of these methods of energy generation.

54. The Sun radiates energy at the rate of 3.77×10^{26} W. Suppose that the net reaction

$$4(^1_1\text{H}) + 2(^0_{-1}\text{e}) \longrightarrow ^4_2\text{He} + 2\nu + \gamma$$

accounts for all the energy released. Calculate the number of protons fused per second.

55. Consider the two nuclear reactions

$$(\text{I}) \qquad \text{A} + \text{B} \longrightarrow \text{C} + \text{E}$$

$$(\text{II}) \qquad \text{C} + \text{D} \longrightarrow \text{F} + \text{G}$$

(a) Show that the net disintegration energy for these two reactions ($Q_{net} = Q_\text{I} + Q_\text{II}$) is identical to the disintegration energy for the net reaction

$$\text{A} + \text{B} + \text{D} \longrightarrow \text{E} + \text{F} + \text{G}$$

(b) One chain of reactions in the proton–proton cycle in the Sun's interior is

$$^1_1\text{H} + ^1_1\text{H} \longrightarrow ^2_1\text{H} + ^0_1\text{e} + \nu$$

$$^0_1\text{e} + ^0_{-1}\text{e} \longrightarrow 2\gamma$$

$$^1_1\text{H} + ^2_1\text{H} \longrightarrow ^3_2\text{He} + \gamma$$

$$^1_1\text{H} + ^3_2\text{He} \longrightarrow ^4_2\text{He} + ^0_1\text{e} + \nu$$

$$^0_1\text{e} + ^0_{-1}\text{e} \longrightarrow 2\gamma$$

Based on part (a), what is Q_{net} for this sequence?

56. Suppose the target in a laser fusion reactor is a sphere of solid hydrogen that has a diameter of 1.50×10^{-4} m and a density of 0.200 g/cm^3. Also assume that half of the nuclei are ^{2}H and half are ^{3}H. (a) If 1.00% of a 200-kJ laser pulse is delivered to this sphere, what temperature does the sphere reach? (b) If all of the hydrogen "burns" according to the D–T reaction, how much energy in joules is released?

57. The carbon cycle, first proposed by Hans Bethe in 1939, is another cycle by which energy is released in stars as hydrogen is converted to helium. The carbon cycle requires higher temperatures than the proton–proton cy-

cle. The series of reactions is

$$^{12}C + {}^{1}H \longrightarrow {}^{13}N + \gamma$$

$$^{13}N \longrightarrow {}^{13}C + e^+ + \nu$$

$$e^+ + e^- \longrightarrow 2\gamma$$

$$^{13}C + {}^{1}H \longrightarrow {}^{14}N + \gamma$$

$$^{14}N + {}^{1}H \longrightarrow {}^{15}O + \gamma$$

$$^{15}O \longrightarrow {}^{15}N + e^+ + \nu$$

$$e^+ + e^- \longrightarrow 2\gamma$$

$$^{15}N + {}^{1}H \longrightarrow {}^{12}C + {}^{4}He$$

(a) If the proton–proton cycle requires a temperature of 1.5×10^7 K, estimate by proportion the temperature required for the carbon cycle. (b) Calculate the Q value for each step in the carbon cycle and the overall energy released. (c) Do you think the energy carried off by the neutrinos is deposited in the star? Explain.

58. When photons pass through matter, the intensity I of the beam (measured in watts per square meter) decreases exponentially according to

$$I = I_0 e^{-\mu x}$$

where I_0 is the intensity of the incident beam, and I is the intensity of the beam that just passed through a thickness x of material. The constant μ is known as the *linear absorption coefficient*, and its value depends on the absorbing material and the wavelength of the photon beam. This wavelength (or energy) dependence allows us to filter out unwanted wavelengths from a broad-spectrum x-ray beam. (a) Two x-ray beams of wavelengths λ_1 and λ_2 and equal incident intensities pass through the same metallic plate. Show that the ratio of the emergent beam intensities is

$$\frac{I_2}{I_1} = e^{-[\mu_2 - \mu_1]x}$$

(b) Compute the ratio of intensities emerging from an aluminum plate 1.00 mm thick if the incident beam contains equal intensities of 50-pm and 100-pm x-rays. The values of μ for aluminum at these two wavelengths are $\mu_1 = 5.4$ cm^{-1} at 50 pm and $\mu_2 = 41.0$ cm^{-1} at 100 pm. (c) Repeat for an aluminum plate 10.0 mm thick.

ANSWERS TO QUICK QUIZZES

45.1 A proton or another neutron. As we learned in Chapter 9, during an elastic collision, the maximum kinetic energy is transferred when the colliding objects have the same mass (see Example 9.8). Consequently, a neutron loses all of its kinetic energy when it collides head-on with a proton, which has approximately the same mass as the neutron, or with another neutron.

45.2 (a). Both forces decrease with increasing separation. However, because the nuclear (attractive) force is an extremely short-range force, it drops off much more rapidly than the electrostatic (repulsive) force between

protons. Because the attractive force becomes small much faster than the repulsive force, the nucleus fissions.

45.3 According to Figure 44.3, the ratio N/Z increases with increasing Z. As a result, when a heavy nucleus fissions to two lighter nuclei, the lighter nuclei tend to have too many neutrons. This leads to instability, and the nuclei return to the line of stability by further decay processes that reduce the number of neutrons.

45.4 Reactions (a) and (b) because in both cases the Z and A values balance on the two sides of the equations. In reaction (c), $Z_{left} = Z_{right}$ but $A_{left} \neq A_{right}$.

Both the circular particle accelerator at Fermilab and the Y-shaped radiotele-scope called the Very Large Array are, in a sense, "time machines." These devices allow us to peer back in time and better understand what the Universe was like soon after it was created. How is this possible? What were things like back then? *(Top, courtesy of Fermilab Visual Media Services; bottom, courtesy of NRAO/AUI, photo by Dave Finley)*

PUZZLER

c h a p t e r

46

Particle Physics and Cosmology

In this concluding chapter, we examine the various known subatomic particles and the fundamental interactions that govern their behavior. We also discuss the current theory of elementary particles, in which all matter is constructed from only two families of particles, quarks and leptons. Finally, we discuss how clarifications of such models might help scientists understand the birth and evolution of the Universe.

The word *atom* comes from the Greek *atomos,* which means "indivisible." The early Greeks believed that atoms were the indivisible constituents of matter; that is, they regarded them as elementary particles. Experiments in the 1890s and the early part of the 20th century showed that this was not the case, however, and after 1932 physicists viewed all matter as consisting of three constituent particles: electrons, protons, and neutrons. Beginning in the 1940s, many "new" particles were discovered in experiments involving high-energy collisions between known particles. The new particles are characteristically very unstable and have very short half-lives, ranging between 10^{-6} s and 10^{-23} s. So far, more than 300 of them have been catalogued.

Until the 1960s, physicists were bewildered by the great number and variety of subatomic particles that were being discovered. They wondered whether the particles had no systematic relationship connecting them, or whether a pattern was emerging that would provide a better understanding of the elaborate structure in the subatomic world. During the last 40 years, many high-energy particle accelerators have been constructed throughout the world, making it possible to observe collisions of highly energetic particles under controlled laboratory conditions and to see the subatomic world in finer detail. In these years, physicists have tremendously advanced our knowledge of the structure of matter by recognizing that all particles except electrons, photons, and a few others are made of smaller particles called quarks. Protons and neutrons, for example, are not truly elementary but are systems of tightly bound quarks.

46.1 ▶ THE FUNDAMENTAL FORCES IN NATURE

As noted in Section 5.1, all natural phenomena can be described by four fundamental forces acting between particles. In order of decreasing strength, they are the nuclear force, the electromagnetic force, the weak force, and the gravitational force.

The nuclear force, as we mentioned in Chapter 44, represents the glue that holds nucleons together. It is very short-range and is negligible for separations greater than about 10^{-15} m (about the size of the nucleus). The electromagnetic force, which binds atoms and molecules together to form ordinary matter, has about 10^{-2} times the strength of the nuclear force. It is a long-range force that decreases in magnitude as the inverse square of the separation between interacting particles. The weak force is a short-range force that tends to produce instability in certain nuclei. It is responsible for decay processes or the conversion of a neutron into a proton, and its strength is only about 10^{-5} times that of the nuclear force. Finally, the gravitational force is a long-range force that has a strength of only about 10^{-39} times that of the nuclear force. Although this familiar interaction is the force that holds the planets, stars, and galaxies together, its effect on elementary particles is negligible.

In modern physics, interactions between particles are often described in terms of the exchange or continuous emission and absorption of entities called **field particles** or **exchange particles.** In the case of the electromagnetic interaction, for instance, the field particles are photons. In the language of modern physics,

TABLE 46.1 Particle Interactions

Interaction	Relative Strength	Range of Force	Mediating Field Particle
Nuclear	1	Short (≈ 1 fm)	Gluon
Electromagnetic	10^{-2}	∞	Photon
Weak	10^{-5}	Short ($\approx 10^{-3}$ fm)	$W^{\pm}$, Z^{0} bosons
Gravitational	10^{-39}	∞	Graviton

the electromagnetic force is said to be *mediated* by photons, and photons are the field particles of the electromagnetic field. Likewise, the nuclear force is mediated by field particles called *gluons* (so called because they "glue" the nucleons together). The weak force is mediated by field particles called W and Z *bosons,* and the gravitational force is mediated by field particles called *gravitons.* These interactions, their ranges, and their relative strengths are summarized in Table 46.1.

46.2 POSITRONS AND OTHER ANTIPARTICLES

In the 1920s, Paul Dirac developed a relativistic quantum-mechanical description of the electron that successfully explained the origin of the electron's spin and its magnetic moment. His theory had one major problem, however: its relativistic wave equation required solutions corresponding to negative energy states, and if negative energy states existed, an electron in a state of positive energy would be expected to make a rapid transition to one of these states, emitting a photon in the process.

Dirac circumvented this difficulty by postulating that all negative energy states are filled. Electrons occupying these negative energy states are said to be in the *Dirac sea* and are not directly observable because the Pauli exclusion principle does not allow them to react to external forces. However, if one of these negative energy states is vacant, leaving a hole in the sea of filled states, the hole can react to external forces and therefore is observable. The way a hole reacts to external forces is similar to the way an electron reacts to the same force except that the hole has a positive charge—it is the *antiparticle* to the electron.

The profound implication of this theory is that *for every particle an antiparticle exists.* The antiparticle for a charged particle has the same mass as the particle but opposite charge. For example, the electron's antiparticle (now called a *positron,* as noted in Section 44.5) has a rest energy of 0.511 MeV and a positive charge of 1.60×10^{-19} C.

Carl Anderson (1905–1991) observed the positron experimentally in 1932, and in 1936 he was awarded a Nobel Prize for his achievement. Anderson discovered the positron while examining tracks created in a cloud chamber by electron-like particles of positive charge (Fig. 46.1). (These early experiments used cosmic rays—mostly energetic protons passing through interstellar space—to initiate high-energy reactions on the order of several GeV.) To discriminate between positive and negative charges, Anderson placed the cloud chamber in a magnetic field, causing moving charges to follow curved paths. He noted that some of the electron-like tracks deflected in a direction corresponding to a positively charged particle.

Paul Adrien Maurice Dirac
British physicist (1902–1984) Dirac won the Nobel Prize for physics in 1933. *(Courtesy of AIP Emilio Segré Visual Archives)*

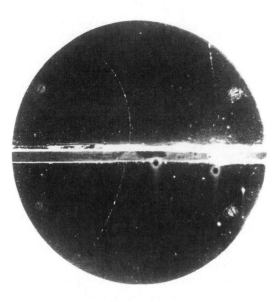

Figure 46.1 The first photograph of a positron track. The particle's track can be seen entering the picture at about the 7 o'clock position and curving upward and to the left. *(Courtesy of Archives, California Institute of Technology.)*

Quick Quiz 46.1

(a) What is the direction of the external magnetic field in Figure 46.1, into or out of the plane of the page? (b) The horizontal line is the edge of a sheet of lead. Why is the curve tighter above the lead than below it?

Since Anderson's discovery, positrons have been observed in a number of experiments. A common source of positrons is **pair production.** In this process, a gamma-ray photon with sufficiently high energy interacts with a nucleus, and an electron–positron pair is created from the photon. (The nucleus is necessary to satisfy the principle of conservation of momentum.) Because the total rest energy of the electron–positron pair is $2m_e c^2 = 1.02$ MeV (where m_e is the mass of the electron), the photon must have at least this much energy to create an electron–positron pair. Thus, electromagnetic energy in the form of a gamma ray is converted to rest energy in accordance with Einstein's famous relationship $E_R = mc^2$. If the gamma-ray photon has energy in excess of the rest energy of the electron and positron, the excess appears as kinetic energy of the two particles. Figure 46.2 shows tracks of electron–positron pairs created by 300-MeV gamma rays striking a lead sheet.

The creation of rest energy from other forms of energy is a general process and occurs in other situations besides pair production. In later sections of this chapter, we shall show how this process can be applied to understand the exchange of field particles between interacting particles.

The reverse process can also occur. Under the proper conditions, an electron and a positron can annihilate each other to produce two gamma-ray photons that have a combined energy of at least 1.02 MeV:

$$e^- + e^+ \longrightarrow 2\gamma$$

Because the initial momentum of the electron–positron system is approximately zero, two gamma rays traveling in opposite directions are necessary in this process to conserve momentum. If all the energy of the system were transformed into one photon, the momentum of the system would be high—momentum would not be conserved. Two photons can move in opposite directions with the result that the

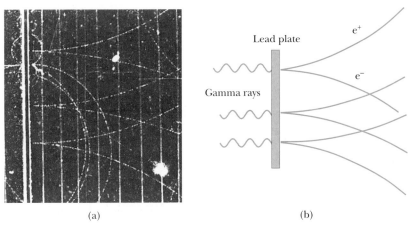

(a)   (b)

Figure 46.2 (a) Bubble-chamber tracks of electron–positron pairs produced by 300-MeV gamma rays striking a lead sheet. *(Courtesy Lawrence Berkeley Laboratory, University of California)* (b) The pertinent pair-production events. The positrons deflect upward and the electrons downward because the direction of the applied magnetic field is into the page.

momentum of the electron–positron system remains small and equal to that of the system before annihilation. Very rarely, a proton and an antiproton also annihilate each other to produce two gamma-ray photons.

Practically every known elementary particle has a distinct antiparticle. Among the exceptions are the photon and the neutral pion (π^0). Following the construction of high-energy accelerators in the 1950s, many other antiparticles were revealed. These included the antiproton, discovered by Emilio Segré (1905–1989) and Owen Chamberlain (b. 1920) in 1955, and the antineutron, discovered shortly thereafter.

Electron–positron annihilation is used in the medical diagnostic technique called *positron emission tomography* (PET). The patient is injected with a glucose solution containing a radioactive substance that decays by positron emission, and the material is carried by the blood throughout the body. A positron emitted during a decay event in one of the radioactive nuclei in the glucose solution annihilates with an electron in the surrounding tissue, resulting in two gamma-ray photons emitted in opposite directions. A gamma detector surrounding the patient pinpoints the source of the photons and, with the assistance of a computer, displays an image of the sites at which the glucose accumulates. (Glucose is metabolized rapidly in cancerous tumors and accumulates at those sites, providing a strong signal for a PET detector system.) The images from a PET scan can indicate a wide variety of disorders in the brain, including Alzheimer's disease (Fig. 46.3). In addi-

Bubble-chamber photograph of electron (green) and positron (red) tracks produced by energetic gamma rays. The highly curved tracks at the top are due to the electron and positron in an electron–positron pair bending in opposite directions in the magnetic field. *(Lawrence Berkeley Laboratory/Science Photo Library/Photo Researchers, Inc.)*

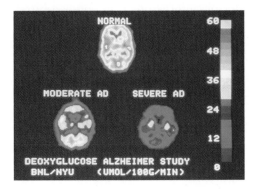

Figure 46.3 PET scans of the brain of a healthy older person and those of patients suffering from Alzheimer's disease. Lighter regions contain higher concentrations of radioactive glucose, indicating higher metabolism rates and therefore increased brain activity. *(Dr. Monty de Leon/New York University Medical Center and National Institute on Aging)*

tion, because glucose metabolizes more rapidly in active areas of the brain, a PET scan can indicate which areas of the brain are involved in the activities in which the patient is engaging at the time of the scan, such as language use, music, and vision.

Hideki Yukawa Japanese physicist (1907–1981) Yukawa was awarded the Nobel Prize in 1949 for predicting the existence of mesons. This photograph of him at work was taken in 1950 in his office at Columbia University. *(UPI/Corbis-Bettman)*

46.3 MESONS AND THE BEGINNING OF PARTICLE PHYSICS

Physicists in the mid-1930s had a fairly simple view of the structure of matter. The building blocks were the proton, the electron, and the neutron. Three other particles were either known or postulated at the time: the photon, the neutrino, and the positron. Together these six particles were considered the fundamental constituents of matter. With this marvelously simple picture of the world, however, no one was able to answer the following important question: In view of the fact that the protons in any nucleus should strongly repel one another due to their like charges, what is the nature of the force that holds the nucleus together? Scientists recognized that this mysterious force must be much stronger than anything encountered in nature up to that time. This is the nuclear force discussed in Section 44.3 and examined in historical perspective in the following paragraphs.[1]

The first theory to explain the nature of the nuclear force was proposed in 1935 by the Japanese physicist Hideki Yukawa—an effort that earned him a Nobel prize. To understand Yukawa's theory, recall the introduction of field particles in Section 46.1, which stated that each fundamental force is mediated by a field particle exchanged between the interacting particles. Yukawa used this idea to explain the nuclear force, proposing the existence of a new particle whose exchange between nucleons in the nucleus causes the nuclear force. He established that the range of the force is inversely proportional to the mass of this particle and predicted the mass to be about 200 times the mass of the electron. (Yukawa's predicted particle is *not* the gluon mentioned in Section 46.1, which is massless and is today considered to be the field particle for the nuclear force.) Because the new particle would have a mass between that of the electron and that of the proton, it was called a **meson** (from the Greek *meso,* "middle").

In efforts to substantiate Yukawa's predictions, physicists began experimental searches for the meson by studying cosmic rays entering the Earth's atmosphere. In 1937, Carl Anderson and his collaborators discovered a particle of mass $106 \text{ MeV}/c^2$, about 207 times the mass of the electron. This was thought to be Yukawa's meson. However, subsequent experiments showed that the particle interacted very weakly with matter and hence could not be the field particle for the nuclear force. That puzzling situation inspired several theoreticians to propose two mesons having slightly different masses equal to about 200 times that of the electron—one having been discovered by Anderson and the other, still undiscovered, predicted by Yukawa. This idea was confirmed in 1947 with the discovery of the **pi**

---

[1] The nuclear force that we discussed in Chapter 44 and continue to discuss here was originally called the *strong* force. Once the quark theory (Section 46.9) was established, the phrase *strong force* was reserved for the force between quarks. We shall follow this convention—the strong force occurs between quarks, and the nuclear force occurs between nucleons. The nuclear force is a secondary result of the strong force, as we shall discuss in Section 46.10. Be prepared, however—because of this historical development of the names for these forces, you may find the nuclear force referred to as the strong force in other books.

meson (π), or simply **pion.** The particle discovered by Anderson in 1937, the one initially thought to be Yukawa's meson, is not really a meson. (We shall discuss the requirements for a particle to be a meson in Section 46.4.) Instead, it takes part in the weak and electromagnetic interactions only and is now called the **muon** (μ).

The pion comes in three varieties, corresponding to three charge states: π^+, π^-, and π^0. The π^+ and π^- particles (π^- is the antiparticle of π^+) each have a mass of 139.6 MeV/c^2, and the π^0 mass is 135.0 MeV/c^2. Two muons exist—μ^- and its antiparticle μ^+.

Pions and muons are very unstable particles. For example, the π^-, which has a mean lifetime of 2.6×10^{-8} s, decays to a muon and an antineutrino. The muon, which has a mean lifetime of 2.2 μs, then decays to an electron, a neutrino, and an antineutrino:

$$\pi^- \longrightarrow \mu^- + \bar{\nu} \qquad \textbf{(46.1)}$$
$$\mu^- \longrightarrow e^- + \nu + \bar{\nu}$$

Note that for chargeless particles (as well as some charged particles, such as the proton), a bar over the symbol indicates an antiparticle, as in beta decay (see Section 44.6).

The interaction between two particles can be represented in a simple diagram called a **Feynman diagram,** developed by the American physicist Richard P. Feynman. Figure 46.4 is such a diagram for the electromagnetic interaction between two electrons. A Feynman diagram is a qualitative graph of time on the vertical axis versus space on the horizontal axis. It is qualitative in the sense that the actual values of time and space are not important, but the overall appearance of the graph provides a representation of the process. The time evolution of the process can be approximated by starting at the bottom of the diagram and moving your eyes upward.

In the simple case of the electron–electron interaction in Figure 46.4, a photon (the field particle) mediates the electromagnetic force between the electrons. Notice that the entire interaction is represented in the diagram as occurring at a single point in time. Thus, the paths of the electrons appear to undergo a discontinuous change in direction at the moment of interaction. This is different from the *actual* paths, which would be curved due to the continuous exchange of large numbers of field particles. This is another aspect of the qualitative nature of Feynman diagrams.

In the electron–electron interaction, the photon, which transfers energy and momentum from one electron to the other, is called a *virtual photon* because it vanishes during the interaction without having been detected. Virtual photons do not violate the law of conservation of energy because they have a very short lifetime Δt that makes the uncertainty in the energy $\Delta E \approx \hbar/2\,\Delta t$ of the system consisting of two electrons and the photon greater than the photon energy.

Now consider a pion mediating the nuclear force between a proton and a neutron, as in Yukawa's model (Fig. 46.5a). We can reason that the rest energy ΔE_R needed to create a pion of mass m_π is given by Einstein's equation $\Delta E_R = m_\pi c^2$. As with the photon in Figure 46.4, the very existence of the pion would violate the law of conservation of energy if the particle existed for a time greater than $\Delta t \approx \hbar/2\,\Delta E_R$ (from the uncertainty principle), where ΔE_R is the rest energy of the pion and Δt is the time it takes the pion to transfer from one nucleon to the other. Therefore,

$$\Delta t \approx \frac{\hbar}{2\,\Delta E_R} = \frac{\hbar}{2m_\pi c^2} \qquad \textbf{(46.2)}$$

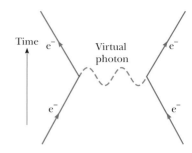

Time $\quad$ e⁻ $\qquad$ Virtual photon $\qquad$ e⁻

e⁻ $\qquad\qquad$ e⁻

Figure 46.4 Feynman diagram representing a photon mediating the electromagnetic force between two electrons.

Richard Feynman **American physicist (1918–1988)** Feynman with his son, Carl, after winning the Nobel Prize for physics in 1965. The prize was shared by Feynman, Julian Schwinger, and Sin Itiro Tomonaga. *(UPI Telephotos)*

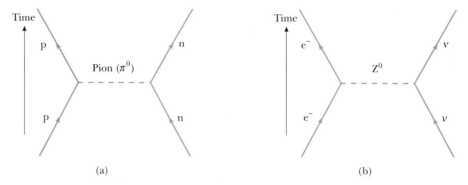

Figure 46.5   (a) Feynman diagram representing a proton and a neutron interacting via the nuclear force with a pion mediating the force. (b) Feynman diagram for an electron and a neutrino interacting via the weak force, with a Z^0 boson mediating the force.

Because the pion cannot travel faster than the speed of light, the maximum distance d it can travel in a time Δt is $c\,\Delta t$. Thus,

$$d = c\,\Delta t \approx \frac{\hbar}{2m_\pi c} \tag{46.3}$$

From Table 46.1, we know that the range of the nuclear force is approximately 1×10^{-15} m. Using this value for d in Equation 46.3, we estimate the rest energy of the pion to be

$$m_\pi c^2 \approx \frac{\hbar c}{2d} = \frac{(1.05 \times 10^{-34}\,\text{J·s})(3.00 \times 10^8\,\text{m/s})}{2(1 \times 10^{-15}\,\text{m})}$$

$$= 1.6 \times 10^{-11}\,\text{J} \approx 100\,\text{MeV}$$

This is the same order of magnitude as the observed masses of the pions, thus boosting confidence in the field-particle model.

The concept we have just described is quite revolutionary. In effect, it says that a system of two nucleons can change into two nucleons plus a pion, as long as it returns to its original state in a very short time. (Remember that this is the older, historical model in which we assume that the pion is the field particle for the nuclear force; keep in mind that the gluon is the actual field particle in current models, as we shall discuss shortly.) Physicists often say that a nucleon undergoes *fluctuations* as it emits and absorbs pions. As we have seen, these fluctuations are a consequence of a combination of quantum mechanics (through the uncertainty principle) and special relativity (through Einstein's energy–mass relationship $E_R = mc^2$).

This section has dealt with the field particles that were originally proposed to mediate the nuclear force (pions) and those that mediate the electromagnetic force (photons). The graviton, the field particle for the gravitational force, has yet to be observed. The $W^\pm$ and Z^0 particles, which mediate the weak force, were discovered in 1983 by the Italian physicist Carlo Rubbia (b. 1934) and his associates, using a proton–antiproton collider. Rubbia and Simon van der Meer (b. 1925), both at CERN (the European Laboratory for Particle Physics), shared the 1984 Nobel Prize for the discovery of the $W^\pm$ and Z^0 particles and the development of the proton–antiproton collider. Figure 46.5b shows a Feynman diagram for a weak interaction mediated by a Z^0 boson.

What does the infinite range of the electromagnetic and gravitational interactions tell you about the masses of the photon and graviton?

46.4 ▶ CLASSIFICATION OF PARTICLES

All particles other than field particles can be classified into two broad categories, hadrons and leptons, according to the interactions in which they take part. Table 46.2 provides a summary of the properties of some of these particles.

Hadrons

Particles that interact through the nuclear force are called **hadrons.** The two classes of hadrons, *mesons* and *baryons,* are distinguished by their masses and spins.

Mesons all have zero or integer spin (0 or 1). As indicated in Section 46.3, the name comes from the expectation that Yukawa's proposed meson mass would lie between the masses of the electron and the proton. Several meson masses do lie in this range, although mesons having masses greater than that of the proton do exist.

All mesons are known to decay finally into electrons, positrons, neutrinos, and photons. The pions are the lightest known mesons; they have masses of about 1.4×10^2 MeV/c^2, and all three pions—π^+, π^-, and π^0—have a spin of 0. (This indicates that the particle discovered by Anderson in 1937, the muon, is not a meson; the muon has spin $\frac{1}{2}$. It belongs in the *lepton* classification, described below.)

Baryons, the second class of hadrons, have masses equal to or greater than the proton mass (the name *baryon* means "heavy" in Greek), and their spin is always a noninteger value ($\frac{1}{2}$ or $\frac{3}{2}$). Protons and neutrons are baryons, as are many other particles. With the exception of the proton, all baryons decay in such a way that the end products include a proton. For example, the baryon called the Ξ hyperon (Greek capital xi) decays to the Λ^0 baryon (Greek capital lambda) in about 10^{-10} s. The Λ^0 then decays to a proton and a π^- in approximately 3×10^{-10} s.

Today it is believed that hadrons are not elementary particles but are composed of more elementary units called quarks, as we shall see in Section 46.9.

Leptons

Leptons (from the Greek *leptos,* meaning "small" or "light") are a group of particles that do not interact by means of the nuclear force. All leptons have spin $\frac{1}{2}$. Whereas hadrons have size and structure, leptons appear to be truly elementary, meaning that they have no structure and are point-like.

Quite unlike the case with hadrons, the number of known leptons is small. Currently, scientists believe that only six leptons exist—the electron, the muon, the tau, and a neutrino associated with each:

$$\begin{pmatrix} e^- \\ \nu_e \end{pmatrix} \qquad \begin{pmatrix} \mu^- \\ \nu_\mu \end{pmatrix} \qquad \begin{pmatrix} \tau^- \\ \nu_\tau \end{pmatrix}$$

The tau lepton, discovered in 1975, has a mass about twice that of the proton. The neutrino associated with the tau has not yet been observed in the laboratory. Each of the six leptons has an antiparticle.

TABLE 46.2 **Some Particles and Their Properties**

Category	Particle Name	Symbol	Antiparticle	Mass (MeV/c²)	B	L_e	L_μ	L_τ	S	Lifetime(s)	Principal Decay Modes[a]
Leptons	Electron	e^-	e^+	0.511	0	+1	0	0	0	Stable	—
	Electron neutrino	ν_e	$\bar{\nu}_e$	$<7\times10^{-6}$	0	+1	0	0	0	Stable	—
	Muon	μ^-	μ^+	105.7	0	0	+1	0	0	2.20×10^{-6}	$e^-\bar{\nu}_e\nu_\mu$
	Muon neutrino	ν_μ	$\bar{\nu}_\mu$	<0.3	0	0	+1	0	0	Stable	—
	Tau	τ^-	τ^+	1784	0	0	0	+1	0	$<4\times10^{-13}$	$\mu^-\bar{\nu}_\mu\nu_\tau,\ e^-\bar{\nu}_e\nu_\tau$
	Tau neutrino	ν_τ	$\bar{\nu}_\tau$	<30	0	0	0	+1	0	Stable	—
Hadrons											
Mesons	Pion	π^+	π^-	139.6	0	0	0	0	0	2.60×10^{-8}	$\mu^+\nu_\mu$
		π^0	Self	135.0	0	0	0	0	0	0.83×10^{-16}	2γ
	Kaon	K^+	K^-	493.7	0	0	0	0	+1	1.24×10^{-8}	$\mu^+\nu_\mu,\ \pi^+\pi^0$
		K_S^0	$\bar{K}_S^0$	497.7	0	0	0	0	+1	0.89×10^{-10}	$\pi^+\pi^-,\ 2\pi^0$
		K_L^0	$\bar{K}_L^0$	497.7	0	0	0	0	+1	5.2×10^{-8}	$\pi^\pm e^\mp \bar{\nu}_e,\ 3\pi^0$ $\pi^\pm \mu^\mp \bar{\nu}_\mu$
	Eta	η	Self	548.8	0	0	0	0	0	$<10^{-18}$	$2\gamma,\ 3\pi$
		η'	Self	958	0	0	0	0	0	2.2×10^{-21}	$\eta\pi^+\pi^-$
Baryons	Proton	p	$\bar{p}$	938.3	+1	0	0	0	0	Stable	—
	Neutron	n	$\bar{n}$	939.6	+1	0	0	0	0	920	$pe^-\bar{\nu}_e$
	Lambda	Λ^0	$\bar{\Lambda}^0$	1115.6	+1	0	0	0	-1	2.6×10^{-10}	$p\pi^-,\ n\pi^0$
	Sigma	Σ^+	$\bar{\Sigma}^-$	1189.4	+1	0	0	0	-1	0.80×10^{-10}	$p\pi^0,\ n\pi^+$
		Σ^0	$\bar{\Sigma}^0$	1192.5	+1	0	0	0	-1	6×10^{-20}	$\Lambda^0\gamma$
		Σ^-	$\bar{\Sigma}^+$	1197.3	+1	0	0	0	-1	1.5×10^{-10}	$n\pi^-$
	Xi	Ξ^0	$\bar{\Xi}^0$	1315	+1	0	0	0	-2	2.9×10^{-10}	$\Lambda^0\pi^0$
		Ξ^-	$\bar{\Xi}^+$	1321	+1	0	0	0	-2	1.64×10^{-10}	$\Lambda^0\pi^-$
	Omega	Ω^-	$\bar{\Omega}^+$	1672	+1	0	0	0	-3	0.82×10^{-10}	$\Xi^0\pi^-,\ \Lambda^0K^-$

[a] Notations in this column such as $p\pi^-$, $n\pi^0$ mean two possible decay modes. In this case, the two possible decays are $\Lambda^0 \rightarrow p + \pi^-$ and $\Lambda^0 \rightarrow n + \pi^0$.

Current studies indicate that neutrinos have a small but nonzero mass. If they do have mass, then they cannot travel at the speed of light. Also, so many neutrinos exist that their combined mass may be sufficient to cause all the matter in the Universe to eventually collapse into a single point, which might then explode and create a completely new Universe! We shall discuss this possibility in more detail in Section 46.12.

46.5   CONSERVATION LAWS

In Chapter 44 we learned that conservation laws are important for understanding why certain radioactive decays and nuclear reactions occur and others do not. In general, the laws of conservation of energy, linear momentum, angular momentum, and electric charge provide us with a set of rules that all processes must follow. In the study of elementary particles, a number of additional conservation laws are important. Although the two described here have no theoretical foundation, they are supported by abundant empirical evidence.

Baryon Number

The law of conservation of baryon number tells us that whenever a baryon is created in a decay or nuclear reaction, an antibaryon is also created. This scheme can be quantified by assigning every particle a quantum number, the **baryon number,** as follows: $B = +1$ for all baryons, $B = -1$ for all antibaryons, and $B = 0$ for all other particles. Thus, the **law of conservation of baryon number** states that **whenever a nuclear reaction or decay occurs, the sum of the baryon numbers before the process must equal the sum of the baryon numbers after the process.**

Conservation of baryon number

If baryon number is absolutely conserved, the proton must be absolutely stable. If it were not for the law of conservation of baryon number, the proton could decay to a positron and a neutral pion. However, such a decay has never been observed. At the present, all we can say is that the proton has a half-life of at least 10^{33} years (the estimated age of the Universe is only 10^{10} years). In one recent theory, however, physicists predicted that the proton is unstable. According to this theory, baryon number is not absolutely conserved.

EXAMPLE 46.1   Checking Baryon Numbers

Use the law of conservation of baryon number to determine whether the following reactions can occur: (a) $p + n \rightarrow p + p + n + \overline{p}$; (b) $p + n \rightarrow p + p + \overline{p}$.

Solution   (a) The left side of the equation gives a total baryon number of $1 + 1 = 2$. The right side gives a total baryon number of $1 + 1 + 1 + (-1) = 2$. Thus, baryon number is conserved and the reaction can occur (provided

the incoming proton has sufficient energy that energy conservation is satisfied).

(b) The left side of the equation gives a total baryon number of $1 + 1 = 2$. However, the right side gives $1 + 1 + (-1) = 1$. Because baryon number is not conserved, the reaction cannot occur.

EXAMPLE 46.2 ▶ Detecting Proton Decay

Measurements taken at the Super Kamiokande neutrino detection facility (Fig. 46.6) indicate that the half-life of protons is at least 10^{33} years. Estimate how long we would have to watch, on average, to observe the decay of a proton in a glass of water.

Solution   Let us estimate that a glass contains about $\frac{1}{4}$ L, or 250 g, of water. The number of molecules of water is

$$\frac{(250\,\text{g})(6.02 \times 10^{23}\,\text{molecules/mol})}{18\,\text{g/mol}} = 8.4 \times 10^{24}\,\text{molecules}$$

Each water molecule contains one proton in each of its two hydrogen atoms plus eight protons in its oxygen atom. Thus, the glass of water contains 8.4×10^{25} protons. The decay constant is given by Equation 44.9:

$$\lambda = \frac{0.693}{T_{1/2}} = \frac{0.693}{10^{33}\,\text{yr}} = 6.9 \times 10^{-34}\,\text{yr}^{-1}$$

This is the probability that any one proton will decay in one year. The probability that any proton in our glass of water will decay in the one-year interval is (Eqs. 44.6 and 44.8)

$$R = (8.4 \times 10^{25})(6.9 \times 10^{-34}\,\text{yr}^{-1}) = 5.8 \times 10^{-8}\,\text{yr}^{-1}$$

So we have to watch our glass of water for $1/R \approx$

17 million years!

Exercise   The Super Kamiokande neutrino facility contains 50 000 tons of water. Estimate the average time between detected proton decays if the half-life is 10^{33} yr.

Answer   Approximately 1 yr.

Figure 46.6   This detector at the Super Kamiokande neutrino facility in Japan is used to study photons and neutrinos. It holds 50 000 tons of highly purified water and 13 000 photomultipliers. The photograph was taken while the detector was being filled. Technicians use a raft to clean the photodetectors before they are submerged. *(Courtesy of KRR [Institute for Cosmic Ray Research], University of Tokyo)*

Lepton Number

We have three conservation laws involving lepton numbers, one for each variety of lepton. The **law of conservation of electron lepton number** states that whenever a nuclear reaction or decay occurs, **the sum of the electron lepton numbers before the process must equal the sum of the electron lepton numbers after the process.**

Conservation of lepton number

The electron and the electron neutrino are assigned an electron lepton number $L_e = +1$, the antileptons e^+ and $\bar{\nu}_e$ are assigned an electron lepton number $L_e = -1$, and all other particles have $L_e = 0$. For example, consider the decay of the neutron:

$$n \longrightarrow p + e^- + \bar{\nu}_e$$

Before the decay, the electron lepton number is $L_e = 0$; after the decay, it is $0 + 1 + (-1) = 0$. Thus, electron lepton number is conserved. (Baryon number must also be conserved, of course, and it is: Before the decay $B = +1$, and after the decay $B = +1 + 0 + 0 = +1$.)

Similarly, when a decay involves muons, the muon lepton number L_μ is conserved. The μ^- and the ν_μ are assigned a muon lepton number $L_\mu = +1$, the antimuon μ^+ and the muon antineutrino $\bar{\nu}_\mu$ are assigned a muon lepton number $L_\mu = -1$, and all other particles have $L_\mu = 0$.

Finally, tau lepton number L_τ is conserved with similar assignments for the tau lepton, its neutrino, and other particles.

EXAMPLE 46.3   Checking Lepton Numbers

Use the law of conservation of lepton numbers to determine which of the following decay schemes can occur: (a) $\mu^- \rightarrow e^- + \bar{\nu}_e + \nu_\mu$; (b) $\pi^+ \rightarrow \mu^+ + \nu_\mu + \nu_e$.

Solution   (a) Because this decay involves a muon and an electron, L_μ and L_e must both be conserved. Before the decay, $L_\mu = +1$ and $L_e = 0$. After the decay, $L_\mu = 0 + 0 + 1 = +1$ and $L_e = +1 + (-1) + 0 = 0$. Thus, both numbers are conserved, and on this basis the decay is possible.

(b) Before the decay, $L_\mu = 0$ and $L_e = 0$. After the decay, $L_\mu = -1 + 1 + 0 = 0$, but $L_e = 0 + 0 + 1 = 1$. Thus, the decay is not possible because electron lepton number is not conserved.

Exercise   Determine whether the decay $\mu^- \rightarrow e^- + \bar{\nu}_e$ can occur.

Answer   No, because L_μ is $+1$ before the decay and 0 after.

Quick Quiz 46.3

A scientist claims to have observed the decay of an electron into two electron neutrinos. Is this believable?

46.6   STRANGE PARTICLES AND STRANGENESS

Many particles discovered in the 1950s were produced by the interaction of pions with protons and neutrons in the atmosphere. A group of these—the kaon (K), lambda (Λ), and sigma (Σ) particles—exhibited unusual properties both as they were created and as they decayed; hence, they were called *strange particles.*

One unusual property of strange particles is that they are always produced in pairs. For example, when a pion collides with a proton, a highly probable result is the production of two neutral strange particles (Fig. 46.7):

$$\pi^- + p \longrightarrow K^0 + \Lambda^0$$

However, the reaction $\pi^- + p \rightarrow K^0 + n^0$, in which only one of the final particles is strange, never occurs, even though no known conservation laws would be violated and even though the energy of the pion is sufficient to initiate the reaction.

The second peculiar feature of strange particles is that, although they are produced in reactions involving the nuclear interaction at a high rate, they do not decay into particles that interact via the nuclear force at a high rate. Instead, they decay very slowly, which is characteristic of the weak interaction. Their half-lives are in the range 10^{-10} s to 10^{-8} s, whereas most other particles that interact via the nuclear force have lifetimes on the order of 10^{-23} s.

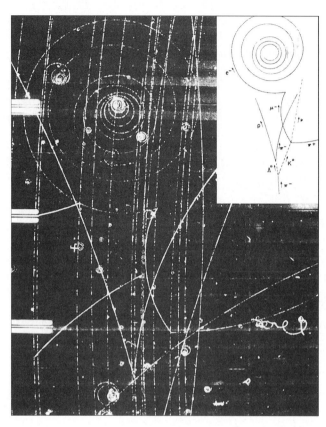

Figure 46.7 This bubble-chamber photograph shows many events, and the inset is a drawing of identified tracks. The strange particles Λ^0 and K^0 are formed at the bottom as a π^- particle interacts with a proton in the reaction $\pi^- + p \rightarrow \Lambda^0 + K^0$. (Note that the neutral particles leave no tracks, as indicated by the dashed lines in the inset.) The Λ^0 then decays in the reaction $\Lambda^0 \rightarrow \pi^- + p$, and the K^0 decays in the reaction $K^0 \rightarrow \pi^+ + \mu^- + \overline{\nu}_\mu$. *(Courtesy Lawrence Berkeley Laboratory, University of California, Photographic Services)*

To explain these unusual properties of strange particles, a new quantum number S, called **strangeness,** was introduced, together with a conservation law. The strangeness numbers for some particles are given in Table 46.2. The production of strange particles in pairs is explained by assigning $S = +1$ to one of the particles, $S = -1$ to the other, and $S = 0$ to all nonstrange particles. The **law of conservation of strangeness** states that **whenever a nuclear reaction or decay occurs, the sum of the strangeness numbers before the process must equal the sum of the strangeness numbers after the process.**

> Conservation of strangeness number

The low decay rate of strange particles can be explained by assuming that the nuclear and electromagnetic interactions obey the law of conservation of strangeness but the weak interaction does not. Because the decay of a strange particle involves the loss of one strange particle, it violates strangeness conservation and hence proceeds slowly via the weak interaction.

EXAMPLE 46.4   **Is Strangeness Conserved?**

(a) Use the law of conservation of strangeness to determine whether the reaction $\pi^0 + n \rightarrow K^+ + \Sigma^-$ occurs.

Solution From Table 46.2, we see that the initial strangeness is $S = 0 + 0 = 0$. Because the strangeness of the K^+ is

$S = +1$ and the strangeness of the Σ^- is $S = -1$, the strangeness of the final products is $+1 - 1 = 0$. Thus, strangeness is conserved, and the reaction is allowed.

(b) Show that the reaction $\pi^- + p \rightarrow \pi^- + \Sigma^+$ does not conserve strangeness.

Solution Before: $S = 0 + 0 = 0$; after: $S = 0 + (-1) = -1$. Thus, strangeness is not conserved.

Exercise Show that the reaction $p + \pi^- \rightarrow K^0 + \Lambda^0$ obeys the law of conservation of strangeness.

46.7 MAKING PARTICLES AND MEASURING THEIR PROPERTIES

The bewildering array of entries in Table 46.2 leaves one yearning for firm ground. For example, it is natural to wonder about an entry for a particle (Σ^0) that exists for 10^{-20} s and has a mass of 1 192.5 MeV/c^2. How is it possible to detect a particle that exists for only 10^{-20} s? Furthermore, how can its mass be measured when it exists for such a short time? If a standard attribute of a particle is some type of permanence or stability, in what sense is such a fleeting entity a particle? In this section we answer such questions and explain how elementary particles are produced and their properties measured.

Most elementary particles are unstable and are created in nature only rarely, in cosmic ray showers. In the laboratory, however, great numbers of these particles are created in controlled collisions between high-energy particles and a suitable target. The incident particles must have very high energy, and it takes considerable time for electromagnetic fields to accelerate particles to high energies. Thus, stable charged particles such as electrons or protons generally make up the incident beam. In addition, targets must be simple and stable, and the simplest target, hydrogen, serves nicely as both a target (the proton) and a detector.

Figure 46.7 documents a typical event in which a bubble chamber served as both target source and detector. Many parallel tracks of negative pions are visible entering the photograph from the bottom. As the labels in the inset drawing show, one of the pions has hit a stationary proton in the hydrogen and produced two strange particles, Λ^0 and K^0, according to the reaction

$$\pi^- + p \longrightarrow \Lambda^0 + K^0$$

Neither neutral strange particle leaves a track, but their subsequent decay into charged particles can be seen in Figure 46.7. A magnetic field directed into the plane of the page causes the track of each charged particle to curve, and from the measured curvature we can determine the particle's charge and linear momentum. If the mass and momentum of the incident particle are known, we can then usually calculate the product particle's mass, kinetic energy, and speed from the laws of conservation of momentum and energy. Finally, combining a product particle's speed with the length of the track it leaves, we can calculate the particle's lifetime. Figure 46.7 shows that sometimes we can use this lifetime technique even for a neutral particle, which leaves no track. As long as the start and finish of the missing track are known, as well as the particle speed, we can infer the missing track length and so determine the lifetime of the neutral particle.

Resonance Particles

With clever experimental technique and much effort, decay track lengths as short as 10^{-6} m can be measured. This means that lifetimes as short as 10^{-16} s can be measured for high-energy particles traveling at about the speed of light. We arrive at this result by assuming that a decaying particle travels 1 μm at a speed of 0.99c in the reference frame of the laboratory, yielding a lifetime of $\Delta t_{\text{lab}} = 1 \times 10^{-6}\,\text{m}/0.99c \approx 3.4 \times 10^{-15}\,\text{s}$. This is not our final result, however, because we must account for the relativistic effects of time dilation. Because the proper lifetime Δt_p as measured in the decaying particle's reference frame is shorter than the laboratory-frame value Δt_{lab} by a factor of $\sqrt{1 - (v^2/c^2)}$ (see Eq. 39.7), we can calculate the proper lifetime:

$$\Delta t_p = \Delta t_{\text{lab}} \sqrt{1 - \frac{v^2}{c^2}} = (3.4 \times 10^{-15}\,\text{s}) \sqrt{1 - \frac{(0.99c)^2}{c^2}} = 4.8 \times 10^{-16}\,\text{s}$$

Unfortunately, even with Einstein's help, the best answer we can obtain with the track-length method is several orders of magnitude away from lifetimes of 10^{-20} s. How, then, can we detect the presence of particles that exist for time intervals such as 10^{-20} s? As we shall soon see, for such short-lived particles, known as **resonance particles,** all we can do is infer their masses, their lifetimes, and their very existence from data on their decay products.

Let us consider this detection process in detail by examining the case of the resonance particle called the delta plus (Δ^+), which has a mass of 1 232 MeV/c^2 and a lifetime of about 10^{-23} s, even shorter than the most short-lived particle listed in Table 46.2. This particle is produced in the reaction

$$\text{e}^- + \text{p} \longrightarrow \text{e}^- + \Delta^+ \tag{46.4}$$

followed in 10^{-23} s by the decay

$$\Delta^+ \longrightarrow \pi^+ + \text{n} \tag{46.5}$$

Because the Δ^+ lifetime is so short, the particle leaves no measurable track in a bubble chamber. It might therefore seem impossible to distinguish the reactions given in Equations 46.4 and 46.5 from the reaction

$$\text{e}^- + \text{p} \longrightarrow \text{e}^- + \pi^+ + \text{n} \tag{46.6}$$

in which the reactants of Equation 46.4 decay directly to e^-, π^+, and n with no intermediate step in which a Δ^+ is produced. Distinguishing between these two possibilities is not impossible, however. If a Δ^+ particle exists, it has a distinct rest energy, which must come from the kinetic energy of the incoming particles. If we imagine firing electrons with increasing kinetic energy at protons, eventually we will provide enough energy to the system to create the Δ^+ particle. This is very similar to firing photons of increasing energy at an atom until you fire them with enough energy to excite the atom to a higher quantum state. In fact, the Δ^+ particle is an excited state of the proton, which we can understand via the quark theory discussed in Section 46.9. After the Δ^+ particle is formed, its rest energy becomes the energies of the outgoing pion and neutron. Equation 39.26 can be solved for the rest energy of the Δ^+ particle in terms of its kinetic energy and linear momentum:

$$(m_{\Delta^+}c^2)^2 = E_{\Delta^+}{}^2 - p_{\Delta^+}{}^2c^2 = E_{\Delta^+}{}^2 - (\mathbf{p}_{\Delta^+})^2 c^2$$

When the Δ^+ particle decays into a pion and a neutron, conservation of energy and momentum requires that

$$E_{\Delta^+} = E_{\pi^+} + E_{\text{n}} \qquad \mathbf{p}_{\Delta^+} = \mathbf{p}_{\pi^+} + \mathbf{p}_{\text{n}}$$

Thus, the rest energy of the Δ^+ particle can be expressed in terms of the energies and momenta of the outgoing particles, which can all be measured in the bubble-

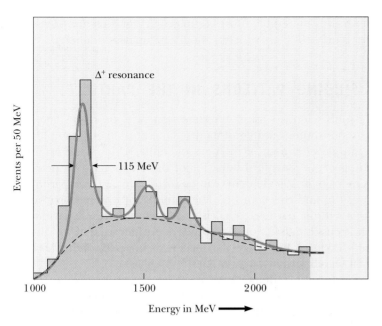

Figure 46.8   Experimental evidence for the existence of the Δ^+ particle. The sharp peak at 1 232 MeV was produced by events in which a Δ^+ formed and promptly decayed to a π^+ and a neutron.

chamber photograph:

$$(m_{\Delta^+}c^2)^2 = (E_{\pi^+} + E_n)^2 - (\mathbf{p}_{\pi^+} + \mathbf{p}_n)^2 c^2$$

Any pions and neutrons that come from the decay of a Δ^+ particle must have energies and momenta that combine in this equation to give the rest energy of the Δ^+ particle. Pions and neutrons coming from the reaction of Equation 46.6 will have a variety of energies and momenta with no particular pattern because the energy of the reactants can divide up in many ways among the three outgoing particles in this reaction. At the energy at which the rest energy of the Δ^+ particle can be created, many reactions occur, as evidenced by the proper combinations of energy and momentum already described.

To show the existence of the Δ^+ particle, we analyze a great number of events in which a π^+ and a neutron are produced. Then the number of events in a given energy range is plotted versus energy. Following this procedure, we obtain a slowly varying curve that has a sharp peak superimposed on it. The peak represents the incident electron energy at which the rest energy of the Δ^+ particle was created, revealing the existence of the particle.

Figure 46.8 is an experimental plot for the Δ^+ particle. The dashed broad curve is produced by direct events in which no Δ^+ was created (see Eq. 46.6). The sharp peak at 1 232 MeV was produced by all the events in which a Δ^+ was formed and decayed to a pion and a neutron. Thus, the rest energy of the Δ^+ particle is 1 232 MeV. Peaks corresponding to two resonance particles with masses greater than that of the Δ^+ particle can also be seen in Figure 46.8.

Graphs such as Figure 46.8 can tell us not only the mass of a short-lived particle but also its lifetime. The width of the resonance peak and the uncertainty relationship $\Delta E \, \Delta t \approx \hbar/2$ are used to infer the lifetime Δt of the particle. The measured width of 115 MeV in Figure 46.8 leads to a lifetime of 0.57×10^{-23} s for the

Δ^+ particle. In this incredibly short lifetime, a Δ^+ particle moving at the limiting speed c travels only 10^{-15} m, which is about one nuclear diameter.

46.8 ▷ FINDING PATTERNS IN THE PARTICLES

One of the tools scientists use is the detection of patterns in data, patterns that contribute to our understanding of nature. One of the best examples of the use of this tool is the development of the periodic table, which provides a fundamental understanding of the chemical behavior of the elements. The periodic table explains how more than 100 elements can be formed from three particles—the electron, the proton, and the neutron. The number of observed particles and resonances observed by particle physicists is even greater than the number of elements. Is it possible that a small number of entities exist from which all of these can be built? Taking a hint from the success of the periodic table, let us explore the historical search for patterns among the particles.

Many classification schemes have been proposed for grouping particles into families. Consider, for instance, the baryons listed in Table 46.2. We can consider these particles as belonging to a group based on the fact that they all have spins of $\frac{1}{2}$. If we plot strangeness versus charge for these baryons using a sloping coordinate system, as in Figure 46.9a, we observe a fascinating pattern: Six of the baryons form a hexagon, and the remaining two are at the hexagon's center.

As a second example, consider the following nine spin-zero mesons listed in Table 46.2: π^+, π^0, π^-, K^+, K^0, K^-, η, η', and the antiparticle $\overline{K}^0$. Figure 46.9b is a plot of strangeness versus charge for this family. Again, a hexagonal pattern emerges. In this case, each particle on the perimeter of the hexagon lies opposite its antiparticle, and the remaining three (which form their own antiparticles) are at the center of the hexagon. These and related symmetric patterns were devel-

Murray Gell-Mann American physicist (b. 1929) Murray Gell-Mann was awarded the Nobel Prize in 1969 for his theoretical studies dealing with subatomic particles.
(Courtesy of Michael R. Dressler)

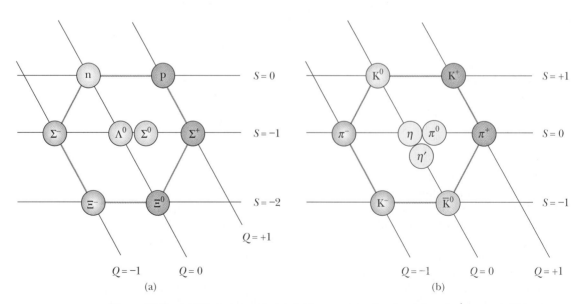

Figure 46.9 (a) The hexagonal eightfold-way pattern for the eight spin-$\frac{1}{2}$ baryons. This strangeness-versus-charge plot uses a sloping axis for charge number Q and a horizontal axis for strangeness S. (b) The eightfold-way pattern for the nine spin-zero mesons.

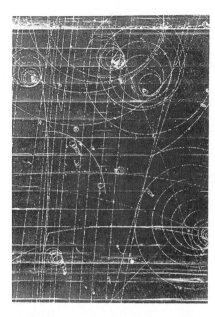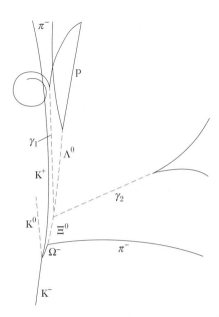

Figure 46.10 Discovery of the Ω^- particle. The K^- particle at the bottom of the photograph collides with a proton to produce the first detected Ω^- particle plus two other particles. *(Courtesy of Brookhaven National Laboratory.)*

oped independently in 1961 by Murray Gell-Mann and Yuval Ne'eman. Gell-Mann called the patterns the **eightfold way,** after the eightfold path to nirvana in Buddhism.

Groups of baryons and mesons can be displayed in many other symmetric patterns within the framework of the eightfold way. For example, the family of spin-$\frac{3}{2}$ baryons contains ten particles arranged in a pattern like that of the pins in a bowling alley. After this pattern was proposed, an empty spot occurred in it, corresponding to a particle that had never been observed. Gell-Mann predicted that the missing particle, which he called the omega minus (Ω^-), should have spin $\frac{3}{2}$, charge -1, strangeness -3, and rest energy $\approx 1\,680$ MeV. Shortly thereafter, in 1964, scientists at the Brookhaven National Laboratory found the missing particle through careful analyses of bubble-chamber photographs (Fig. 46.10) and confirmed all its predicted properties.

The prediction of the particle missing from the eightfold way has much in common with the prediction of missing elements in the periodic table. Whenever a vacancy occurs in an organized pattern of information, experimentalists have a guide for their investigations.

46.9 QUARKS—FINALLY

As we have noted, leptons appear to be truly elementary particles because there are only a few types of them, and they have no measurable size or internal structure. Hadrons, on the other hand, are complex particles having size and structure. The existence of the strangeness-charge patterns of the eightfold way suggests that hadrons have substructure. Furthermore, we know that hundreds of types of hadrons exist and that many of them decay into other hadrons.

The Original Quark Model

In 1963 Gell-Mann and George Zweig independently proposed a model for the substructure of hadrons. According to their model, all hadrons are composite systems of two or three elementary constituents called **quarks.** (Gell-Mann borrowed the word *quark* from the passage "Three quarks for Muster Mark" in James Joyce's *Finnegans Wake.*) The model had three types of quarks, designated by the symbols u, d, and s. These are given the arbitrary names **up, down,** and **strange.** The various types of quarks are called **flavors.** Figure 46.11 is a pictorial representation of the quark compositions of several hadrons.

An unusual property of quarks is that they carry a fractional electronic charge. The u, d, and s quarks have charges of $+2e/3$, $-e/3$, and $-e/3$, respectively. These and other properties of quarks and antiquarks are given in Table 46.3. Notice that quarks have spin $\frac{1}{2}$, which means that all quarks are **fermions,** defined as any particle having half integral spin. As Table 46.3 shows, associated with each quark is an antiquark of opposite charge, baryon number, and strangeness.

The compositions of all hadrons known when Gell-Mann and Zweig presented their model could be completely specified by three simple rules:

- A meson consists of one quark and one antiquark, giving it a baryon number of 0, as required.
- A baryon consists of three quarks.
- An antibaryon consists of three antiquarks.

The theory put forth by Gell-Mann and Zweig is referred to as the *original quark model.*

Quick Quiz 46.4

We have seen a law of lepton number conservation and a law of baryon number conservation. Why is there no law of meson number conservation? (*Hint:* Imagine creating particle-antiparticle pairs from energy, and focus on the creation of a quark–antiquark pair.)

Charm and Other Developments

Although the original quark model was highly successful in classifying particles into families, some discrepancies occurred between its predictions and certain experimental decay rates. Consequently, several physicists proposed a fourth quark flavor in 1967. They argued that if four types of leptons exist (as was thought at the time), then there should also be four flavors of quarks because of an underlying symmetry in nature. The fourth quark, designated c, was assigned a property called **charm.** A *charmed* quark has charge $+2e/3$, just as the up quark does, but its charm distinguishes it from the other three quarks. This introduces a new quantum number C, representing charm. The new quark has charm $C = +1$, its antiquark has charm $C = -1$, and all other quarks have $C = 0$. Charm, like strangeness, is conserved in nuclear and electromagnetic interactions but not in weak interactions.

Evidence that the charmed quark exists began to accumulate in 1974, when a heavy meson called the J/Ψ particle (or simply Ψ, uppercase Greek psi) was discovered independently by two groups, one led by Burton Richter (b. 1931) at the Stanford Linear Accelerator (SLAC), and the other led by Samuel Ting (b. 1936) at the Brookhaven National Laboratory. In 1976 Richter and Ting were awarded a Nobel Prize for this work. The J/Ψ particle does not fit into the three-quark

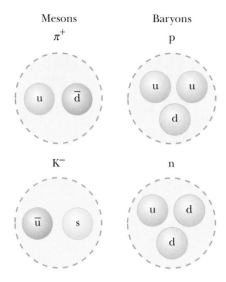

Figure 46.11 Quark composition of two mesons and two baryons.

model; instead, it has properties of a combination of the proposed charmed quark and its antiquark ($c\bar{c}$). It is much more massive than the other known mesons ($\sim 3\,100$ MeV/c^2), and its lifetime is much longer than the lifetimes of particles that interact via the nuclear force. Soon, related mesons were discovered, corresponding to such quark combinations as $\bar{c}d$ and $c\bar{d}$, all of which have great masses

TABLE 46.3 Properties of Quarks and Antiquarks

				Quarks				
Name	**Symbol**	**Spin**	**Charge**	**Baryon Number**	**Strangeness**	**Charm**	**Bottomness**	**Topness**
Up	u	$\frac{1}{2}$	$+\frac{2}{3}e$	$\frac{1}{3}$	0	0	0	0
Down	d	$\frac{1}{2}$	$-\frac{1}{3}e$	$\frac{1}{3}$	0	0	0	0
Strange	s	$\frac{1}{2}$	$-\frac{1}{3}e$	$\frac{1}{3}$	-1	0	0	0
Charmed	c	$\frac{1}{2}$	$+\frac{2}{3}e$	$\frac{1}{3}$	0	$+1$	0	0
Bottom	b	$\frac{1}{2}$	$-\frac{1}{3}e$	$\frac{1}{3}$	0	0	$+1$	0
Top	t	$\frac{1}{2}$	$+\frac{2}{3}e$	$\frac{1}{3}$	0	0	0	$+1$

				Antiquarks				
Name	**Symbol**	**Spin**	**Charge**	**Baryon Number**	**Strangeness**	**Charm**	**Bottomness**	**Topness**
Anti-up	$\bar{u}$	$\frac{1}{2}$	$-\frac{2}{3}e$	$-\frac{1}{3}$	0	0	0	0
Anti-down	$\bar{d}$	$\frac{1}{2}$	$+\frac{1}{3}e$	$-\frac{1}{3}$	0	0	0	0
Anti-strange	$\bar{s}$	$\frac{1}{2}$	$+\frac{1}{3}e$	$-\frac{1}{3}$	$+1$	0	0	0
Anti-charmed	$\bar{c}$	$\frac{1}{2}$	$-\frac{2}{3}e$	$-\frac{1}{3}$	0	-1	0	0
Anti-bottom	$\bar{b}$	$\frac{1}{2}$	$+\frac{1}{3}e$	$-\frac{1}{3}$	0	0	-1	0
Anti-top	$\bar{t}$	$\frac{1}{2}$	$-\frac{2}{3}e$	$-\frac{1}{3}$	0	0	0	-1

TABLE 46.4 **Quarkᵃ Composition of Mesons**

		b		c		s		d		u	
	b	Υ^-	$(\overline{b}b)$								
	c	B_c^+	$(\overline{b}c)$	J/Ψ	$(\overline{c}c)$						
Quarks	s	B_s^0	$(\overline{b}s)$	D_s^-	$(\overline{c}s)$	η, η'	$(\overline{s}s)$	$\overline{K}_0$	$(\overline{d}s)$	K^-	$(\overline{u}s)$
	d	B^0	$(\overline{b}d)$	D^-	$(\overline{c}d)$	K^0	$(\overline{s}d)$	π^0, η, η'	$(\overline{d}d)$	π^-	$(\overline{u}d)$
	u	B^+	$(\overline{b}u)$	$\overline{D}^0$	$(\overline{c}u)$	K^+	$(\overline{s}u)$	π^+	$(\overline{d}u)$	π^0, η, η'	$(\overline{u}u)$

From D. Kestenbaum, "Physicists Find the Last of the Mesons," *Science* 280:35, 1998.

ᵃ The top quark does not form mesons because it decays too quickly.

and long lifetimes. The existence of these new mesons provided firm evidence for the fourth quark flavor.

In 1975, researchers at Stanford University reported strong evidence for the tau (τ) lepton, mass 1 784 MeV/c^2. This was the fifth type of lepton, which led physicists to propose that more flavors of quarks might exist, on the basis of symmetry arguments similar to those leading to the proposal of the charmed quark. These proposals led to more elaborate quark models and the prediction of two new quarks, **top** (t) and **bottom** (b). (Some physicists prefer *truth* and *beauty*.) To distinguish these quarks from the others, quantum numbers called *topness* and *bottomness* (with allowed values $+1, 0, -1$) were assigned to all quarks and antiquarks (see Table 46.3). In 1977, researchers at the Fermi National Laboratory, under the direction of Leon Lederman (b. 1922), reported the discovery of a very massive new meson Υ^- (Greek capital upsilon), whose composition is considered to be $b\overline{b}$, providing evidence for the bottom quark. In March 1995, researchers at Fermilab announced the discovery of the top quark (supposedly the last of the quarks that will be found), which has a mass of 173 GeV/c^2.

Table 46.4 lists the quark compositions of mesons formed from the up, down, strange, charmed, and bottom quarks. The meson formed by the combination of the bottom antiquark and the charmed quark ($\overline{b}c$) was the last to be found. It was discovered in 1998.[2] Table 46.5 shows the quark combinations for the baryons

TABLE 46.5 **Quark Composition of Several Baryons**

Particle	Quark Composition
p	uud
n	udd
Λ^0	uds
Σ^+	uus
Σ^0	uds
Σ^-	dds
Ξ^0	uss
Ξ^-	dss
Ω^-	sss

[2] For information about the discovery of the B_c^+ meson, read D. Kestenbaum, "Physicists Find the Last of the Mesons," *Science* 280:35, 1998.

TABLE 46.6	The Elementary Particles and Their Rest Energies and Charges	
Particle	Rest Energy	Charge
Quarks		
u	360 MeV	$+\frac{2}{3}e$
d	360 MeV	$-\frac{1}{3}e$
c	1 500 MeV	$+\frac{2}{3}e$
s	540 MeV	$-\frac{1}{3}e$
t	173 GeV	$+\frac{2}{3}e$
b	5 GeV	$-\frac{1}{3}e$
Leptons		
e^-	511 keV	$-e$
μ^-	105.7 MeV	$-e$
τ^-	1 784 MeV	$-e$
ν_e	<7 eV	0
ν_μ	<0.3 MeV	0
ν_τ	<30 MeV	0

listed in Table 46.2. Note that only two flavors of quarks, u and d, are contained in all hadrons encountered in ordinary matter (protons and neutrons).

You are probably wondering whether the discoveries of elementary particles will ever end. How many "building blocks" of matter really exist? At the present, physicists believe that the elementary particles in nature are six quarks and six leptons, together with their antiparticles. Table 46.6 lists their rest energies and their charges.

Despite extensive experimental effort, no isolated quark has ever been observed. Physicists now believe that quarks are permanently confined inside ordinary particles because of an exceptionally strong force that prevents them from escaping, called (appropriately) the **strong force**[3] (discussed in Section 46.10). This force increases with separation distance, similar to the force exerted by a stretched spring. One author described its great magnitude as follows:[4]

Quarks are slaves, . . . bound like prisoners of a chain gang. . . . Any locksmith can break the chain between two prisoners, but no locksmith is expert enough to break the gluon chains between quarks. Quarks remain slaves forever.

46.10 MULTICOLORED QUARKS

Shortly after the concept of quarks was proposed, scientists recognized that certain particles had quark compositions that violated the exclusion principle applied to quarks. In Section 42.6, we applied the exclusion principle to electrons in atoms. The

[3] As a reminder, the original meaning of the term *strong force* was the short-range attractive force between nucleons, which we have called the *nuclear force*. As we shall discuss in Section 46.10, the nuclear force between nucleons is a secondary effect of the strong force between quarks.

[4] H. Fritzsch, *Quarks, the Stuff of Matter*. London, Allen & Lane, 1983.

principle is more general, however, and applies to all particles having half-integral spin ($\frac{1}{2}$, $\frac{3}{2}$, and so on), which we collectively call fermions, as noted in the previous section. Because all quarks are fermions with spin $\frac{1}{2}$, they are expected to follow the exclusion principle. One example of a particle that appears to violate the exclusion principle is the Ω^- (sss) baryon, which contains three strange quarks with parallel spins, giving it a total spin of $\frac{3}{2}$. All three quarks have the same spin quantum number, in violation of the exclusion principle. Other examples of baryons made up of identical quarks having parallel spins are the Δ^{++} (uuu) and the Δ^- (ddd).

To resolve this problem, it was suggested that quarks possess an additional property called **color charge,** not to be confused with the color associated with visible light. This property is similar in many respects to electric charge except that it occurs in three varieties rather than two. The colors assigned to quarks are red, green, and blue, and antiquarks have the colors antired, antigreen, and antiblue. Thus, the colors red, green, and blue serve as the "quantum numbers" for the color of the quark. To satisfy the exclusion principle, the three quarks in any baryon must all have different colors. The three colors "neutralize" to white, as the electric charges + and − neutralize to zero net charge. The quark and antiquark in any meson must be of a color and the corresponding anticolor. The result is that baryons and mesons are always colorless (or white). Thus, the apparent violation of the exclusion principle in the Ω^- baryon is removed because the three quarks in the particle have different colors.

Note that the new property of color increases the number of quarks by a factor of three, since each of the six quarks comes in three colors. Although the concept of color in the quark model was originally conceived to satisfy the exclusion principle, it also provided a better theory for explaining certain experimental results. For example, the modified theory correctly predicts the lifetime of the π^0 meson.

The theory of how quarks interact with each other is called **quantum chromodynamics,** or QCD, to parallel the name *quantum electrodynamics* (the theory of interaction between electric charges). In QCD, each quark is said to carry a color charge, in analogy to electric charge. The strong force between quarks is often called the **color force.** Thus, the terms *strong force* and *color force* are used interchangeably.

In Section 46.1 we stated that the nuclear interaction between hadrons is mediated by massless field particles called **gluons** (analogous to photons for the electromagnetic force). As we shall discuss in greater detail shortly, however, the nuclear force is actually a secondary effect of the strong force between quarks, and so the gluons are actually mediators of the strong force. Let us first investigate this as the primary effect and then investigate how the gluons also mediate the nuclear force. When a quark emits or absorbs a gluon, the quark's color may change. For example, a blue quark that emits a gluon may become a red quark, and a red quark that absorbs this gluon becomes a blue quark.

The color force between quarks is analogous to the electric force between charges: likes repel, and opposites attract. Therefore, two red quarks repel each other, but a red quark is attracted to an antired quark. The attraction between quarks of opposite color to form a meson ($q\bar{q}$) is indicated in Figure 46.12a. Differently colored quarks also attract one another, although with less intensity than the oppositely colored quark and antiquark. For example, a cluster of red, blue, and green quarks all attract one another to form a baryon, as in Figure 46.12b. Thus, every baryon contains three quarks of three different colors.

Although the nuclear force between two colorless hadrons is negligible at large separations, the net strong force between their constituent quarks is not exactly zero at small separations. This residual strong force is the nuclear force that

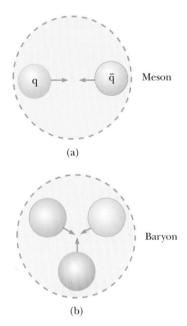

Figure 46.12   (a) A red quark is attracted to an antired quark to form a meson whose quark structure is ($q\bar{q}$). (b) Three quarks of different colors attract one another to form a baryon.

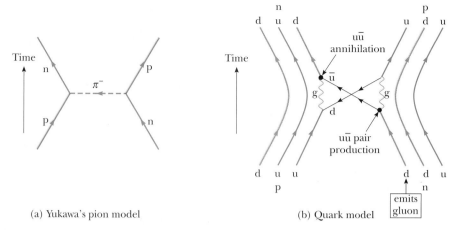

Figure 46.13 (a) A nuclear interaction between a proton and a neutron explained in terms of Yukawa's pion-exchange model. Because the pion carries charge, the proton and neutron switch identities. (b) The same interaction explained in terms of quarks and gluons. Note that the exchanged $\bar{u}d$ quark pair makes a π^- meson.

binds protons and neutrons to form nuclei. It is similar to the force between two electric dipoles. Each dipole is electrically neutral. An electric field surrounds the dipoles, however, because of the separation of the positive and negative charges (see Section 23.6). As a result, an electric force occurs between the dipoles, albeit weaker than the force between single charges.

According to QCD, a more basic explanation of the nuclear force can be given in terms of quarks and gluons. Figure 46.13a shows the nuclear interaction between a neutron and a proton by means of Yukawa's pion, in this case a π^-. This drawing differs from Figure 46.5a, in which the field particle is a π^0; thus, there is no transfer of charge from one nucleon to the other. In Figure 46.13a, the charged pion carries charge from one nucleon to the other, so the nucleons change identities—the proton becomes a neutron and the neutron becomes a proton.

Let us look at the same interaction from the viewpoint of the quark model, shown in Figure 46.13b. In this Feynman diagram, the proton and neutron are represented by their quark constituents. Each quark in the neutron and proton is continuously emitting and absorbing gluons. The energy of a gluon can result in the creation of quark–antiquark pairs. This is similar to the creation of electron–positron pairs in pair production, which we investigated in Section 46.2. When the neutron and proton approach to within 1 fm of each other, these gluons and quarks can be exchanged between the two nucleons, and such exchanges produce the nuclear force. Figure 46.13b depicts one possibility for the process shown in Figure 46.13a. A down quark in the neutron on the right emits a gluon (represented by the wavy line labeled g on the right side). The energy of the gluon is then transformed to create a $u\bar{u}$ pair. The u quark stays within the nucleon (which has now changed to a proton), and the recoiling d quark and the $\bar{u}$ antiquark are transmitted to the proton on the left side of the diagram. Here the $\bar{u}$ annihilates a u quark within the proton (with the creation of a gluon), and the d is captured. Thus, the net effect is to change a u quark to a d quark, and the proton on the left has changed to a neutron.

As the d quark and $\bar{u}$ antiquark in Figure 46.13b transfer between the nucleons, the d and $\bar{u}$ exchange gluons with each other and can be considered to be bound to each other by means of the strong force. If we look back at Table 46.4, we see that this combination is a π^- —Yukawa's field particle! Thus, the quark model of interactions between nucleons is consistent with the pion-exchange model.

46.11 THE STANDARD MODEL

Scientists now believe that there are three classifications of truly elementary particles: leptons, quarks, and field particles. These three particles are further classified as either **fermions** (quarks and leptons) or **bosons** (field particles). Note that quarks and leptons have spin $\frac{1}{2}$ and hence are fermions, while the field particles have spin 1 or higher and are bosons.

Recall from Section 46.1 that the weak force is believed to be mediated by the W^+, W^-, and Z^0 bosons. These particles are said to have *weak charge,* just as quarks have color charge. Thus, each elementary particle can have mass, electric charge, color charge, and weak charge. Of course, one or more of these could be zero.

In 1979, Sheldon Glashow (b. 1932), Abdus Salam (1926–1996), and Steven Weinberg (b. 1933) won a Nobel Prize for developing a theory that unifies the electromagnetic and weak interactions. This **electroweak theory** postulates that the weak and electromagnetic interactions have the same strength when the particles involved have very high energies. Thus, the two interactions are viewed as different manifestations of a single unifying electroweak interaction. The theory makes many concrete predictions, but perhaps the most spectacular is the prediction of the masses of the W and Z particles at about 82 GeV/c^2 and 93 GeV/c^2, respectively—predictions that are borne out by experiment.

The combination of the electroweak theory and QCD for the strong interaction is referred to in high-energy physics as the **Standard Model.** Although the details of the Standard Model are complex, its essential ingredients can be summarized with the help of Figure 46.14. (The Standard Model does not include the gravitational force at present; however, we include gravity in Figure 46.14 because physicists hope to eventually incorporate this force into a unified theory.) This dia-

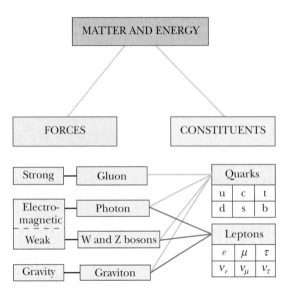

Figure 46.14 The Standard Model of particle physics.

gram shows that quarks participate in all of the fundamental forces and that leptons participate in all except the strong force.

The Standard Model does not answer all questions. A major question that is still unanswered is why, of the two mediators of the electroweak interaction, the photon has no mass but the W and Z bosons do. Because of this mass difference, the electromagnetic and weak forces are quite distinct at low energies but become similar at very high energies, at which the rest energy is negligible relative to the total energy. The behavior as one goes from high to low energies is called *symmetry breaking* because the forces are similar, or symmetric, at high energies but are very different at low energies. The nonzero rest energies of the W and Z bosons raise the question of the origin of particle masses. To resolve this problem, a hypothetical particle called the **Higgs boson,** which provides a mechanism for breaking the electroweak symmetry, has been proposed. The Standard Model modified to include the Higgs mechanism provides a logically consistent explanation of the massive nature of the W and Z bosons. Unfortunately, the Higgs boson has not yet been found, but physicists know that its rest energy should be less than 1 TeV. For us to determine whether the Higgs boson exists, two quarks each having at least 1 TeV of energy must collide, but calculations show that such a collision requires injecting 40 TeV of energy within the volume of a proton.

Scientists are convinced that, because of the limited energy available in conventional accelerators using fixed targets, it is necessary to build colliding-beam accelerators called **colliders.** The concept of colliders is straightforward. Particles that have equal masses and equal kinetic energies, traveling in opposite directions in an accelerator ring, collide head-on to produce the required reaction and form new particles. Because the total momentum of the interacting particles is zero, all of their kinetic energy is available for the reaction. The Large Electron–Positron (LEP) Collider at CERN (Fig. 46.15) and the Stanford Linear Collider collide both electrons and positrons. The Super Proton Synchrotron at CERN accelerates

A technician works on one of the particle detectors at CERN, the European center for particle physics near Geneva, Switzerland. Electrons and positrons accelerated to an energy of 50 GeV collide in a circular tunnel 2 km in circumference, located 100 m underground. *(David Parker/Science Photo Library/Photo Researchers, Inc.)*

Figure 46.15 A view from inside the Large Electron–Positron (LEP) Collider tunnel, which is 27 km in circumference. *(Courtesy of CERN)*

protons and antiprotons to energies of 270 GeV, whereas the world's highest-energy proton accelerator, the Tevatron at the Fermi National Laboratory in Illinois, produces protons at almost 1 000 GeV (1 TeV). The Superconducting Super Collider (SSC), which was being built in Texas, was an accelerator designed to produce 20-TeV protons in a ring 52 mi in circumference. After much debate in Congress and an investment of almost 2 billion dollars, the U.S. Department of Energy canceled the SSC project in October 1993. CERN expects a 2005 completion date for the Large Hadron Collider (LHC), a proton–proton collider that will provide a center-of-mass energy of 14 TeV and enable exploration of Higgs-boson physics. The accelerator will be constructed in the same 27-km circumference tunnel now housing the LEP Collider, and many countries are expected to participate in the project.

Following the success of the electroweak theory, scientists attempted to combine it with QCD in a **grand unification theory** (GUT). In this model, the electroweak force is merged with the strong force to form a grand unified force. One version of the theory considers leptons and quarks as members of the same family that can change into each other by exchanging an appropriate field particle.

web

Visit the Conseil Européen de Recherche Nucléair (now called the European Laboratory for Particle Physics) at **www.CERN.ch**

The World Wide Web was invented at CERN in 1991 as a way for physicists to easily share data.

46.12 ▶ THE COSMIC CONNECTION

In this section we describe one of the most fascinating theories in all of science—the Big Bang theory of the creation of the Universe—and the experimental evidence that supports it. This theory of cosmology states that the Universe had a beginning and, furthermore, that the beginning was so cataclysmic that it is impossible to look back beyond it. According to this theory, the Universe erupted from an infinitely dense singularity about 15 to 20 billion years ago. The first few minutes after the Big Bang saw such extremely high energy that it is believed that all four interactions of physics were unified and all matter was contained in an undifferentiated "quark soup."

The evolution of the four fundamental forces from the Big Bang to the present is shown in Figure 46.16. During the first 10^{-43} s (the ultrahot epoch, $T \sim 10^{32}$ K),

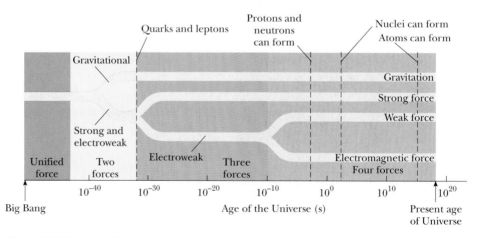

Figure 46.16 A brief history of the Universe from the Big Bang to the present. The four forces became distinguishable during the first nanosecond. Following this, all the quarks combined to form particles that interact via the nuclear force. However, the leptons remained separate and to this day exist as individual, observable particles.

it is presumed that the strong, electroweak, and gravitational forces were joined to form a completely unified force. In the first 10^{-35} s following the Big Bang (the hot epoch, $T \sim 10^{29}$ K), gravity broke free of this unification while the strong and electroweak forces remained as one, described by a grand unification theory. This was a period when particle energies were so great ($> 10^{16}$ GeV) that very massive particles as well as quarks, leptons, and their antiparticles existed. Then, after 10^{-35} s, the Universe rapidly expanded and cooled (the warm epoch, $T \sim 10^{29}$ to 10^{15} K), the strong and electroweak forces parted company, and the grand unification scheme was broken. As the Universe continued to cool, the electroweak force split into the weak force and the electromagnetic force about 10^{-10} s after the Big Bang.

After a few minutes, protons condensed out of the hot soup. For half an hour the Universe underwent thermonuclear detonation, exploding as a hydrogen bomb and producing most of the helium nuclei that exist now. The Universe continued to expand, and its temperature dropped. Until about 700 000 years after the Big Bang, the Universe was dominated by radiation. Energetic radiation prevented matter from forming single hydrogen atoms because collisions would instantly ionize any atoms that happened to form. Photons experienced continuous Compton scattering from the vast numbers of free electrons, resulting in a Universe that was opaque to radiation. By the time the Universe was about 700 000 years old, it had expanded and cooled to about 3 000 K, and protons could bind to electrons to form neutral hydrogen atoms. Because of the quantized energies of the atoms, far more wavelengths of radiation were not absorbed by atoms than were absorbed, and the Universe suddenly became transparent to photons. Radiation no longer dominated the Universe, and clumps of neutral matter steadily grew—first atoms, then molecules, gas clouds, stars, and finally galaxies.

Observation of Radiation from the Primordial Fireball

In 1965, Arno A. Penzias (b. 1933) and Robert W. Wilson (b. 1936) of Bell Laboratories were testing a sensitive microwave receiver and made an amazing discovery. A pesky signal producing a faint background hiss was interfering with their satellite communications experiments. In spite of their valiant efforts, the signal remained. Ultimately, it became clear that they were perceiving microwave background radiation (at a wavelength of 7.35 cm), which represented the leftover "glow" from the Big Bang.

The microwave horn that served as their receiving antenna is shown in Figure 46.17. The intensity of the detected signal remained unchanged as the antenna was pointed in different directions. The fact that the radiation had equal strengths in all directions suggested that the entire Universe was the source of this radiation. Evicting a flock of pigeons from the 20-ft horn and cooling the microwave detector both failed to remove the signal. Through a casual conversation, Penzias and Wilson discovered that a group at Princeton had predicted the residual radiation from the Big Bang and were planning an experiment to attempt to confirm the theory. The excitement in the scientific community was high when Penzias and Wilson announced that they had already observed an excess microwave background compatible with a 3-K blackbody source.

Because Penzias and Wilson made their measurements at a single wavelength, they did not completely confirm the radiation as 3-K blackbody radiation. Subsequent experiments by other groups added intensity data at different wavelengths, as shown in Figure 46.18. The results confirm that the radiation is that of a black body at 2.7 K. This figure is perhaps the most clear-cut evidence for the Big Bang theory. The 1978 Nobel Prize was awarded to Penzias and Wilson for this most important discovery.

Figure 46.17   Robert W. Wilson *(left)* and Arno A. Penzias with the Bell Telephone Laboratories horn-reflector antenna. *(AT&T Bell Laboratories)*

The discovery of the cosmic background radiation brought with it a problem, however—the radiation was too uniform. Scientists believed that slight fluctuations in this background had to occur in order for such objects as galaxies to form. In 1989, NASA launched a satellite called COBE (KOH-bee), for Cosmic Background Explorer, to study this radiation in greater detail. In 1992, George Smoot

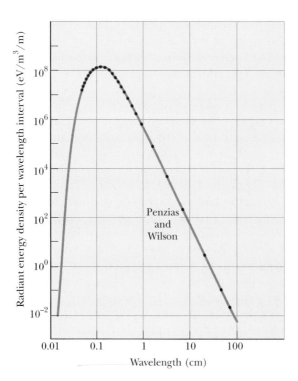

Figure 46.18   Theoretical black-body (red curve) and measured radiation spectra (black points) of the Big Bang. Most of the data were collected from the Cosmic Background Explorer (COBE) satellite. The data of Wilson and Penzias are indicated.

(b. 1945) at the Lawrence Berkeley Laboratory found, on the basis of the data collected, that the background was not perfectly uniform but instead contained irregularities that corresponded to temperature variations of 0.000 3 K. It is these small variations that provided nucleation sites for the formation of the galaxies and other objects we now see in the sky.

Other Evidence for an Expanding Universe

The Big Bang theory of cosmology predicts that the Universe is expanding. Most of the key discoveries supporting the theory of an expanding Universe were made in the 20th century. Vesto Melvin Slipher (1875–1969), an American astronomer, reported in 1912 that most nebulae are receding from the Earth at speeds up to several million miles per hour. Slipher was one of the first scientists to use Doppler shifts (see Section 17.5) in spectral lines to measure velocities.

In the late 1920s, Edwin P. Hubble made the bold assertion that the whole Universe is expanding. From 1928 to 1936, until they reached the limits of the 100-inch telescope, Hubble and Milton Humason (1891–1972) worked at Mount Wilson in California to prove this assertion. The results of that work and of its continuation with the use of a 200-inch telescope in the 1940s showed that the speeds at which galaxies are receding from the Earth increase in direct proportion to their distance R from us (Fig. 46.19). This linear relationship, known as **Hubble's law,** may be written

$$v = HR \qquad \textbf{(46.7)}$$

where H, called the **Hubble parameter,** has the approximate value

$$H \approx 17 \times 10^{-3} \, \text{m/s} \cdot \text{ly}$$

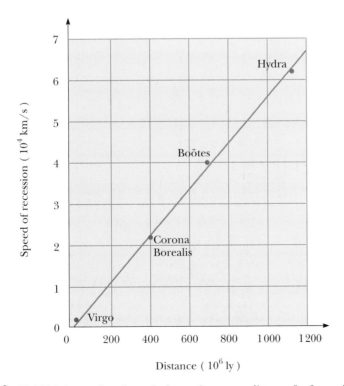

Figure 46.19 Hubble's law: a plot of speed of recession versus distance for four galaxies.

EXAMPLE 46.5   Recession of a Quasar

A quasar is an object that appears similar to a star and is very distant from the Earth. Its speed can be determined from Doppler-shift measurements in the light it emits. A certain quasar recedes from the Earth at a speed of 0.55c. How far away is it?

Exercise   Assuming that the quasar has moved at this speed ever since the Big Bang, estimate the age of the Universe.

Answer $t = R/v = 1/H \approx 18$ billion years, which is in reasonable agreement with other calculations.

Solution   We can find the distance through Hubble's law:

$$R = \frac{v}{H} = \frac{(0.55)(3.00 \times 10^8 \text{ m/s})}{17 \times 10^{-3} \text{ m/s} \cdot \text{ly}} = \boxed{9.7 \times 10^9 \text{ ly}}$$

Will the Universe Expand Forever?

In the 1950s and 1960s, Allan R. Sandage used the 200-inch telescope at Mount Palomar to measure the speeds of galaxies at distances of up to 6 billion lightyears away from the Earth. These measurements showed that these very distant galaxies were moving about 10 000 km/s faster than Hubble's law predicted. According to this result, the Universe must have been expanding more rapidly 1 billion years ago, and consequently we conclude that the expansion rate is slowing[5] (Fig. 46.20). Today, astronomers and physicists are trying to determine the rate of slow-

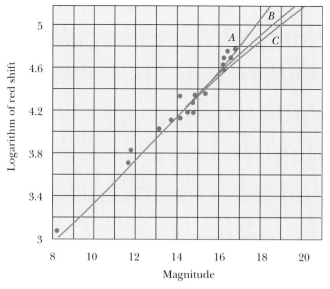

Figure 46.20   Red shift, or speed of recession, versus magnitude (which is related to brightness) of 18 faint galaxy clusters. Significant scatter of the data occurs, so the extrapolation of the curve to the upper right is uncertain. Curve A is the trend suggested by the six faintest clusters. Curve C corresponds to a Universe having a constant rate of expansion. If more data are taken and the complete set of data indicates a curve that falls between B and C, the expansion will slow but never stop. If the data fall to the left of B, expansion will eventually stop and the Universe will begin to contract.

---

[5] The data at large distances have large observational uncertainties and may be systematically in error from effects such as abnormal brightness in the most distant visible clusters.

ing. If the average mass density of the Universe is less than some critical value $(\rho_c \approx 3 \text{ atoms/m}^3)$, the galaxies will slow in their outward rush but still escape to infinity. If the average density exceeds the critical value, the expansion will eventually stop and contraction will begin, possibly leading to a superdense state followed by another expansion. In this case, we have an oscillating Universe.

EXAMPLE 46.6 ▶ **The Critical Density of the Universe**

(a) Starting from energy conservation, derive an expression for the critical mass density of the Universe ρ_c in terms of the Hubble parameter H and the universal gravitational constant G.

Solution Figure 46.21 shows a large section of the Universe, contained within a sphere of radius R. The total mass of the galaxies in this volume is M. A galaxy of mass $m \ll M$ that has a speed v at a distance R from the center of the sphere will escape to infinity (at which its speed will approach zero) if the sum of its kinetic energy and the gravitational potential energy of the system—galaxy plus rest of the Universe—is zero at any time. The Universe may be infinite in spatial extent, but Gauss's law implies that only the mass M inside the sphere contributes to the gravitational potential energy of the galaxy:

$$E_{\text{total}} = 0 = K + U = \tfrac{1}{2}mv^2 - \frac{GmM}{R}$$

We substitute for the mass M contained within the sphere the product of the critical density and the volume of the sphere:

$$\tfrac{1}{2}mv^2 = \frac{Gm\tfrac{4}{3}\pi R^3 \rho_c}{R}$$

Solving for the critical density gives

$$\rho_c = \frac{3v^2}{8\pi G R^2}$$

From Hubble's law, the ratio of v to R is $v/R = H$, so this expression becomes

$$\boxed{\rho_c = \frac{3H^2}{8\pi G}}$$

(b) Estimate a numerical value for the critical density in grams per cubic centimeter.

Solution Using $H = 17 \times 10^{-3} \text{ m/s} \cdot \text{ly}$, where $1 \text{ ly} = 9.46 \times 10^{15}$ m, we find for the critical density

$$\rho_c = \frac{3H^2}{8\pi G} = \frac{3(17 \times 10^{-3} \text{ m/s} \cdot \text{ly})^2}{8\pi(6.67 \times 10^{-11} \text{ N} \cdot \text{m}^2/\text{kg}^2)}\left(\frac{1 \text{ ly}}{9.46 \times 10^{15} \text{m}}\right)^2$$

$$= 6 \times 10^{-27} \text{ kg/m}^3$$

Converting this to the requested units, we have $\rho_c = 6 \times 10^{-30} \text{ g/cm}^3$. Because the mass of a hydrogen atom is 1.67×10^{-24} g, this value of ρ_c corresponds to 3×10^{-6} hydrogen atoms per cubic centimeter, or 3 atoms per cubic meter.

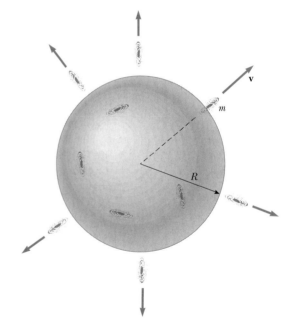

Figure 46.21 The galaxy marked with mass m is escaping from a large cluster of galaxies contained within a spherical volume of radius R. Only the mass within R slows the mass m.

Missing Mass in the Universe?

The luminous matter in galaxies averages out to a Universe density of 5×10^{-33} g/cm^3. The radiation in the Universe has a mass equivalent of approximately 2% of the luminous matter. The total mass of all nonluminous matter (such as interstellar gas and black holes) may be estimated from the speeds of galaxies orbiting each other in a cluster. The higher the galaxy speeds, the more mass in the cluster. Measurements on the Coma cluster of galaxies indicate, surprisingly, that the amount of nonluminous matter is 20 to 30 times the amount of luminous matter present in stars and luminous gas clouds. Yet even this large invisible component of *dark matter,* if extrapolated to the Universe as a whole, leaves the observed mass density a factor of 10 less than ρ_c. The deficit, called *missing mass,* has been the subject of intense theoretical and experimental work, with exotic particles such as axions, photinos, and superstring particles suggested as candidates for the missing mass. Some researchers have made the more mundane proposal that the missing mass is present in neutrinos. In fact, neutrinos are so abundant that a tiny neutrino rest energy on the order of only 20 eV would furnish the missing mass and "close" the Universe. Thus, current experiments designed to measure the rest energy of the neutrino will have an impact on predictions for the future of the Universe.

Although we have some degree of certainty about the beginning of the Universe, we are uncertain about how the story will end. Will the Universe keep on expanding forever, or will it someday collapse and then expand again, perhaps in an endless series of oscillations? Results and answers to these questions remain inconclusive, and the exciting controversy continues.

46.13 ▷ PROBLEMS AND PERSPECTIVES

While particle physicists have been exploring the realm of the very small, cosmologists have been exploring cosmic history back to the first microsecond of the Big Bang. Observation of the events that occur when two particles collide in an accelerator is essential for reconstructing the early moments in cosmic history. For this reason, perhaps the key to understanding the early Universe is to first understand the world of elementary particles. Cosmologists and physicists now find that they have many common goals and are joining hands in an attempt to understand the physical world at its most fundamental level.

Our understanding of physics at short distances is far from complete. Particle physics is faced with many questions. Why does so little antimatter exist in the Universe? Is it possible to unify the strong and electroweak theories in a logical and consistent manner? Why do quarks and leptons form three similar but distinct families? Are muons the same as electrons apart from their difference in mass, or do they have other subtle differences that have not been detected? Why are some particles charged and others neutral? Why do quarks carry a fractional charge? What determines the masses of the elementary constituents of matter? Can isolated quarks exist? The questions go on and on. Because of the rapid advances and new discoveries in the field of particle physics, by the time you read this book, some of these questions may be resolved and new ones may emerge.

An important and obvious question that remains is whether leptons and quarks have an underlying structure. If they do, we can envision an infinite number of deeper structure levels. However, if leptons and quarks are indeed the ultimate constituents of matter, as physicists today tend to believe, we should be able to construct a final theory of the structure of matter, just as Einstein dreamed of doing. This theory, whimsically called the Theory of Everything, is a combination of GUT and a quantum theory of gravity. In the view of many physicists, the end of the road is in sight, but how long it will take to reach it is anyone's guess.

SUMMARY

Before quark theory was developed, the four fundamental forces in nature were identified as nuclear, electromagnetic, weak, and gravitational. All the interactions in which these forces take part are mediated by **field particles.** The electromagnetic interaction is mediated by the photon; the weak interaction is mediated by the $W^{\pm}$ and Z^0 bosons; the gravitational interaction is mediated by gravitons; the nuclear interaction is mediated by gluons.

A charged particle and its **antiparticle** have the same mass but opposite charge, and other properties may have opposite values, such as lepton number and baryon number. It is possible to produce particle–antiparticle pairs in nuclear reactions if the available energy is greater than $2mc^2$, where m is the mass of the particle (or antiparticle).

Particles other than field particles are classified as hadrons or leptons. **Hadrons** interact via all four fundamental forces. They have size and structure and are not elementary particles. There are two types—**baryons** and **mesons.** Baryons, which generally are the most massive particles, have nonzero **baryon number** and a spin of $\frac{1}{2}$ or $\frac{3}{2}$. Mesons have baryon number zero and either zero or integral spin.

Leptons have no structure or size and are considered truly elementary. They interact only via the weak, gravitational, and electromagnetic forces. Six types of leptons exist: the electron e^-; the muon μ^-; the tau τ^-; and their neutrinos ν_e, ν_μ, and ν_τ.

In all reactions and decays, quantities such as energy, linear momentum, angular momentum, electric charge, baryon number, and lepton number are strictly conserved. Certain particles have properties called **strangeness** and **charm.** These unusual properties are conserved only in the decays and nuclear reactions that occur via the strong force.

Theorists in elementary particle physics have postulated that all hadrons are composed of smaller units known as **quarks.** Quarks have fractional electric charge and come in six **flavors:** up (u), down (d), strange (s), charmed (c), top (t), and bottom (b). Each baryon contains three quarks, and each meson contains one quark and one antiquark.

According to the theory of **quantum chromodynamics,** quarks have a property called **color,** and the force between quarks is referred to as the **strong force** or the **color force.** The strong force is now considered to be a fundamental force. The nuclear force, which was originally considered to be fundamental, is now understood to be a secondary effect of the strong force, due to gluon exchanges between hadrons.

The electromagnetic and weak forces are now considered to be manifestations of a single force called the **electroweak force.** The combination of quantum chromodynamics and the electroweak theory is called the **Standard Model.**

The background microwave radiation discovered by Penzias and Wilson strongly suggests that the Universe started with a Big Bang 12 to 15 billion years ago. The background radiation is equivalent to that of a black body at 3 K. Various astronomical measurements strongly suggest that the Universe is expanding. According to **Hubble's law,** distant galaxies are receding from the Earth at a speed $v = HR$, where R is the distance from the Earth to the galaxy and H is the **Hubble parameter,** $H \approx 17 \times 10^{-3}$ m/s·ly.

QUESTIONS

1. Name the four fundamental interactions and the field particle that mediates each.
2. Describe the quark model of hadrons, including the properties of quarks.
3. What are the differences between hadrons and leptons?
4. Describe the properties of baryons and mesons and the important differences between them.
5. Particles known as resonances have very short lifetimes, of the order of 10^{-23} s. From this information, would you guess that they are hadrons or leptons? Explain.
6. Kaons all decay into final states that contain no protons or neutrons. What is the baryon number of kaons?
7. The Ξ^0 particle decays by the weak interaction according to the decay mode $\Xi^0 \rightarrow \Lambda^0 + \pi^0$. Would you expect this decay to be fast or slow? Explain.
8. Identify the particle decays listed in Table 46.2 that occur by the weak interaction. Justify your answers.
9. Identify the particle decays listed in Table 46.2 that occur by the electromagnetic interaction. Justify your answers.
10. Two protons in a nucleus interact via the nuclear interaction. Are they also subject to the weak interaction?
11. Discuss the following conservation laws: energy, linear momentum, angular momentum, electric charge, baryon number, lepton number, and strangeness. Are all of these laws based on fundamental properties of nature? Explain.
12. An antibaryon interacts with a meson. Can a baryon be produced in such an interaction? Explain.
13. Describe the essential features of the Standard Model of particle physics.
14. How many quarks are in each of the following: (a) a baryon, (b) an antibaryon, (c) a meson, (d) an antimeson? How do you account for the fact that baryons have half-integral spins while mesons have spins of 0 or 1? (*Hint:* Quarks have spin $\frac{1}{2}$.)
15. In the theory of quantum chromodynamics, quarks come in three colors. How would you justify the statement that "all baryons and mesons are colorless"?
16. Which baryon did Murray Gell-Mann predict in 1961? What is the quark composition of this particle?
17. What is the quark composition of the Ξ^- particle? (See Table 46.5.)
18. The W and Z bosons were first produced at CERN in 1983 (by causing a beam of protons and a beam of antiprotons to meet at high energy). Why was this an important discovery?
19. How did Edwin Hubble (in 1928) determine that the Universe is expanding?
20. **Review Question.** A girl and her grandmother grind corn while the woman tells the girl some illuminating stories. A boy keeps crows away from ripening corn while his grandfather sits in the shade and explains to him the Universe and his place in it. What the children do not understand this summer they will better understand next year. Now you must take the part of the adults. State the most general, most fundamental, most universal truths that you know. If you find yourself repeating someone else's ideas, get the best version of those ideas that you can, and state your source. If there is something you do not understand, make a plan to understand it better within the next year.

PROBLEMS

1, 2, 3 = straightforward, intermediate, challenging   ☐ = full solution available in the *Student Solutions Manual and Study Guide*
WEB = solution posted at **http://www.saunderscollege.com/physics/**   🖥 = Computer useful in solving problem   📻 = Interactive Physics
☐ = paired numerical/symbolic problems

Section 46.1   The Fundamental Forces in Nature
Section 46.2   Positrons and Other Antiparticles

1. A photon produces a proton–antiproton pair according to the reaction $\gamma \rightarrow p + \bar{p}$. What is the minimum possible frequency of the photon? What is its wavelength?
2. Two photons are produced when a proton and an antiproton annihilate each other. What are the minimum frequency and corresponding wavelength of each photon?
3. A photon with an energy $E_\gamma = 2.09$ GeV creates a proton–antiproton pair in which the proton has a ki-netic energy of 95.0 MeV. What is the kinetic energy of the antiproton? ($m_p c^2 = 938.3$ MeV.)

Section 46.3   Mesons and the Beginning of Particle Physics

4. Occasionally, high-energy muons collide with electrons and produce two neutrinos according to the reaction $\mu^+ + e^- \rightarrow 2\nu$. What kind of neutrinos are these?
5. One of the mediators of the weak interaction is the Z^0 boson, with mass 93 GeV/c^2. Use this information to find the order of magnitude of the range of the weak interaction.

6. A free neutron actually beta-decays by creating a proton, an electron, and an antineutrino according to the reaction $n \rightarrow p + e^- + \bar{\nu}$. Assume, nevertheless, that a free neutron decays by creating a proton and an electron according to the reaction

$$n \longrightarrow p + e^-$$

and assume that the neutron is initially at rest in the laboratory. (a) Determine the energy released in this reaction. (b) Determine the speeds of the proton and electron after the reaction. (Energy and momentum are conserved in the reaction.) (c) Is either of these particles moving at relativistic speeds? Explain.

7. When a high-energy proton or pion traveling near the speed of light collides with a nucleus, it travels an average distance of 3×10^{-15} m before interacting. From this information, find the order of magnitude of the time for the strong interaction to occur.

8. Calculate the range of the force that might be produced by the virtual exchange of a proton.

WEB 9. A neutral pion at rest decays into two photons according to

$$\pi^0 \longrightarrow \gamma + \gamma$$

Find the energy, momentum, and frequency of each photon.

Section 46.4 Classification of Particles

10. Identify the unknown particle on the left side of the following reaction:

$$? + p \longrightarrow n + \mu^+$$

11. Name one possible decay mode (see Table 46.2) for Ω^+, $\overline{K_S^0}$, $\overline{\Lambda^0}$, and $\bar{n}$.

Section 46.5 Conservation Laws

12. Each of the following reactions is forbidden. Determine a conservation law that is violated for each reaction.
(a) $p + \bar{p} \rightarrow \mu^+ + e^-$
(b) $\pi^- + p \rightarrow p + \pi^+$
(c) $p + p \rightarrow p + \pi^+$
(d) $p + p \rightarrow p + p + n$
(e) $\gamma + p \rightarrow n + \pi^0$

13. (a) Show that baryon number and charge are conserved in the following reactions of a pion with a proton.

$$\pi^+ + p \longrightarrow K^+ + \Sigma^+ \qquad (1)$$

$$\pi^+ + p \longrightarrow \pi^+ + \Sigma^+ \qquad (2)$$

(b) The first reaction is observed, but the second never occurs. Explain.

14. The first of the following two reactions may occur, but the second cannot. Explain.

$$K_S^0 \longrightarrow \pi^+ + \pi^- \qquad \text{(can occur)}$$

$$\Lambda^0 \longrightarrow \pi^+ + \pi^- \qquad \text{(cannot occur)}$$

WEB 15. The following reactions or decays involve one or more neutrinos. In each case, supply the missing neutrino (ν_e, ν_μ, or ν_τ).
(a) $\pi^- \rightarrow \mu^- + ?$
(b) $K^+ \rightarrow \mu^+ + ?$
(c) $? + p \rightarrow n + e^+$
(d) $? + n \rightarrow p + e^-$
(e) $? + n \rightarrow p + \mu^-$
(f) $\mu^- \rightarrow e^- + ? + ?$

16. A K_S^0 particle at rest decays into a π^+ and a π^-. What will be the speed of each of the pions? The mass of the K_S^0 is 497.7 MeV/c^2, and the mass of each π is 139.6 MeV/c^2.

WEB 17. Determine which of the following reactions can occur. For those that cannot occur, determine the conservation law (or laws) violated.
(a) $p \rightarrow \pi^+ + \pi^0$
(b) $p + p \rightarrow p + p + \pi^0$
(c) $p + p \rightarrow p + \pi^+$
(d) $\pi^+ \rightarrow \mu^+ + \nu_\mu$
(e) $n \rightarrow p + e^- + \bar{\nu}_e$
(f) $\pi^+ \rightarrow \mu^+ + n$

18. (a) Show that the proton-decay reaction

$$p \longrightarrow e^+ + \gamma$$

cannot occur because it violates conservation of baryon number. (b) Imagine that this reaction does occur, and that the proton is initially at rest. Determine the energy and momentum of the positron and photon after the reaction. (*Hint:* Recall that energy and momentum must be conserved in the reaction.) (c) Determine the speed of the positron after the reaction.

19. Determine the type of neutrino or antineutrino involved in each of the following processes.
(a) $\pi^+ \rightarrow \pi^0 + e^+ + ?$
(b) $? + p \rightarrow \mu^- + p + \pi^+$
(c) $\Lambda^0 \rightarrow p + \mu^- + ?$
(d) $\tau^+ \rightarrow \mu^+ + ? + ?$

Section 46.6 Strange Particles and Strangeness

20. The neutral ρ meson decays by the strong interaction into two pions: $\rho^0 \rightarrow \pi^+ + \pi^-$, half-life 10^{-23} s. The neutral kaon also decays into two pions: $K_S^0 \rightarrow \pi^+ + \pi^-$, half-life 10^{-10} s. How do you explain the difference in half-lives?

21. Determine whether strangeness is conserved in the following decays and reactions.
(a) $\Lambda^0 \rightarrow p + \pi^-$
(b) $\pi^- + p \rightarrow \Lambda^0 + K^0$
(c) $\bar{p} + p \rightarrow \overline{\Lambda^0} + \Lambda^0$
(d) $\pi^- + p \rightarrow \pi^- + \Sigma^+$
(e) $\Xi^- \rightarrow \Lambda^0 + \pi^-$
(f) $\Xi^0 \rightarrow p + \pi^-$

22. For each of the following forbidden decays, determine which conservation law is violated.
(a) $\mu^- \rightarrow e^- + \gamma$
(b) $n \rightarrow p + e^- + \nu_e$
(c) $\Lambda^0 \rightarrow p + \pi^0$
(d) $p \rightarrow e^+ + \pi^0$
(e) $\Xi^0 \rightarrow n + \pi^0$

23. Which of the following processes are allowed by the strong interaction, the electromagnetic interaction, the weak interaction, or no interaction at all?
(a) $\pi^- + p \rightarrow 2\eta$

(b) $K^- + n \rightarrow \Lambda^0 + \pi^-$
(c) $K^- \rightarrow \pi^- + \pi^0$
(d) $\Omega^- \rightarrow \Xi^- + \pi^0$
(e) $\eta \rightarrow 2\gamma$

24. Identify the conserved quantities in the following processes.
(a) $\Xi^- \rightarrow \Lambda^0 + \mu^- + \nu_\mu$
(b) $K_S^0 \rightarrow 2\pi^0$
(c) $K^- + p \rightarrow \Sigma^0 + n$
(d) $\Sigma^0 \rightarrow \Lambda^0 + \gamma$
(e) $e^+ + e^- \rightarrow \mu^+ + \mu^-$
(f) $\bar{p} + n \rightarrow \Lambda^0 + \Sigma^-$

25. Fill in the missing particle. Assume that (a) occurs via the strong interaction and that (b) and (c) involve the weak interaction.
(a) $K^+ + p \rightarrow ? + p$
(b) $\Omega^- \rightarrow ? + \pi^-$
(c) $K^+ \rightarrow ? + \mu^+ + \nu_\mu$

Section 46.7 Making Particles and Measuring Their Properties

Section 46.8 Finding Patterns in the Particles

Section 46.9 Quarks—Finally

26. The quark composition of the proton is uud, and that of the neutron is udd. Show that in each case the charge, baryon number, and strangeness of the particle equal, respectively, the sums of these numbers for the quark constituents.

27. (a) Find the number of electrons and the number of each species of quark in 1 L of water. (b) Make an order-of-magnitude estimate of the number of each kind of fundamental matter particle in your body. State your assumptions and the quantities you take as data.

28. The quark compositions of the K^0 and Λ^0 particles are $\bar{s}d$ and uds, respectively. Show that the charge, baryon number, and strangeness of these particles equal, respectively, the sums of these numbers for the quark constituents.

29. Assuming that binding energies can be neglected, find the masses of the u and d quarks from the masses of the proton and neutron.

30. The text stated that the reaction $\pi^- + p \rightarrow K^0 + \Lambda^0$ occurs with high probability, whereas the reaction $\pi^- + p \rightarrow K^0 + n$ never occurs. Analyze these reactions at the quark level. Show that the first reaction conserves the total number of each type of quark, and the second reaction does not.

31. Analyze each reaction in terms of constituent quarks.
(a) $\pi^- + p \rightarrow K^0 + \Lambda^0$
(b) $\pi^+ + p \rightarrow K^+ + \Sigma^+$
(c) $K^- + p \rightarrow K^+ + K^0 + \Omega^-$
(d) $p + p \rightarrow K^0 + p + \pi^+ + ?$
In the last reaction, identify the mystery particle.

32. A Σ^0 particle traveling through matter strikes a proton; then a Σ^+ and a gamma ray emerge, as well as a third

particle. Use the quark model of each to determine the identity of the third particle.

33. Identify the particles corresponding to the quark states (a) suu, (b) $\bar{u}d$, (c) $\bar{s}d$, and (d) ssd.

34. What is the electrical charge of the baryons with the quark compositions (a) $\bar{u}\,\bar{u}\,d$ and (b) $\bar{u}\,d\,d$? What are these baryons called?

Section 46.10 Multicolored Quarks

Section 46.11 The Standard Model

Section 46.12 The Cosmic Connection

35. **Review Problem.** Review Section 39.4. Prove that the Doppler shift in wavelength of electromagnetic waves is given by

$$\lambda' = \lambda \sqrt{\frac{1 + v/c}{1 - v/c}}$$

where λ' is the wavelength measured by an observer moving at speed v away from a source radiating waves of wavelength λ.

36. Using Hubble's law (Eq. 46.7), find the wavelength of the 590-nm sodium line emitted from galaxies (a) 2.00×10^6 ly away from Earth, (b) 2.00×10^8 ly away, and (c) 2.00×10^9 ly away. You may use the result of Problem 35.

WEB 37. A distant quasar is moving away from Earth at such high speed that the blue 434-nm hydrogen line is observed at 650 nm, in the red portion of the spectrum. (a) How fast is the quasar receding? You may use the result of Problem 35. (b) Using Hubble's law, determine the distance from Earth to this quasar.

38. The various spectral lines observed in the light from a distant quasar have longer wavelengths λ'_n than the wavelengths λ_n measured in light from a stationary source. The fractional change in wavelength toward the red is the same for all spectral lines. That is, the redshift parameter Z defined by

$$Z = (\lambda'_n - \lambda_n)/\lambda_n$$

is common to all spectral lines for one object. In terms of Z, determine (a) the speed of recession of the quasar and (b) the distance from Earth to this quasar. Use the result of Problem 35 and Hubble's law.

39. It is mostly your roommate's fault. Nosy astronomers have discovered enough junk and clutter in your dorm room to constitute the missing mass required to close the Universe. After observing your floor, closet, bed, and computer files, they calculate the average density of the observable Universe as 1.20 ρ_c. How many times larger will the Universe become before it begins to collapse? That is, by what factor will the distance between remote galaxies increase in the future?

40. The early Universe was dense with gamma-ray photons of energy $\sim k_B T$ and at such a high temperature that protons and antiprotons were created by the process

$\gamma \rightarrow p + \bar{p}$ as rapidly as they annihilated each other. As the Universe cooled in adiabatic expansion, its temperature fell below a certain value, and proton-pair production became rare. At that time, slightly more protons than antiprotons existed, and essentially all of the protons in the Universe today date from that time. (a) Estimate the order of magnitude of the temperature of the Universe when protons condensed out. (b) Estimate the order of magnitude of the temperature of the Universe when electrons condensed out.

41. **Review Problem.** The cosmic background radiation is blackbody radiation at a temperature of 2.73 K. (a) Determine the wavelength at which this radiation has its maximum intensity. (b) In what part of the electromagnetic spectrum is the peak of the distribution?

Section 46.13 Problems and Perspectives

42. Classical general relativity views the structure of space–time as deterministic and well defined down to arbitrarily small distances. On the other hand, quantum general relativity forbids distances less than the Planck length given by $L = (\hbar G / c^3)^{1/2}$. (a) Calculate the value of the Planck length. The quantum limitation suggests that after the Big Bang, when all of the presently observable section of the Universe was reduced to a point-like singularity, nothing could be observed until that singularity grew larger than the Planck length. Because the size of the singularity grew at the speed of light, we can infer that no observations were possible during the time it took for light to travel the Planck length. (b) Calculate this time, known as the Planck time T, and compare it with the ultrahot epoch mentioned in the text. (c) Does this suggest that we may never know what happened between the time $t = 0$ and the time $t = T$?

ADDITIONAL PROBLEMS

43. The nuclear force can be attributed to the exchange of an elementary particle between protons and neutrons if they are sufficiently close. Take the range of the nuclear force as approximately 1.4×10^{-15} m. (a) Use the uncertainty principle $\Delta E \, \Delta t \geq \hbar/2$ to estimate the mass of the elementary particle if it moves at nearly the speed of light. (b) Using Table 46.2, identify the particle.

44. Name at least one conservation law that prevents each of the following reactions.
    (a) $\pi^- + p \rightarrow \Sigma^+ + \pi^0$
    (b) $\mu^- \rightarrow \pi^- + \nu_e$
    (c) $p \rightarrow \pi^+ + \pi^+ + \pi^-$

WEB 45. The energy flux carried by neutrinos from the Sun is estimated to be on the order of 0.4 W/m^2 at Earth's surface. Estimate the fractional mass loss of the Sun over 10^9 years due to the radiation of neutrinos. (The mass of the Sun is 2×10^{30} kg. The Earth–Sun distance is 1.5×10^{11} m.)

46. Two protons approach each other with 70.4 MeV of kinetic energy and engage in a reaction in which a proton and a positive pion emerge at rest. What third particle, obviously uncharged and therefore difficult to detect, must have been created?

47. **Review Problem.** Supernova 1987A, located about 170 000 ly from Earth, is estimated to have emitted a burst of $\sim 10^{46}$ J of neutrinos. Suppose that the average neutrino energy was 6 MeV and your body presented cross-sectional area 5 000 cm^2. To an order of magnitude, how many of these neutrinos passed through you?

48. A gamma-ray photon strikes a stationary electron. Determine the minimum gamma-ray energy to make this reaction go:

$$\gamma + e^- \longrightarrow e^- + e^- + e^+$$

49. Determine the kinetic energies of the proton and pion resulting from the decay of a Λ^0 at rest:

$$\Lambda^0 \longrightarrow p + \pi^-$$

50. An unstable particle, initially at rest, decays into a proton (rest energy 938.3 MeV) and a negative pion (rest energy 139.5 MeV). A uniform magnetic field of 0.250 T exists perpendicular to the velocities of the created particles. The radius of curvature of each track is found to be 1.33 m. What is the mass of the original unstable particle?

51. A Σ^0 particle at rest decays according to

$$\Sigma^0 \longrightarrow \Lambda^0 + \gamma$$

Find the gamma-ray energy.

52. Two protons approach each other with equal and opposite velocities. What is the minimum kinetic energy of each of the protons if they are to produce a π^+ meson at rest in the following reaction?

$$p + p \longrightarrow p + n + \pi^+$$

53. If a K_S^0 meson at rest decays in 0.900×10^{-10} s, how far will a K_S^0 meson travel if it is moving at $0.960 c$ through a bubble chamber?

54. A π-meson at rest decays according to $\pi^- \rightarrow \mu^- + \bar{\nu}_\mu$. What energy is carried off by the neutrino? (Assume a massless neutrino that moves off with the speed of light.) $m_\pi c^2 = 139.5$ MeV, $m_\mu c^2 = 105.7$ MeV.

55. **Review Problem.** Use the Boltzmann distribution function $e^{-E/k_B T}$ to calculate the temperature at which 1.00% of a population of photons will have energy greater than 1.00 eV. The energy required to excite an atom is on the order of 1 eV. Thus, as the temperature of the Universe fell below the value you calculate, neutral atoms could form from plasma, and the Universe became transparent. The cosmic background radiation represents our vastly red-shifted view of the opaque fireball of the Big Bang as it was at that time and temperature. The fireball surrounds us; we are embers.

56. What processes are described by the Feynman diagrams in Figure P46.56? What is the exchange particle in each process?

57. Identify the mediators for the two interactions described in the Feynman diagrams shown in Figure P46.57.

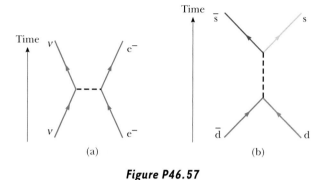

(a)   (b)

Figure P46.57

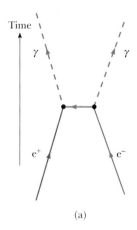

(a)

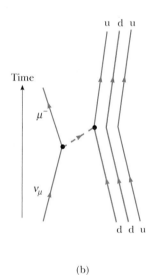

(b)

Figure P46.56

ANSWERS TO QUICK QUIZZES

46.1 (a) Into the plane. The right-hand rule for the positive particle tells you that this is the direction that leads to a force directed toward the center of curvature of the path. (b) The particle must have been slowed (by collisions) during its encounter with the lead, causing it to move in a tighter circular path (see Eq. 29.13, $r = mv/qB$).

46.2 Equation 46.3 indicates that the masses of both of these field particles must be zero; otherwise d would be less than infinity.

46.3 No, because several conservation laws are violated. Electric charge is not conserved because the negative charge on the electron disappears. Electron lepton number is not conserved because an electron with $L_e = 1$ exists before the decay and two neutrinos, each with $L_e = 1$, exist afterward. Angular momentum is not conserved because one spin-$\frac{1}{2}$ particle exists before the decay, and two spin-$\frac{1}{2}$ particles exist afterward.

46.4 We can argue this from the point of view of creating particle–antiparticle pairs from available energy. If energy is converted to rest energy of a lepton–antilepton pair, no net change occurs in lepton number because $L_e = 1$ for the lepton and $L_e = -1$ for the antilepton. Energy can also be transformed to rest energy of a baryon–antibaryon pair. The baryon has $B = +1$ and the antibaryon has $B = -1$, so no net change occurs in baryon number.

Now suppose energy is transformed to rest energy of a quark–antiquark pair. By definition, such a pair is a meson. Thus, a meson has been created from energy—no meson existed before; now one does. Thus, meson number is not conserved. With more energy, we can create more mesons, with no restriction from a conservation law other than the law of conservation of energy.

"Particles, particles, particles."

The Meaning of Success

To earn the respect of intelligent people and to win the affection of children;
To appreciate the beauty in nature and all that surrounds us;
To seek out and nurture the best in others;
To give the gift of yourself to others without the slightest thought of return, for it is in giving that we receive;
To have accomplished a task, whether it be saving a lost soul, healing a sick child, writing a book, or risking your life for a friend;
To have celebrated and laughed with great joy and enthusiasm and sung with exaltation;
To have hope even in times of despair, for as long as you have hope, you have life;
To love and be loved;
To be understood and to understand;
To know that even one life has breathed easier because you have lived;
This is the meaning of success.

RALPH WALDO EMERSON
Modified by Ray Serway, December 1989

TABLE A.1 Conversion Factors

Length

	m	cm	km	in.	ft	mi
1 meter	1	10^2	10^{-3}	39.37	3.281	6.214×10^{-4}
1 centimeter	10^{-2}	1	10^{-5}	0.393 7	3.281×10^{-2}	6.214×10^{-6}
1 kilometer	10^3	10^5	1	3.937×10^4	3.281×10^3	0.621 4
1 inch	2.540×10^{-2}	2.540	2.540×10^{-5}	1	8.333×10^{-2}	1.578×10^{-5}
1 foot	0.304 8	30.48	3.048×10^{-4}	12	1	1.894×10^{-4}
1 mile	1 609	1.609×10^5	1.609	6.336×10^4	5 280	1

Mass

	kg	g	slug	u
1 kilogram	1	10^3	6.852×10^{-2}	6.024×10^{26}
1 gram	10^{-3}	1	6.852×10^{-5}	6.024×10^{23}
1 slug	14.59	1.459×10^4	1	8.789×10^{27}
1 atomic mass unit	1.660×10^{-27}	1.660×10^{-24}	1.137×10^{-28}	1

Note: 1 metric ton = 1 000 kg.

Time

	s	min	h	day	yr
1 second	1	1.667×10^{-2}	2.778×10^{-4}	1.157×10^{-5}	3.169×10^{-8}
1 minute	60	1	1.667×10^{-2}	6.994×10^{-4}	1.901×10^{-6}
1 hour	3 600	60	1	4.167×10^{-2}	1.141×10^{-4}
1 day	8.640×10^4	1 440	24	1	2.738×10^{-5}
1 year	3.156×10^7	5.259×10^5	8.766×10^3	365.2	1

Speed

	m/s	cm/s	ft/s	mi/h
1 meter per second	1	10^2	3.281	2.237
1 centimeter per second	10^{-2}	1	3.281×10^{-2}	2.237×10^{-2}
1 foot per second	0.304 8	30.48	1	0.681 8
1 mile per hour	0.447 0	44.70	1.467	1

Note: 1 mi/min = 60 mi/h = 88 ft/s.

continued

TABLE A.1   *Continued*

Force

	N	lb
1 newton	1	0.224 8
1 pound	4.448	1

Work, Energy, Heat

	J	ft·lb	eV
1 joule	1	0.737 6	6.242×10^{18}
1 ft·lb	1.356	1	8.464×10^{18}
1 eV	1.602×10^{-19}	1.182×10^{-19}	1
1 cal	4.186	3.087	2.613×10^{19}
1 Btu	1.055×10^{3}	7.779×10^{2}	6.585×10^{21}
1 kWh	3.600×10^{6}	2.655×10^{6}	2.247×10^{25}

	cal	Btu	kWh
1 joule	0.238 9	9.481×10^{-4}	2.778×10^{-7}
1 ft·lb	0.323 9	1.285×10^{-3}	3.766×10^{-7}
1 eV	3.827×10^{-20}	1.519×10^{-22}	4.450×10^{-26}
1 cal	1	3.968×10^{-3}	1.163×10^{-6}
1 Btu	2.520×10^{2}	1	2.930×10^{-4}
1 kWh	8.601×10^{5}	3.413×10^{2}	1

Pressure

	Pa	atm
1 pascal	1	9.869×10^{-6}
1 atmosphere	1.013×10^{5}	1
1 centimeter mercury[a]	1.333×10^{3}	1.316×10^{-2}
1 pound per inch²	6.895×10^{3}	6.805×10^{-2}
1 pound per foot²	47.88	4.725×10^{-4}

	cm Hg	lb/in.²	lb/ft²
1 newton per meter²	7.501×10^{-4}	1.450×10^{-4}	2.089×10^{-2}
1 atmosphere	76	14.70	2.116×10^{3}
1 centimeter mercury[a]	1	0.194 3	27.85
1 pound per inch²	5.171	1	144
1 pound per foot²	3.591×10^{-2}	6.944×10^{-3}	1

[a] At 0°C and at a location where the acceleration due to gravity has its "standard" value, 9.806 65 m/s².

TABLE A.2 Symbols, Dimensions, and Units of Physical Quantities

Quantity	Common Symbol	Unit[a]	Dimensions[b]	Unit in Terms of Base SI Units
Acceleration	$\mathbf{a}$	m/s^2	L/T^2	m/s^2
Amount of substance	n	mole		mol
Angle	θ, ϕ	radian (rad)	1	
Angular acceleration	$\boldsymbol{\alpha}$	rad/s^2	T^{-2}	s^{-2}
Angular frequency	ω	rad/s	T^{-1}	s^{-1}
Angular momentum	$\mathbf{L}$	$kg \cdot m^2/s$	ML^2/T	$kg \cdot m^2/s$
Angular velocity	$\boldsymbol{\omega}$	rad/s	T^{-1}	s^{-1}
Area	A	m^2	L^2	m^2
Atomic number	Z			
Capacitance	C	farad (F)	Q^2T^2/ML^2	$A^2 \cdot s^4/kg \cdot m^2$
Charge	q, Q, e	coulomb (C)	Q	$A \cdot s$
Charge density				
Line	λ	C/m	Q/L	$A \cdot s/m$
Surface	σ	C/m^2	Q/L^2	$A \cdot s/m^2$
Volume	ρ	C/m^3	Q/L^3	$A \cdot s/m^3$
Conductivity	σ	$1/\Omega \cdot m$	Q^2T/ML^3	$A^2 \cdot s^3/kg \cdot m^3$
Current	I	AMPERE	Q/T	A
Current density	$\mathbf{J}$	A/m^2	Q/T^2	A/m^2
Density	ρ	kg/m^3	M/L^3	kg/m^3
Dielectric constant	κ			
Displacement	$\mathbf{r, s}$	METER	L	m
Distance	d, h			
Length	ℓ, L			
Electric dipole moment	$\mathbf{p}$	$C \cdot m$	QL	$A \cdot s \cdot m$
Electric field	$\mathbf{E}$	V/m	ML/QT^2	$kg \cdot m/A \cdot s^3$
Electric flux	Φ_E	$V \cdot m$	ML^3/QT^2	$kg \cdot m^3/A \cdot s^3$
Electromotive force	$\boldsymbol{\varepsilon}$	volt (V)	ML^2/QT^2	$kg \cdot m^2/A \cdot s^3$
Energy	E, U, K	joule (J)	ML^2/T^2	$kg \cdot m^2/s^2$
Entropy	S	J/K	$ML^2/T^2 \cdot K$	$kg \cdot m^2/s^2 \cdot K$
Force	$\mathbf{F}$	newton (N)	ML/T^2	$kg \cdot m/s^2$
Frequency	f	hertz (Hz)	T^{-1}	s^{-1}
Heat	Q	joule (J)	ML^2/T^2	$kg \cdot m^2/s^2$
Inductance	L	henry (H)	ML^2/Q^2	$kg \cdot m^2/A^2 \cdot s^2$
Magnetic dipole moment	$\boldsymbol{\mu}$	$N \cdot m/T$	QL^2/T	$A \cdot m^2$
Magnetic field	$\mathbf{B}$	tesla (T) $(= Wb/m^2)$	M/QT	$kg/A \cdot s^2$
Magnetic flux	Φ_B	weber (Wb)	ML^2/QT	$kg \cdot m^2/A \cdot s^2$
Mass	m, M	KILOGRAM	M	kg
Molar specific heat	C	$J/mol \cdot K$		$kg \cdot m^2/s^2 \cdot mol \cdot K$
Moment of inertia	I	$kg \cdot m^2$	ML^2	$kg \cdot m^2$
Momentum	$\mathbf{p}$	$kg \cdot m/s$	ML/T	$kg \cdot m/s$
Period	T	s	T	s
Permeability of space	μ_0	$N/A^2 (= H/m)$	ML/Q^2T	$kg \cdot m/A^2 \cdot s^2$
Permittivity of space	ϵ_0	$C^2/N \cdot m^2 (= F/m)$	Q^2T^2/ML^3	$A^2 \cdot s^4/kg \cdot m^3$
Potential	V	volt (V) $(= J/C)$	ML^2/QT^2	$kg \cdot m^2/A \cdot s^3$
Power	$\mathscr{P}$	watt (W) $(= J/s)$	ML^2/T^3	$kg \cdot m^2/s^3$

continued

TABLE A.2 *Continued*

Quantity	Common Symbol	Unit[a]	Dimensions[b]	Unit in Terms of Base SI Units
Pressure	P	pascal (Pa) $= (N/m^2)$	M/LT^2	$kg/m \cdot s^2$
Resistance	R	ohm $(\Omega)(=V/A)$	ML^2/Q^2T	$kg \cdot m^2/A^2 \cdot s^3$
Specific heat	c	$J/kg \cdot K$	$L^2/T^2 \cdot K$	$m^2/s^2 \cdot K$
Speed	v	m/s	L/T	m/s
Temperature	T	KELVIN	K	K
Time	t	SECOND	T	s
Torque	τ	$N \cdot m$	ML^2/T^2	$kg \cdot m^2/s^2$
Volume	V	m^3	L^3	m^3
Wavelength	λ	m	L	m
Work	W	joule $(J)(=N \cdot m)$	ML^2/T^2	$kg \cdot m^2/s^2$

[a] The base SI units are given in uppercase letters.

[b] The symbols M, L, T, and Q denote mass, length, time, and charge, respectively.

TABLE A.3   Table of Atomic Masses[a]

Atomic Number Z	Element	Symbol	Chemical Atomic Mass (u)	Mass Number (* Indicates Radioactive) A	Atomic Mass (u)	Percent Abundance	Half-Life (If Radioactive) $T_{1/2}$
0	(Neutron)	n		1*	1.008 665		10.4 min
1	Hydrogen	H	1.007 9	1	1.007 825	99.985	
	Deuterium	D		2	2.014 102	0.015	
	Tritium	T		3*	3.016 049		12.33 yr
2	Helium	He	4.002 60	3	3.016 029	0.000 14	
				4	4.002 602	99.999 86	
				6*	6.018 886		0.81 s
3	Lithium	Li	6.941	6	6.015 121	7.5	
				7	7.016 003	92.5	
				8*	8.022 486		0.84 s
4	Beryllium	Be	9.012 2	7*	7.016 928		53.3 days
				9	9.012 174	100	
				10*	10.013 534		1.5×10^6 yr
5	Boron	B	10.81	10	10.012 936	19.9	
				11	11.009 305	80.1	
				12*	12.014 352		0.020 2 s
6	Carbon	C	12.011	10*	10.016 854		19.3 s
				11*	11.011 433		20.4 min
				12	12.000 000	98.90	
				13	13.003 355	1.10	
				14*	14.003 242		5 730 yr
				15*	15.010 599		2.45 s
7	Nitrogen	N	14.006 7	12*	12.018 613		0.011 0 s
				13*	13.005 738		9.96 min
				14	14.003 074	99.63	
				15	15.000 108	0.37	
				16*	16.006 100		7.13 s
				17*	17.008 450		4.17 s

TABLE A.3   Continued

Atomic Number Z	Element	Symbol	Chemical Atomic Mass (u)	Mass Number (* Indicates Radioactive) A	Atomic Mass (u)	Percent Abundance	Half-Life (If Radioactive) $T_{1/2}$
8	Oxygen	O	15.999 4	14*	14.008 595		70.6 s
				15*	15.003 065		122 s
				16	15.994 915	99.761	
				17	16.999 132	0.039	
				18	17.999 160	0.20	
				19*	19.003 577		26.9 s
9	Fluorine	F	18.998 40	17*	17.002 094		64.5 s
				18*	18.000 937		109.8 min
				19	18.998 404	100	
				20*	19.999 982		11.0 s
				21*	20.999 950		4.2 s
10	Neon	Ne	20.180	18*	18.005 710		1.67 s
				19*	19.001 880		17.2 s
				20	19.992 435	90.48	
				21	20.993 841	0.27	
				22	21.991 383	9.25	
				23*	22.994 465		37.2 s
11	Sodium	Na	22.989 87	21*	20.997 650		22.5 s
				22*	21.994 434		2.61 yr
				23	22.989 770	100	
				24*	23.990 961		14.96 h
12	Magnesium	Mg	24.305	23*	22.994 124		11.3 s
				24	23.985 042	78.99	
				25	24.985 838	10.00	
				26	25.982 594	11.01	
				27*	26.984 341		9.46 min
13	Aluminum	Al	26.981 54	26*	25.986 892		7.4×10^5 yr
				27	26.981 538	100	
				28*	27.981 910		2.24 min
14	Silicon	Si	28.086	28	27.976 927	92.23	
				29	28.976 495	4.67	
				30	29.973 770	3.10	
				31*	30.975 362		2.62 h
				32*	31.974 148		172 yr
15	Phosphorus	P	30.973 76	30*	29.978 307		2.50 min
				31	30.973 762	100	
				32*	31.973 908		14.26 days
				33*	32.971 725		25.3 days
16	Sulfur	S	32.066	32	31.972 071	95.02	
				33	32.971 459	0.75	
				34	33.967 867	4.21	
				35*	34.969 033		87.5 days
				36	35.967 081	0.02	
17	Chlorine	Cl	35.453	35	34.968 853	75.77	
				36*	35.968 307		3.0×10^5 yr
				37	36.965 903	24.23	

continued

TABLE A.3 *Continued*

Atomic Number Z	Element	Symbol	Chemical Atomic Mass (u)	Mass Number (* Indicates Radioactive) A	Atomic Mass (u)	Percent Abundance	Half-Life (If Radioactive) $T_{1/2}$
18	Argon	Ar	39.948	36	35.967 547	0.337	
				37*	36.966 776		35.04 days
				38	37.962 732	0.063	
				39*	38.964 314		269 yr
				40	39.962 384	99.600	
				42*	41.963 049		33 yr
19	Potassium	K	39.098 3	39	38.963 708	93.258 1	
				40*	39.964 000	0.011 7	1.28×10^9 yr
				41	40.961 827	6.730 2	
20	Calcium	Ca	40.08	40	39.962 591	96.941	
				41*	40.962 279		1.0×10^5 yr
				42	41.958 618	0.647	
				43	42.958 767	0.135	
				44	43.955 481	2.086	
				46	45.953 687	0.004	
				48	47.952 534	0.187	
21	Scandium	Sc	44.955 9	41*	40.969 250		0.596 s
				45	44.955 911	100	
22	Titanium	Ti	47.88	44*	43.959 691		49 yr
				46	45.952 630	8.0	
				47	46.951 765	7.3	
				48	47.947 947	73.8	
				49	48.947 871	5.5	
				50	49.944 792	5.4	
23	Vanadium	V	50.941 5	48*	47.952 255		15.97 days
				50*	49.947 161	0.25	1.5×10^{17} yr
				51	50.943 962	99.75	
24	Chromium	Cr	51.996	48*	47.954 033		21.6 h
				50	49.946 047	4.345	
				52	51.940 511	83.79	
				53	52.940 652	9.50	
				54	53.938 883	2.365	
25	Manganese	Mn	54.938 05	54*	53.940 361		312.1 days
				55	54.938 048	100	
26	Iron	Fe	55.847	54	53.939 613	5.9	
				55*	54.938 297		2.7 yr
				56	55.934 940	91.72	
				57	56.935 396	2.1	
				58	57.933 278	0.28	
				60*	59.934 078		1.5×10^6 yr
27	Cobalt	Co	58.933 20	59	58.933 198	100	
				60*	59.933 820		5.27 yr
28	Nickel	Ni	58.693	58	57.935 346	68.077	
				59*	58.934 350		7.5×10^4 yr
				60	59.930 789	26.223	
				61	60.931 058	1.140	
				62	61.928 346	3.634	
				63*	62.929 670		100 yr
				64	63.927 967	0.926	

TABLE A.3 Continued

Atomic Number Z	Element	Symbol	Chemical Atomic Mass (u)	Mass Number (* Indicates Radioactive) A	Atomic Mass (u)	Percent Abundance	Half-Life (If Radioactive) $T_{1/2}$
29	Copper	Cu	63.54	63	62.929 599	69.17	
				65	64.927 791	30.83	
30	Zinc	Zn	65.39	64	63.929 144	48.6	
				66	65.926 035	27.9	
				67	66.927 129	4.1	
				68	67.924 845	18.8	
				70	69.925 323	0.6	
31	Gallium	Ga	69.723	69	68.925 580	60.108	
				71	70.924 703	39.892	
32	Germanium	Ge	72.61	70	69.924 250	21.23	
				72	71.922 079	27.66	
				73	72.923 462	7.73	
				74	73.921 177	35.94	
				76	75.921 402	7.44	
33	Arsenic	As	74.921 6	75	74.921 594	100	
34	Selenium	Se	78.96	74	73.922 474	0.89	
				76	75.919 212	9.36	
				77	76.919 913	7.63	
				78	77.917 307	23.78	
				79*	78.918 497		$\leq 6.5 \times 10^4$ yr
				80	79.916 519	49.61	
				82*	81.916 697	8.73	1.4×10^{20} yr
35	Bromine	Br	79.904	79	78.918 336	50.69	
				81	80.916 287	49.31	
36	Krypton	Kr	83.80	78	77.920 400	0.35	
				80	79.916 377	2.25	
				81*	80.916 589		2.1×10^5 yr
				82	81.913 481	11.6	
				83	82.914 136	11.5	
				84	83.911 508	57.0	
				85*	84.912 531		10.76 yr
				86	85.910 615	17.3	
37	Rubidium	Rb	85.468	85	84.911 793	72.17	
				87*	86.909 186	27.83	4.75×10^{10} yr
38	Strontium	Sr	87.62	84	83.913 428	0.56	
				86	85.909 266	9.86	
				87	86.908 883	7.00	
				88	87.905 618	82.58	
				90*	89.907 737		29.1 yr
39	Yttrium	Y	88.905 8	89	88.905 847	100	
40	Zirconium	Zr	91.224	90	89.904 702	51.45	
				91	90.905 643	11.22	
				92	91.905 038	17.15	
				93*	92.906 473		1.5×10^6 yr
				94	93.906 314	17.38	
				96	95.908 274	2.80	

continued

TABLE A.3 Continued

Atomic Number Z	Element	Symbol	Chemical Atomic Mass (u)	Mass Number (* Indicates Radioactive) A	Atomic Mass (u)	Percent Abundance	Half-Life (If Radioactive) $T_{1/2}$
41	Niobium	Nb	92.906 4	91*	90.906 988		6.8×10^2 yr
				92*	91.907 191		3.5×10^7 yr
				93	92.906 376	100	
				94*	93.907 280		2×10^4 yr
42	Molybdenum	Mo	95.94	92	91.906 807	14.84	
				93*	92.906 811		3.5×10^3 yr
				94	93.905 085	9.25	
				95	94.905 841	15.92	
				96	95.904 678	16.68	
				97	96.906 020	9.55	
				98	97.905 407	24.13	
				100	99.907 476	9.63	
43	Technetium	Tc		97*	96.906 363		2.6×10^6 yr
				98*	97.907 215		4.2×10^6 yr
				99*	98.906 254		2.1×10^5 yr
44	Ruthenium	Ru	101.07	96	95.907 597	5.54	
				98	97.905 287	1.86	
				99	98.905 939	12.7	
				100	99.904 219	12.6	
				101	100.905 558	17.1	
				102	101.904 348	31.6	
				104	103.905 428	18.6	
45	Rhodium	Rh	102.905 5	103	102.905 502	100	
46	Palladium	Pd	106.42	102	101.905 616	1.02	
				104	103.904 033	11.14	
				105	104.905 082	22.33	
				106	105.903 481	27.33	
				107*	106.905 126		6.5×10^6 yr
				108	107.903 893	26.46	
				110	109.905 158	11.72	
47	Silver	Ag	107.868	107	106.905 091	51.84	
				109	108.904 754	48.16	
48	Cadmium	Cd	112.41	106	105.906 457	1.25	
				108	107.904 183	0.89	
				109*	108.904 984		462 days
				110	109.903 004	12.49	
				111	110.904 182	12.80	
				112	111.902 760	24.13	
				113*	112.904 401	12.22	9.3×10^{15} yr
				114	113.903 359	28.73	
				116	115.904 755	7.49	
49	Indium	In	114.82	113	112.904 060	4.3	
				115*	114.903 876	95.7	4.4×10^{14} yr
50	Tin	Sn	118.71	112	111.904 822	0.97	
				114	113.902 780	0.65	
				115	114.903 345	0.36	
				116	115.901 743	14.53	
				117	116.902 953	7.68	

TABLE A.3 *Continued*

Atomic Number Z	Element	Symbol	Chemical Atomic Mass (u)	Mass Number (* Indicates Radioactive) A	Atomic Mass (u)	Percent Abundance	Half-Life (If Radioactive) $T_{1/2}$
(50)	(Tin)			118	117.901 605	24.22	
				119	118.903 308	8.58	
				120	119.902 197	32.59	
				121*	120.904 237		55 yr
				122	121.903 439	4.63	
				124	123.905 274	5.79	
51	Antimony	Sb	121.76	121	120.903 820	57.36	
				123	122.904 215	42.64	
				125*	124.905 251		2.7 yr
52	Tellurium	Te	127.60	120	119.904 040	0.095	
				122	121.903 052	2.59	
				123*	122.904 271	0.905	1.3×10^{13} yr
				124	123.902 817	4.79	
				125	124.904 429	7.12	
				126	125.903 309	18.93	
				128*	127.904 463	31.70	$> 8 \times 10^{24}$ yr
				130*	129.906 228	33.87	$\leq 1.25 \times 10^{21}$ yr
53	Iodine	I	126.904 5	127	126.904 474	100	
				129*	128.904 984		1.6×10^{7} yr
54	Xenon	Xe	131.29	124	123.905 894	0.10	
				126	125.904 268	0.09	
				128	127.903 531	1.91	
				129	128.904 779	26.4	
				130	129.903 509	4.1	
				131	130.905 069	21.2	
				132	131.904 141	26.9	
				134	133.905 394	10.4	
				136*	135.907 215	8.9	$\geq 2.36 \times 10^{21}$ yr
55	Cesium	Cs	132.905 4	133	132.905 436	100	
				134*	133.906 703		2.1 yr
				135*	134.905 891		2×10^{6} yr
				137*	136.907 078		30 yr
56	Barium	Ba	137.33	130	129.906 289	0.106	
				132	131.905 048	0.101	
				133*	132.905 990		10.5 yr
				134	133.904 492	2.42	
				135	134.905 671	6.593	
				136	135.904 559	7.85	
				137	136.905 816	11.23	
				138	137.905 236	71.70	
57	Lanthanum	La	138.905	137*	136.906 462		6×10^{4} yr
				138*	137.907 105	0.090 2	1.05×10^{11} yr
				139	138.906 346	99.909 8	
58	Cerium	Ce	140.12	136	135.907 139	0.19	
				138	137.905 986	0.25	
				140	139.905 434	88.43	
				142*	141.909 241	11.13	$> 5 \times 10^{16}$ yr
59	Praseodymium	Pr	140.907 6	141	140.907 647	100	

continued

TABLE A.3 *Continued*

Atomic Number Z	Element	Symbol	Chemical Atomic Mass (u)	Mass Number (* Indicates Radioactive) A	Atomic Mass (u)	Percent Abundance	Half-Life (If Radioactive) $T_{1/2}$
60	Neodymium	Nd	144.24	142	141.907 718	27.13	
				143	142.909 809	12.18	
				144*	143.910 082	23.80	2.3×10^{15} yr
				145	144.912 568	8.30	
				146	145.913 113	17.19	
				148	147.916 888	5.76	
				150*	149.920 887	5.64	$>1 \times 10^{18}$ yr
61	Promethium	Pm		143*	142.910 928		265 days
				145*	144.912 745		17.7 yr
				146*	145.914 698		5.5 yr
				147*	146.915 134		2.623 yr
62	Samarium	Sm	150.36	144	143.911 996	3.1	
				146*	145.913 043		1.0×10^{8} yr
				147*	146.914 894	15.0	1.06×10^{11} yr
				148*	147.914 819	11.3	7×10^{15} yr
				149*	148.917 180	13.8	$>2 \times 10^{15}$ yr
				150	149.917 273	7.4	
				151*	150.919 928		90 yr
				152	151.919 728	26.7	
				154	153.922 206	22.7	
63	Europium	Eu	151.96	151	150.919 846	47.8	
				152*	151.921 740		13.5 yr
				153	152.921 226	52.2	
				154*	153.922 975		8.59 yr
				155*	154.922 888		4.7 yr
64	Gadolinium	Gd	157.25	148*	147.918 112		75 yr
				150*	149.918 657		1.8×10^{6} yr
				152*	151.919 787	0.20	1.1×10^{14} yr
				154	153.920 862	2.18	
				155	154.922 618	14.80	
				156	155.922 119	20.47	
				157	156.923 957	15.65	
				158	157.924 099	24.84	
				160	159.927 050	21.86	
65	Terbium	Tb	158.925 3	159	158.925 345	100	
66	Dysprosium	Dy	162.50	156	155.924 277	0.06	
				158	157.924 403	0.10	
				160	159.925 193	2.34	
				161	160.926 930	18.9	
				162	161.926 796	25.5	
				163	162.928 729	24.9	
				164	163.929 172	28.2	
67	Holmium	Ho	164.930 3	165	164.930 316	100	
				166*	165.932 282		1.2×10^{3} yr
68	Erbium	Er	167.26	162	161.928 775	0.14	
				164	163.929 198	1.61	
				166	165.930 292	33.6	

TABLE A.3 *Continued*

Atomic Number Z	Element	Symbol	Chemical Atomic Mass (u)	Mass Number (* Indicates Radioactive) A	Atomic Mass (u)	Percent Abundance	Half-Life (If Radioactive) $T_{1/2}$
(68)	(Erbium)			167	166.932 047	22.95	
				168	167.932 369	27.8	
				170	169.935 462	14.9	
69	Thulium	Tm	168.934 2	169	168.934 213	100	
				171*	170.936 428		1.92 yr
70	Ytterbium	Yb	173.04	168	167.933 897	0.13	
				170	169.934 761	3.05	
				171	170.936 324	14.3	
				172	171.936 380	21.9	
				173	172.938 209	16.12	
				174	173.938 861	31.8	
				176	175.942 564	12.7	
71	Lutecium	Lu	174.967	173*	172.938 930		1.37 yr
				175	174.940 772	97.41	
				176*	175.942 679	2.59	3.78×10^{10} yr
72	Hafnium	Hf	178.49	174*	173.940 042	0.162	2.0×10^{15} yr
				176	175.941 404	5.206	
				177	176.943 218	18.606	
				178	177.943 697	27.297	
				179	178.945 813	13.629	
				180	179.946 547	35.100	
73	Tantalum	Ta	180.947 9	180	179.947 542	0.012	
				181	180.947 993	99.988	
74	Tungsten (Wolfram)	W	183.85	180	179.946 702	0.12	
				182	181.948 202	26.3	
				183	182.950 221	14.28	
				184	183.950 929	30.7	
				186	185.954 358	28.6	
75	Rhenium	Re	186.207	185	184.952 951	37.40	
				187*	186.955 746	62.60	4.4×10^{10} yr
76	Osmium	Os	190.2	184	183.952 486	0.02	
				186*	185.953 834	1.58	2.0×10^{15} yr
				187	186.955 744	1.6	
				188	187.955 832	13.3	
				189	188.958 139	16.1	
				190	189.958 439	26.4	
				192	191.961 468	41.0	
				194*	193.965 172		6.0 yr
77	Iridium	Ir	192.2	191	190.960 585	37.3	
				193	192.962 916	62.7	
78	Platinum	Pt	195.08	190*	189.959 926	0.01	6.5×10^{11} yr
				192	191.961 027	0.79	
				194	193.962 655	32.9	
				195	194.964 765	33.8	
				196	195.964 926	25.3	
				198	197.967 867	7.2	
79	Gold	Au	196.966 5	197	196.966 543	100	

continued

TABLE A.3 Continued

Atomic Number Z	Element	Symbol	Chemical Atomic Mass (u)	Mass Number (* Indicates Radioactive) A	Atomic Mass (u)	Percent Abundance	Half-Life (If Radioactive) $T_{1/2}$
80	Mercury	Hg	200.59	196	195.965 806	0.15	
				198	197.966 743	9.97	
				199	198.968 253	16.87	
				200	199.968 299	23.10	
				201	200.970 276	13.10	
				202	201.970 617	29.86	
				204	203.973 466	6.87	
81	Thallium	Tl	204.383	203	202.972 320	29.524	
				204*	203.973 839		3.78 yr
				205	204.974 400	70.476	
		(Ra E″)		206*	205.976 084		4.2 min
		(Ac C″)		207*	206.977 403		4.77 min
		(Th C″)		208*	207.981 992		3.053 min
		(Ra C″)		210*	209.990 057		1.30 min
82	Lead	Pb	207.2	202*	201.972 134		5×10^4 yr
				204*	203.973 020	1.4	$\geq 1.4 \times 10^{17}$ yr
				205*	204.974 457		1.5×10^7 yr
				206	205.974 440	24.1	
				207	206.975 871	22.1	
				208	207.976 627	52.4	
		(Ra D)		210*	209.984 163		22.3 yr
		(Ac B)		211*	210.988 734		36.1 min
		(Th B)		212*	211.991 872		10.64 h
		(Ra B)		214*	213.999 798		26.8 min
83	Bismuth	Bi	208.980 3	207*	206.978 444		32.2 yr
				208*	207.979 717		3.7×10^5 yr
				209	208.980 374	100	
		(Ra E)		210*	209.984 096		5.01 days
		(Th C)		211*	210.987 254		2.14 min
				212*	211.991 259		60.6 min
		(Ra C)		214*	213.998 692		19.9 min
				215*	215.001 836		7.4 min
84	Polonium	Po		209*	208.982 405		102 yr
		(Ra F)		210*	209.982 848		138.38 days
		(Ac C′)		211*	210.986 627		0.52 s
		(Th C′)		212*	211.988 842		0.30 μs
		(Ra C′)		214*	213.995 177		164 μs
		(Ac A)		215*	214.999 418		0.001 8 s
		(Th A)		216*	216.001 889		0.145 s
		(Ra A)		218*	218.008 965		3.10 min
85	Astatine	At		215*	214.998 638		≈ 100 μs
				218*	218.008 685		1.6 s
				219*	219.011 294		0.9 min
86	Radon	Rn					
		(An)		219*	219.009 477		3.96 s
		(Tn)		220*	220.011 369		55.6 s
		(Rn)		222*	222.017 571		3.823 days
87	Francium	Fr					
		(Ac K)		223*	223.019 733		22 min

TABLE A.3   Continued

Atomic Number Z	Element	Symbol	Chemical Atomic Mass (u)	Mass Number (* Indicates Radioactive) A	Atomic Mass (u)	Percent Abundance	Half-Life (If Radioactive) $T_{1/2}$
88	Radium	Ra					
		(Ac X)		223*	223.018 499		11.43 days
		(Th X)		224*	224.020 187		3.66 days
		(Ra)		226*	226.025 402		1 600 yr
		(Ms Th$_1$)		228*	228.031 064		5.75 yr
89	Actinium	Ac		227*	227.027 749		21.77 yr
		(Ms Th$_2$)		228*	228.031 015		6.15 h
90	Thorium	Th	232.038 1				
		(Rd Ac)		227*	227.027 701		18.72 days
		(Rd Th)		228*	228.028 716		1.913 yr
				229*	229.031 757		7 300 yr
		(Io)		230*	230.033 127		75.000 yr
		(UY)		231*	231.036 299		25.52 h
		(Th)		232*	232.038 051	100	1.40×10^{10} yr
		(UX$_1$)		234*	234.043 593		24.1 days
91	Protactinium	Pa		231*	231.035 880		32.760 yr
		(Uz)		234*	234.043 300		6.7 h
92	Uranium	U	238.028 9	232*	232.037 131		69 yr
				233*	233.039 630		1.59×10^5 yr
				234*	234.040 946	0.005 5	2.45×10^5 yr
		(Ac U)		235*	235.043 924	0.720	7.04×10^8 yr
				236*	236.045 562		2.34×10^7 yr
		(UI)		238*	238.050 784	99.274 5	4.47×10^9 yr
93	Neptunium	Np		235*	235.044 057		396 days
				236*	236.046 560		1.15×10^5 yr
				237*	237.048 168		2.14×10^6 yr
94	Plutonium	Pu		236*	236.046 033		2.87 yr
				238*	238.049 555		87.7 yr
				239*	239.052 157		2.412×10^4 yr
				240*	240.053 808		6 560 yr
				241*	241.056 846		14.4 yr
				242*	242.058 737		3.73×10^6 yr
				244*	244.064 200		8.1×10^7 yr

[a] The masses in the sixth column are atomic masses, which include the mass of Z electrons. Data are from the National Nuclear Data Center, Brookhaven National Laboratory, prepared by Jagdish K. Tuli, July 1990. The data are based on experimental results reported in *Nuclear Data Sheets* and *Nuclear Physics* and also from *Chart of the Nuclides*, 14th ed. Atomic masses are based on those by A. H. Wapstra, G. Audi, and R. Hoekstra. Isotopic abundances are based on those by N. E. Holden.

These appendices in mathematics are intended as a brief review of operations and methods. Early in this course, you should be totally familiar with basic algebraic techniques, analytic geometry, and trigonometry. The appendices on differential and integral calculus are more detailed and are intended for those students who have difficulty applying calculus concepts to physical situations.

B.1 SCIENTIFIC NOTATION

Many quantities that scientists deal with often have very large or very small values. For example, the speed of light is about 300 000 000 m/s, and the ink required to make the dot over an i in this textbook has a mass of about 0.000 000 001 kg. Obviously, it is very cumbersome to read, write, and keep track of numbers such as these. We avoid this problem by using a method dealing with powers of the number 10:

$$10^0 = 1$$
$$10^1 = 10$$
$$10^2 = 10 \times 10 = 100$$
$$10^3 = 10 \times 10 \times 10 = 1000$$
$$10^4 = 10 \times 10 \times 10 \times 10 = 10\,000$$
$$10^5 = 10 \times 10 \times 10 \times 10 \times 10 = 100\,000$$

and so on. The number of zeros corresponds to the power to which 10 is raised, called the **exponent** of 10. For example, the speed of light, 300 000 000 m/s, can be expressed as 3×10^8 m/s.

In this method, some representative numbers smaller than unity are

$$10^{-1} = \frac{1}{10} = 0.1$$
$$10^{-2} = \frac{1}{10 \times 10} = 0.01$$
$$10^{-3} = \frac{1}{10 \times 10 \times 10} = 0.001$$
$$10^{-4} = \frac{1}{10 \times 10 \times 10 \times 10} = 0.000\,1$$
$$10^{-5} = \frac{1}{10 \times 10 \times 10 \times 10 \times 10} = 0.000\,01$$

In these cases, the number of places the decimal point is to the left of the digit 1 equals the value of the (negative) exponent. Numbers expressed as some power of 10 multiplied by another number between 1 and 10 are said to be in **scientific notation.** For example, the scientific notation for 5 943 000 000 is 5.943×10^9 and that for 0.000 083 2 is 8.32×10^{-5}.

When numbers expressed in scientific notation are being multiplied, the following general rule is very useful:

$$10^n \times 10^m = 10^{n+m} \qquad \text{(B.1)}$$

where n and m can be *any* numbers (not necessarily integers). For example, $10^2 \times 10^5 = 10^7$. The rule also applies if one of the exponents is negative: $10^3 \times 10^{-8} = 10^{-5}$.

When dividing numbers expressed in scientific notation, note that

$$\frac{10^n}{10^m} = 10^n \times 10^{-m} = 10^{n-m} \qquad \text{(B.2)}$$

EXERCISES

With help from the above rules, verify the answers to the following:

1. $86\ 400 = 8.64 \times 10^4$
2. $9\ 816\ 762.5 = 9.816\ 762\ 5 \times 10^6$
3. $0.000\ 000\ 039\ 8 = 3.98 \times 10^{-8}$
4. $(4 \times 10^8)(9 \times 10^9) = 3.6 \times 10^{18}$
5. $(3 \times 10^7)(6 \times 10^{-12}) = 1.8 \times 10^{-4}$
6. $\dfrac{75 \times 10^{-11}}{5 \times 10^{-3}} = 1.5 \times 10^{-7}$
7. $\dfrac{(3 \times 10^6)(8 \times 10^{-2})}{(2 \times 10^{17})(6 \times 10^5)} = 2 \times 10^{-18}$

B.2 ► ALGEBRA

Some Basic Rules

When algebraic operations are performed, the laws of arithmetic apply. Symbols such as x, y, and z are usually used to represent quantities that are not specified, what are called the **unknowns.**

First, consider the equation

$$8x = 32$$

If we wish to solve for x, we can divide (or multiply) each side of the equation by the same factor without destroying the equality. In this case, if we divide both sides by 8, we have

$$\frac{8x}{8} = \frac{32}{8}$$

$$x = 4$$

Next consider the equation

$$x + 2 = 8$$

In this type of expression, we can add or subtract the same quantity from each side. If we subtract 2 from each side, we get

$$x + 2 - 2 = 8 - 2$$

$$x = 6$$

In general, if $x + a = b$, then $x = b - a$.

Now consider the equation

$$\frac{x}{5} = 9$$

If we multiply each side by 5, we are left with x on the left by itself and 45 on the right:

$$\left(\frac{x}{5}\right)(5) = 9 \times 5$$

$$x = 45$$

In all cases, *whatever operation is performed on the left side of the equality must also be performed on the right side.*

The following rules for multiplying, dividing, adding, and subtracting fractions should be recalled, where a, b, and c are three numbers:

	Rule	**Example**
Multiplying	$\left(\dfrac{a}{b}\right)\left(\dfrac{c}{d}\right) = \dfrac{ac}{bd}$	$\left(\dfrac{2}{3}\right)\left(\dfrac{4}{5}\right) = \dfrac{8}{15}$
Dividing	$\dfrac{(a/b)}{(c/d)} = \dfrac{ad}{bc}$	$\dfrac{2/3}{4/5} = \dfrac{(2)(5)}{(4)(3)} = \dfrac{10}{12}$
Adding	$\dfrac{a}{b} \pm \dfrac{c}{d} = \dfrac{ad \pm bc}{bd}$	$\dfrac{2}{3} - \dfrac{4}{5} = \dfrac{(2)(5) - (4)(3)}{(3)(5)} = -\dfrac{2}{15}$

EXERCISES

In the following exercises, solve for x:

        Answers

1. $a = \dfrac{1}{1 + x}$ $x = \dfrac{1 - a}{a}$

2. $3x - 5 = 13$ $x = 6$

3. $ax - 5 = bx + 2$ $x = \dfrac{7}{a - b}$

4. $\dfrac{5}{2x + 6} = \dfrac{3}{4x + 8}$ $x = -\dfrac{11}{7}$

Powers

When powers of a given quantity x are multiplied, the following rule applies:

$$x^n x^m = x^{n+m} \tag{B.3}$$

For example, $x^2x^4 = x^{2+4} = x^6$.

When dividing the powers of a given quantity, the rule is

$$\frac{x^n}{x^m} = x^{n-m} \tag{B.4}$$

For example, $x^8/x^2 = x^{8-2} = x^6$.

A power that is a fraction, such as $\frac{1}{3}$, corresponds to a root as follows:

$$x^{1/n} = \sqrt[n]{x} \tag{B.5}$$

For example, $4^{1/3} = \sqrt[3]{4} = 1.5874$. (A scientific calculator is useful for such calculations.)

Finally, any quantity x^n raised to the mth power is

$$(x^n)^m = x^{nm} \tag{B.6}$$

Table B.1 summarizes the rules of exponents.

TABLE B.1
Rules of Exponents

$x^0 = 1$
$x^1 = x$
$x^nx^m = x^{n+m}$
$x^n/x^m = x^{n-m}$
$x^{1/n} = \sqrt[n]{x}$
$(x^n)^m = x^{nm}$

EXERCISES

Verify the following:

1. $3^2 \times 3^3 = 243$
2. $x^5x^{-8} = x^{-3}$
3. $x^{10}/x^{-5} = x^{15}$
4. $5^{1/3} = 1.709\,975$ (Use your calculator.)
5. $60^{1/4} = 2.783\,158$ (Use your calculator.)
6. $(x^4)^3 = x^{12}$

Factoring

Some useful formulas for factoring an equation are

$ax + ay + az = a(x + y + x)$ common factor

$a^2 + 2ab + b^2 = (a + b)^2$ perfect square

$a^2 - b^2 = (a + b)(a - b)$ differences of squares

Quadratic Equations

The general form of a quadratic equation is

$$ax^2 + bx + c = 0 \tag{B.7}$$

where x is the unknown quantity and a, b, and c are numerical factors referred to as **coefficients** of the equation. This equation has two roots, given by

$$x = \frac{-b \pm \sqrt{b^2 - 4ac}}{2a} \tag{B.8}$$

If $b^2 \geq 4ac$, the roots are real.

EXAMPLE 1

The equation $x^2 + 5x + 4 = 0$ has the following roots corresponding to the two signs of the square-root term:

$$x = \frac{-5 \pm \sqrt{5^2 - (4)(1)(4)}}{2(1)} = \frac{-5 \pm \sqrt{9}}{2} = \frac{-5 \pm 3}{2}$$

$$x_+ = \frac{-5 + 3}{2} = \boxed{-1} \qquad x_- = \frac{-5 - 3}{2} = \boxed{-4}$$

where x_+ refers to the root corresponding to the positive sign and x_- refers to the root corresponding to the negative sign.

EXERCISES

Solve the following quadratic equations:

Answers

1. $x^2 + 2x - 3 = 0$ $x_+ = 1$ $x_- = -3$
2. $2x^2 - 5x + 2 = 0$ $x_+ = 2$ $x_- = \frac{1}{2}$
3. $2x^2 - 4x - 9 = 0$ $x_+ = 1 + \sqrt{22}/2$ $x_- = 1 - \sqrt{22}/2$

Linear Equations

A linear equation has the general form

$$y = mx + b \qquad \text{(B.9)}$$

where m and b are constants. This equation is referred to as being linear because the graph of y versus x is a straight line, as shown in Figure B.1. The constant b, called the **y-intercept,** represents the value of y at which the straight line intersects the y axis. The constant m is equal to the **slope** of the straight line and is also equal to the tangent of the angle that the line makes with the x axis. If any two points on the straight line are specified by the coordinates (x_1, y_1) and (x_2, y_2), as in Figure B.1, then the slope of the straight line can be expressed as

$$\text{Slope} = \frac{y_2 - y_1}{x_2 - x_1} = \frac{\Delta y}{\Delta x} = \tan \theta \qquad \text{(B.10)}$$

Note that m and b can have either positive or negative values. If $m > 0$, the straight line has a *positive* slope, as in Figure B1. If $m < 0$, the straight line has a *negative* slope. In Figure B.1, both m and b are positive. Three other possible situations are shown in Figure B.2.

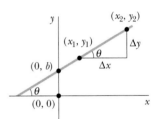

Figure B.1

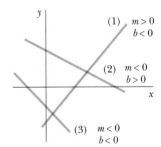

Figure B.2

EXERCISES

1. Draw graphs of the following straight lines:
   (a) $y = 5x + 3$ (b) $y = -2x + 4$ (c) $y = -3x - 6$
2. Find the slopes of the straight lines described in Exercise 1.

Answers   (a) 5   (b) -2 (c) -3

3. Find the slopes of the straight lines that pass through the following sets of points:
   (a) $(0, -4)$ and $(4, 2)$, (b) $(0, 0)$ and $(2, -5)$, and (c) $(-5, 2)$ and $(4, -2)$

Answers   (a) $3/2$ (b) $-5/2$ (c) $-4/9$

Solving Simultaneous Linear Equations

Consider the equation $3x + 5y = 15$, which has two unknowns, x and y. Such an equation does not have a unique solution. For example, note that $(x = 0, y = 3)$, $(x = 5, y = 0)$, and $(x = 2, y = 9/5)$ are all solutions to this equation.

If a problem has two unknowns, a unique solution is possible only if we have *two* equations. In general, if a problem has n unknowns, its solution requires n equations. In order to solve two simultaneous equations involving two unknowns, x and y, we solve one of the equations for x in terms of y and substitute this expression into the other equation.

EXAMPLE 2

Solve the following two simultaneous equations:

$$(1) \quad 5x + y = -8$$

$$(2) \quad 2x - 2y = 4$$

Solution   From (2), $x = y + 2$. Substitution of this into (1) gives

$$5(y + 2) + y = -8$$

$$6y = -18$$

$$y = -3$$

$$x = y + 2 = \boxed{-1}$$

Alternate Solution   Multiply each term in (1) by the factor 2 and add the result to (2):

$$10x + 2y = -16$$

$$\underline{2x - 2y = 4}$$

$$12x = -12$$

$$x = -1$$

$$y = x - 2 = \boxed{-3}$$

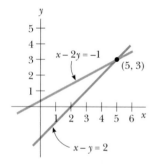

Figure B.3

Two linear equations containing two unknowns can also be solved by a graphical method. If the straight lines corresponding to the two equations are plotted in a conventional coordinate system, the intersection of the two lines represents the solution. For example, consider the two equations

$$x - y = 2$$

$$x - 2y = -1$$

These are plotted in Figure B.3. The intersection of the two lines has the coordinates $x = 5$, $y = 3$. This represents the solution to the equations. You should check this solution by the analytical technique discussed above.

EXERCISES

Solve the following pairs of simultaneous equations involving two unknowns:

                                Answers

1. $x + y = 8$ $x = 5, y = 3$
   $x - y = 2$

2. $98 - T = 10a$ $\qquad$ $T = 65, a = 3.27$
   $T - 49 = 5a$
3. $6x + 2y = 6$ $\qquad$ $x = 2, y = -3$
   $8x - 4y = 28$

Logarithms

Suppose that a quantity x is expressed as a power of some quantity a:

$$x = a^y \tag{B.11}$$

The number a is called the **base** number. The **logarithm** of x with respect to the base a is equal to the exponent to which the base must be raised in order to satisfy the expression $x = a^y$:

$$y = \log_a x \tag{B.12}$$

Conversely, the **antilogarithm** of y is the number x:

$$x = \text{antilog}_a y \tag{B.13}$$

In practice, the two bases most often used are base 10, called the *common* logarithm base, and base $e = 2.718\ldots$, called Euler's constant or the *natural* logarithm base. When common logarithms are used,

$$y = \log_{10} x \qquad (\text{or } x = 10^y) \tag{B.14}$$

When natural logarithms are used,

$$y = \ln_e x \qquad (\text{or } x = e^y) \tag{B.15}$$

For example, $\log_{10} 52 = 1.716$, so that $\text{antilog}_{10} 1.716 = 10^{1.716} = 52$. Likewise, $\ln_e 52 = 3.951$, so $\text{antiln}_e 3.951 = e^{3.951} = 52$.

In general, note that you can convert between base 10 and base e with the equality

$$\ln_e x = (2.302\ 585) \log_{10} x \tag{B.16}$$

Finally, some useful properties of logarithms are

$$\log(ab) = \log a + \log b$$
$$\log(a/b) = \log a - \log b$$
$$\log(a^n) = n \log a$$
$$\ln e = 1$$
$$\ln e^a = a$$
$$\ln\left(\frac{1}{a}\right) = -\ln a$$

B.3 GEOMETRY

The **distance** d between two points having coordinates (x_1, y_1) and (x_2, y_2) is

$$d = \sqrt{(x_2 - x_1)^2 + (y_2 - y_1)^2} \tag{B.17}$$

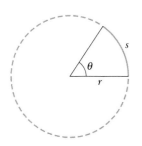

Figure B.4

Radian measure: The arc length s of a circular arc (Fig. B.4) is proportional to the radius r for a fixed value of θ (in radians):

$$s = r\theta$$

$$\theta = \frac{s}{r}$$

(B.18)

Table B.2 gives the areas and volumes for several geometric shapes used throughout this text:

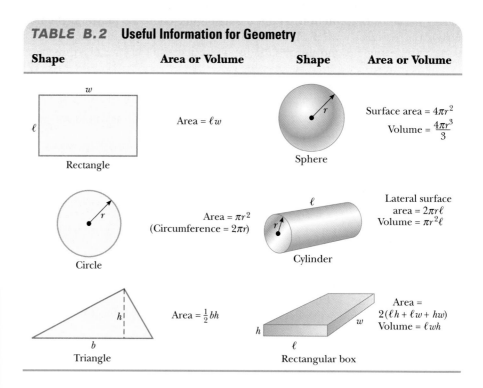

TABLE B.2	Useful Information for Geometry		
Shape	**Area or Volume**	**Shape**	**Area or Volume**
Rectangle	Area $= \ell w$	Sphere	Surface area $= 4\pi r^2$ Volume $= \frac{4\pi r^3}{3}$
Circle	Area $= \pi r^2$ (Circumference $= 2\pi r$)	Cylinder	Lateral surface area $= 2\pi r\ell$ Volume $= \pi r^2 \ell$
Triangle	Area $= \frac{1}{2}bh$	Rectangular box	Area $= 2(\ell h + \ell w + hw)$ Volume $= \ell w h$

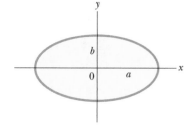

Figure B.5

The equation of a **straight line** (Fig. B.5) is

$$y = mx + b$$

(B.19)

where b is the y-intercept and m is the slope of the line.

The equation of a **circle** of radius R centered at the origin is

$$x^2 + y^2 = R^2$$

(B.20)

The equation of an **ellipse** having the origin at its center (Fig. B.6) is

$$\frac{x^2}{a^2} + \frac{y^2}{b^2} = 1$$

(B.21)

where a is the length of the semi-major axis (the longer one) and b is the length of the semi-minor axis (the shorter one).

Figure B.6

The equation of a **parabola** the vertex of which is at $y = b$ (Fig. B.7) is

$$y = ax^2 + b \qquad \text{(B.22)}$$

The equation of a **rectangular hyperbola** (Fig. B.8) is

$$xy = \text{constant} \qquad \text{(B.23)}$$

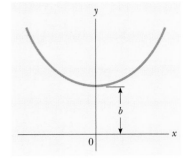

Figure B.7

B.4 ▶ TRIGONOMETRY

That portion of mathematics based on the special properties of the right triangle is called trigonometry. By definition, a right triangle is one containing a 90° angle. Consider the right triangle shown in Figure B.9, where side a is opposite the angle θ, side b is adjacent to the angle θ, and side c is the hypotenuse of the triangle. The three basic trigonometric functions defined by such a triangle are the sine (sin), cosine (cos), and tangent (tan) functions. In terms of the angle θ, these functions are defined by

$$\sin \theta \equiv \frac{\text{side opposite } \theta}{\text{hypotenuse}} = \frac{a}{c} \qquad \text{(B.24)}$$

$$\cos \theta \equiv \frac{\text{side adjacent to } \theta}{\text{hypotenuse}} = \frac{b}{c} \qquad \text{(B.25)}$$

$$\tan \theta \equiv \frac{\text{side opposite } \theta}{\text{side adjacent to } \theta} = \frac{a}{b} \qquad \text{(B.26)}$$

The Pythagorean theorem provides the following relationship between the sides of a right triangle:

$$c^2 = a^2 + b^2 \qquad \text{(B.27)}$$

From the above definitions and the Pythagorean theorem, it follows that

$$\sin^2 \theta + \cos^2 \theta = 1$$

$$\tan \theta = \frac{\sin \theta}{\cos \theta}$$

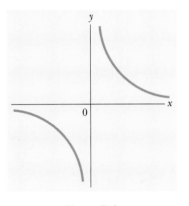

Figure B.8

The cosecant, secant, and cotangent functions are defined by

$$\csc \theta \equiv \frac{1}{\sin \theta} \qquad \sec \theta \equiv \frac{1}{\cos \theta} \qquad \cot \theta \equiv \frac{1}{\tan \theta}$$

The relationships below follow directly from the right triangle shown in Figure B.9:

$$\sin \theta = \cos(90° - \theta)$$

$$\cos \theta = \sin(90° - \theta)$$

$$\cot \theta = \tan(90° - \theta)$$

Some properties of trigonometric functions are

$$\sin(-\theta) = -\sin \theta$$

$$\cos(-\theta) = \cos \theta$$

$$\tan(-\theta) = -\tan \theta$$

The following relationships apply to *any* triangle, as shown in Figure B.10:

$$\alpha + \beta + \gamma = 180°$$

a = opposite side
b = adjacent side
c = hypotenuse

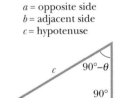

Figure B.9

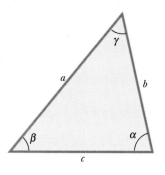

Figure B.10

$$a^2 = b^2 + c^2 - 2bc \cos \alpha$$
Law of cosines $\quad b^2 = a^2 + c^2 - 2ac \cos \beta$
$$c^2 = a^2 + b^2 - 2ab \cos \gamma$$

Law of sines $\quad \dfrac{a}{\sin \alpha} = \dfrac{b}{\sin \beta} = \dfrac{c}{\sin \gamma}$

Table B.3 lists a number of useful trigonometric identities.

TABLE B.3   Some Trigonometric Identities

$\sin^2 \theta + \cos^2 \theta = 1$	$\csc^2 \theta = 1 + \cot^2 \theta$
$\sec^2 \theta = 1 + \tan^2 \theta$	$\sin^2 \dfrac{\theta}{2} = \tfrac{1}{2}(1 - \cos \theta)$
$\sin 2\theta = 2 \sin \theta \cos \theta$	$\cos^2 \dfrac{\theta}{2} = \tfrac{1}{2}(1 + \cos \theta)$
$\cos 2\theta = \cos^2 \theta - \sin^2 \theta$	$1 - \cos \theta = 2 \sin^2 \dfrac{\theta}{2}$
$\tan 2\theta = \dfrac{2 \tan \theta}{1 - \tan^2 \theta}$	$\tan \dfrac{\theta}{2} = \sqrt{\dfrac{1 - \cos \theta}{1 + \cos \theta}}$

$\sin(A \pm B) = \sin A \cos B \pm \cos A \sin B$
$\cos(A \pm B) = \cos A \cos B \mp \sin A \sin B$
$\sin A \pm \sin B = 2 \sin[\tfrac{1}{2}(A \pm B)]\cos[\tfrac{1}{2}(A \mp B)]$
$\cos A + \cos B = 2 \cos[\tfrac{1}{2}(A + B)]\cos[\tfrac{1}{2}(A - B)]$
$\cos A - \cos B = 2 \sin[\tfrac{1}{2}(A + B)]\sin[\tfrac{1}{2}(B - A)]$

EXAMPLE 3

Consider the right triangle in Figure B.11, in which $a = 2$, $b = 5$, and c is unknown. From the Pythagorean theorem, we have

$$c^2 = a^2 + b^2 = 2^2 + 5^2 = 4 + 25 = 29$$

$$c = \sqrt{29} = \boxed{5.39}$$

To find the angle θ, note that

$$\tan \theta = \frac{a}{b} = \frac{2}{5} = 0.400$$

From a table of functions or from a calculator, we have

$$\theta = \tan^{-1}(0.400) = \boxed{21.8°}$$

where $\tan^{-1}(0.400)$ is the notation for "angle whose tangent is 0.400," sometimes written as arctan (0.400).

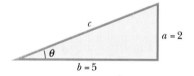

Figure B.11

EXERCISES

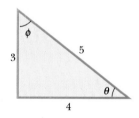

Figure B.12

1. In Figure B.12, identify (a) the side opposite θ and (b) the side adjacent to ϕ and then find (c) $\cos \theta$, (d) $\sin \phi$, and (e) $\tan \phi$.

   Answers   (a) 3, (b) 3, (c) $\frac{4}{5}$, (d) $\frac{4}{5}$, and (e) $\frac{4}{3}$

2. In a certain right triangle, the two sides that are perpendicular to each other are 5 m and 7 m long. What is the length of the third side?

   Answer   8.60 m

3. A right triangle has a hypotenuse of length 3 m, and one of its angles is 30°. What is the length of (a) the side opposite the 30° angle and (b) the side adjacent to the 30° angle?

Answers (a) 1.5 m, (b) 2.60 m

B.5 SERIES EXPANSIONS

$$(a + b)^n = a^n + \frac{n}{1!}a^{n-1}b + \frac{n(n-1)}{2!}a^{n-2}b^2 + \cdots$$

$$(1 + x)^n = 1 + nx + \frac{n(n-1)}{2!}x^2 + \cdots$$

$$e^x = 1 + x + \frac{x^2}{2!} + \frac{x^3}{3!} + \cdots$$

$$\ln(1 \pm x) = \pm x - \tfrac{1}{2}x^2 \pm \tfrac{1}{3}x^3 - \cdots$$

$$\left.\begin{array}{l} \sin x = x - \dfrac{x^3}{3!} + \dfrac{x^5}{5!} - \cdots \\[2ex] \cos x = 1 - \dfrac{x^2}{2!} + \dfrac{x^4}{4!} - \cdots \\[2ex] \tan x = x + \dfrac{x^3}{3} + \dfrac{2x^5}{15} + \cdots \qquad |x| < \pi/2 \end{array}\right\} \quad x \text{ in radians}$$

For $x \ll 1$, the following approximations can be used[1]:

$$(1 + x)^n \approx 1 + nx \qquad \sin x \approx x$$

$$e^x \approx 1 + x \qquad \cos x \approx 1$$

$$\ln(1 \pm x) \approx \pm x \qquad \tan x \approx x$$

B.6 DIFFERENTIAL CALCULUS

In various branches of science, it is sometimes necessary to use the basic tools of calculus, invented by Newton, to describe physical phenomena. The use of calculus is fundamental in the treatment of various problems in Newtonian mechanics, electricity, and magnetism. In this section, we simply state some basic properties and "rules of thumb" that should be a useful review to the student.

First, a **function** must be specified that relates one variable to another (such as a coordinate as a function of time). Suppose one of the variables is called y (the dependent variable), the other x (the independent variable). We might have a function relationship such as

$$y(x) = ax^3 + bx^2 + cx + d$$

If a, b, c, and d are specified constants, then y can be calculated for any value of x. We usually deal with continuous functions, that is, those for which y varies "smoothly" with x.

---

[1] The approximations for the functions $\sin x$, $\cos x$, and $\tan x$ are for $x \le 0.1$ rad.

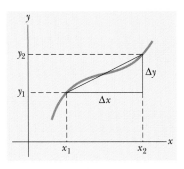

Figure B.13

The **derivative** of y with respect to x is defined as the limit, as Δx approaches zero, of the slopes of chords drawn between two points on the y versus x curve. Mathematically, we write this definition as

$$\frac{dy}{dx} = \lim_{\Delta x \to 0} \frac{\Delta y}{\Delta x} = \lim_{\Delta x \to 0} \frac{y(x + \Delta x) - y(x)}{\Delta x} \qquad \textbf{(B.28)}$$

where Δy and Δx are defined as $\Delta x = x_2 - x_1$ and $\Delta y = y_2 - y_1$ (Fig. B.13). It is important to note that dy/dx *does not* mean dy divided by dx, but is simply a notation of the limiting process of the derivative as defined by Equation B.28.

A useful expression to remember when $y(x) = ax^n$, where a is a *constant* and n is *any* positive or negative number (integer or fraction), is

$$\frac{dy}{dx} = nax^{n-1} \qquad \textbf{(B.29)}$$

If $y(x)$ is a polynomial or algebraic function of x, we apply Equation B.29 to *each* term in the polynomial and take $d[\text{constant}]/dx = 0$. In Examples 4 through 7, we evaluate the derivatives of several functions.

EXAMPLE 4

Suppose $y(x)$ (that is, y as a function of x) is given by

$$y(x) = ax^3 + bx + c$$

where a and b are constants. Then it follows that

$$y(x + \Delta x) = a(x + \Delta x)^3$$
$$+ b(x + \Delta x) + c$$
$$y(x + \Delta x) = a(x^3 + 3x^2\Delta x + 3x\Delta x^2 + \Delta x^3)$$
$$+ b(x + \Delta x) + c$$

so

$$\Delta y = y(x + \Delta x) - y(x) = a(3x^2\Delta x + 3x\Delta x^2 + \Delta x^3)$$
$$+ b\Delta x$$

Substituting this into Equation B.28 gives

$$\frac{dy}{dx} = \lim_{\Delta x \to 0} \frac{\Delta y}{\Delta x} = \lim_{\Delta x \to 0} [3ax^2 + 3x\Delta x + \Delta x^2] + b$$

$$\frac{dy}{dx} = 3ax^2 + b$$

EXAMPLE 5

$$y(x) = 8x^5 + 4x^3 + 2x + 7$$

Solution Applying Equation B.29 to each term independently, and remembering that d/dx (constant) $= 0$, we have

$$\frac{dy}{dx} = 8(5)x^4 + 4(3)x^2 + 2(1)x^0 + 0$$

$$\frac{dy}{dx} = 40x^4 + 12x^2 + 2$$

Special Properties of the Derivative

A. Derivative of the product of two functions If a function $f(x)$ is given by the product of two functions, say, $g(x)$ and $h(x)$, then the derivative of $f(x)$ is defined as

$$\frac{d}{dx} f(x) = \frac{d}{dx}[g(x)h(x)] = g\frac{dh}{dx} + h\frac{dg}{dx} \qquad \textbf{(B.30)}$$

B. Derivative of the sum of two functions If a function $f(x)$ is equal to the sum of two functions, then the derivative of the sum is equal to the sum of the derivatives:

$$\frac{d}{dx} f(x) = \frac{d}{dx}[g(x) + h(x)] = \frac{dg}{dx} + \frac{dh}{dx} \qquad \textbf{(B.31)}$$

C. Chain rule of differential calculus If $y = f(x)$ and $x = g(z)$, then dy/dx can be written as the product of two derivatives:

$$\frac{dy}{dz} = \frac{dy}{dx} \frac{dx}{dz} \qquad \textbf{(B.32)}$$

D. The second derivative The second derivative of y with respect to x is defined as the derivative of the function dy/dx (the derivative of the derivative). It is usually written

$$\frac{d^2y}{dx^2} = \frac{d}{dx}\left(\frac{dy}{dx}\right) \qquad \textbf{(B.33)}$$

EXAMPLE 6

Find the derivative of $y(x) = x^3/(x + 1)^2$ with respect to x.

Solution We can rewrite this function as $y(x) = x^3(x + 1)^{-2}$ and apply Equation B.30:

$$\frac{dy}{dx} = (x + 1)^{-2}\frac{d}{dx}(x^3) + x^3\frac{d}{dx}(x + 1)^{-2}$$

$$= (x + 1)^{-2}3x^2 + x^3(-2)(x + 1)^{-3}$$

$$\frac{dy}{dx} = \frac{3x^2}{(x + 1)^2} - \frac{2x^3}{(x + 1)^3}$$

EXAMPLE 7

A useful formula that follows from Equation B.30 is the derivative of the quotient of two functions. Show that

$$\frac{d}{dx}\left[\frac{g(x)}{h(x)}\right] = \frac{h\dfrac{dg}{dx} - g\dfrac{dh}{dx}}{h^2}$$

Solution We can write the quotient as gh^{-1} and then apply Equations B.29 and B.30:

$$\frac{d}{dx}\left(\frac{g}{h}\right) = \frac{d}{dx}(gh^{-1}) = g\frac{d}{dx}(h^{-1}) + h^{-1}\frac{d}{dx}(g)$$

$$= -gh^{-2}\frac{dh}{dx} + h^{-1}\frac{dg}{dx}$$

$$= \frac{h\dfrac{dg}{dx} - g\dfrac{dh}{dx}}{h^2}$$

Some of the more commonly used derivatives of functions are listed in Table B.4.

B.7 INTEGRAL CALCULUS

We think of integration as the inverse of differentiation. As an example, consider the expression

$$f(x) = \frac{dy}{dx} = 3ax^2 + b \qquad \textbf{(B.34)}$$

which was the result of differentiating the function

$$y(x) = ax^3 + bx + c$$

$$\frac{d}{dx}(a) = 0$$

$$\frac{d}{dx}(ax^n) = nax^{n-1}$$

$$\frac{d}{dx}(e^{ax}) = ae^{ax}$$

$$\frac{d}{dx}(\sin ax) = a\cos ax$$

$$\frac{d}{dx}(\cos ax) = -a\sin ax$$

$$\frac{d}{dx}(\tan ax) = a\sec^2 ax$$

$$\frac{d}{dx}(\cot ax) = -a\csc^2 ax$$

$$\frac{d}{dx}(\sec x) = \tan x\sec x$$

$$\frac{d}{dx}(\csc x) = -\cot x\csc x$$

$$\frac{d}{dx}(\ln ax) = \frac{1}{x}$$

Note: The letters a and n are constants.

in Example 4. We can write Equation B.34 as $dy = f(x)\,dx = (3ax^2 + b)\,dx$ and obtain $y(x)$ by "summing" over all values of x. Mathematically, we write this inverse operation

$$y(x) = \int f(x)\,dx$$

For the function $f(x)$ given by Equation B.34, we have

$$y(x) = \int (3ax^2 + b)\,dx = ax^3 + bx + c$$

where c is a constant of the integration. This type of integral is called an *indefinite integral* because its value depends on the choice of c.

A general **indefinite integral** $I(x)$ is defined as

$$I(x) = \int f(x)\,dx \tag{B.35}$$

where $f(x)$ is called the *integrand* and $f(x) = \dfrac{dI(x)}{dx}$.

For a *general continuous* function $f(x)$, the integral can be described as the area under the curve bounded by $f(x)$ and the x axis, between two specified values of x, say, x_1 and x_2, as in Figure B.14.

The area of the blue element is approximately $f(x_i)\Delta x_i$. If we sum all these area elements from x_1 and x_2 and take the limit of this sum as $\Delta x_i \to 0$, we obtain the *true* area under the curve bounded by $f(x)$ and x, between the limits x_1 and x_2:

$$\text{Area} = \lim_{\Delta x_i \to 0} \sum_i f(x_i)\,\Delta x_i = \int_{x_1}^{x_2} f(x)\,dx \tag{B.36}$$

Integrals of the type defined by Equation B.36 are called **definite integrals.**

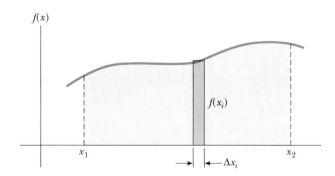

Figure B.14

One common integral that arises in practical situations has the form

$$\int x^n\,dx = \frac{x^{n+1}}{n+1} + c \qquad (n \neq -1) \tag{B.37}$$

This result is obvious, being that differentiation of the right-hand side with respect to x gives $f(x) = x^n$ directly. If the limits of the integration are known, this integral becomes a *definite integral* and is written

$$\int_{x_1}^{x_2} x^n\,dx = \frac{x_2^{\,n+1} - x_1^{\,n+1}}{n+1} \qquad (n \neq -1) \tag{B.38}$$

EXAMPLES

1. $\int_0^a x^2\, dx = \dfrac{x^3}{3}\Big]_0^a = \dfrac{a^3}{3}$

2. $\int_0^b x^{3/2}\, dx = \dfrac{x^{5/2}}{5/2}\Big]_0^b = \dfrac{2}{5} b^{5/2}$

3. $\int_3^5 x\, dx = \dfrac{x^2}{2}\Big]_3^5 = \dfrac{5^2 - 3^2}{2} = 8$

Partial Integration

Sometimes it is useful to apply the method of *partial integration* (also called "integrating by parts") to evaluate certain integrals. The method uses the property that

$$\int u\, dv = uv - \int v\, du \tag{B.39}$$

where u and v are *carefully* chosen so as to reduce a complex integral to a simpler one. In many cases, several reductions have to be made. Consider the function

$$I(x) = \int x^2 e^x\, dx$$

This can be evaluated by integrating by parts twice. First, if we choose $u = x^2$, $v = e^x$, we get

$$\int x^2 e^x\, dx = \int x^2\, d(e^x) = x^2 e^x - 2\int e^x x\, dx + c_1$$

Now, in the second term, choose $u = x$, $v = e^x$, which gives

$$\int x^2 e^x\, dx = x^2 e^x - 2x e^x + 2\int e^x\, dx + c_1$$

or

$$\int x^2 e^x\, dx = x^2 e^x - 2x e^x + 2e^x + c_2$$

The Perfect Differential

Another useful method to remember is the use of the *perfect differential,* in which we look for a change of variable such that the differential of the function is the differential of the independent variable appearing in the integrand. For example, consider the integral

$$I(x) = \int \cos^2 x \sin x\, dx$$

This becomes easy to evaluate if we rewrite the differential as $d(\cos x) = -\sin x\, dx$. The integral then becomes

$$\int \cos^2 x \sin x\, dx = -\int \cos^2 x\, d(\cos x)$$

If we now change variables, letting $y = \cos x$, we obtain

$$\int \cos^2 x \sin x\, dx = -\int y^2 dy = -\dfrac{y^3}{3} + c = -\dfrac{\cos^3 x}{3} + c$$

Table B.5 lists some useful indefinite integrals. Table B.6 gives Gauss's probability integral and other definite integrals. A more complete list can be found in various handbooks, such as *The Handbook of Chemistry and Physics,* CRC Press.

TABLE B.5 Some Indefinite Integrals (An arbitrary constant should be added to each of these integrals.)

$$\int x^n \, dx = \frac{x^{n+1}}{n+1} \qquad (\text{provided } n \neq -1)$$

$$\int \frac{dx}{x} = \int x^{-1} \, dx = \ln x$$

$$\int \frac{dx}{a+bx} = \frac{1}{b} \ln(a+bx)$$

$$\int \frac{x \, dx}{a+bx} = \frac{x}{b} - \frac{a}{b^2} \ln(a+bx)$$

$$\int \frac{dx}{x(x+a)} = -\frac{1}{a} \ln \frac{x+a}{x}$$

$$\int \frac{dx}{(a+bx)^2} = -\frac{1}{b(a+bx)}$$

$$\int \frac{dx}{a^2+x^2} = \frac{1}{a} \tan^{-1} \frac{x}{a}$$

$$\int \frac{dx}{a^2-x^2} = \frac{1}{2a} \ln \frac{a+x}{a-x} \qquad (a^2-x^2 > 0)$$

$$\int \frac{dx}{x^2-a^2} = \frac{1}{2a} \ln \frac{x-a}{x+a} \qquad (x^2-a^2 > 0)$$

$$\int \frac{x \, dx}{a^2 \pm x^2} = \pm \tfrac{1}{2} \ln(a^2 \pm x^2)$$

$$\int \frac{dx}{\sqrt{a^2-x^2}} = \sin^{-1} \frac{x}{a} = -\cos^{-1} \frac{x}{a} \qquad (a^2-x^2 > 0)$$

$$\int \frac{dx}{\sqrt{x^2 \pm a^2}} = \ln(x + \sqrt{x^2 \pm a^2})$$

$$\int \frac{x \, dx}{\sqrt{a^2-x^2}} = -\sqrt{a^2-x^2}$$

$$\int \frac{x \, dx}{\sqrt{x^2 \pm a^2}} = \sqrt{x^2 \pm a^2}$$

$$\int \sqrt{a^2-x^2} \, dx = \tfrac{1}{2}\left(x\sqrt{a^2-x^2} + a^2 \sin^{-1} \frac{x}{a} \right)$$

$$\int x\sqrt{a^2-x^2} \, dx = -\tfrac{1}{3}(a^2-x^2)^{3/2}$$

$$\int \sqrt{x^2 \pm a^2} \, dx = \tfrac{1}{2}[x\sqrt{x^2 \pm a^2} \pm a^2 \ln(x + \sqrt{x^2 \pm a^2})]$$

$$\int x(\sqrt{x^2 \pm a^2}) \, dx = \tfrac{1}{3}(x^2 \pm a^2)^{3/2}$$

$$\int e^{ax} \, dx = \frac{1}{a} e^{ax}$$

$$\int \ln ax \, dx = (x \ln ax) - x$$

$$\int xe^{ax} \, dx = \frac{e^{ax}}{a^2}(ax-1)$$

$$\int \frac{dx}{a+be^{cx}} = \frac{x}{a} - \frac{1}{ac} \ln(a+be^{cx})$$

$$\int \sin ax \, dx = -\frac{1}{a} \cos ax$$

$$\int \cos ax \, dx = \frac{1}{a} \sin ax$$

$$\int \tan ax \, dx = \frac{1}{a} \ln(\cos ax) = \frac{1}{a} \ln(\sec ax)$$

$$\int \cot ax \, dx = \frac{1}{a} \ln(\sin ax)$$

$$\int \sec ax \, dx = \frac{1}{a} \ln(\sec ax + \tan ax) = \frac{1}{a} \ln\left[\tan\left(\frac{ax}{2} + \frac{\pi}{4} \right) \right]$$

$$\int \csc ax \, dx = \frac{1}{a} \ln(\csc ax - \cot ax) = \frac{1}{a} \ln\left(\tan \frac{ax}{2} \right)$$

$$\int \sin^2 ax \, dx = \frac{x}{2} - \frac{\sin 2ax}{4a}$$

$$\int \cos^2 ax \, dx = \frac{x}{2} + \frac{\sin 2ax}{4a}$$

$$\int \frac{dx}{\sin^2 ax} = -\frac{1}{a} \cot ax$$

$$\int \frac{dx}{\cos^2 ax} = \frac{1}{a} \tan ax$$

$$\int \tan^2 ax \, dx = \frac{1}{a}(\tan ax) - x$$

$$\int \cot^2 ax \, dx = -\frac{1}{a}(\cot ax) - x$$

$$\int \sin^{-1} ax \, dx = x(\sin^{-1} ax) + \frac{\sqrt{1-a^2x^2}}{a}$$

$$\int \cos^{-1} ax \, dx = x(\cos^{-1} ax) - \frac{\sqrt{1-a^2x^2}}{a}$$

$$\int \frac{dx}{(x^2+a^2)^{3/2}} = \frac{x}{a^2\sqrt{x^2+a^2}}$$

$$\int \frac{x \, dx}{(x^2+a^2)^{3/2}} = -\frac{1}{\sqrt{x^2+a^2}}$$

TABLE B.6 Gauss's Probability Integral and Other Definite Integrals

$$\int_0^\infty x^n\, e^{-ax}\, dx = \frac{n!}{a^{n+1}}$$

$$I_0 = \int_0^\infty e^{-ax^2}\, dx = \frac{1}{2}\sqrt{\frac{\pi}{a}} \qquad \text{(Gauss's probability integral)}$$

$$I_1 = \int_0^\infty x e^{-ax^2}\, dx = \frac{1}{2a}$$

$$I_2 = \int_0^\infty x^2 e^{-ax^2}\, dx = -\frac{dI_0}{da} = \frac{1}{4}\sqrt{\frac{\pi}{a^3}}$$

$$I_3 = \int_0^\infty x^3 e^{-ax^2}\, dx = -\frac{dI_1}{da} = \frac{1}{2a^2}$$

$$I_4 = \int_0^\infty x^4 e^{-ax^2}\, dx = \frac{d^2 I_0}{da^2} = \frac{3}{8}\sqrt{\frac{\pi}{a^5}}$$

$$I_5 = \int_0^\infty x^5 e^{-ax^2}\, dx = \frac{d^2 I_1}{da^2} = \frac{1}{a^3}$$

$$\cdot$$
$$\cdot$$
$$\cdot$$

$$I_{2n} = (-1)^n \frac{d^n}{da^n} I_0$$

$$I_{2n+1} = (-1)^n \frac{d^n}{da^n} I_1$$

Group I	Group II	Transition elements						
H 1 1.008 0 $1s^1$								
Li 3 6.94 $2s^1$	**Be** 4 9.012 $2s^2$							
Na 11 22.99 $3s^1$	**Mg** 12 24.31 $3s^2$							
K 19 39.102 $4s^1$	**Ca** 20 40.08 $4s^2$	**Sc** 21 44.96 $3d^14s^2$	**Ti** 22 47.90 $3d^24s^2$	**V** 23 50.94 $3d^34s^2$	**Cr** 24 51.996 $3d^54s^1$	**Mn** 25 54.94 $3d^54s^2$	**Fe** 26 55.85 $3d^64s^2$	**Co** 27 58.93 $3d^74s^2$
Rb 37 85.47 $5s^1$	**Sr** 38 87.62 $5s^2$	**Y** 39 88.906 $4d^15s^2$	**Zr** 40 91.22 $4d^25s^2$	**Nb** 41 92.91 $4d^45s^1$	**Mo** 42 95.94 $4d^55s^1$	**Tc** 43 (99) $4d^55s^2$	**Ru** 44 101.1 $4d^75s^1$	**Rh** 45 102.91 $4d^85s^1$
Cs 55 132.91 $6s^1$	**Ba** 56 137.34 $6s^2$	57-71*	**Hf** 72 178.49 $5d^26s^2$	**Ta** 73 180.95 $5d^36s^2$	**W** 74 183.85 $5d^46s^2$	**Re** 75 186.2 $5d^56s^2$	**Os** 76 190.2 $5d^66s^2$	**Ir** 77 192.2 $5d^76s^2$
Fr 87 (223) $7s^1$	**Ra** 88 (226) $7s^2$	89-103**	**Rf** 104 (261) $6d^27s^2$	**Db** 105 (262) $6d^37s^2$	**Sg** 106 (263)	**Bh** 107 (262)	**Hs** 108 (265)	**Mt** 109 (266)

Symbol — **Ca** 20 — Atomic number
Atomic mass † — 40.08
$4s^2$ — Electron configuration

*Lanthanide series

La 57 138.91 $5d^16s^2$	**Ce** 58 140.12 $5d^14f^16s^2$	**Pr** 59 140.91 $4f^36s^2$	**Nd** 60 144.24 $4f^46s^2$	**Pm** 61 (147) $4f^56s^2$	**Sm** 62 150.4 $4f^66s^2$

**Actinide series

Ac 89 (227) $6d^17s^2$	**Th** 90 (232) $6d^27s^2$	**Pa** 91 (231) $5f^26d^17s^2$	**U** 92 (238) $5f^36d^17s^2$	**Np** 93 (239) $5f^46d^17s^2$	**Pu** 94 (239) $5f^66d^07s^2$

◻ Atomic mass values given are averaged over isotopes in the percentages in which they exist in nature.
† For an unstable element, mass number of the most stable known isotope is given in parentheses.
†† Elements 110, 111, 112, and 114 have not yet been named.
††† For a description of the atomic data, visit **physics.nist.gov/atomic**

			Group III	Group IV	Group V	Group VI	Group VII	Group 0
							H 1 1.008 0 $1s^1$	**He** 2 4.002 6 $1s^2$
			B 5 10.81 $2p^1$	**C** 6 12.011 $2p^2$	**N** 7 14.007 $2p^3$	**O** 8 15.999 $2p^4$	**F** 9 18.998 $2p^5$	**Ne** 10 20.18 $2p^6$
			Al 13 26.98 $3p^1$	**Si** 14 28.09 $3p^2$	**P** 15 30.97 $3p^3$	**S** 16 32.06 $3p^4$	**Cl** 17 35.453 $3p^5$	**Ar** 18 39.948 $3p^6$
Ni 28 58.71 $3d^8 4s^2$	**Cu** 29 63.54 $3d^{10}4s^1$	**Zn** 30 65.37 $3d^{10}4s^2$	**Ga** 31 69.72 $4p^1$	**Ge** 32 72.59 $4p^2$	**As** 33 74.92 $4p^3$	**Se** 34 78.96 $4p^4$	**Br** 35 79.91 $4p^5$	**Kr** 36 83.80 $4p^6$
Pd 46 106.4 $4d^{10}$	**Ag** 47 107.87 $4d^{10}5s^1$	**Cd** 48 112.40 $4d^{10}5s^2$	**In** 49 114.82 $5p^1$	**Sn** 50 118.69 $5p^2$	**Sb** 51 121.75 $5p^3$	**Te** 52 127.60 $5p^4$	**I** 53 126.90 $5p^5$	**Xe** 54 131.30 $5p^6$
Pt 78 195.09 $5d^9 6s^1$	**Au** 79 196.97 $5d^{10}6s^1$	**Hg** 80 200.59 $5d^{10}6s^2$	**Tl** 81 204.37 $6p^1$	**Pb** 82 207.2 $6p^2$	**Bi** 83 208.98 $6p^3$	**Po** 84 (210) $6p^4$	**At** 85 (218) $6p^5$	**Rn** 86 (222) $6p^6$
110†† (269)	111†† (272)	112†† (277)	114†† (289)					

Eu 63 152.0 $4f^7 6s^2$	**Gd** 64 157.25 $5d^1 4f^7 6s^2$	**Tb** 65 158.92 $5d^1 4f^8 6s^2$	**Dy** 66 162.50 $4f^{10}6s^2$	**Ho** 67 164.93 $4f^{11}6s^2$	**Er** 68 167.26 $4f^{12}6s^2$	**Tm** 69 168.93 $4f^{13}6s^2$	**Yb** 70 173.04 $4f^{14}6s^2$	**Lu** 71 174.97 $5d^1 4f^{14}6s^2$
Am 95 (243) $5f^7 6d^0 7s^2$	**Cm** 96 (245) $5f^7 6d^1 7s^2$	**Bk** 97 (247) $5f^8 6d^1 7s^2$	**Cf** 98 (249) $5f^{10}6d^0 7s^2$	**Es** 99 (254) $5f^{11}6d^0 7s^2$	**Fm** 100 (253) $5f^{12}6d^0 7s^2$	**Md** 101 (255) $5f^{13}6d^0 7s^2$	**No** 102 (255) $6d^0 7s^2$	**Lr** 103 (257) $6d^1 7s^2$

APPENDIX D · *SI Units*

TABLE D.1 SI Units

Base Quantity	SI Base Unit	
	Name	Symbol
Length	Meter	m
Mass	Kilogram	kg
Time	Second	s
Electric current	Ampere	A
Temperature	Kelvin	K
Amount of substance	Mole	mol
Luminous intensity	Candela	cd

TABLE D.2 Some Derived SI Units

Quantity	Name	Symbol	Expression in Terms of Base Units	Expression in Terms of Other SI Units
Plane angle	radian	rad	m/m	
Frequency	hertz	Hz	s^{-1}	
Force	newton	N	$kg \cdot m/s^2$	J/m
Pressure	pascal	Pa	$kg/m \cdot s^2$	N/m^2
Energy; work	joule	J	$kg \cdot m^2/s^2$	$N \cdot m$
Power	watt	W	$kg \cdot m^2/s^3$	J/s
Electric charge	coulomb	C	$A \cdot s$	
Electric potential	volt	V	$kg \cdot m^2/A \cdot s^3$	W/A
Capacitance	farad	F	$A^2 \cdot s^4/kg \cdot m^2$	C/V
Electric resistance	ohm	Ω	$kg \cdot m^2/A^2 \cdot s^3$	V/A
Magnetic flux	weber	Wb	$kg \cdot m^2/A \cdot s^2$	$V \cdot s$
Magnetic field intensity	tesla	T	$kg/A \cdot s^2$	
Inductance	henry	H	$kg \cdot m^2/A^2 \cdot s^2$	$T \cdot m^2/A$

All Nobel Prizes in physics are listed (and marked with a P), as well as relevant Nobel Prizes in Chemistry (C). The key dates for some of the scientific work are supplied; they often antedate the prize considerably.

1901 (P) *Wilhelm Roentgen* for discovering x-rays (1895).

1902 (P) *Hendrik A. Lorentz* for predicting the Zeeman effect and *Pieter Zeeman* for discovering the Zeeman effect, the splitting of spectral lines in magnetic fields.

1903 (P) *Antoine-Henri Becquerel* for discovering radioactivity (1896) and *Pierre* and *Marie Curie* for studying radioactivity.

1904 (P) *Lord Rayleigh* for studying the density of gases and discovering argon.
(C) *William Ramsay* for discovering the inert gas elements helium, neon, xenon, and krypton, and placing them in the periodic table.

1905 (P) *Philipp Lenard* for studying cathode rays, electrons (1898–1899).

1906 (P) *J. J. Thomson* for studying electrical discharge through gases and discovering the electron (1897).

1907 (P) *Albert A. Michelson* for inventing optical instruments and measuring the speed of light (1880s).

1908 (P) *Gabriel Lippmann* for making the first color photographic plate, using interference methods (1891).
(C) *Ernest Rutherford* for discovering that atoms can be broken apart by alpha rays and for studying radioactivity.

1909 (P) *Guglielmo Marconi* and *Carl Ferdinand Braun* for developing wireless telegraphy.

1910 (P) *Johannes D. van der Waals* for studying the equation of state for gases and liquids (1881).

1911 (P) *Wilhelm Wien* for discovering Wien's law giving the peak of a blackbody spectrum (1893).
(C) *Marie Curie* for discovering radium and polonium (1898) and isolating radium.

1912 (P) *Nils Dalén* for inventing automatic gas regulators for lighthouses.

1913 (P) *Heike Kamerlingh Onnes* for the discovery of superconductivity and liquefying helium (1908).

1914 (P) *Max T. F. von Laue* for studying x-rays from their diffraction by crystals, showing that x-rays are electromagnetic waves (1912).
(C) *Theodore W. Richards* for determining the atomic weights of sixty elements, indicating the existence of isotopes.

1915 (P) *William Henry Bragg* and *William Lawrence Bragg*, his son, for studying the diffraction of x-rays in crystals.

1917 (P) *Charles Barkla* for studying atoms by x-ray scattering (1906).

1918 (P) *Max Planck* for discovering energy quanta (1900).

1919 (P) *Johannes Stark*, for discovering the Stark effect, the splitting of spectral lines in electric fields (1913).

1920 (P) *Charles-Édouard Guillaume* for discovering invar, a nickel-steel alloy with low coefficient of expansion.

(C) *Walther Nernst* for studying heat changes in chemical reactions and formulating the third law of thermodynamics (1918).

1921 (P) *Albert Einstein* for explaining the photoelectric effect and for his services to theoretical physics (1905).

(C) *Frederick Soddy* for studying the chemistry of radioactive substances and discovering isotopes (1912).

1922 (P) *Niels Bohr* for his model of the atom and its radiation (1913).

(C) *Francis W. Aston* for using the mass spectrograph to study atomic weights, thus discovering 212 of the 287 naturally occurring isotopes.

1923 (P) *Robert A. Millikan* for measuring the charge on an electron (1911) and for studying the photoelectric effect experimentally (1914).

1924 (P) *Karl M. G. Siegbahn* for his work in x-ray spectroscopy.

1925 (P) *James Franck* and *Gustav Hertz* for discovering the Franck-Hertz effect in electron-atom collisions.

1926 (P) *Jean-Baptiste Perrin* for studying Brownian motion to validate the discontinuous structure of matter and measure the size of atoms.

1927 (P) *Arthur Holly Compton* for discovering the Compton effect on x-rays, their change in wavelength when they collide with matter (1922), and *Charles T. R. Wilson* for inventing the cloud chamber, used to study charged particles (1906).

1928 (P) *Owen W. Richardson* for studying the thermionic effect and electrons emitted by hot metals (1911).

1929 (P) *Louis Victor de Broglie* for discovering the wave nature of electrons (1923).

1930 (P) *Chandrasekhara Venkata Raman* for studying Raman scattering, the scattering of light by atoms and molecules with a change in wavelength (1928).

1932 (P) *Werner Heisenberg* for creating quantum mechanics (1925).

1933 (P) *Erwin Schrödinger* and *Paul A. M. Dirac* for developing wave mechanics (1925) and relativistic quantum mechanics (1927).

(C) *Harold Urey* for discovering heavy hydrogen, deuterium (1931).

1935 (P) *James Chadwick* for discovering the neutron (1932).

(C) *Irène* and *Frédéric Joliot-Curie* for synthesizing new radioactive elements.

1936 (P) *Carl D. Anderson* for discovering the positron in particular and antimatter in general (1932) and *Victor F. Hess* for discovering cosmic rays.

(C) *Peter J. W. Debye* for studying dipole moments and diffraction of x-rays and electrons in gases.

1937 (P) *Clinton Davisson* and *George Thomson* for discovering the diffraction of electrons by crystals, confirming de Broglie's hypothesis (1927).

1938 (P) *Enrico Fermi* for producing the transuranic radioactive elements by neutron irradiation (1934–1937).

1939 (P) *Ernest O. Lawrence* for inventing the cyclotron.

1943 (P) *Otto Stern* for developing molecular-beam studies (1923), and using them to discover the magnetic moment of the proton (1933).

1944 (P) *Isidor I. Rabi* for discovering nuclear magnetic resonance in atomic and molecular beams.

(C) *Otto Hahn* for discovering nuclear fission (1938).

1945 (P) *Wolfgang Pauli* for discovering the exclusion principle (1924).

1946 (P) *Percy W. Bridgman* for studying physics at high pressures.

1947 (P) *Edward V. Appleton* for studying the ionosphere.

1948 (P) *Patrick M. S. Blackett* for studying nuclear physics with cloud-chamber photographs of cosmic-ray interactions.

1949 (P) *Hideki Yukawa* for predicting the existence of mesons (1935).

1950 (P) *Cecil F. Powell* for developing the method of studying cosmic rays with photographic emulsions and discovering new mesons.

1951 (P) *John D. Cockcroft* and *Ernest T. S. Walton* for transmuting nuclei in an accelerator (1932).

(C) *Edwin M. McMillan* for producing neptunium (1940) and *Glenn T. Seaborg* for producing plutonium (1941) and further transuranic elements.

1952 (P) *Felix Bloch* and *Edward Mills Purcell* for discovering nuclear magnetic resonance in liquids and gases (1946).

1953 (P) *Frits Zernike* for inventing the phase-contrast microscope, which uses interference to provide high contrast.

1954 (P) *Max Born* for interpreting the wave function as a probability (1926) and other quantum-mechanical discoveries and *Walther Bothe* for developing the coincidence method to study subatomic particles (1930–1931), producing, in particular, the particle interpreted by Chadwick as the neutron.

1955 (P) *Willis E. Lamb, Jr.,* for discovering the Lamb shift in the hydrogen spectrum (1947) and *Polykarp Kusch* for determining the magnetic moment of the electron (1947).

1956 (P) *John Bardeen, Walter H. Brattain,* and *William Shockley* for inventing the transistor (1956).

1957 (P) *T.-D. Lee* and *C.-N. Yang* for predicting that parity is not conserved in beta decay (1956).

1958 (P) *Pavel A. Čerenkov* for discovering Čerenkov radiation (1935) and *Ilya M. Frank* and *Igor Tamm* for interpreting it (1937).

1959 (P) *Emilio G. Segrè* and *Owen Chamberlain* for discovering the antiproton (1955).

1960 (P) *Donald A. Glaser* for inventing the bubble chamber to study elementary particles (1952).

(C) *Willard Libby* for developing radiocarbon dating (1947).

1961 (P) *Robert Hofstadter* for discovering internal structure in protons and neutrons and *Rudolf L. Mössbauer* for discovering the Mössbauer effect of recoilless gamma-ray emission (1957).

1962 (P) *Lev Davidovich Landau* for studying liquid helium and other condensed matter theoretically.

1963 (P) *Eugene P. Wigner* for applying symmetry principles to elementary-particle theory and *Maria Goeppert Mayer* and *J. Hans D. Jensen* for studying the shell model of nuclei (1947).

1964 (P) *Charles H. Townes, Nikolai G. Basov,* and *Alexandr M. Prokhorov* for developing masers (1951–1952) and lasers.

1965 (P) *Sin-itiro Tomonaga, Julian S. Schwinger,* and *Richard P. Feynman* for developing quantum electrodynamics (1948).

1966 (P) *Alfred Kastler* for his optical methods of studying atomic energy levels.

1967 (P) *Hans Albrecht Bethe* for discovering the routes of energy production in stars (1939).

1968 (P) *Luis W. Alvarez* for discovering resonance states of elementary particles.

1969 (P) *Murray Gell-Mann* for classifying elementary particles (1963).

1970 (P) *Hannes Alfvén* for developing magnetohydrodynamic theory and *Louis Eugène Félix Néel* for discovering antiferromagnetism and ferrimagnetism (1930s).

1971 (P) *Dennis Gabor* for developing holography (1947).

(C) *Gerhard Herzberg* for studying the structure of molecules spectroscopically.

1972 (P) *John Bardeen, Leon N. Cooper,* and *John Robert Schrieffer* for explaining superconductivity (1957).

1973 (P) *Leo Esaki* for discovering tunneling in semiconductors, *Ivar Giaever* for discovering tunneling in superconductors, and *Brian D. Josephson* for predicting the Josephson effect, which involves tunneling of paired electrons (1958–1962).

1974 (P) *Anthony Hewish* for discovering pulsars and *Martin Ryle* for developing radio interferometry.

1975 (P) *Aage N. Bohr, Ben R. Mottelson,* and *James Rainwater* for discovering why some nuclei take asymmetric shapes.

1976 (P) *Burton Richter* and *Samuel C. C. Ting* for discovering the J/psi particle, the first charmed particle (1974).

1977 (P) *John H. Van Vleck, Nevill F. Mott,* and *Philip W. Anderson* for studying solids quantum-mechanically.

(C) *Ilya Prigogine* for extending thermodynamics to show how life could arise in the face of the second law.

1978 (P) *Arno A. Penzias* and *Robert W. Wilson* for discovering the cosmic background radiation (1965) and *Pyotr Kapitsa* for his studies of liquid helium.

1979 (P) *Sheldon L. Glashow, Abdus Salam,* and *Steven Weinberg* for developing the theory that unified the weak and electromagnetic forces (1958–1971).

1980 (P) *Val Fitch* and *James W. Cronin* for discovering CP (charge-parity) violation (1964), which possibly explains the cosmological dominance of matter over antimatter.

1981 (P) *Nicolaas Bloembergen* and *Arthur L. Schawlow* for developing laser spectroscopy and *Kai M. Siegbahn* for developing high-resolution electron spectroscopy (1958).

1982 (P) *Kenneth G. Wilson* for developing a method of constructing theories of phase transitions to analyze critical phenomena.

1983 (P) *William A. Fowler* for theoretical studies of astrophysical nucleosynthesis and *Subramanyan Chandrasekhar* for studying physical processes of importance to stellar structure and evolution, including the prediction of white dwarf stars (1930).

1984 (P) *Carlo Rubbia* for discovering the W and Z particles, verifying the electroweak unification, and *Simon van der Meer,* for developing the method of stochastic cooling of the CERN beam that allowed the discovery (1982–1983).

1985 (P) *Klaus von Klitzing* for the quantized Hall effect, relating to conductivity in the presence of a magnetic field (1980).

1986 (P) *Ernst Ruska* for inventing the electron microscope (1931), and *Gerd Binnig* and *Heinrich Rohrer* for inventing the scanning-tunneling electron microscope (1981).

1987 (P) *J. Georg Bednorz* and *Karl Alex Müller* for the discovery of high temperature superconductivity (1986).

1988 (P) *Leon M. Lederman, Melvin Schwartz,* and *Jack Steinberger* for a collaborative experiment that led to the development of a new tool for studying the weak nuclear force, which affects the radioactive decay of atoms.

1989 (P) *Norman Ramsay* (U.S.) for various techniques in atomic physics; and *Hans Dehmelt* (U.S.) and *Wolfgang Paul* (Germany) for the development of techniques for trapping single charge particles.

1990 (P) *Jerome Friedman, Henry Kendall* (both U.S.), and *Richard Taylor* (Canada) for experiments important to the development of the quark model.

1991 (P) *Pierre-Gilles de Gennes* for discovering that methods developed for studying order phenomena in simple systems can be generalized to more complex forms of matter, in particular to liquid crystals and polymers.

1992 (P) *George Charpak* for developing detectors that trace the paths of evanescent subatomic particles produced in particle accelerators.

1993 (P) *Russell Hulse* and *Joseph Taylor* for discovering evidence of gravitational waves.

1994 (P) *Bertram N. Brockhouse* and *Clifford G. Shull* for pioneering work in neutron scattering.

1995 (P) *Martin L. Perl* and *Frederick Reines* for discovering the tau particle and the neutrino, respectively.

1996 (P) *David M. Lee, Douglas C. Osheroff,* and *Robert C. Richardson* for developing a superfluid using helium-3.

1997 (P) *Steven Chu, Claude Cohen-Tannoudji,* and *William D. Phillips* for developing methods to cool and trap atoms with laser light.

1998 (P) *Robert B. Laughlin, Horst L. Störmer,* and *Daniel C. Tsui* for discovering a new form of quantum fluid with fractionally charged excitations.

Answers to Odd-Numbered Problems

Chapter 1

1. 2.15×10^4 kg/m^3

3. 184 g

5. (a) 7.10 cm^3 (b) 1.18×10^{-29} m^3 (c) 0.228 nm
(d) 12.7 cm^3, 2.11×10^{-29} m^3, 0.277 nm

7. (a) 4.00 u = 6.64×10^{-24} g (b) 55.9 u =
9.29×10^{-23} g (c) 207 u = 3.44×10^{-22} g

9. (a) 9.83×10^{-16} g (b) 1.06×10^7 atoms

11. (a) 4.01×10^{25} molecules (b) 3.65×10^4 molecules

13. no

15. (b) only

17. $0.579t$ ft^3/s + $1.19 \times 10^{-9}t^2$ ft^3/s^2

19. 1.39×10^3 m^2

21. (a) 0.071 4 gal/s (b) 2.70×10^{-4} m^3/s (c) 1.03 h

23. 4.05×10^3 m^2

25. 11.4×10^3 kg/m^3

27. 1.19×10^{57} atoms

29. (a) 190 y (b) 2.32×10^4 times

31. 151 μm

33. 1.00×10^{10} lb

35. 3.08×10^4 m^3

37. 5.0 m

39. 2.86 cm

41. $\sim 10^6$ balls

43. $\sim 10^7$ or 10^8 rev

45. $\sim 10^7$ or 10^8 blades

47. $\sim 10^2$ kg; $\sim 10^3$ kg

49. $\sim 10^2$ tuners

51. (a) (346 ± 13) m^2 (b) (66.0 ± 1.3) m

53. $(1.61 \pm 0.17) \times 10^3$ kg/m^3

55. 115.9 m

57. 316 m

59. 4.50 m^2

61. 3.41 m

63. 0.449%

65. (a) 0.529 cm/s (b) 11.5 cm/s

67. 1×10^{10} gal/yr

69. $\sim 10^{11}$ stars

Chapter 2

1. (a) 2.30 m/s (b) 16.1 m/s (c) 11.5 m/s

3. (a) 5 m/s (b) 1.2 m/s (c) -2.5 m/s (d) -3.3 m/s
(e) 0

5. (a) 3.75 m/s (b) 0

7. (a)

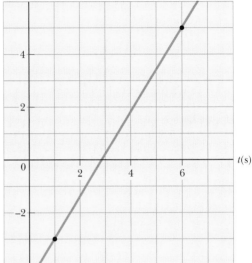

(b) 1.60 m/s

9. (a) -2.4 m/s (b) -3.8 m/s (c) 4.0 s

11. (a) 5.0 m/s (b) -2.5 m/s (c) 0 (d) 5.0 m/s

13. 1.34×10^4 m/s^2

15. (a)

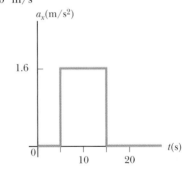

(b) 1.6 m/s^2 and 0.80 m/s^2

17. (a) 2.00 m (b) -3.00 m/s (c) -2.00 m/s^2

19. (a) 1.3 m/s^2 (b) 2.0 m/s^2 at 3 s (c) at $t = 6$ s and for
$t > 10$ s (d) -1.5 m/s^2 at 8 s

21. 2.74×10^5 m/s^2, which is 2.79×10^4 g

23. (a) 6.61 m/s (b) -0.448 m/s^2

25. -16.0 cm/s^2

27. (a) 2.56 m (b) -3.00 m/s

29. (a) 8.94 s (b) 89.4 m/s

31. (a) 20.0 s (b) no

33. $x_f - x_i = v_{xf}\,t - a_x\,t^2/2$; $v_{xf} = 3.10$ m/s

35. (a) 35.0 s (b) 15.7 m/s
37. (a) -202 m/s^2 (b) 198 m
39. (a) 3.00 m/s (b) 6.00 s (c) -0.300 m/s^2
    (d) 2.05 m/s
41. (a) -4.90 m, -19.6 m, -44.1 m (b) -9.80 m/s,
    -19.6 m/s, -29.4 m/s
43. (a) 10.0 m/s up (b) 4.68 m/s down
45. No. In 0.2 s the bill falls out from between David's fin-
    gers.
47. (a) 29.4 m/s (b) 44.1 m
49. (a) 7.82 m (b) 0.782 s
51. (a) 1.53 s (b) 11.5 m (c) -4.60 m/s, -9.80 m/s^2
53. (a) $a_x = a_{xi} + Jt$, $v_x = v_{xi} + a_{xi}t + \frac{1}{2}Jt^2$,
    $x = x_i + v_{xi}t + \frac{1}{2}a_{xi}t^2 + \frac{1}{6}Jt^3$
55. 0.222 s
57. 0.509 s
59. (a) 41.0 s (b) 1.73 km (c) -184 m/s
61. $v_{xi}\,t + at^2/2$, in agreement with Equation 2.11
63. (a) 5.43 m/s^2 and 3.83 m/s^2 (b) 10.9 m/s and 11.5 m/s
    (c) Maggie by 2.62 m
65. (a) 45.7 s (b) 574 m (c) 12.6 m/s (d) 765 s
67. (a) 2.99 s (b) -15.4 m/s (c) 31.3 m/s down and
    34.9 m/s down
69. (a) 5.46 s (b) 73.0 m (c) $v_{\text{Stan}} = 22.6$ m/s, $v_{\text{Kathy}} =$
    26.7 m/s
71. (a) See top of next column.
    (b) See top of next column.
73. $0.577v$

Chapter 3

1. $(-2.75, -4.76)$ m
3. 1.15; 2.31
5. (a) 2.24 m (b) 2.24 m at 26.6° from the positive x axis.
7. (a) 484 m (b) 18.1° north of west
9. 70.0 m
11. (a) approximately 6.1 units at 112° (b) approximately
    14.8 units at 22°
13. (a) 10.0 m (b) 15.7 m (c) 0
15. (a) 5.2 m at 60° (b) 3.0 m at 330° (c) 3.0 m at 150°
    (d) 5.2 m at 300°
17. approximately 420 ft at $-3°$
19. 5.83 m at 59.0° to the right of his initial direction
21. 1.31 km north and 2.81 km east
23. (a) 10.4 cm (b) 35.5°
25. 47.2 units at 122° from the positive x axis.
27. $(-25.0\mathbf{i})$m $+ (43.3\mathbf{j})$m
29. 7.21 m at 56.3° from the positive x axis.
31. (a) $2.00\mathbf{i} - 6.00\mathbf{j}$ (b) $4.00\mathbf{i} + 2.00\mathbf{j}$ (c) 6.32 (d) 4.47
    (e) 288°; 26.6° from the positive x axis.
33. (a) $(-11.1\mathbf{i} + 6.40\mathbf{j})$ m (b) $(1.65\mathbf{i} + 2.86\mathbf{j})$ cm
    (c) $(-18.0\mathbf{i} - 12.6\mathbf{j})$ in.
35. 9.48 m at 166°
37. (a) 185 N at 77.8° from the positive x axis
    (b) $(-39.3\mathbf{i} - 181\mathbf{j})$ N
39. $\mathbf{A} + \mathbf{B} = (2.60\mathbf{i} + 4.50\mathbf{j})$ m

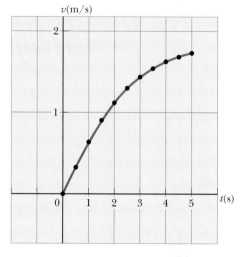

Chapter 2, Problem 71(a)

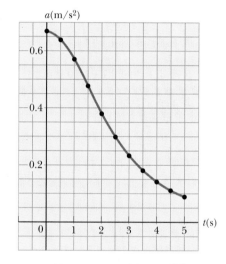

Chapter 2, Problem 71(b)

41. 196 cm at $-14.7°$ from the positive x axis.
43. (a) $8.00\mathbf{i} + 12.0\mathbf{j} - 4.00\mathbf{k}$ (b) $2.00\mathbf{i} + 3.00\mathbf{j} - 1.00\mathbf{k}$
    (c) $-24.0\mathbf{i} - 36.0\mathbf{j} + 12.0\mathbf{k}$
45. (a) 5.92 m is the magnitude of $(5.00\mathbf{i} - 1.00\mathbf{j} - 3.00\mathbf{k})$ m
    (b) 19.0 m is the magnitude of $(4.00\mathbf{i} - 11.0\mathbf{j} + 15.0\mathbf{k})$ m
47. 157 km
49. (a) $-3.00\mathbf{i} + 2.00\mathbf{j}$ (b) 3.61 at 146° from the positive
    x axis. (c) $3.00\mathbf{i} - 6.00\mathbf{j}$
51. (a) $49.5\mathbf{i} + 27.1\mathbf{j}$ (b) 56.4 units at 28.7° from the posi-
    tive x axis.
53. 1.15°
55. (a) 2.00, 1.00, 3.00 (b) 3.74 (c) $\theta_x = 57.7°$, $\theta_y = 74.5°$,
    $\theta_z = 36.7°$
57. 240 m at 237°
59. 390 mi/h at 7.37° north of east
61. $\mathbf{R}_1 = a\mathbf{i} + b\mathbf{j}$; $R_1 = \sqrt{a^2 + b^2}$ (b) $\mathbf{R}_2 = a\mathbf{i} + b\mathbf{j} + c\mathbf{k}$

Chapter 4

1. (a) 4.87 km at 209° from east (b) 23.3 m/s
(c) 13.5 m/s at 209°
3. (a) $(18.0t)\mathbf{i} + (4.00t - 4.90t^2)\mathbf{j}$
(b) $18.0\mathbf{i} + (4.00 - 9.80t)\mathbf{j}$ (c) $-9.80\mathbf{j}$
(d) $(54.0\mathbf{i} - 32.1\mathbf{j})$ m
(e) $(18.0\mathbf{i} - 25.4\mathbf{j})$ m/s (f) $(-9.80\mathbf{j})$ m/s^2
5. (a) $(2.00\mathbf{i} + 3.00\mathbf{j})$ m/s^2
(b) $(3.00t + t^2)\mathbf{i}$ m, $(1.50\,t^2 - 2.00t)\mathbf{j}$ m
7. (a) $(0.800\mathbf{i} - 0.300\mathbf{j})$ m/s^2 (b) 339°
(c) $(360\mathbf{i} - 72.7\mathbf{j})$ m, $-15.2°$
9. (a) $(3.34\mathbf{i})$ m/s (b) $-50.9°$
11. (a) 20.0° (b) 3.05 s
13. $x = 7.23$ km $y = 1.68$ km
15. 53.1°
17. 22.4° or 89.4°
19. (a) The ball clears by 0.889 m (b) while descending
21. $d \tan \theta_i - gd^2/(2v_i^2 \cos^2\theta_i)$
23. (a) 0.852 s (b) 3.29 m/s (c) 4.03 m/s (d) 50.8°
(e) 1.12 s
25. 377 m/s^2
27. 10.5 m/s, 219 m/s^2
29. (a) 6.00 rev/s (b) 1.52 km/s^2 (c) 1.28 km/s^2
31. 1.48 m/s^2 inward at 29.9° behind the radius
33. (a) 13.0 m/s^2 (b) 5.70 m/s (c) 7.50 m/s^2
35. (a)

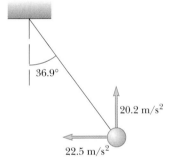

(b) 29.7 m/s^2 (c) 6.67 m/s at 36.9° above the horizontal
37. 2.02×10^3 s; 21.0% longer
39. 153 km/h at 11.3° north of west
41. (a) 36.9° (b) 41.6° (c) 3.00 min
43. 15.3 m
45. $2\,v_i t \cos \theta_i$
47. (b) $45° + \phi/2$; $v_i^2(1 - \sin \phi)/g\cos^2\phi$
49. (a) 41.7 m/s (b) 3.81 s (c) $(34.1\mathbf{i} - 13.4\mathbf{j})$ m/s; 36.6 m/s
51. (a) 25.0 m/s^2 (radial); 9.80 m/s^2 (tangential)
(b)

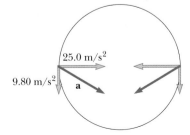

(c) 26.8 m/s^2 inward at 21.4° below the horizontal
53. 8.94 m/s at $-63.4°$ relative to the positive x axis.
55. 20.0 m
57. (a) 0.600 m (b) 0.402 m (c) 1.87 m/s^2 toward center
(d) 9.80 m/s^2 down
59. (a) 6.80 km (b) 3.00 km vertically above the impact point (c) 66.2°
61. (a) 46.5 m/s (b) $-77.6°$ (c) 6.34 s
63. (a) 1.53 km (b) 36.2 s (c) 4.04 km
65. (a) 20.0 m/s, 5.00 s (b) $(16.0\mathbf{i} - 27.1\mathbf{j})$ m/s (c) 6.54 s
(d) $(24.6\mathbf{i})$ m
67. (a) 43.2 m (b) $(9.66\mathbf{i} - 25.5\mathbf{j})$ m/s
69. Imagine you are shaking down the mercury in a fever thermometer. Starting with your hand at the level of your shoulder, move your hand down as fast as you can and snap it around an arc at the bottom. ~ 100 m/s$^2 \approx 10\,g$

Chapter 5

1. (a) 1/3 (b) 0.750 m/s^2
3. $(6.00\mathbf{i} + 15.0\mathbf{j})$ N; 16.2 N
5. 312 N
7. (a) $x = vt/2$ (b) $F_g v\mathbf{i}/gt + F_g\mathbf{j}$
9. (a) $(2.50\mathbf{i} + 5.00\mathbf{j})$ N (b) 5.59 N
11. (a) 3.64×10^{-18} N (b) 8.93×10^{-30} N is 408 billion times smaller.
13. 2.38 kN
15. (a) 5.00 m/s^2 at 36.9° (b) 6.08 m/s^2 at 25.3°
17. (a) $\sim 10^{-22}$ m/s^2 (b) $\sim 10^{-23}$ m
19. (a) 0.200 m/s^2 forward (b) 10.0 m (c) 2.00 m/s
21. (a) 15.0 lb up (b) 5.00 lb up (c) 0
23. 613 N

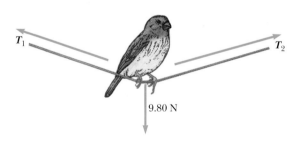

27. (a) 49.0 N (b) 98.0 N (c) 24.5 N
29. 8.66 N east
31. 100 N and 204 N
33. 3.73 m
35. $a = F/(m_1 + m_2)$; $T = F m_1/(m_1 + m_2)$
37. (a) $F_x > 19.6$ N (b) $F_x \leq -78.4$ N
(c) See top of next page.
39. (a) 706 N (b) 814 N (c) 706 N (d) 648 N
41. $\mu_s = 0.306$; $\mu_k = 0.245$
43. (a) 256 m (b) 42.7 m
45. (a) 1.78 m/s^2 (b) 0.368 (c) 9.37 N (d) 2.67 m/s
47. (a) 0.161 (b) 1.01 m/s^2
49. 37.8 N

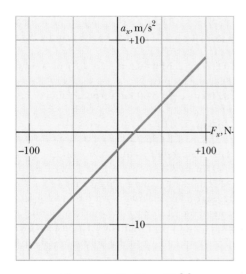

Chapter 5, Problem 37(c)

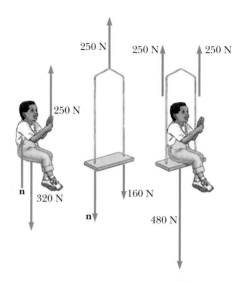

Chapter 5, Problem 55(a)

51. (a)

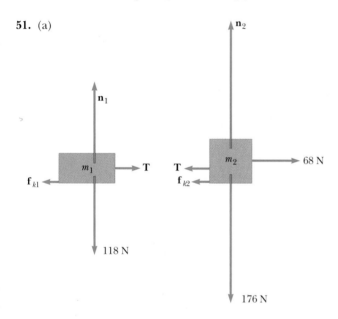

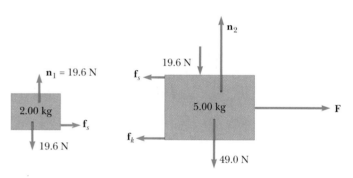

Chapter 5, Problem 67(a)

71. (a)

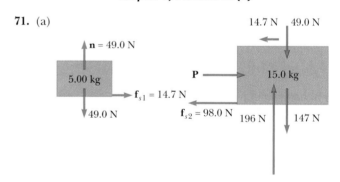

(b) 27.2 N, 1.29 m/s²

53. Any value between 31.7 N and 48.6 N

55. (a) See top of next column.
(b) 0.408 m/s²   (c) 83.3 N

57. 1.18 kN

59. (a) *Mg*/2, *Mg*/2, *Mg*/2, 3*Mg*/2, *Mg*   (b) *Mg*/2

61. (b)

θ	0	15.0°	30.0°	45.0°	60.0°
P(N)	40.0	46.4	60.1	94.3	260

63. (a) 19.3°   (b) 4.21 N

65. (a) 2.13 s   (b) 1.67 m

67. (a) See next column.
Static friction between the two blocks accelerates the upper block.   (b) 34.7 N   (c) 0.306

69. $(M + m_1 + m_2)(m_2g/m_1)$

(b) 113 N   (c) 0.980 m/s² and 1.96 m/s²

73. (a) 0.087 1   (b) 27.4 N

75. (a) 30.7°   (b) 0.843 N

77. (a) 3.34   (b) Either the car would flip over backwards, or the wheels would skid, spinning in place, and the time would increase.

Chapter 6

1. (a) 8.00 m/s   (b) 3.02 N

3. Any speed up to 8.08 m/s

5. 6.22×10^{-12} N

7. (a) 1.52 m/s^2 (b) 1.66 km/s (c) $6\,820$ s

9. (a) static friction (b) $0.085\,0$

11. $v \leq 14.3$ m/s

13. (a) 68.6 N toward the center of the circle and 784 N up (b) 0.857 m/s^2

15. No. The jungle lord needs a vine of tensile strength 1.38 kN.

17. (a) 4.81 m/s (b) 700 N up

19. 3.13 m/s

21. (a) 2.49×10^4 N up (b) 12.1 m/s

23. (a) 0.822 m/s^2 (b) 37.0 N (c) 0.0839

25. (a) $17.0°$ (b) 5.12 N

27. (a) 491 N (b) 50.1 kg (c) 2.00 m/s^2

29. $0.0927°$

31. (a) 32.7 s^{-1} (b) 9.80 m/s^2 (c) 4.90 m/s^2

33. 3.01 N

35. (a) 1.47 N·s/m (b) 2.04×10^{-3} s (c) 2.94×10^{-2} N

37. (a) 0.0347 s^{-1} (b) 2.50 m/s (c) $a = -cv$

39. $\sim 10^1$ N

41. (a) 13.7 m/s down

(b)

t (s)	x (m)	v (m/s)
0	0	0
0.2	0	-1.96
0.4	-0.392	-3.88
. . .	. . .	. . .
1.0	-3.77	-8.71
. . . 2.0	-14.4	-12.56
. . . 4.0	-41.0	-13.67

43. (a) 49.5 m/s and 4.95 m/s

(b)

t (s)	y (m)	v (m/s)
0	$1\,000$	0
. . . 1	995	-9.7
. . . 2	980	-18.6
. . . 10	674	-47.7
. . . 10.1	671	-16.7
. . . 12	659	-4.95
. . . 145	0	-4.95

45. (a) 2.33×10^{-4} kg/m (b) 53 m/s (c) 42 m/s. The second trajectory is higher and shorter. In both, the ball attains maximum height when it has covered about 57% of its horizontal range, and it attains minimum speed somewhat later. The impact speeds also are both about 30 m/s.

47. (a) $mg - mv^2/R$ (b) $\sqrt{gR}$

49. (a) 2.63 m/s^2 (b) 201 m (c) 17.7 m/s

51. (a) 9.80 N (b) 9.80 N (c) 6.26 m/s

53. (b) 732 N down at the equator and 735 N down at the poles

59. (a) 1.58 m/s^2 (b) 455 N (c) 329 N (d) 397 N upward and $9.15°$ inward

61. (a) 5.19 m/s (b) Child + seat:

$T \cos 28.0°$

$T \sin 28.0°$

490 N

$T = 555$ N

63. (b) 2.54 s; 23.6 rev/min

65. 215 N horizontally inward

67. (a) either $70.4°$ or $0°$ (b) $0°$

69. 12.8 N

71. (a)

t (s)	d (m)
0	0
1	4.9
2	18.9
. . . 5	112.6
. . . 10	347.0
11	399.1
. . . 15	611.3
. . . 20	876.5

(b)

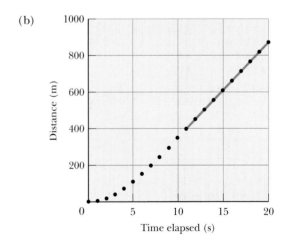

(c) The graph is straight for $11\,\text{s} < t < 20\,\text{s}$, with slope 53.0 m/s.

Chapter 7

1. 15.0 MJ

3. (a) 32.8 mJ (b) -32.8 mJ

5. (a) 31.9 J (b) 0 (c) 0 (d) 31.9 J

7. 4.70 kJ

9. 14.0

11. (a) 16.0 J (b) $36.9°$

13. (a) $11.3°$ (b) $156°$ (c) $82.3°$

15. (a) 24.0 J   (b) − 3.00 J   (c) 21.0 J
17. (a) 7.50 J   (b) 15.0 J   (c) 7.50 J   (d) 30.0 J
19. (a) 0.938 cm   (b) 1.25 J
21. 0.299 m/s
23. 12.0 J
25. (b) mgR
27. (a) 1.20 J   (b) 5.00 m/s   (c) 6.30 J
29. (a) 60.0 J   (b) 60.0 J
31. (a) $\sqrt{2W/m}$ (b) W/d
33. (a) 650 J   (b) − 588 J   (c) 0   (d) 0   (e) 62.0 J
(f) 1.76 m/s
35. (a) − 168 J   (b) − 184 J   (c) 500 J   (d) 148 J
(e) 5.64 m/s
37. 2.04 m
39. (a) 22 500 N   (b) 1.33×10^{-4} s
41. (a) 0.791 m/s   (b) 0.531 m/s
43. 875 W
45. 830 N
47. (a) 5 910 W   (b) It is 53.0% of 11 100 W
49. (a) 0.013 5 gal   (b) 73.8   (c) 8.08 kW
51. 5.90 km/L
53. (a) 5.37×10^{-11} J   (b) 1.33×10^{-9} J
55. 90.0 J
59. (a) $(2 + 24t^2 + 72t^4)$ J   (b) $12t$ m/s^2; $48t$ N
(c) $(48t + 288t^3)$ W   (d) 1 250 J
61. − 0.047 5 J
63. 878 kN
65. (b) 240 W
67. (a) $\mathbf{F}_1 = (20.5\mathbf{i} + 14.3\mathbf{j})$ N; $\mathbf{F}_2 = (− 36.4\mathbf{i} + 21.0\mathbf{j})$ N
(b) $(− 15.9\mathbf{i} + 35.3\mathbf{j})$ N   (c) $(− 3.18\mathbf{i} + 7.07\mathbf{j})$ m/s^2
(d) $(− 5.54\mathbf{i} + 23.7\mathbf{j})$ m/s   (e) $(− 2.30\mathbf{i} + 39.3\mathbf{j})$ m
(f) 1 480 J   (g) 1 480 J
69. (a) 4.12 m   (b) 3.35 m
71. 1.68 m/s
73. (a) 14.5 m/s   (b) 1.75 kg   (c) 0.350 kg
75. 0.799 J

Chapter 8

1. (a) 259 kJ, 0, − 259 kJ   (b) 0, − 259 kJ, − 259 kJ
3. (a) − 196 J   (b) − 196 J   (c) − 196 J. The force is conservative.
5. (a) 125 J   (b) 50.0 J   (c) 66.7 J   (d) Nonconservative. The results differ.
7. (a) 40.0 J   (b) − 40.0 J   (c) 62.5 J
9. (a) $Ax^2/2 − Bx^3/3$ (b) $\Delta U = 5A/2 − 19B/3$; $\Delta K = − 5A/2 + 19B/3$
11. 0.344 m
13. (a) $v_B = 5.94$ m/s; $v_C = 7.67$ m/s   (b) 147 J
15. $v = (3gR)^{1/2}$, 0.098 0 N down
17. 10.2 m
19. (a) 19.8 m/s   (b) 78.4 J   (c) 1.00
21. (a) 4.43 m/s   (b) 5.00 m
23. (a) 18.5 km, 51.0 km   (b) 10.0 MJ
25. (b) 60.0°
27. 5.49 m/s

29. 2.00 m/s, 2.79 m/s, 3.19 m/s
31. 3.74 m/s
33. (a) − 160 J   (b) 73.5 J   (c) 28.8 N   (d) 0.679
35. 489 kJ
37. (a) 1.40 m/s   (b) 4.60 cm after release   (c) 1.79 m/s
39. 1.96 m
41. (A/r^2) away from the other particle
43. (a) $r = 1.5$ mm, stable; 2.3 mm, unstable; 3.2 mm, stable; $r \rightarrow \infty$ neutral   (b) − 5.6 J $< E <$ 1 J
(c) 0.6 mm $< r <$ 3.6 mm   (d) 2.6 J   (e) 1.5 mm
(f) 4 J
45. (a) + at ⓑ, − at ⓓ, 0 at ⓐ, ©, and Ⓔ   (b) © stable;
ⓐ and Ⓔ unstable
(c)

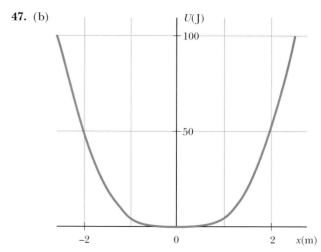

47. (b)

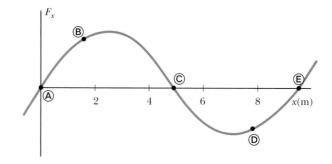

Equilibrium at $x = 0$ (c) $v = \sqrt{0.800\,\mathrm{J}/m}$
49. (a) 1.50×10^{-10} J   (b) 1.07×10^{-9} J   (c) 9.15×10^{-10} J
51. 48.2° Note that the answer is independent of the pumpkin's mass and of the radius of the dome.
53. (a) 0.225 J   (b) $\Delta E_f = − 0.363$ J   (c) No; the normal force changes in a complicated way.
55. $\sim 10^2$ W sustainable power
57. 0.327
59. (a) 23.6 cm   (b) 5.90 m/s^2 up the incline; no.
(c) Gravitational potential energy turns into kinetic energy plus elastic potential energy and then entirely into elastic potential energy.
61. 1.25 m/s

63. (a) 0.400 m   (b) 4.10 m/s   (c) The block stays on the track.
65. (b) 2.06 m/s
67. (b) 1.44 m   (c) 0.400 m   (d) No. A very strong wind pulls the string out horizontally (parallel to the ground). The largest possible equilibrium height is equal to L.
71. (a) 6.15 m/s   (b) 9.87 m/s
73. 0.923 m/s

Chapter 9

1. (a) $(9.00\mathbf{i} - 12.0\mathbf{j})$ kg·m/s   (b) 15.0 kg·m/s at 307°
3. 6.25 cm/s west
5. $\sim 10^{-23}$ m/s
7. (b) $p = \sqrt{2mK}$
9. (a) 13.5 N·s   (b) 9.00 kN   (c) 18.0 kN
11. 260 N normal to the wall
13. 15.0 N in the direction of the initial velocity of the exiting water stream
15. 65.2 m/s
17. 301 m/s
19. (a) $v_{gx} = 1.15$ m/s   (b) $v_{px} = -0.346$ m/s
21. (a) 20.9 m/s east   (b) 8.68 kJ into internal energy
23. (a) 2.50 m/s   (b) 37.5 kJ   (c) Each process is the time-reversal of the other. The same momentum conservation equation describes both.
25. (a) 0.284   (b) 115 fJ and 45.4 fJ
27. 91.2 m/s
29. (a) 2.88 m/s at 32.3° north of east   (b) 783 J into internal energy
31. No; his speed was 41.5 mi/h.
33. 2.50 m/s at $-60.0°$
35. $(3.00\mathbf{i} - 1.20\mathbf{j})$ m/s
37. Orange: $v_i \cos\theta$; yellow: $v_i \sin\theta$
39. (a) $(-9.33\mathbf{i} - 8.33\mathbf{j})$ Mm/s   (b) 439 fJ
41. $\mathbf{r}_{CM} = (11.7\mathbf{i} + 13.3\mathbf{j})$ cm
43. 0.006 73 nm from the oxygen nucleus along the bisector of the angle
45. (a) 15.9 g   (b) 0.153 m
47. 0.700 m
49. (a) $(1.40\mathbf{i} + 2.40\mathbf{j})$ m/s   (b) $(7.00\mathbf{i} + 12.0\mathbf{j})$ kg·m/s
51. (a) 39.0 MN up   (b) 3.20 m/s² up
53. (a) 442 metric tons   (b) 19.2 metric tons
55. (a) $(1.33\mathbf{i})$ m/s   (b) $(-235\mathbf{i})$ N   (c) 0.680 s
   (d) $(-160\mathbf{i})$ N·s and $(+160\mathbf{i})$ N·s   (e) 1.81 m
   (f) 0.454 m   (g) -427 J   (h) $+107$ J
   (i) Equal friction forces act through different distances on person and cart to do different amounts of work on them. The total work on both together, -320 J, becomes $+320$ J of internal energy in this perfectly inelastic collision.
57. 1.39 km/s
59. 240 s
61. 0.980 m
63. (a) 6.81 m/s   (b) 1.00 m
65. $(3Mgx/L)\mathbf{j}$

67. (a) 3.75 kg·m/s²   (b) 3.75 N   (c) 3.75 N   (d) 2.81 J
   (e) 1.41 J   (f) Friction between sand and belt converts half of the input work into internal energy.
69. (a) As the child walks to the right, the boat moves to the left and the center of mass remains fixed.   (b) 5.55 m from the pier (c) No, since 6.55 m is less than 7.00 m.
71. (a) 100 m/s   (b) 374 J
73. (a) $\sqrt{2}\, v_i$ for m and $\sqrt{2/3}\, v_i$ for $3m$ (b) 35.3°
75. (a) 3.73 km/s   (b) 153 km

Chapter 10

1. (a) 4.00 rad/s²   (b) 18.0 rad
3. (a) 1 200 rad/s   (b) 25.0 s
5. (a) 5.24 s   (b) 27.4 rad
7. (a) 5.00 rad, 10.0 rad/s, 4.00 rad/s²   (b) 53.0 rad, 22.0 rad/s, 4.00 rad/s²
9. 13.7 rad/s²
11. $\sim 10^7$ rev/y
13. (a) 0.180 rad/s   (b) 8.10 m/s² toward the center of the track
15. (a) 8.00 rad/s   (b) 8.00 m/s, $a_r = -64.0$ m/s², $a_t = 4.00$ m/s²   (c) 9.00 rad
17. (a) 54.3 rev   (b) 12.1 rev/s
19. (a) 126 rad/s   (b) 3.78 m/s   (c) 1.27 km/s²
   (d) 20.2 m
21. (a) $-2.73\mathbf{i}$ m $+ 1.24\mathbf{j}$ m   (b) second quadrant, 156°
   (c) $-1.85\mathbf{i}$ m/s $- 4.10\mathbf{j}$ m/s
   (d) into the third quadrant at 246°

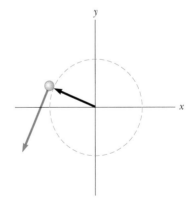

   (e) $6.15\mathbf{i}$ m/s² $- 2.78\mathbf{j}$ m/s²
   (f) $24.6\mathbf{i}$ N $- 11.1\mathbf{j}$ N
23. (a) 92.0 kg·m², 184 J   (b) 6.00 m/s, 4.00 m/s, 8.00 m/s, 184 J
25. (a) 143 kg·m²   (b) 2.57 kJ
29. 1.28 kg·m²
31. $\sim 10^0 = 1$ kg·m²
33. -3.55 N·m
35. 882 N·m
37. (a) 24.0 N·m   (b) 0.035 6 rad/s²   (c) 1.07 m/s²
39. (a) 0.309 m/s²   (b) 7.67 N and 9.22 N
41. (a) 872 N   (b) 1.40 kN

43. 2.36 m/s

45. (a) 11.4 N, 7.57 m/s^2, 9.53 m/s down   (b) 9.53 m/s

49. (a) $2(Rg/3)^{1/2}$ (b) $4(Rg/3)^{1/2}$ (c) $(Rg)^{1/2}$

51. $\frac{1}{3}\ell$

53. (a) 1.03 s   (b) 10.3 rev

55. (a) 4.00 J   (b) 1.60 s   (c) yes

57. (a) 12.5 rad/s   (b) 128 rad

59. (a) $(3g/L)^{1/2}$ (b) $3g/2L$ (c) $-\frac{3}{2}g\mathbf{i} - \frac{3}{4}g\mathbf{j}$

     (d) $-\frac{3}{2}Mg\mathbf{i} + \frac{1}{4}Mg\mathbf{j}$

61. $\alpha = g(h_2 - h_1)/2\pi R^2$

63. (b) $2gM(\sin\theta - \mu\cos\theta)(m + 2M)^{-1}$

65. 139 m/s

67. 5.80 kg·m^2; the height makes no difference.

69. (a) 2 160 N·m   (b) 439 W

71. (a) 118 N and 156 N   (b) 1.19 kg·m^2

73. (a) $\alpha = -0.176$ rad/s^2 (b) 1.29 rev   (c) 9.26 rev

Chapter 11

1. (a) 500 J   (b) 250 J   (c) 750 J

3. $\frac{7}{10} Mv^2$

5. (a) $\frac{2}{3}g\sin\theta$ for the disk, larger than $\frac{1}{2}g\sin\theta$ for the hoop

     (b) $\frac{1}{3}\tan\theta$

7. 1.21×10^{-4} kg·m^2. The height is unnecessary.

9. $-7.00\mathbf{i} + 16.0\mathbf{j} - 10.0\mathbf{k}$

11. (a) $-17.0\mathbf{k}$ (b) 70.5°

13. (a) 2.00 N·m   (b) $\mathbf{k}$

15. (a) negative z direction   (b) positive z direction

17. 45.0°

19. $(17.5\mathbf{k})$ kg·m^2/s

21. $(60.0\mathbf{k})$ kg·m^2/s

23. $mvR[\cos(vt/R) + 1]\mathbf{k}$

25. (a) zero   (b) $(-mv_i^3\sin^2\theta\cos\theta/2g)\mathbf{k}$

     (c) $(-2mv_i^3\sin^2\theta\cos\theta/g)\mathbf{k}$ (d) The downward force of gravity exerts a torque in the $-z$ direction.

27. $-m\ell\,gt\cos\theta\,\mathbf{k}$

29. 4.50 kg·m^2/s up

31. (a) 0.433 kg·m^2/s   (b) 1.73 kg·m^2/s

33. (a) $\omega_f = \omega_i I_1/(I_1 + I_2)$ (b) $I_1/(I_1 + I_2)$

35. (a) 1.91 rad/s   (b) 2.53 J, 6.44 J

37. (a) 0.360 rad/s counterclockwise   (b) 99.9 J

39. (a) $mv\ell$ down   (b) $M/(M + m)$

41. (a) $\omega = 2mv_i d/(M + 2m)R^2$ (b) No; some mechanical energy changes into internal energy.

43. (a) 2.19×10^6 m/s   (b) 2.18×10^{-18} J

     (c) 4.13×10^{16} rad/s

45. $[10Rg(1 - \cos\theta)/7r^2]^{1/2}$

51. (a) $2.70R$ (b) $F_x = -\frac{20}{7}mg$, $F_y = -mg$

53. 0.632

55. (a) $v_i r_i/r$ (b) $T = (mv_i^2 r_i^2)r^{-3}$ (c) $\frac{1}{2}mv_i^2(r_i^2/r^2 - 1)$

     (d) 4.50 m/s, 10.1 N, 0.450 J

57. 54.0°

59. (a) 3 750 kg·m^2/s   (b) 1.88 kJ   (c) 3 750 kg·m^2/s

     (d) 10.0 m/s   (e) 7.50 kJ   (f) 5.62 kJ

61. $(M/m)[3ga(\sqrt{2} - 1)]^{1/2}$

63. (c) $(8Fd/3M)^{1/2}$

67. (a) 0.800 m/s^2, 0.400 m/s^2 (b) 0.600 N backward on the plank and forward on the roller, at the top of each roller; 0.200 N forward on each roller and backward on the floor, at the bottom of each roller.

Chapter 12

1. 10.0 N up; 6.00 N·m counterclockwise

3. $[(m_1 + m_b)d + m_1\ell/2]/m_2$

5. -0.429 m

7. (3.85 cm, 6.85 cm)

9. $(-1.50$ m, -1.50 m)

11. (a) 859 N   (b) 1 040 N left and upward at 36.9°

13. (a) $f_s = 268$ N, $n = 1\,300$ N   (b) 0.324

15. (a) 1.04 kN at 60.0°   (b) $(370\mathbf{i} + 900\mathbf{j})$ N

17. 2.94 kN on each rear wheel and 4.41 kN on each front wheel

19. (a) 29.9 N   (b) 22.2 N

21. (a) 35.5 kN   (b) 11.5 kN   (c) -4.19 kN

23. 88.2 N and 58.8 N

25. 4.90 mm

27. 0.023 8 mm

29. 0.912 mm

31. $\dfrac{8m_1 m_2 g L_i}{\pi d^2 Y(m_1 + m_2)}$

33. (a) 3.14×10^4 N   (b) 6.28×10^4 N

35. 1.80×10^8 N/m^2

37. $n_A = 5.98 \times 10^5$ N, $n_B = 4.80 \times 10^5$ N

39. (a) 0.400 mm   (b) 40.0 kN   (c) 2.00 mm   (d) 2.40 mm

     (e) 48.0 kN

41. (a)

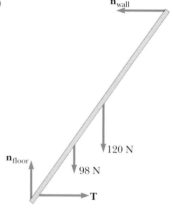

     (b) 69.8 N   (c) 0.877L

43. (a) 160 N right   (b) 13.2 N right   (c) 292 N up

     (d) 192 N

45. (a) $T = F_g(L + d)/\sin\theta(2L + d)$

     (b) $R_x = F_g(L + d)\cot\theta/(2L + d)$; $R_y = F_g L/(2L + d)$

47. 0.789 L

49. 5.08 kN, $R_x = 4.77$ kN, $R_y = 8.26$ kN

51. $T = 2.71$ kN, $R_x = 2.65$ kN

53. (a) $\mu_k = 0.571$; the normal force acts 20.1 cm to the left of the front edge of the sliding cabinet.   (b) 0.501 m

55. (b) 60.0°

57. (a) $M = (m/2)(2\mu_s \sin\theta - \cos\theta)(\cos\theta - \mu_s \sin\theta)^{-1}$
(b) $R = (m + M)g(1 + \mu_s^2)^{1/2}$,
$F = g[M^2 + \mu_s^2(m + M)^2]^{1/2}$

59. (a) 133 N (b) $n_A = 429$ N and $n_B = 257$ N
(c) $R_x = 133$ N and $R_y = -257$ N

61. 66.7 N

65. 1.09 m

67. (a) 4 500 N (b) 4.50×10^6 N/m² (c) yes.

69. (a) $P_y = (F_g/L)(d - ah/g)$ (b) 0.306 m
(c) $\mathbf{P} = (-306\mathbf{i} + 553\mathbf{j})$ N

71. $n_A = n_E = 6.66$ kN; $F_{AB} = 10.4$ kN $= F_{BC} = F_{DC} = F_{DE}$;
$F_{AC} = 7.94$ kN $= F_{CE}$; $F_{BD} = 15.9$ kN

Chapter 13

1. (a) 1.50 Hz, 0.667 s (b) 4.00 m (c) π rad (d) 2.83 m

3. (a) 20.0 cm (b) 94.2 cm/s as the particle passes through equilibrium (c) 17.8 m/s² at the maximum displacement from equilibrium

5. (b) 18.8 cm/s, 0.333 s (c) 178 cm/s², 0.500 s
(d) 12.0 cm

7. 0.627 s

9. (a) 40.0 cm/s, 160 cm/s² (b) 32.0 cm/s, -96.0 cm/s²
(c) 0.232 s

11. 40.9 N/m

13. (a) 0.750 m (b) $x = -(0.750 \text{ m})\sin(2.00t/s)$

15. 0.628 m/s

17. 2.23 m/s

19. (a) 28.0 mJ (b) 1.02 m/s (c) 12.2 mJ (d) 15.8 mJ

21. (a) 2.61 m/s (b) 2.38 m/s

23. 2.60 cm and -2.60 cm

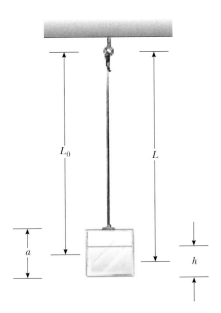

Chapter 13, Problem 57(a)

25. (a) 35.7 m (b) 29.1 s

27. $\sim 10^0$ s

29. (a) 0.817 m/s (b) 2.54 rad/s² (c) 0.634 N

33. 0.944 kg·m²

37. (a) 5.00×10^{-7} kg·m² (b) 3.16×10^{-4} N·m/rad

39. The x coordinate of the crank pin is $A \cos\omega t$.

41. 1.00×10^{-3} s^{-1}

43. (a) 2.95 Hz (b) 2.85 cm

47. Either 1.31 Hz or 0.641 Hz

49. 6.58 kN/m

51. (a) $2Mg$; $Mg(1 + y/L)$ (b) $T = (4\pi/3)(2L/g)^{1/2}$; 2.68 s

53. 6.62 cm

55. 9.19×10^{13} Hz

57. (a) See bottom of preceding column.
(b) $\dfrac{dT}{dt} = \dfrac{\pi(dM/dt)}{2\rho a^2 g^{1/2}[L_i + (dM/dt)t/2\rho a^2]^{1/2}}$
(c) $T = 2\pi g^{-1/2}[L_i + (dM/dt)t/2\rho a^2]^{1/2}$

59. $f = (2\pi L)^{-1}(gL + kh^2/M)^{1/2}$

61. (a) 3.56 Hz (b) 2.79 Hz (c) 2.10 Hz

63. (a) 3.00 s (b) 14.3 J (c) 25.5°

65. 0.224 rad/s

Chapter 14

1. $\sim 10^{-7}$ N toward you

3. $\mathbf{g} = (Gm/\ell^2)(\frac{1}{2} + \sqrt{2})$ toward the opposite corner

5. $(-100\mathbf{i} + 59.3\mathbf{j})$ pN

7. (a) 4.39×10^{20} N (b) 1.99×10^{20} N (c) 3.55×10^{22} N

9. 0.613 m/s² toward the Earth

11.

1.000 m $-$ 61.3 nm

15. 12.6×10^{31} kg

17. 1.27

19. 1.90×10^{27} kg

21. 8.92×10^7 m

25. $g = 2MGr(r^2 + a^2)^{-3/2}$ toward the center of mass

27. (a) -4.77×10^9 J (b) 569 N (c) 569 N up

29. (a) 1.84×10^9 kg/m³ (b) 3.27×10^6 m/s²
(c) -2.08×10^{13} J

31. (a) -1.67×10^{-14} J (b) At the center

33. 1.58×10^{10} J

35. (a) 1.48 h (b) 7.79 km/s (c) 6.43×10^9 J

37. 1.66×10^4 m/s

41. 15.6 km/s

43. $GM_E m/12R_E$

45. $2GmM/\pi R^2$ straight up in the picture

47. (a) 7.41×10^{-10} N   (b) 1.04×10^{-8} N
     (c) 5.21×10^{-9} N

49. 2.26×10^{-7}

51. (b) 1.10×10^{32} kg

53. (b) $GMm/2R$

55. 7.79×10^{14} kg

57. 7.41×10^{-10} N

59. $v_{esc} = (8\pi G\rho/3)^{1/2} R$

61. (a) $v_1 = m_2(2G/d)^{1/2}(m_1 + m_2)^{-1/2}$
     $v_2 = m_1(2G/d)^{1/2}(m_1 + m_2)^{-1/2}$
     $v_{rel} = (2G/d)^{1/2}(m_1 + m_2)^{1/2}$
     (b) $K_1 = 1.07 \times 10^{32}$ J, $K_2 = 2.67 \times 10^{31}$ J

63. (a) $A = M/\pi R^4$ (b) $F = GmM/r^2$ toward the center
     (c) $F = GmMr^2/R^4$ toward the center

65. 119 km

67. (a) -36.7 MJ   (b) 9.24×10^{10} kg·m²/s
     (c) 5.58 km/s, 10.4 Mm   (d) 8.69 Mm   (e) 134 min

71.

t (s)	x (m)	y (m)	v_x (m/s)	v_y (m/s)
0	0	12 740 000	5 000	0
10	50 000	12 740 000	4 999.9	− 24.6
20	99 999	12 739 754	4 999.7	− 49.1
30	149 996	12 739 263	4 999.4	− 73.7 . . .

The object does not hit the Earth; its minimum radius is $1.33R_E$. Its period is 1.09×10^4 s. A circular orbit would require speed 5.60 km/s.

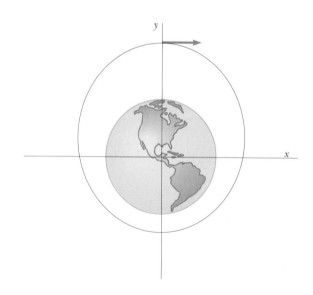

Chapter 15

1. 0.111 kg

3. 6.24 MPa

5. 5.27×10^{18} kg

7. 1.62 m

9. 7.74×10^{-3} m²

11. 271 kN horizontally backward

13. $P_0 + (\rho d/2)(g^2 + a^2)^{1/2}$

15. 0.722 mm

17. 10.5 m; no, some alcohol and water evaporate.

19. 12.6 cm

21. 1.07 m²

23. (a) 9.80 N   (b) 6.17 N

25. (a) 7.00 cm   (b) 2.80 kg

27. $\rho_{oil} = 1\,250$ kg/m³; $\rho_{sphere} = 500$ kg/m³

29. 1 430 m³

31. 2.67×10^3 kg

33. (a) 1.06 m/s   (b) 4.24 m/s

35. (a) 17.7 m/s   (b) 1.73 mm

37. 31.6 m/s

39. 68.0 kPa

41. 103 m/s

43. (a) 4.43 m/s   (b) The siphon can be no higher than 10.3 m.

45. $2\sqrt{h(h_0 - h)}$

47. 0.258 N

49. 1.91 m

53. 709 kg/m³

55. top scale 17.3 N; bottom scale 31.7 N

59. 90.04%

61. 4.43 m/s

63. (a) 10.3 m   (b) 0

65. (a) 18.3 mm   (b) 14.3 mm   (c) 8.56 mm

67. (a) 2.65 m/s   (b) 2.31×10^4 Pa

69. (a) 1.25 cm   (b) 13.8 m/s

Chapter 16

1. $y = 6[(x - 4.5t)^2 + 3]^{-1}$

3. (a) left   (b) 5.00 m/s

5. (a) longitudinal   (b) 665 s

7. (a) 156°   (b) 0.058 4 cm

9. (a) y_1 in $+x$ direction, y_2 in $-x$ direction   (b) 0.750 s
     (c) 1.00 m

11. 30.0 N

13. 1.64 m/s²

15. 13.5 N

17. 586 m/s

19. 32.9 ms

21. 0.329 s

23. (a) See top of next page   (b) 0.125 s

25. 0.319 m

27. 2.40 m/s

29. (a) 0.250 m   (b) 40.0 rad/s   (c) 0.300 rad/m
     (d) 20.9 m   (e) 133 m/s   (f) $+x$

31. (a) $y = (8.00 \text{ cm}) \sin(7.85x + 6\pi t)$
     (b) $y = (8.00 \text{ cm}) \sin(7.85x + 6\pi t - 0.785)$

33. (a) 0.500 Hz, 3.14 rad/s   (b) 3.14 rad/m
     (c) $(0.100 \text{ m}) \sin(3.14x/\text{m} - 3.14t/\text{s})$

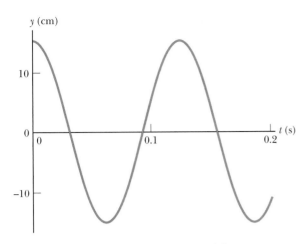

Chapter 16, Problem 23(a)

(d) $(0.100 \text{ m}) \sin(-3.14t/\text{s})$
(e) $(0.100 \text{ m}) \sin(4.71 \text{ rad} - 3.14t/\text{s})$ (f) 0.314 m/s
35. 2.00 cm, 2.98 m, 0.576 Hz, 1.72 m/s
37. (b) 3.18 Hz
41. 55.1 Hz
43. (a) 62.5 m/s (b) 7.85 m (c) 7.96 Hz (d) 21.1 W
45. (a) $A = 40.0$ (b) $A = 7.00, B = 0, C = 3.00$. One can
take the dot product of the given equation with each one
of $\mathbf{i}, \mathbf{j}$, and $\mathbf{k}$. (c) By inspection, $A = 0, B = 7.00$ mm,
$C = 3.00/\text{m}, D = 4.00/\text{s}, E = 2.00$. Consider the average
value of both sides of the given equation to find A. Then
consider the maximum value of both sides to find B. You
can evaluate the partial derivative of both sides of the
given equation with respect to x and separately with respect to t to obtain equations yielding C and D upon chosen substitutions for x and t. Then substitute $x = 0$
and $t = 0$ to obtain E.
47. It is if $v = (T/\mu)^{1/2}$
49. ~1 min
51. (a) $3.33\mathbf{i}$ m/s (b) -5.48 cm (c) 0.667 m, 5.00 Hz
(d) 11.0 m/s
53. $(Lm/Mg \sin \theta)^{1/2}$
55. (a) 39.2 N (b) 0.892 m (c) 83.6 m/s
57. 14.7 kg
61. (a) $(0.707)2(L/g)^{1/2}$ (b) $L/4$
63. 3.86×10^{-4}
65. (a) $v = (2T_0/\mu_0)^{1/2} = v_0 2^{1/2}$
$v' = (2T_0/3\mu_0)^{1/2} = v_0 (2/3)^{1/2}$
(b) $0.966t_0$
67. 130 m/s, 1.73 km

Chapter 17

1. 5.56 km
3. 7.82 m
5. (a) 27.2 s (b) 25.7 s; the interval in (a) is longer
7. (a) 153 m/s (b) 614 m
9. (a) amplitude 2.00 μm, wavelength 40.0 cm,
speed 54.6 m/s (b) -0.433 μm (c) 1.72 mm/s

11. $\Delta P = (0.2 \text{ Pa}) \sin(62.8x/\text{m} - 2.16 \times 10^4 \ t/\text{s})$
13. (a) 6.52 mm (b) 20.5 m/s
15. 5.81 m
17. 66.0 dB
19. (a) 3.75 W/m^2 (b) 0.600 W/m^2
21. (a) $1.32 \times 10^{-4} \text{ W/m}^2$ (b) 81.2 dB
23. 65.6 dB
25. (a) 65.0 dB (b) 67.8 dB (c) 69.6 dB
27. 1.13 μW
29. (a) 30.0 m (b) 9.49×10^5 m
31. (a) 332 J (b) 46.4 dB
33. (a) 75.7-Hz drop (b) 0.948 m
35. 26.4 m/s
37. 19.3 m
39. (a) 338 Hz (b) 483 Hz
41. 56.4°
43. (a) 56.3 s (b) 56.6 km farther along
45. 400 m; 27.5%
47. (a) 23.2 cm (b) 8.41×10^{-8} m (c) 1.38 cm
49. (a) 0.515/min (b) 0.614/min
51. 7.94 km
53. (a) 55.8 m/s (b) 2 500 Hz
55. Bat is gaining on the insect at the rate of 1.69 m/s.
57. (a)

(b) 0.343 m (c) 0.303 m (d) 0.383 m
(e) 1.03 kHz
59. (a) 0.691 m (b) 691 km
61. 1204.2 Hz
63. (a) 0.948° (b) 4.40°
65. 1.34×10^4 N
67. 95.5 s
69. (b) 531 Hz
71. (a) 6.45 (b) 0
73. ~10^{11} Hz

Chapter 18

1. (a) 9.24 m (b) 600 Hz
3. 5.66 cm
5. 91.3°
7. (a) 2 (b) 9.28 m and 1.99 m
9. 15.7 m, 31.8 Hz, 500 m/s
11. At 0.089 1 m, 0.303 m, 0.518 m, 0.732 m, 0.947 m, and
1.16 m from one speaker
13. (a) 4.24 cm (b) 6.00 cm (c) 6.00 cm (d) 0.500 cm,
1.50 cm, and 2.50 cm
17. 0.786 Hz, 1.57 Hz, 2.36 Hz, and 3.14 Hz
19. (a) 163 N (b) 660 Hz
21. 19.976 kHz

23. 31.2 cm from the bridge; 3.84%
25. (a) 350 Hz   (b) 400 kg
27. 0.352 Hz
29. (a) 3.66 m/s   (b) 0.200 Hz
31. (a) 0.357 m   (b) 0.715 m
33. (a) 531 Hz   (b) 42.5 mm
35. around 3 kHz
37. n(206 Hz) for $n = 1$ to 9, and n(84.5 Hz) for $n = 2$ to 23
39. 239 s
41. 0.502 m and 0.837 m
43. (a) 350 m/s   (b) 1.14 m
45. (a) 19.5 cm   (b) 841 Hz
47. (a) 1.59 kHz   (b) odd-numbered harmonics
   (c) 1.11 kHz
49. 5.64 beats/s
51. (a) 1.99 beats/s   (b) 3.38 m/s
53. The second harmonic of E is close to the third harmonic of A, and the fourth harmonic of C# is close to the fifth harmonic of A.
55. (a) 3.33 rad   (b) 283 Hz
57. 3.85 m/s away from the station or 3.77 m/s toward the station
59. 85.7 Hz
61. 31.1 N
63. (a) 59.9 Hz   (b) 20.0 cm
65. (a) 1/2   (b) $[n/(n + 1)]^2 T$ (c) 9/16
67. 50.0 Hz, 1.70 m
69. (a) $2A \sin(2\pi x/\lambda) \cos(2\pi vt/\lambda)$
   (b) $2A \sin(\pi x/L) \cos(\pi vt/L)$
   (c) $2A \sin(2\pi x/L) \cos(2\pi vt/L)$
   (d) $2A \sin(n\pi x/L) \cos(n\pi vt/L)$

Chapter 19

1. (a) $37.0°C = 310$ K   (b) $-20.6°C = 253$ K
3. (a) $-274°C$ (b) 1.27 atm   (c) 1.74 atm
5. (a) $-320°F$ (b) 77.3 K
7. (a) $810°F$ (b) 450 K
9. 3.27 cm
11. (a) 3.005 8 m   (b) 2.998 6 m
13. 55.0°C
15. (a) 0.109 cm^2 (b) increase
17. (a) 0.176 mm   (b) 8.78 μm   (c) 0.093 0 cm^3
19. (a) 2.52 MN/m^2 (b) It will not break.
21. 1.14°C
23. (a) 99.4 cm^3 (b) 0.943 cm
25. (a) 3.00 mol   (b) 1.80×10^{24} molecules
27. 1.50×10^{29} molecules
29. 472 K
31. (a) 41.6 mol   (b) 1.20 kg, in agreement with the tabulated density
33. (a) 400 kPa   (b) 449 kPa
35. 2.27 kg
37. 3.67 cm^3
39. 4.39 kg
43. (a) 94.97 cm   (b) 95.03 cm

45. 208°C
47. 3.55 cm
49. (a) Expansion makes density drop.   (b) 5×10^{-5} (°C)$^{-1}$
51. (a) $h = nRT/(mg + P_0 A)$ (b) 0.661 m
53. $\alpha \, \Delta T$ is much less than 1.
55. (a) 9.49×10^{-5} s   (b) 57.4 s lost
57. (a) $\rho g P_0 V_i (P_0 + \rho g d)^{-1}$ (b) decrease   (c) 10.3 m
61. (a) 5.00 MPa   (b) 9.58×10^{-3}
63. 2.74 m
65. $L_c = 9.17$ cm, $L_s = 14.2$ cm
67. (a) $L_f = L_i e^{\alpha \Delta T}$ (b) 2.00×10^{-4}%; 59.4%
69. (a) 6.17×10^{-3} kg/m   (b) 632 N   (c) 580 N; 192 Hz

Chapter 20

1. $(10.0 + 0.117)$°C
3. 0.234 kJ/kg · °C
5. 29.6°C
7. (a) 0.435 cal/g · °C   (b) beryllium
9. (a) 25.8°C   (b) No
11. 50.7 ks
13. 0.294 g
15. 0.414 kg
17. (a) 0°C   (b) 114 g
19. 59.4°C
21. 1.18 MJ
23. (a) $4P_i V_i$ (b) $T = (P_i/nRV_i) V^2$
25. 466 J
27. 810 J, 506 J, 203 J
29. $Q = -720$ J
31.

	Q	W	ΔE_{int}
BC	$-$	0	$-$
CA	$-$	$-$	$-$
AB	$+$	$+$	$+$

33. (a) 7.50 kJ   (b) 900 K
35. 3.10 kJ; 37.6 kJ
37. (a) 0.041 0 m^3 (b) -5.48 kJ   (c) -5.48 kJ
41. 2.40×10^6 cal/s
43. 10.0 kW
45. 51.2°C
47. (a) 0.89 ft^2 · °F · h/Btu   (b) 1.85 ft^2 · °F · h/Btu   (c) 2.08
49. (a) $\sim 10^3$ W   (b) decreasing at $\sim 10^{-1}$ K/s
51. 364 K
53. 47.7 g
55. (a) 16.8 L   (b) 0.351 L/s
57. 2.00 kJ/kg · °C
59. 1.87 kJ
61. (a) $4P_i V_i$ (b) $4P_i V_i$ (c) 9.08 kJ
63. 5.31 h
65. 872 g
67. (a) 15.0 mg. Block: $Q = 0$, $W = +5.00$ J, $\Delta E_{int} = 0$, $\Delta K = -5.00$ J; Ice: $Q = 0$, $W = -5.00$ J, $\Delta E_{int} = 5.00$ J, $\Delta K = 0$.

(b) 15.0 mg. Block: $Q = 0$, $W = 0$, $\Delta E_{int} = 5.00$ J,
$\Delta K = -5.00$ J; Metal: $Q = 0$, $W = 0$, $\Delta E_{int} = 0$, $\Delta K = 0$.
(c) 0.004 04°C. Moving slab: $Q = 0$, $W = +2.50$ J,
$\Delta E_{int} = 2.50$ J, $\Delta K = -5.00$ J; Stationary slab: $Q = 0$,
$W = -2.50$ J, $\Delta E_{int} = 2.50$ J, $\Delta K = 0$

69. 10.2 h
71. 9.32 kW

Chapter 21

1. 6.64×10^{-27} kg
3. 0.943 N; 1.57 Pa
5. 17.6 kPa
7. 3.32 mol
9. (a) 3.53×10^{23} atoms   (b) 6.07×10^{-21} J
   (c) 1.35 km/s
11. (a) 8.76×10^{-21} J for both   (b) 1.62 km/s for helium;
   514 m/s for argon
13. 75.0 J
15. (a) 3.46 kJ   (b) 2.45 kJ   (c) 1.01 kJ
17. (a) 118 kJ   (b) 6.03×10^3 kg
19. Between 10^{-2}°C and 10^{-3}°C
21. (a) 316 K   (b) 200 J
23. $9\, P_i V_i$
25. (a) 1.39 atm   (b) 366 K, 253 K   (c) 0, 4.66 kJ, -4.66 kJ
27. 227 K
29. (a)

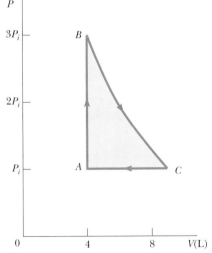

(b) 8.79 L   (c) 900 K   (d) 300 K   (e) 336 J
31. 25.0 kW
33. (a) 9.95 cal/K, 13.9 cal/K   (b) 13.9 cal/K, 17.9 cal/K
35. 2.33×10^{-21} J
37. The ratio of oxygen to nitrogen molecules decreases to
   85.5% of its sea-level value.
39. (a) 6.80 m/s   (b) 7.41 m/s   (c) 7.00 m/s
43. 819°C
45. (a) 3.21×10^{12} molecules   (b) 778 km
   (c) 6.42×10^{-4} s^{-1}
49. (a) 9.36×10^{-8} m   (b) 9.36×10^{-8} atm   (c) 302 atm
51. (a) 100 kPa, 66.5 L, 400 K, 5.82 kJ, 7.48 kJ, 1.66 kJ

(b) 133 kPa, 49.9 L, 400 K, 5.82 kJ, 5.82 kJ, 0
(c) 120 kPa, 41.6 L, 300 K, 0, -910 J, -910 J
(d) 120 kPa, 43.3 L, 312 K, 722 J, 0, -722 J
55. 0.625
57. (a) Pressure increases as volume decreases.
   (d) 0.500 atm^{-1}, 0.300 atm^{-1}
59. 1.09×10^{-3}; 2.69×10^{-2}; 0.529; 1.00; 0.199;
   1.01×10^{-41}; 1.25×10^{-1082}
61. (a) Larger-mass molecules settle to the outside.
63. (a) 0.203 mol   (b) $T_B = T_C = 900$ K; $V_C = 15.0$ L

(c, d)	P (atm)	V (L)	T (K)	E_{int} (kJ)
A	1	5	300	0.760
B	3	5	900	2.28
C	1	15	900	2.28

(e) For $A \rightarrow B$, lock the piston in place and put the cylinder into an oven at 900 K. For $B \rightarrow C$, keep the gas in the oven while gradually letting the gas expand to lift a load on the piston as far as it can. For $C \rightarrow A$, move the cylinder from the oven back to the 300-K room and let the gas cool and contract.

(f, g)	Q (kJ)	W (kJ)	ΔE_{int} (kJ)
$A \rightarrow B$	1.52	0	1.52
$B \rightarrow C$	1.67	1.67	0
$C \rightarrow A$	-2.53	-1.01	-1.52
$ABCA$	0.656	0.656	0

65. (a) 3.34×10^{26} molecules   (b) during the 27th day
   (c) 2.53×10^6
67. (a) 0.510 m/s   (b) 20 ms

Chapter 22

1. (a) 6.94%   (b) 335 J
3. (a) 10.7 kJ   (b) 0.533 s
5. (a) 1.00 kJ   (b) 0
7. (a) 67.2%   (b) 58.8 kW
9. (a) 869 MJ   (b) 330 MJ
11. (a) 741 J   (b) 459 J
13. 0.330 or 33.0%
15. (a) 5.12%   (b) 5.27 TJ/h   (c) As conventional energy
   sources become more expensive, or as their true costs are
   recognized, alternative sources become economically viable.
17. (a) 214 J, 64.3 J
   (b) -35.7 J, -35.7 J. The net effect is the transport of
   energy from the cold to the hot reservoir without expenditure of external work.
   (c) 333 J, 233 J
   (d) 83.3 J, 83.3 J, 0. The net effect is the expulsion of the
   energy entering the system by heat, entirely by work, in a
   cyclic process.
   (e) -0.111 J/K. The entropy of the Universe has
   decreased.

19. (a) 244 kPa   (b) 192 J
21. 146 kW, 70.8 kW
23. 9.00
27. 72.2 J
29. (a) 24.0 J   (b) 144 J
31. -610 J/K
33. 195 J/K
35. 3.27 J/K
37. 1.02 kJ/K
39. 5.76 J/K. Temperature is constant if the gas is ideal.
41. 0.507 J/K
43. 18.4 J/K
45. (a) 1   (b) 6
47. (a)

Macrostate	Possible Microstates	Total Number of Microstates
All R	RRR	1
2R, 1G	RRG, RGR, GRR	3
1R, 2G	GRR, GRG, RGG	3
All G	GGG	1

(b)

Macrostate	Possible Microstates	Total Number of Microstates
All R	RRRRR	1
4R, 1G	RRRRG, RRRGR, RRGRR, RGRRR, GRRRR	5
3R, 2G	RRRGG, RRGRG, RGRRG, GRRRG, RRGGR, RGRGR, GRRGR, RGGRR, GRGRR, GGRRR	10
2R, 3G	GGGRR, GGRGR, GRGGR, RGGGR, GGRRG, GRGRG, RGGRG, GRRGG, RGRGG, RRGGG	10
1R, 4G	GGGGR, GGGRG, GGRGG, GRGGG, RGGGG	5
All G	GGGGG	1

49. 1.86
51. (a) 5.00 kW   (b) 763 W
53. (a) $2nRT_i \ln 2$ (b) 0.273
55. 23.1 mW
57. 5.97×10^4 kg/s
59. (a) 3.19 cal/K   (b) 98.19°F, 2.59 cal/K
61. 1.18 J/K
63. (a) $10.5nRT_i$ (b) $8.50nRT_i$ (c) 0.190   (d) 0.833
65. $nC_P \ln 3$
69. (a) 96.9 W $= 8.33 \times 10^4$ cal/hr
  (b) 1.19°C/h = 2.14°F/h

Chapter 23

1. (a) 2.62×10^{24} electrons   (b) 2.38 electrons
3. The force is $\sim 10^{26}$ N.
5. 514 kN
7. 0.873 N at 330°
9. (a) 82.2 nN   (b) 2.19 Mm/s
11. (a) 55.8 pN/C down   (b) 102 nN/C up
13. 1.82 m to the left of the negative charge
15. (a) $(18.0\mathbf{i} - 218\mathbf{j})$ kN/C   (b) $(36.0\mathbf{i} - 436\mathbf{j})$ mN
17. (a) The field is zero at the center of the triangle.

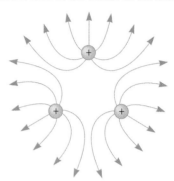

(b) $(1.73\, k_e q/a^2)\mathbf{j}$
19. (a) $5.91 k_e q/a^2$ at 58.8°   (b) $5.91 k_e\, q^2/a^2$ at 58.8°
23. $-(\pi^2 k_e q/6a^2)\,\mathbf{i}$
25. $-(k_e \lambda_0/x_0)\,\mathbf{i}$
27. (a) $6.64\mathbf{i}$ MN/C   (b) $24.1\mathbf{i}$ MN/C   (c) $6.40\mathbf{i}$ MN/C
  (d) $0.664\mathbf{i}$ MN/C, taking the axis of the ring as the x axis
29. (a) 383 MN/C away   (b) 324 MN/C away
  (c) 80.7 MN/C away   (d) 6.68 MN/C away
33. $-21.6\mathbf{i}$ MN/C
37. (a) 86.4 pC for each
  (b) 324 pC, 459 pC, 459 pC, 432 pC
  (c) 57.6 pC, 106 pC, 154 pC, 96.0 pC
39.

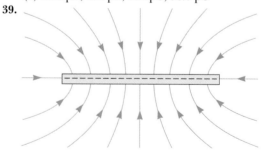

41. 4.39 Mm/s and 2.39 km/s
43. (a) 61.4 Gm/s^2 (b) 19.5 μs   (c) 11.7 m   (d) 1.20 fJ
45. K/ed in the direction of motion
47. (a) 111 ns   (b) 5.67 mm   (c) $(450\mathbf{i} + 102\mathbf{j})$ km/s
49. (a) 36.9°, 53.1°   (b) 167 ns, 221 ns
51. (a) 21.8 μm   (b) 2.43 cm
53. (a) 10.9 nC   (b) 5.43 mN
55. 40.9 N at 263°
57. 26.7 μC
61. $-707\mathbf{j}$ mN

63. (a) $\theta_1 = \theta_2$

65. (b) The object's acceleration is a negative constant times its displacement from equilibrium.
$T = (\pi 8^{-1/4})(mL^3/k_e qQ)^{1/2}$

67. (a) $-(4k_e q/3a^2)\mathbf{j}$ (b) $(0, 2.00 \text{ m})$

69. (a) $\mathbf{F} = 1.90(k_e q^2/s^2)(\mathbf{i} + \mathbf{j} + \mathbf{k})$
(b) $\mathbf{F} = 3.29(k_e q^2/s^2)$ in the direction away from the diagonally opposite vertex

71. $(-1.36\mathbf{i} + 1.96\mathbf{j}) \text{ kN/C}$

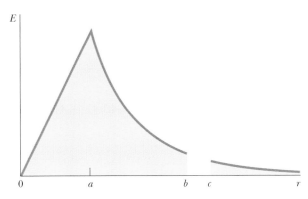

Chapter 24, Problem 53

Chapter 24

1. (a) $858 \text{ N} \cdot \text{m}^2/\text{C}$ (b) 0   (c) $657 \text{ N} \cdot \text{m}^2/\text{C}$

3. 4.14 MN/C

5. (a) $-2.34 \text{ kN} \cdot \text{m}^2/\text{C}$ (b) $+2.34 \text{ kN} \cdot \text{m}^2/\text{C}$ (c) 0

7. q/ϵ_0

9. EhR

11. (a) $-6.89 \text{ MN} \cdot \text{m}^2/\text{C}$ (b) The number of lines entering exceeds the number leaving by 2.91 times or more.

13. (a) $q/2\epsilon_0$ (b) $q/2\epsilon_0$ (c) Plane and square look the same to the charge.

15. (a) $+Q/2\epsilon_0$ (b) $-Q/2\epsilon_0$

17. $5.22 \text{ kN} \cdot \text{m}^2/\text{C}$

19. $-18.8 \text{ kN} \cdot \text{m}^2/\text{C}$

21. 0 when $R < d$ and $\lambda L/\epsilon_0$ when $R > d$

23. (a) $3.20 \text{ MN} \cdot \text{m}^2/\text{C}$ (b) $19.2 \text{ MN} \cdot \text{m}^2/\text{C}$ (c) The answer to part (a) could change, but the answer to part (b) would stay the same.

25. $-q/24\epsilon_0$

27. (a) 0   (b) 366 kN/C (c) 1.46 MN/C (d) 650 kN/C

29. $\mathbf{E} = \rho r/2\epsilon_0$ away from the axis

31. (a) 0   (b) 7.19 MN/C away from the center

33. (a) $\sim 1 \text{ mN}$ (b) $\sim 100 \text{ nC}$ (c) $\sim 10 \text{ kN/C}$
(d) $\sim 10 \text{ kN} \cdot \text{m}^2/\text{C}$

35. (a) 51.4 kN/C outward   (b) $646 \text{ N} \cdot \text{m}^2/\text{C}$

37. 508 kN/C up

39. (a) 0   (b) $5\,400 \text{ N/C}$ outward   (c) 540 N/C outward

41. (a) $+708 \text{ nC/m}^2$ and -708 nC/m^2 (b) $+177 \text{ nC}$ and -177 nC

43. 2.00 N

45. (a) $-\lambda, +3\lambda$ (b) $3\lambda/2\pi\epsilon_0 r$ radially outward

47. (a) 80.0 nC/m^2 on each face   (b) $(9.04\mathbf{k}) \text{ kN/C}$
(c) $(-9.04\mathbf{k}) \text{ kN/C}$

49. (a) 0   (b) 79.9 MN/C radially outward   (c) 0
(d) 7.35 MN/C radially outward

51. (b) $Q/2\epsilon_0$ (c) Q/ϵ_0

53. (a) $+2Q$ (b) radially outward   (c) $2k_e Q/r^2$
(d) 0   (e) 0   (f) $3Q$ (g) $3k_e Q/r^2$ radially outward
(h) $3Qr^3/a^3$ (i) $3k_e Qr/a^3$ radially outward   (j) $-3Q$
(k) $+2Q$ (l) See top of next column

55. (a) $\rho r/3\epsilon_0$; $Q/4\pi\epsilon_0 r^2$; 0; $Q/4\pi\epsilon_0 r^2$, all radially outward
(b) $-Q/4\pi b^2$ and $+Q/4\pi c^2$

57. For $r < a$, $\mathbf{E} = \lambda/2\pi\epsilon_0 r$ radially outward. For $a < r < b$,
$\mathbf{E} = [\lambda + \rho\pi(r^2 - a^2)]/2\pi\epsilon_0 r$ radially outward. For
$r > b$, $\mathbf{E} = [\lambda + \rho\pi(b^2 - a^2)]/2\pi\epsilon_0 r$ radially outward

59. (a) σ/ϵ_0 away from both plates   (b) 0   (c) σ/ϵ_0 away from both plates

61. (c) $f = (1/2\pi)(k_e e^2/m_e R^3)^{1/2}$ (d) 102 pm

65. $\mathbf{E} = a/2\epsilon_0$ radially outward

69. (b) $\mathbf{g} = GM_E r/R_E^3$ radially inward

Chapter 25

1. 1.35 MJ

3. (a) 152 km/s (b) 6.49 Mm/s

5. -0.502 V

7. 1.67 MN/C

9. -38.9 V; the origin

11. (a) 0.500 m (b) 0.250 m (c) 1.26 s (d) 0.343 m

13. 40.2 kV

15. 0.300 m/s

17. (a) $\mathbf{F} = 0$ (b) $\mathbf{E} = 0$ (c) 45.0 kV

19. (a) -27.3 eV (b) -6.81 eV (c) 0

21. -11.0 MV

23. $-5k_e q/R$

27. (a) 10.8 m/s and 1.55 m/s (b) larger

29. $0.720 \text{ m}, 1.44 \text{ m}, 2.88 \text{ m}$. No. The radii of the equipotentials are inversely proportional to the potential.

31. 27.4 fm

33. -3.96 J

35. $22.8k_e q^2/s$

37. $\mathbf{E} = (-5 + 6xy)\mathbf{i} + (3x^2 - 2z^2)\mathbf{j} - 4yz\mathbf{k}$; 7.08 N/C

39. $E_y = \dfrac{k_e Q}{\ell y}\left[1 - \dfrac{y^2}{\ell^2 + y^2 + \ell\sqrt{\ell^2 + y^2}}\right]$

41. $-0.553k_e Q/R$

43. (a) C/m^2 (b) $k_e \alpha[L - d\ln(1 + L/d)]$

45. $(\sigma/2\epsilon_0)[(x^2 + b^2)^{1/2} - (x^2 + a^2)^{1/2}]$

47. 1.56×10^{12} electrons removed

49. (a) $0, 1.67 \text{ MV}$ (b) 5.85 MN/C away, 1.17 MV
(c) 11.9 MN/C away, 1.67 MV

51. (a) 450 kV (b) $7.50 \text{ }\mu\text{C}$

53. 253 MeV

57. (a) 6.00 m (b) $-2.00 \text{ }\mu\text{C}$

59. (a) $\dfrac{k_e Q}{h}\ln\left(\dfrac{d + h + \sqrt{(d + h)^2 + R^2}}{d + \sqrt{d^2 + R^2}}\right)$

(b) $\dfrac{k_eQ}{R^2h}\left[(d+h)\sqrt{(d+h)^2+R^2}-d\sqrt{d^2+R^2}\right.$

$\left.+\,R^2\ln\!\left(\dfrac{d+h+\sqrt{(d+h)^2+R^2}}{d+\sqrt{d^2+R^2}}\right)-2dh-h^2\right]$

61. $k_eQ^2/2R$

63. $V_2-V_1=(-\lambda/2\pi\epsilon_0)\ln(r_2/r_1)$

65. (a) The fields are radial. $E_A=0$; $E_B=(89.9\text{ V/m})/r^2$;
$E_C=(-45\text{ V/m})/r^2$
(b) $V_A=150$ V; $V_B=-450$ V $+\,89.9$ V/r; $V_C=-45$ V/r

67. (a) 1.00 kV $-\,(1.41$ kV/m$)\,x-(1.44$ kV$)\ln\!\left(1-\dfrac{x}{3\text{ m}}\right)$
(b) $+633$ nJ

69. (a) $E_r=2k_ep\cos\theta/r^3$; $E_\theta=k_ep\sin\theta/r^3$; yes; no
(b) $V=k_epy(x^2+y^2)^{-3/2}$; $\mathbf{E}=3k_epxy\,(x^2+y^2)^{-5/2}\,\mathbf{i}$
$+\,k_ep(2y^2-x^2)(x^2+y^2)^{-5/2}\,\mathbf{j}$

71. $V=\pi k_eC\left[R\sqrt{x^2+R^2}+x^2\ln\!\left(\dfrac{x}{R+\sqrt{x^2+R^2}}\right)\right]$

73. (a) 1.42 mm   (b) 9.20 kV/m

Chapter 26

1. (a) $48.0\ \mu$C   (b) $6.00\ \mu$C
3. (a) $1.33\ \mu$C/m^2 (b) 13.3 pF
5. (a) $5.00\ \mu$C on the larger and $2.00\ \mu$C on the smaller
sphere   (b) 89.9 kV
7. (a) 11.1 kV/m toward the negative plate
(b) 98.3 nC/m^2 (c) 3.74 pF   (d) 74.8 pC
9. $4.42\ \mu$m
11. (a) 2.68 nF   (b) 3.02 kV
13. 1.23 kV
15. (a) 15.6 pF   (b) 256 kV
17. (a) $17.0\ \mu$F   (b) 9.00 V   (c) $45.0\ \mu$C and $108\ \mu$C
19. 6.00 pF and 3.00 pF
21. (a) $5.96\ \mu$F   (b) $89.5\ \mu$C on the 20-μF capacitor,
$63.2\ \mu$C on the 6-μF capacitor, and $26.3\ \mu$C on the 15-μF
and 3-μF capacitors
23. $120\ \mu$C; $80.0\ \mu$C and $40.0\ \mu$C
25. (a) $400\ \mu$C   (b) 2.50 kN/m
27. 10
29. $83.6\ \mu$C
31. (a) $216\ \mu$J   (b) $54.0\ \mu$J
33. Stored energy doubles.
37. 9.79 kg
39. 1.40 fm
41. (a) 8.13 nF   (b) 2.40 kV
43. 1.04 m
45. (a) $4.00\ \mu$F   (b) $8.40\ \mu$F   (c) 5.71 V and $48.0\ \mu$C
47. $4\pi\kappa_1\kappa_2abc\epsilon_0/[\kappa_2bc-\kappa_1ab+(\kappa_1-\kappa_2)ac]$
49. 22.5 V
51. (b) -8.78 MN/C$\cdot$m; $-55.3\mathbf{i}$ mN
57. 0.188 m^2
59. $\epsilon_0A/(s-d)$
61. $1+q/q_0$
63. (a) $Q_0^2d(\ell-x)/(2\ell^3\epsilon_0)$ (b) $Q_0^2d/(2\ell^3\epsilon_0)$ to the
right   (c) $Q_0^2/(2\ell^4\epsilon_0)$ (d) $Q_0^2/(2\ell^4\epsilon_0)$
65. $4.29\ \mu$F

67. $3.00\ \mu$F
69. (b) $Q/Q_0=\kappa$
71. $2/3$
73. 19.0 kV
75. $3.00\ \mu$F

Chapter 27

1. 7.50×10^{15} electrons
3. (a) $0.632\ I_0\tau$ (b) $0.999\ 95\ I_0\tau$ (c) $I_0\tau$
5. 400 nA
7. (a) 17.0 A   (b) 85.0 kA/m^2
9. (a) 99.5 kA/m^2 (b) 0.800 cm
11. (a) 2.55 A/m^2 (b) 5.31×10^{10} m^{-3} (c) 1.20×10^{10} s
13. 500 mA
15. 6.43 A
17. (a) 1.82 m   (b) $280\ \mu$m
19. (a) 777 nΩ (b) $3.28\ \mu$m/s
21. $1.56\ R$
23. $6.00\times10^{-15}/\Omega\cdot$m
25. 0.180 V/m
27. 21.2 nm
29. $1.44\times10^3\ ^\circ$C
31. (a) 31.5 n$\Omega\cdot$m   (b) 6.35 MA/m^2 (c) 49.9 mA
(d) $659\ \mu$m/s   (e) 0.400 V
33. 0.125
35. 67.6°C
37. 5.00 A; $24.0\ \Omega$
39. $28.9\ \Omega$
41. 36.1%
43. (a) 5.97 V/m   (b) 74.6 W   (c) 66.1 W
45. 0.833 W
47. 26.9 cents/day
49. (a) 184 W   (b) 461°C
51. $\sim\$1$
53. 25.5 yr
57. Experimental resistivity $=1.47\ \mu\Omega\cdot$m $\pm\,4\%$,
in agreement with $1.50\ \mu\Omega\cdot$m
59. (a) $8.00\mathbf{i}$ V/m   (b) $0.637\ \Omega$ (c) 6.28 A
(d) $200\mathbf{i}$ MA/m^2
61. $2\ 020^\circ$C
63. (a) 667 A   (b) 50.0 km
65.

Material	$\alpha'=\alpha/(1-20\alpha)$
Silver	$4.1\times10^{-3}/^\circ$C
Copper	$4.2\times10^{-3}/^\circ$C
Gold	$3.6\times10^{-3}/^\circ$C
Aluminum	$4.2\times10^{-3}/^\circ$C
Tungsten	$4.9\times10^{-3}/^\circ$C
Iron	$5.6\times10^{-3}/^\circ$C
Platinum	$4.25\times10^{-3}/^\circ$C
Lead	$4.2\times10^{-3}/^\circ$C
Nichrome	$0.4\times10^{-3}/^\circ$C
Carbon	$-0.5\times10^{-3}/^\circ$C
Germanium	$-24\times10^{-3}/^\circ$C
Silicon	$-30\times10^{-3}/^\circ$C

67. No. The fuses should pass no more than 3.87 A.

71. The graphs are as follows:

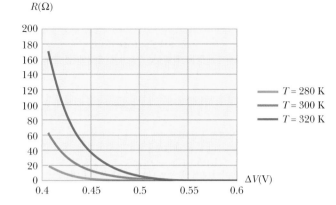

R(Ω)

T = 280 K
T = 300 K
T = 320 K

ΔV(V)

65. Place in parallel with the galvanometer a branch consisting of three resistors in series, with contacts between them as follows:

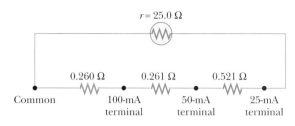

$r = 25.0\ \Omega$

$0.260\ \Omega$ $0.261\ \Omega$ $0.521\ \Omega$

Common   100-mA terminal   50-mA terminal   25-mA terminal

67. (a) 0 in the 3-kΩ resistor and 333 μA in the others
   (b) 50.0 μC   (c) $278\ \mu A e^{-t/180\ \text{ms}}$ (d) 290 ms

71. 48.0 W

73. (a) $\mathcal{E}^2/3R$ (b) $3\mathcal{E}^2/R$ (c) In the parallel connection

Chapter 28

1. (a) 6.73 Ω   (b) 1.97 Ω

3. (a) 4.59 Ω   (b) 8.16%

5. 12.0 Ω

7. He can put the three resistors in parallel.

9. (a) 227 mA   (b) 5.68 V

11. (a) 75.0 V   (b) 25.0 W, 6.25 W, and 6.25 W; 37.5 W

13. 1.00 kΩ

15. 14.3 W, 28.5 W, 1.33 W, 4.00 W

17. (a) The 11-Ω resistor   (b) 148 W = 148 W
   (c) The 22-Ω resistor   (d) 33.0 W = 33.0 W
   (e) The parallel configuration

19. 846 mA down in the 8-Ω resistor; 462 mA down in the middle branch; 1.31 A up in the right-hand branch

21. (a) − 222 J and 1.88 kJ   (b) 687 J, 128 J, 25.6 J, 616 J, 205 J   (c) 1.66 kJ

23. 50.0 mA from *a* to *e*

25. Starter, 171 A; battery, 0.283 A

27. (a) 909 mA   (b) − 1.82 V

29. (a) 5.00 s   (b) 150 μC   (c) 4.06 μA

31. $U_0/4$

33. (a) 6.00 V   (b) 8.29 μs

35. 1.60 MΩ

37. 0.982 s

39. 16.6 kΩ

41. 0.302 Ω

43. 0.588 A

45. 1.36 V

47. (a) Heater, 12.5 A; toaster, 6.25 A; grill, 8.33 A
   (b) No; together they would require 27.1 A.

49. 15.5 A

51. 2.22 h

53. 4.00 V, with *a* at higher potential

55. 6.00 Ω; 3.00 Ω

57. (a) $R \to \infty$ (b) $R \to 0$ (c) $R = r$

59. (a) $R \le 1050\ \Omega$ (b) $R \ge 10.0\ \Omega$

61. (a) 9.93 μC   (b) 33.7 nA   (c) 334 nW   (d) 337 nW

63. (a) 40.0 W   (b) 80.0 V, 40.0 V, 40.0 V

Chapter 29

1. (a) up   (b) toward you, out of the plane of the paper
   (c) no deflection   (d) into the plane of the paper

3. negative *z* direction

5. $(- 20.9\mathbf{j})$ mT

7. 8.93×10^{-30} N down, 1.60×10^{-17} N up, 4.80×10^{-17} N down

9. 48.8° or 131°

11. 2.34 aN

13. 0.245 T east

15. (a) 4.73 N   (b) 5.46 N   (c) 4.73 N

17. 196 A east

19. 1.07 m/s

21. $2\pi r I B \sin\theta$ up

23. (a) 5.41 mA·m²   (b) 4.33 mN·m

25. 9.98 N·m; clockwise as seen looking down from above

27. (a) 80.1 mN·m   (b) 0.104 N·m   (c) 0.132 N·m
   (d) The circular loop experiences largest torque.

29. (a) minimum: with its north end pointing north at 48.0° below the horizontal; maximum: with its north end pointing south at 48.0° above the horizontal   (b) 1.07 μJ

31. (a) 49.7 aN south   (b) 1.29 km

33. 115 keV

35. $r_\alpha = r_d = \sqrt{2}r_p$

37. 4.99×10^8 rad/s

39. 7.88 pT

41. 244 kV/m

43. 0.278 m

45. (a) 4.31×10^7 rad/s   (b) 51.7 Mm/s

47. 70.1 mT

49. 3.70×10^{-9} m³/C

51. 43.2 μT

53. (a) 179 ps   (b) 351 keV

55. (a) The electric current experiences a magnetic force.

57. (a) B_x is indeterminate; B_y is zero; B_z is $-F_i/ev_i$.
   (b) $-F_i\mathbf{j}$ (c) $-F_i\mathbf{j}$

59. (a) $(3.52\mathbf{i} - 1.60\mathbf{j})$ aN   (b) 24.4°

61. 0.588 T

63. 19.6 mT
65. 438 kHz
67. 3.70×10^{-24} N·m
69. (a) 0.501 m   (b) 45.0°
71. (a) 1.33 m/s   (b) No. Positive ions moving toward you in the magnetic field to the right feel an upward magnetic force and migrate upward in the blood vessel. Negative ions moving toward you feel a downward magnetic force and accumulate at the bottom of this section of the vessel. Thus, both species can participate in the generation of the emf.

Chapter 30

1. 12.5 T
3. (a) 28.3 μT into the page   (b) 24.7 μT into the page
5. $\mu_0 I/4\pi x$ into the page
7. 58.0 μT into the page
9. 26.2 μT into the page
11. $\dfrac{\mu_0 I}{12}\left(\dfrac{1}{a} - \dfrac{1}{b}\right)$ out of the page
13. $0.475\mu_0 I/R$ into the page
15. $-13.0\mathbf{j}$ μT
17. $-27.0\mathbf{i}$ μN
19. 20.0 μT toward the bottom of the page
21. 200 μT toward the top of the page; 133 μT toward the bottom of the page
23. (a) 3.60 T   (b) 1.94 T
25. (a) 6.34 mN/m inward   (b) greater
27. (a) $\mu_0 b r_1^2/3$ (b) $\mu_0 b R^3/3r_2$
29. 31.8 mA
31. 464 mT
33. (a) 3.13 mWb   (b) 0
35. (a) $B\pi R^2 \cos\theta$ (b) $B\pi R^2 \cos\theta$
37. (a) 11.3 GV·m/s   (b) 0.100 A
39. 0.191 T
41. 150 μT·m^2
43. 2.62 MA/m
45. (b) 6.45×10^4 K·A/T·m
47. (a) 8.63×10^{45} electrons   (b) 4.01×10^{20} kg
49. 2.00 GA west
51. ~10^{-5} T or 10^{-6} T; on the order of one-tenth as large
53. $\dfrac{\mu_0 I}{2\pi w}\ln\left(1 + \dfrac{w}{b}\right)\mathbf{k}$
55. 143 pT away along the axis
59. (a) 2.46 N up   (b) 107 m/s^2 up
61. (a) 274 μT   (b) $-274\mathbf{j}$ μT   (c) $1.15\mathbf{i}$ mN
    (d) $0.384\mathbf{i}$ m/s^2 (e) acceleration is constant.
    (f) $0.999\mathbf{i}$ m/s
63. (a) $\mu_0\sigma v/2$ (b) out of the plane of the page, parallel to the roller axes
65. 28.8 mT
67. $4\sqrt{2}/\pi^2$
71. $\dfrac{\mu_0 I}{4\pi}\left(1 - e^{-2\pi}\right)$ out of the plane of the page

73. $\rho\mu_0\omega R^2/3$
75. (a) $\dfrac{\mu_0 I}{\pi r}\dfrac{(2r^2 - a^2)}{(4r^2 - a^2)}$ to the left

    (b) $\dfrac{\mu_0 I}{\pi r}\dfrac{(2r^2 + a^2)}{(4r^2 + a^2)}$ toward the top of the page

Chapter 31

1. 500 mV
3. 9.82 mV
5. 160 A
7. (a) 1.60 A counterclockwise   (b) 20.1 μT   (c) up
9. (a) $(\mu_0 IL/2\pi)\ln(1 + w/h)$ (b) -4.80 μV; current is counterclockwise
11. 283 μA upward
13. $(68.2$ mV$)e^{-1.6t}$ counterclockwise
15. 272 m
17. $(0.422$ V$)\cos\omega t$
19. 0.880 C
21. (a) 3.00 N to the right   (b) 6.00 W
23. 0.763 V with the left-hand wingtip positive
25. 2.83 mV
27. (a) $F = N^2 B^2 w^2 v/R$ to the left   (b) 0
    (c) $F = N^2 B^2 w^2 v/R$ to the left
29. negative
31. 145 μA
33. 1.80 mN/C upward and to the left, perpendicular to r_1
35. (a) $(9.87$ mV/m$)\cos(100\pi t)$ (b) clockwise
37. (a) 7.54 kV   (b) The plane of the coil is parallel to **B**.
39. $(28.6$ mV$)\sin(4\pi t)$
41. (a) 0.640 N·m   (b) 241 W
43. (a) $(8.00$ mT·m$^2)\cos(377t)$ (b) $(3.02$ V$)\sin(377t)$
    (c) $(3.02$ A$)\sin(377t)$ (d) $(9.10$ W$)\sin^2(377t)$
    (e) $(24.1$ mN·m$)\sin^2(377t)$
45. (b) Larger R makes current smaller, so the loop must travel faster to maintain equality of magnetic force and weight. (c) The magnetic force is proportional to the product of the magnetic field and current, while the current is proportional to the magnetic field. If B is halved, the speed must be quadrupled to compensate.
47. $(-2.87\mathbf{j} + 5.75\mathbf{k})$ Gm/s^2
49. -7.22 mV $\cos(2\pi\,523\,t/s)$
51. (a) 43.8 A   (b) 38.3 W
53. (a) 3.50 A and 1.40 A   (b) 34.3 W   (c) 4.29 N
57. 1.20 μC
59. (a) 0.900 A   (b) 0.108 N   (c) b (d) no
61. (a) $a\pi r^2$ (b) $-b\pi r^2$ (c) $-b\pi r^2/R$ (d) $b^2\pi^2 r^4/R$
63. (a) 36.0 V   (b) 600 mWb/s   (c) 35.9 V   (d) 4.32 N·m
65. -10.2 μV
67. $\mu_0 I\ell vw/2\pi Rr(r + w)$
69. 6.00 A
71. (a) $(1.19$ V$)\cos(120\pi t)$ (b) 88.5 mW
73. $(-87.1$ mV$)\cos(200\pi t + \phi)$
75. -6.75 V

Chapter 32

1. 19.5 mV
3. 100 V
5. $240\ \text{nT}\cdot\text{m}^2$
7. $(18.8\ \text{V})\cos(377t)$
9. $-0.421\ \text{A/s}$
11. (a) 360 mV   (b) 180 mV   (c) 3.00 s
13. (a) 15.8 μH   (b) 12.6 mH
15. $\mathcal{E}_0/k^2\,L$
17. (a) 0.139 s   (b) 0.461 s
19. (a) 2.00 ms   (b) 0.176 A   (c) 1.50 A   (d) 3.22 ms
21. (a) 0.800   (b) 0
23. (a) 6.67 A/s   (b) 0.332 A/s
25. $(500\ \text{mA})(1 - e^{-10t/s})$, $1.50\ \text{A} - (0.25\ \text{A})e^{-10t/s}$
27. 0 for $t < 0$; $(10\ \text{A})(1 - e^{-10\,000t})$ for $0 < t < 200\ \mu$s; $(63.9\ \text{A})e^{-10\,000t}$ for $t > 200\ \mu$s
29. (a) 5.66 ms   (b) 1.22 A   (c) 58.1 ms
31. 0.064 8 J
33. 2.44 μJ
35. $44.2\ \text{nJ/m}^3$ for the **E** field and $995\ \mu\text{J/m}^3$ for the **B** field
37. (a) 20.0 W   (b) 20.0 W   (c) 0   (d) 20.0 J
39. $2\pi B_0^2 R^3/\mu_0 = 2.70 \times 10^{18}\ \text{J}$
41. 1.00 V
43. (a) 18.0 mH   (b) 34.3 mH   (c) -9.00 mV
45. (b) 3.95 nH
47. $(L_1L_2 - M^2)/(L_1 + L_2 - 2M)$
49. 20.0 V
51. 608 pF
53. (a) 135 Hz   (b) 119 μC   (c) -114 mA
55. (a) 6.03 J   (b) 0.529 J   (c) 6.56 J
57. (a) 4.47 krad/s   (b) 4.36 krad/s   (c) 2.53%
59. $0.693(2L/R)$ (b) $0.347(2L/R)$
61. (a) -20.0 mV   (b) $-(10.0\ \text{MV/s}^2)t^2$ (c) 63.2 μs
63. $\dfrac{Q}{2N}\sqrt{\dfrac{3L}{C}}$
65. (a) $L \approx (\pi/2)N^2\mu_0 R$ (b) ~ 100 nH   (c) ~ 1 ns
71. (a) 72.0 V; b
(b)

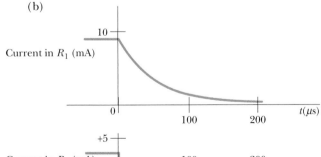

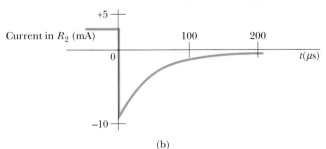

(b)

73. 300 Ω
(c) 75.2 μs
75. (a) It creates a magnetic field.   (b) The long narrow rectangular area between the conductors encloses all of the magnetic flux.
77. (a) 62.5 GJ   (b) 2 000 N
79. (a) 2.93 mT up   (b) 3.42 Pa   (c) clockwise   (d) up   (e) 1.30 mN

Chapter 33

1. $\Delta v(t) = (283\ \text{V})\sin(628t)$
3. 2.95 A, 70.7 V
5. 14.6 Hz
7. 3.38 W
9. (a) 42.4 mH   (b) 942 rad/s
11. 5.60 A
13. $0.450\ \text{T}\cdot\text{m}^2$
15. (a) 141 mA   (b) 235 mA
17. 100 mA
19. (a) 194 V   (b) current leads by 49.9°
21. (a) 78.5 Ω (b) 1.59 kΩ (c) 1.52 kΩ (d) 138 mA   (e) $-84.3°$
23. (a) 17.4°   (b) voltage leads the current
25. (a) 146 V   (b) 213 V   (c) 179 V   (d) 33.4 V
27. (a) 124 nF   (b) 51.5 kV
29. 8.00 W
31. (a) 16.0 Ω (b) $-12.0\ \Omega$
33. 132 mm
35. $11(\Delta V)^2/14R$
37. 1.82 pF
39. 242 mJ
41. 0.591 and 0.987; the circuit in Problem 23
43. 687 V
45. 87.5 Ω
47. (a) 29.0 kW   (b) 5.80×10^{-3} (c) If the generator were limited to 4 500 V, no more than 17.5 kW could be delivered to the load, never 5 000 kW.
49. (a) 613 μF   (b) 0.756
51. (a) 580 μH and 54.6 μF   (b) 1   (c) 894 Hz
(d) ΔV_{out} leads ΔV_{in} by 60.0° at 200 Hz; ΔV_{out} and ΔV_{in} are in phase at 894 Hz; ΔV_{out} lags ΔV_{in} by 60.0° at 4 000 Hz.   (e) 1.56 W, 6.25 W, 1.56 W   (f) 0.408
53. 0.317
55. 56.7 W
57. (a) 225 mA   (b) 450 mA
59. (a) Circuit (a) is a high-pass filter, and circuit (b) is a low-pass filter.
(b) $\dfrac{\Delta V_{\text{out}}}{\Delta V_{\text{in}}} = \dfrac{(R_L^2 + X_L^2)^{1/2}}{[R_L^2 + (X_L - X_C)^2]^{1/2}}$ for circuit (a)
$\dfrac{\Delta V_{\text{out}}}{\Delta V_{\text{in}}} = \dfrac{X_C}{[R_L^2 + (X_L - X_C)^2]^{1/2}}$ for circuit (b)
61. $\sim 10^2$ or 10^3 A
63. (a) 200 mA; voltage leads by 36.8°   (b) 40.0 V; $\phi = 0°$
(c) 20.0 V; $\phi = -90.0°$ (d) 50.0 V; $\phi = +90.0°$
65. (b) 31.6

67. (a) 919 Hz   (b) 1.50 A, 1.60 A, 6.73 mA   (c) 2.19 A
(d) current lagging by $\phi = -46.7°$

69. (a)

f (Hz)	X_L (Ω)	X_C (Ω)	Z (Ω)
300	282	12 600	12 300
600	565	6 290	5 720
800	754	4 710	3 960
1 000	942	3 770	2 830
1 500	1 410	2 510	1 100
2 000	1 880	1 880	40.0
3 000	2 830	1 260	1 570
4 000	3 770	942	2 830
6 000	5 660	629	5 030
10 000	9 420	377	9 040

(b)

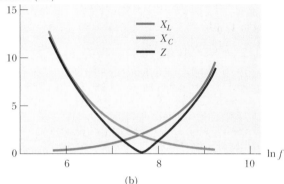

(b)

71. (a) 1.84 kHz
(b)

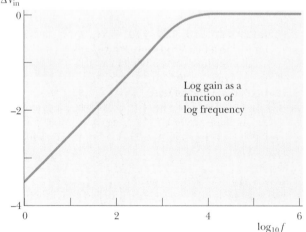

Log gain as a function of log frequency

Chapter 34

1. 2680 A.D.
3. 733 nT
5. (a) 6.00 MHz   (b) 73.3 nT $(-\mathbf{k})$
(c) $\mathbf{B} = (-73.3 \text{ nT}) \cos(0.126x - 3.77 \times 10^7 t)\mathbf{k}$

7. (a) 0.333 μT   (b) 0.628 μm   (c) 477 THz
9. 75.0 MHz
11. 3.33 μJ/m^3
13. 307 μW/m^2
15. 3.33×10^3 m^2
17. (a) 332 kW/m^2 radially inward   (b) 1.88 kV/m and 222 μT
19. 29.5 nT
21. (a) 2.33 mT   (b) 650 MW/m^2 (c) 510 W
23. (a) 4.97 kW/m^2 (b) 16.6 μJ/m^3
25. 83.3 nPa
27. (a) 5.36 N   (b) 8.93×10^{-4} m/s^2 (c) 10.7 days
29. (a) 1.90 kN/C   (b) 50.0 pJ   (c) 1.67×10^{-19} kg·m/s
31. (a) 11.3 kJ   (b) 1.13×10^{-4} kg·m/s
33. (a) 2.26 kW   (b) 4.71 kW/m^2
35. (a) away along the perpendicular bisector of the line segment joining the antennas   (b) along the extensions of the line segment joining the antennas
37. 56.2 m
39. (a) radio, radio, radio, microwave, infrared, ultraviolet, x-ray, γ-ray, γ-ray;   (b) radio, radio, microwave, infrared, ultraviolet or x-ray, x- or γ-ray, γ-ray, γ-ray
41. 545 THz
43. (a) 6.00 pm   (b) 7.50 cm
45. The radio audience hears it first, 8.41 ms before the people in the newsroom.
47. (a) 186 m to 556 m   (b) 2.78 m to 3.41 m
49. (a) 3.77×10^{26} W   (b) 1.01 kV/m and 3.35 μT
51. (a) $2\pi^2 r^2 f B_{max} \cos \theta$, where θ is the angle between the magnetic field and the normal to the loop   (b) The loop should be in the vertical plane that contains the line of sight to the transmitter.
53. (a) 6.67×10^{-16} T   (b) 5.31×10^{-17} W/m^2
(c) 1.67×10^{-14} W   (d) 5.56×10^{-23} N
55. (a) $\frac{1}{2}\mu_0 n^2 r i \dfrac{di}{dt}$ radially inward   (b) $\mu_0 \pi n^2 r^2 \ell\, i \dfrac{di}{dt}$
(c) $(\Delta V) i$
57. 95.1 mV/m
59. (a) $B_{max} = 583$ nT, $k = 419$ rad/m, $\omega = 12.6$ Trad/s, the xz plane   (b) $S_{av} = \frac{1}{2}S_{max} = 40.6$ W/m^2 (c) 271 nPa
(d) 406 nm/s^2
61. (a) 22.6 h   (b) 30.6 s
63. (a) 8.32×10^7 W/m^2 (b) 1.05 kW
65. (a) 625 kW/m^2 (b) 21.7 kN/C, 72.4 μT   (c) 17.8 min
67. (b) 17.6 Tm/s^2, 1.75×10^{-27} W   (c) 1.80×10^{-24} W
69. 3.00×10^{-2} degrees
71. $\epsilon_0 E^2 A / 2m$

Chapter 35

1. 299.5 Mm/s
3. 114 rad/s
5. 25.5°, 442 nm
7. 19.5° above the horizon
9. (a) 181 Mm/s   (b) 225 Mm/s   (c) 136 Mm/s
11. 30.0° and 19.5° at entry; 19.5° and 30.0° at exit

13. 3.39 m

15. six times from mirror 1 and five times from mirror 2

17. 106 ps

19. 6.39 ns

21. (a) 58.9°   (b) only if $\theta_1 = 0$

23. (a) 66.8 μs   (b) 0.250% longer

25. 0.171°

27. 86.8°

29. 4.61°

31. 27.9°

33. 18.4°

35. (a) 24.4°   (b) 37.0°   (c) 49.8°

37. 1.000 08

39. 62.4°

41. 1.08 cm $< d <$ 1.17 cm

43. Skylight incident from above travels down the plastic. If the index of refraction of the plastic is greater than 1.41, the rays close in direction to the vertical are totally reflected from the side walls of the slab and from both facets at the bottom of the plastic, where it is not immersed in gasoline. This light returns up inside the plastic and makes it look bright. Where the plastic is immersed in gasoline, total internal reflection is frustrated, and the downward-propagating light passes from the plastic out into the gasoline. Little light is reflected upward, and the gauge looks dark.

45. 77.5°

47. 2.27 m

49. (a) 0.172 mm/s   (b) 0.345 mm/s   (c) Northward at 50.0° below the horizontal   (d) Northward at 50.0° below the horizontal

51. (a) 1.76×10^7 (b) 3.25×10^{-6} degree

53. 62.2%

55. 82 reflections

57. (b) 68.5%

59. 27.5°

61. (a) It always happens.   (b) 30.3°   (c) It cannot happen.

63. 2.37 cm

67. (a) $n = [1 + (4t/d)^2]^{1/2}$ (b) 2.10 cm   (c) violet

69. (a) 1.20   (b) 3.40 ns

Chapter 36

1. $\sim 10^{-9}$ s younger

3. 2′11″

5. 10.0 ft, 30.0 ft, 40.0 ft

7. 0.267 m behind the mirror; virtual, upright, diminished; $M = 0.026\,7$

9. (a) -12.0 cm; 0.400   (b) -15.0 cm; 0.250   (c) upright

11. (a) $q = 45.0$ cm; $M = -0.500$ (b) $q = -60.0$ cm; $M = 3.00$ (c) Image (a) is real, inverted, and diminished. Image (b) is virtual, upright, and enlarged. The ray diagrams are like Figures 36.15a and 36.15b, respectively.

13. (a) a concave mirror with radius of curvature 2.08 m   (b) 1.25 m from the object

15. (a) 15.0 cm   (b) 60.0 cm

17. (a) A real image moves from $+0.600$ m to infinity, then a virtual image moves from $-\infty$ to 0.   (b) 0.639 s and 0.782 s

19. 38.2 cm below the top surface of the ice

21. 8.57 cm

23. (a) 45.0 cm   (b) -90.0 cm   (c) -6.00 cm

25. inside the bowl at -9.01 cm

27. (a) 16.4 cm   (b) 16.4 cm

29. (a) 650 cm from the lens on the opposite side from the object; real, inverted, enlarged   (b) 600 cm from the lens on the same side as the object; virtual, upright, enlarged

31. 2.84 cm

33. (a) either 9.63 cm or 3.27 cm   (b) 2.10 cm

37. (a) -12.3 cm, to the left of the lens   (b) 0.615   (c)

39. (a) 4.00 m and 1.00 m   (b) Whereas both images are real and inverted, the first has magnification -0.250, and the second, -4.00.

41. 2.18 mm away from the film plane

43. 21.3 cm

45. -4.00 diopters; a diverging lens

47. 3.50

49. -575

51. (a) -800 (b) upside down

53. (a) virtual   (b) at infinity   (c) 15.0 cm, -5.00 cm

55. -40.0 cm

57. 160 cm to the left of the lens; inverted; $M = -0.800$

59. 25.3 cm to the right of the mirror; virtual; upright; enlarged 8.05 times

61. 0.107 m to the right of the vertex of the hemispherical face

63. 8.00 cm

65. 1.50 m in front of the mirror; 1.40 cm (inverted)
67. (a) 30.0 cm and 120 cm   (b) 24.0 cm   (c) real, inverted, diminished, with $M = -0.250$
69. -75.0
71. (a) 44.6 diopters   (b) 3.03 diopters
73. (a) 20.0 cm to the right of the second lens; $M = -6.00$
(b) inverted   (c) 6.67 cm to the right of the second lens; $M = -2.00$; inverted

Chapter 37

1. 1.58 cm
3. (a) 55.7 m   (b) 124 m
5. 1.54 mm
7. (a) 2.62 mm   (b) 2.62 mm
9. 11.3 m
11. (a) 13.2 rad   (b) 6.28 rad   (c) 0.012 7 degree
(d) 0.059 7 degree
13. (a) 1.93 μm   (b) 3.00 λ (c) maximum
15. 48.0 μm
17. (a) 7.95 rad   (b) 0.453
19. (a) and (b) 19.7 kN/C at 35.0°   (c) 9.36 kN/C at 169°
21. $10.0 \sin(100\pi t + 0.927)$
23. $26.2 \sin(\omega t + 36.6°)$
25. $\pi/2$
27. $360°/N$
29. (a) green   (b) violet
31. 0.500 cm
33. No reflection maxima in the visible spectrum
35. 290 nm
37. 4.35 μm
39. 1.20 mm
41. 39.6 μm
43. 1.000 369
45. 1.25 m
47. $\lambda/2(n-1)$
49. 5.00 km^2
51. 3.58°
53. 421 nm
55. 113 dark fringes
59. (a) $2(4h^2 + d^2)^{1/2} - 2d$ (b) $(4h^2 + d^2)^{1/2} - d$
61. $y' = (n-1)tL/d$
63. (a) 70.6 m   (b) 136 m
65. $7.99 \sin(\omega t + 4.44 \text{ rad})$
69. 0.505 mm

Chapter 38

1. 4.22 mm
3. 0.230 mm
5. Three maxima, at 0° and near 46° to the left and right
7. 0.016 2
11. 1.00 mrad
13. 3.09 m
15. 13.1 m

17. 1.90 m if the predominant wavelength is 650 nm
19. 105 m
21. 2.10 m
23. 7.35°
25. 5.91° in first order, 13.2° in second order, 26.5° in third order
27. (a) 478.7 nm, 647.6 nm, and 696.6 nm   (b) 20.51°, 28.30°, and 30.66°
29. (a) 12 000, 24 000, 36 000   (b) 11.1 pm
31. (a) 2 800 lines   (b) 4.72 μm
33. (a) 5 orders   (b) 10 orders in the short-wavelength region
35. 93.4 pm
37. 14.4°
39. 31.9°
41. 3/8
43. (a) 54.7°   (b) 63.4°   (c) 71.6°
45. 60.5°
47. 36.9° above the horizon
49. (a) 6   (b) 7.50°
51. 632.8 nm
53. (a) 25.6°   (b) 19.0°
55. 0.244 rad = 14.0°
57. (a) 3.53×10^3 lines per centimeter   (b) 11 maxima
59. 15.4
61. (a) 41.8°   (b) 0.593   (c) 0.262 m
63. (b) 15.3 μm
65. $a = 99.5$ μm $\pm 1\%$

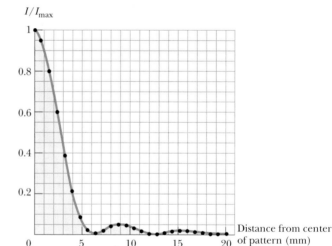

67. $\phi = 1.391\ 557\ 4$ after 17 steps or fewer
69. (b) 0.428 mm

Chapter 39

5. $0.866c$
7. 64.9/min; 10.6/min
9. 1.54 ns
11. $0.800c$

13. (a) 2.18 μs   (b) 649 m

15. 0.789c

17. (a) 20.0 m   (b) 19.0 m   (c) 0.312c

19. 1.13 × 10^4 Hz

21. (b) 0.050 4c

23. 0.960c

25. 0.357c

27. (a) 2.73 × 10^{-24} kg·m/s   (b) 1.58 × 10^{-22} kg·m/s
(c) 5.64 × 10^{-22} kg·m/s

29. 4.50 × 10^{-14}

31. 0.285c

33. 1.63 × 10^3 MeV/c

35. (a) 939 MeV   (b) 3.01 GeV   (c) 2.07 GeV

37. 18.4 g/cm^3

41. (a) 0.302c (b) 4.00 fJ

43. 3.88 MeV and 28.8 MeV

45. 3.18 × 10^{-12} kg, not detectable

47. 0.842 kg

49. 4.19 × 10^9 kg/s

53. (a) a few hundred seconds   (b) ~ 10^8 km

55. 0.712%

57. (a) 0.946c (b) 0.160 ly   (c) 0.114 yr   (d) 7.50 × 10^{22} J

59. (a) 76.0 min   (b) 52.1 min

61. Yes, with 18.8 m to spare

63. (b) For u small compared with c, the relativistic acceleration agrees with the classical expression. As u approaches
c, the acceleration approaches zero; thus, the object's
speed can never reach or surpass the speed of light.
(c) Perform $\int(1 - u^2/c^2)^{-3/2}du = (qE/m)\int dt$ to
obtain $u = qEct(m^2c^2 + q^2E^2t^2)^{-1/2}$ and then $\int dx =$
$\int qEct(m^2c^2 + q^2E^2t^2)^{-1/2}dt$ to obtain $x =$
$(c/qE)[(m^2c^2 + q^2E^2t^2)^{1/2} - mc]$.

65. (a) 6.67 ks   (b) 4.00 ks

67. (a) 0.800c (b) 7.50 ks   (c) 1.44 Tm, − 0.385c
(d) 12.5 ks

69. (a) 0.544c, 0.866c (b) 0.833 m

71. 0.185c = 55.4 Mm/s

73. 6.71 × 10^8 kg

Chapter 40

1. 5.18 × 10^3 K

3. (a) 999 nm   (b) The infrared region of the spectrum is
much wider than the visible region, and the spectral distribution function is highest in the infrared.

5. (a) 70.9 kW   (b) 580 nm   (c) 7.99 × 10^{10} W/m
(d) 9.42 × 10^{-1226} W/m   (e) 1.00 × 10^{-227} W/m
(f) 5.44 × 10^{10} W/m   (g) 7.38 × 10^{10} W/m
(h) 0.260 W/m   (i) 2.60 × 10^{-9} W/m   (j) ≈20 kW

7. (a) 2.57 eV   (b) 12.8 μeV   (c) 191 neV
(d) 484 nm (visible), 9.68 cm and 6.52 m (radio waves)

9. 2.27 × 10^{30} photons/s

11. 1.34 × 10^{31}

15. (a) 296 nm, 1.01 PHz   (b) 2.71 V

17. (a) only lithium   (b) 0.808 eV

19. (a) 1.90 eV   (b) 0.216 V

21. 148 days; absurdly large

23. (a) The Doppler effect increases the frequency incident
on the metal.   (b) 3.87 eV   (c) 8.78 eV

25. (a) 488 fm   (b) 268 keV   (c) 31.5 keV

27. 70.1°

29. (a) 43.0°   (b) 602 keV, 3.21 × 10^{-22} kg·m/s
(c) 278 keV, 3.21 × 10^{-22} kg·m/s

31. (a) 33.0°   (b) 0.785c

33. (a) 2.88 pm   (b) 101°

37. (a) 5   (b) no; no

39. 634 nm, red

41. (a) 0.212 nm   (b) 9.95 × 10^{-25} kg·m/s
(c) 2.11 × 10^{-34} kg·m^2/s   (d) 3.40 eV   (e) − 6.80 eV
(f) − 3.40 eV

43. (a) 3.03 eV   (b) 410 nm   (c) 732 THz

47. 97.5 nm

49. (a) $E_n = − 54.4$ eV/n^2 for $n = 1, 2, 3, \ldots$
(b) 54.4 eV

53. 397 fm

55. (a) 0.709 nm   (b) 414 nm

57. (a) ~ 100 MeV or more   (b) No. With kinetic energy
much larger than the magnitude of its negative electric
potential energy, the electron would immediately escape.

59. (b) No. $\lambda^{-2} + \lambda_C{}^{-2}$ cannot be equal to λ^{-2}.

61. $c/\sqrt{2} = 212$ Mm/s

63. 1.36 eV

65. (a) See bottom of page.
(b) 6.4 × 10^{-34} J·s ± 8%   (c) 1.4 eV

67. The particles are separated by $r_n = (0.106$ nm $)n^2$, and
$E_n = −6.80$ eV/n^2, for $n = 1, 2, 3, \ldots$

69. The classical frequency is $4\pi^2 m_e k_e{}^2 e^4/h^3 n^3$.

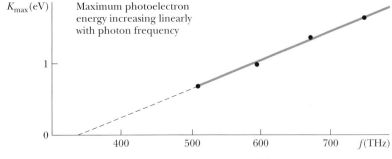

Chapter 40, Problem 65(a)

Chapter 41

1. (a) 993 nm　　(b) 4.97 mm　　(c) If its detection forms part of an interference pattern, the neutron must have passed through both slits. If we test to see which slit a particular neutron passes through, it will not form part of the interference pattern.
3. (b) 907 fm
5. (a) 15.1 keV　(b) 124 keV
7. At $1.00°$ on both sides of the central maximum
9. Within 1.16 mm for the electron, 5.28×10^{-32} m for the bullet
11. 1.16 Mm/s
13. (b) 0.519 fm
15. (a) 126 pm　(b) 5.27×10^{-24} kg·m/s　(c) 95.5 eV
17. (a) n

$$
\begin{array}{ll}
4 \text{————————} & 603 \text{ eV} \\
\\
\\
3 \text{————————} & 339 \text{ eV} \\
\\
2 \text{————————} & 151 \text{ eV} \\
\\
1 \text{————————} & 37.7 \text{ eV}
\end{array}
$$

(b) 2.20 nm, 2.75 nm, 4.12 nm, 4.71 nm, 6.60 nm, 11.0 nm
19. 0.793 nm
21. 202 fm, 6.14 MeV, a gamma ray
23. 0.513 MeV, 2.05 MeV, 4.62 MeV; yes
27. At $L/4$ and at $3L/4$
29. (a) 5.13×10^{-3} eV　(b) 9.41 eV　(c) The electron has much higher energy because it is much less massive.

33. (a) $E = \hbar^2/mL^2$
(b) Requiring $\int_{-L}^{L} A^2(1 - x^2/L^2)^2 \, dx = 1$ gives
$A = (15/16L)^{1/2}$.　(c) $47/81 = 0.580$

35. (a)

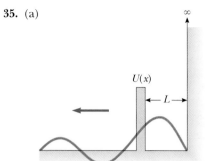

(b) $2L$
37. (a) 0.010 3　(b) 0.990
39. By 0.959 nm, to 1.91 nm
41. 3.92%
43. (b) $b = m\omega/2\hbar$, $E = 3\hbar\omega/2$　(c) first excited state
45. (a) $B = (m\omega/\pi\hbar)^{1/4}$　(b) $(m\omega/\pi\hbar)^{1/2}\delta$
47. $\sim 10^{-10^{30}}$
49. See bottom of page.
(c) The wave function is continuous. It shows localization by approaching zero as $x \rightarrow \pm\infty$. It is everywhere finite and can be normalized.　(d) $A = \sqrt{\alpha}$　(e) 0.632
51. 0.029 4
53. (a) 2.82×10^{-37} m　(b) 1.06×10^{-32} J
(c) $2.87 \times 10^{-35}\%$ or more
55. (a) 434 THz　(b) 691 nm　(c) 165 peV or more
59. (a)

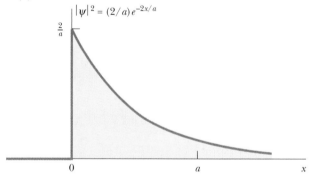

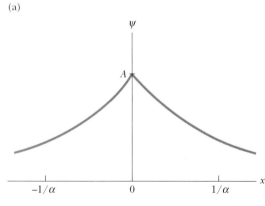

(a)

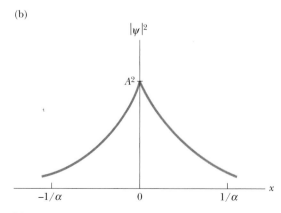

(b)

Chapter 41, Problem 49

(b) 0   (c) 0.865

61. (a) $\Delta p \geq \hbar/2r$
(b) Choosing $p \cong \hbar/r$, $E = \hbar^2/2m_e r^2 - k_e e^2/r$
(c) $r = \hbar^2/m_e k_e e^2 = a_0$ and $E = -13.6$ eV, in agreement with the Bohr theory.

63. (a) $-7k_e e^2/3d$ (b) $\hbar^2/36m_e d^2$ (c) 49.9 pm
(d) Li atom spacing is 280 pm, which is 5.62 times greater than answer (c).

65. (a) $A = (2/17L)^{1/2}$ (b) $|A|^2 + |B|^2 = 1/a$

67. (a) The light is unpolarized. It contains both horizontal and vertical electric field oscillations.   (b) The interference pattern appears, but with diminished overall intensity.   (c) The results are the same in each case.   (d) The interference pattern appears and disappears as the polarizer turns, with alternately increasing and decreasing contrast between the bright and dark fringes. The intensity on the screen is precisely zero at the center of a dark fringe four times in each revolution, when the filter axis has turned by 45°, 135°, 225°, and 315° from the vertical.   (e) Looking at the overall light energy arriving at the screen, we see a low-contrast interference pattern. After we sort out the individual photon runs into those for trial 1, those for trial 2, and those for trial 3, we have the original results replicated: The runs for trials 1 and 2 form the two blue graphs in Figure 41.3, and the runs for trial 3 build up the red graph.

Chapter 42

1. (a) 56.8 fm   (b) 11.3 N
3. (a) 3   (b) 520 km/s
5. (a) 1.31 μm   (b) 164 nm
7. (a)

n	ℓ	m_ℓ	m_s
3	2	2	1/2
3	2	2	$-1/2$
3	2	1	1/2
3	2	1	$-1/2$
3	2	0	1/2
3	2	0	$-1/2$
3	2	-1	1/2
3	2	-1	$-1/2$
3	2	-2	1/2
3	2	-2	$-1/2$

(b)

n	ℓ	m_ℓ	m_s
3	1	1	1/2
3	1	1	$-1/2$
3	1	0	1/2
3	1	0	$-1/2$
3	1	-1	1/2
3	1	-1	$-1/2$

9. (b) 0.497
11. It does, with $E = -k_e e^2/2a_0$.
13. (a) $\sqrt{6}\hbar$ (b) $\sqrt{12}\hbar$
15. $\sqrt{6}\hbar$

17. (a) 2   (b) 8   (c) 18   (d) 32   (e) 50
19. (a) 3.99×10^{17} kg/m^3 (b) 81.7 am   (c) 1.77 Tm/s
(d) 5.91×10^3 c
21. (a) 5.05×10^{-27} J/T $= 31.6$ neV/T   (b) Here μ_n is 1 836 times smaller than μ_B, because a proton is 1 836 times more massive than an electron. The electron has a greater charge-to-mass ratio than any other particle, which gives it a bigger "handle" for a magnetic field to twist.
23. $n = 3$; $\ell = 2$; $m_\ell = -2, -1, 0, 1,$ or 2; $s = 1$; $m_s = -1, 0,$ or 1
25. The $4s$ subshell is filled first. We would expect $[\text{Ar}]3d^4 4s^2$ to have lower energy, but $[\text{Ar}]3d^5 4s^1$ has more unpaired spins and lower energy, as suggested by Hund's rule. It is the ground-state configuration of chromium.
27. (a) Zn or Cu   (b) $1s^2 2s^2 2p^6 3s^2 3p^6 4s^2 3d^{10} 5s^2$ or $1s^2 2s^2 2p^6 3s^2 3p^6 4s^2 3d^{10} 5s^1$
29. (a) $1s$, $2s$, $2p$, $3s$, $3p$, $4s$, $3d$, $4p$, $5s$, $4d$, $5p$, $6s$, $4f$, $5d$, $6p$, $7s$ (b) Element 15 should have valence $+3$ or -5, and it does. Element 47 should have valence -1, but it has valence $+1$. Element 86 should be inert, and it is.
31. (a) $\ell = 0$ and $m_\ell = 0$; or $\ell = 1$ and $m_\ell = -1, 0,$ or 1; or $\ell = 2$ and $m_\ell = -2, -1, 0, 1,$ or 2   (b) -6.05 eV
33. 0.031 0 nm
35. 0.072 5 nm
37. (a) 14 keV   (b) 89 pm
39. (a) 1   (b) 0.69
41. 6.21×10^{-14} J$\cdot$s/m^3
43. 3.49×10^{16} photons
45. (a) 1.22×10^{-33} (b) $10^{-2\,253}$
47. (a) 4.24 PW/m^2 (b) 1.20 pJ $= 7.50$ MeV
51. (a) 1.57×10^{14} m$^{-3/2}$ (b) 2.47×10^{28} m^{-3}
(c) 8.69×10^8 m^{-1}
53. (a) 4.20 mm   (b) 1.05×10^{19} photons
(c) 8.82×10^{16}/mm^3
57. 5.39 keV
59. (a) For Al, about 0.255 nm ~ 0.1 nm; for U, about 0.276 nm ~ 0.1 nm.   (b) For an outer electron, the nuclear charge is screened by the interior electrons. For an inner-shell electron, the nuclear charge is unscreened. The distance scale of the wave function, representing the orbital size, is proportional to a_0/Z.
61. 0.125
63. (b) 0.846 ns
65. 9.79 GHz
67. (a) 137   (b) $\lambda_C/r_e = 2\pi/\alpha$
(c) $a_0/\lambda_C = 1/2\pi\alpha$ (d) $1/R_H a_0 = 4\pi/\alpha$

Chapter 43

1. (a) 921 pN toward the other ion   (b) -2.88 eV
3. (a) $(2A/B)^{1/6}$ (b) $B^2/4A$ (c) 74.2 pm, 4.46 eV
5. (a) 40.0 μeV, 9.66 GHz   (b) 20% too large if r is 10% too small
7. 5.69 Trad/s
9. (a) 1.81×10^{-45} kg$\cdot$m^2 (b) 1.62 cm
11. 0.358 eV
13. (a) 0, 364 μeV, 1.09 meV   (b) 98.2 meV, 295 meV, 491 meV

15. 558

17. 6.25×10^9

19. -7.84 eV

23. An average atom contributes 0.981 electron to the conduction band.

25. (a) 2.54×10^{28} electrons/m^3 (b) 3.15 eV
(c) 1.05 Mm/s

27. 5.28 eV

31. (a) 1.10 (b) 1.55×10^{-25}; much smaller

33. All of the Balmer lines are absorbed, except for the red line at 656 nm, which is transmitted.

35. 1.24 eV or less; yes

37. (a) 59.5 mV (b) -59.5 mV

39. 4.19 mA

41. 203 A to produce a magnetic field in the direction of the original field

43. (a) 61.5 THz (b) 1.59×10^{-46} kg·m^2
(c) 4.79 μm or 4.96 μm

45. 7

49. (a) 0.350 nm (b) -7.02 eV (c) $-1.20\mathbf{i}$ nN

51. (a) r_0 (b) B (c) $(a/\pi)[B/2\mu]^{1/2}$
(d) $B - (ha/\pi)[B/8\mu]^{1/2}$

53. (b) No. The fourth term is larger than the sum of the first three.

Chapter 44

1. $\sim 10^{28}$, $\sim 10^{28}$, $\sim 10^{28}$

3. (a) 27.6 N (b) 4.17×10^{27} m/s^2 away from the nucleus
(c) 1.73 MeV

5. (a) 455 fm (b) 6.04 Mm/s

7. (a) 1.90 fm (b) 7.44 fm

9. 16.0 km

11. Z magic: He, O, Ca, Ni, Sn, Pb; N magic: isotopes of H, He, N, O, Cl, K, Ca, V, Cr, Sr, Y, Zr, Xe, Ba, La, Ce, Pr, Nd, Pb, Bi, Po

13.

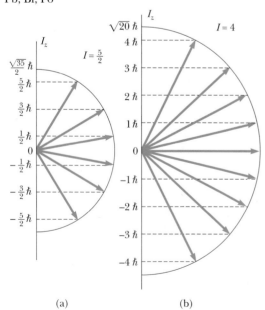

(a)   (b)

15. (a) 1.11 MeV/nucleon (b) 7.07 MeV/nucleon
(c) 8.79 MeV/nucleon (d) 7.57 MeV/nucleon

17. (a) Cs (b) La (c) Cs

19. greater for N by 3.54 MeV

21. 7.93 MeV

23. 200 MeV

25. 1.16 ks

27. (a) 1.55×10^{-5}/s, 12.4 h (b) 2.39×10^{13} atoms
(c) 1.87 mCi

29. 9.47×10^9 nuclei

31. 41.8 TBq or 4.18×10^{13} decays/s

33. 4.27 MeV

35. 18.6 keV

37. (a) $e^- + p \rightarrow n + \nu$ (b) $^{15}_8$O atom $\rightarrow$ $^{15}_7$N atom $+ \nu$
(c) 2.75 MeV

39.

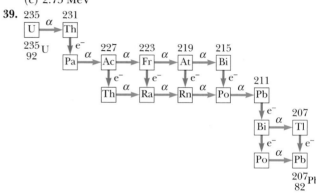

41. (a) 0.281 (b) 1.65×10^{-29} (c) Radium-226 continuously creates radon

43. (a) $^{197}_{79}$Au $+ ^1_0$n $\rightarrow$ $^{198}_{80}$Hg $+ ^{\;\;0}_{-1}$e $+ \bar{\nu}$ (b) 7.89 MeV

45. 1 MeV

47. 8.005 3 u, 10.013 5 u

49. $\sqrt{2}$

51. (a) 2.52×10^{24} (b) 2.29 TBq (c) 1.07 Myr

53. (a) 2.75 fm (b) 152 N (c) 2.62 MeV
(d) 7.44 fm, 379 N, 17.6 MeV

55. (a) Conservation of energy (b) Electrostatic energy of the nucleus (c) 1.20 MeV

57. (b) 4.78 MeV

59. (b) 0.001 94 eV

61. (a) $\sim 10^{-1349}$ (b) 0.892

63. (a) 4.28 pJ (b) 1.19×10^{57} atoms (c) 107 Gyr

65. 2.20 μeV

69. 0.400%

71. 2.64 min

Chapter 45

1. 192 MeV

3. n $+ ^{232}$Th $\rightarrow ^{233}$Th $\rightarrow ^{233}$Pa $+ e^- + \bar{\nu}$ and
^{233}Pa $\rightarrow ^{233}$U $+ e^- + \bar{\nu}$

5. (a) 201 MeV (b) 0.091 3%

7. 5.80 Mm

9. about 3 000 yr

11. 6.25×10^{19} Bq

13. (a) $2.30 \times 10^{-14} \, Z_1 Z_2 \, \text{J}$ (b) 0.144 MeV for both
15. (a) 2.22 Mm/s (b) $\sim 10^{-7}$ s
17. (a) 1.7×10^7 J (b) 7.3 kg
19. 1.30×10^{25} ^{6}Li; 1.61×10^{26} ^{7}Li
21. 1.66×10^3 yr
23. (a) 2.5 mrem per x-ray (b) 5 rem/yr is 38 times 0.13 rem/yr
25. 2.09×10^6 s
27. 1.14 rad
29. (a) 3.12×10^7 (b) 3.12×10^{10} electrons
31. (a) 10 (b) 10^6 (c) 1.00×10^8 eV
33. 4.45×10^{-8} kg/h
35. (a) $\sim 10^6$ (b) $\sim 10^{-15}$ g
37. (a) 8×10^4 eV (b) 4.62 MeV and 13.9 MeV
(c) 1.03×10^7 kWh
39. 0.375% for D–T fusion, which is about four times larger than 0.095 0% for ^{235}U fission.
41. 482 Ci, less than the fission inventory by on the order of 100 million times
43. 2.56×10^4 kg
45. (a) 2.65 GJ (b) The fusion energy is 78.0 times larger.
47. (a) 4.91×10^8 kg/h $= 4.91 \times 10^5$ m^3/h (b) 0.141 kg/h
49. (a) 15.5 cm (b) 51.7 MeV (c) The number of decays per second is the decay rate R and the energy released in each decay is Q. Then the energy released per time is $\mathcal{P} = QR$. (d) 227 kJ/yr (e) 3.18 J/yr
51. 14.1 MeV
53. (a) 2.24×10^7 kWh (b) 17.6 MeV (c) 2.34×10^8 kWh (d) 9.36 kWh (e) Coal is cheap at this moment in human history. We hope that safety and waste disposal problems can be solved so that nuclear energy can be affordable before scarcity drives up the price of fossil fuels.
55. (b) 26.7 MeV
57. (a) 5×10^7 K

(b)

Reaction	Q (MeV)
$^{12}\text{C} + {}^1\text{H} \rightarrow {}^{13}\text{N} + \gamma$	1.94
$^{13}\text{N} \rightarrow {}^{13}\text{C} + e^+ + \nu$	1.20
$e^+ + e^- \rightarrow 2\gamma$	1.02
$^{13}\text{C} + {}^1\text{H} \rightarrow {}^{14}\text{N} + \gamma$	7.55
$^{14}\text{N} + {}^1\text{H} \rightarrow {}^{15}\text{O} + \gamma$	7.30
$^{15}\text{O} \rightarrow {}^{15}\text{N} + e^+ + \nu$	1.73
$e^+ + e^- \rightarrow 2\gamma$	1.02
$^{15}\text{N} + {}^1\text{H} \rightarrow {}^{12}\text{C} + {}^4\text{He}$	4.97
Overall	26.7

(c) Most of the neutrinos leave the star directly after their creation, without interacting with any other particles.

Chapter 46

1. 453 ZHz; 662 am
3. 118 MeV
5. $\sim 10^{-18}$ m
7. $\sim 10^{-23}$ s
9. 67.5 MeV, 67.5 MeV/c, 16.3 ZHz
11. $\Omega^+ \rightarrow \overline{\Lambda}^0 + K^+$ $\overline{K}_S^0 \rightarrow \pi^+ + \pi^-$ $\overline{\Lambda}^0 \rightarrow \overline{p} + \pi^+$
$\overline{n} \rightarrow \overline{p} + e^+ + \nu_e$
13. (b) The second reaction violates strangeness conservation.
15. (a) $\overline{\nu}_\mu$ (b) ν_μ (c) $\overline{\nu}_e$ (d) ν_e (e) ν_μ (f) $\overline{\nu}_e + \nu_\mu$
17. (a), (c), and (f) violate baryon number conservation. (b), (d), and (e) can occur. (f) violates muon–lepton number conservation
19. (a) ν_e (b) ν_μ (c) $\overline{\nu}_\mu$ (d) $\nu_\mu + \overline{\nu}_\tau$
21. (b) and (c) conserve strangeness. (a), (d), (e), and (f) violate strangeness conservation.
23. (a) not allowed; violates conservation of baryon number (b) strong interaction (c) weak interaction (d) weak interaction (e) electromagnetic interaction
25. (a) K^+ (b) Ξ^0 (c) π^0
27. (a) 3.34×10^{26} e$^-$, 9.36×10^{26} u, 8.70×10^{26} d (b) $\sim 10^{28}$ e$^-$, $\sim 10^{29}$ u, $\sim 10^{29}$ d. My strangeness, charm, truth, and beauty are zero.
29. $m_u = 312$ MeV/c^2 $m_d = 314$ MeV/c^2
31. (a) The reaction $\overline{u}d + uud \rightarrow \overline{s}d + uds$ has a total of 1 u, 2 d, and 0 s quarks originally and finally. (b) The reaction $\overline{d}u + uud \rightarrow \overline{s}u + uus$ has a net of 3 u, 0 d, and 0 s before and after. (c) $\overline{u}s + uud \rightarrow \overline{s}u + \overline{s}d + sss$ shows conservation at 1 u, 1 d, and 1 s quark. (d) The process $uud + uud \rightarrow \overline{s}d + uud + \overline{d}u + uds$ nets 4 u, 2 d, 0 s initially and finally; the mystery particle is a Λ^0.
33. (a) Σ^+ (b) π^- (c) K^0 (d) Ξ^-
37. (a) $0.383c$ (b) 6.76×10^9 ly
39. 6.00
41. (a) 1.06 mm (b) microwave
43. (a) ~ 100 MeV/c^2 (b) charged or neutral pion
45. one part in 50 000 000
47. $\sim 10^{14}$
49. 5.35 MeV and 32.3 MeV
51. 74.4 MeV
53. 9.26 cm
55. 2.52×10^3 K
57. (a) Z^0 boson (b) gluon

Page numbers in *italics* indicate illustrations; page numbers followed by an "n" indicate footnotes, page numbers followed by "t" indicate tables.